Ecological Studies
Analysis and Synthesis

Edited by

W.D. Billings, Durham (USA) F. Golley, Athens (USA)

O.L. Lange, Würzburg (FRG) J.S. Olson, Oak Ridge (USA)

H. Remmert, Marburg (FRG)

Volume 97

Ecological Studies

R.K. Olson D. Binkley
M. Böhm Editors

The Response of Western Forests to Air Pollution

Contributors

M. Arbaugh D. Binkley M. Böhm L. Brubaker A. Bytnerowicz
S. Cline E. Cook T. Droessler D. Duriscoe C. Earle D. Ford
D. Graybill N. Grulke J. Miller P. Miller R. Olson D. Peterson
C. Ribic M. Rose G. Segura K. Stolte S. Vega-Gonzalez

With 123 Illustrations

Springer-Verlag
New York Berlin Heidelberg London Paris
Tokyo Hong Kong Barcelona Budapest

Richard K. Olson
ManTech Environmental Technology, Inc.
US EPA Environmental Research Laboratory
Corvallis, OR 97333 USA

Dan Binkley
Department of Forest and Wood Science
Colorado State University
Fort Collins, CO 80523 USA

Margi Böhm
ManTech Environmental Technology, Inc.
US EPA Environmental Research Laboratory
Corvallis, OR 97333 USA

Library of Congress Cataloging-in-Publication Data
The response of western forests to air pollution / R.K. Olson, D.
  Binkley, M. Böhm, editors.
      p.   cm. — (Ecological studies ; 97)
  Includes bibliographical references and index.

  ISBN-13: 978-1-4612-7734-7   e-ISBN-13: 978-1-4612-2960-5
  DOI: 10.1007/978-1-4612-2960-5

  1. Trees—West (U.S.)—Effect of air pollution on.   2. Conifers-
-West (U.S.)—Effect of air pollution on.   3. Forest ecology—West
(U.S.)   4. Conifers—West (U.S.)—Growth.   5. Trees—Wounds and
injuries—West (U.S.)   6. Conifers—Wounds and injuries—West (U.S.)
I. Olson, Richard K.   II. Binkley, Dan.   III. Böhm, M. (Margi)
IV. Series: Ecological studies ; v. 97.
SB745.R48   1992
634.9'619—dc20   92-2192

Printed on acid-free paper.

© 1992 Springer-Verlag New York, Inc.

Softcover reprint of the hardcover 1st edition 1992

Production managed by Henry Krell; manufacturing supervised by Jacqui Ashri.
Negatives supplied by the editors.

9 8 7 6 5 4 3 2 1

PREFACE

Passage of the Acid Precipitation Act of 1980 (Title VII, Energy Security Act, P.L. 96-294) led to the establishment of the National Acid Precipitation Assessment Program (NAPAP). A coalition of 12 federal agencies, NAPAP's goal was to conduct research and assessments in support of regulatory and policy decisions on controlling acid rain. The culmination of NAPAP's efforts came in 1990 with the publication of 27 State of Science and Technology reports. A wide range of topics was addressed including emissions, transport, and effects of acid rain and its precursors.

The Forest Response Program (FRP) was developed in 1985 as part of NAPAP. A joint effort of the US Environmental Protection Agency (EPA) and the USDA Forest Service, the FRP evaluated the effects of acid rain and associated pollutants on US forests. Research was conducted within four regional cooperatives, and coniferous forests in the eleven Western states fell under the purview of the Western Conifers Research Cooperative (WCRC), which operated from 1986–1990. Research by the WCRC was conducted at scales ranging from small (e.g., open-top chambers) to large (e.g., regional growth trends), and at a variety of biological levels of organization including seedlings, trees, and stands. Individual studies were conducted as components of an integrated program of research on the effects of air pollution on Western forests. In addition to EPA and Forest Service researchers, scientists from eleven universities, state agencies, and industry groups worked within the Cooperative.

This book is the final product of the WCRC. Most of the chapter authors conducted research for the Cooperative. Results from the WCRC form the core of the book, but much of the information presented comes from other sources. The goal is to describe air quality and deposition patterns for the western United States, and to describe the effects of this pollution on Western forests.

Support and encouragement for this book came from numerous people at the US EPA Environmental Research Laboratory, Corvallis, Oregon. Roger Blair and Charley Peterson provided programmatic guidance; Ann Hairston and Marcia Bollman editorial support; and Deborah Coffey and Kate Dwire quality assurance review. Terralyn Vandetta conducted much of the data analysis in support of Chapter 3. Over 50 extramural reviewers provided comments on chapter drafts.

We thank Pris Hardin, Arlene Kovash, Paul Bissex, Julie Borden, Sue Moore, Sam Olfano and Dave Zaworski of Page Craft, Corvallis, Oregon for producing the book and for preparing many of the figures. For permission to reprint figures, we thank the following:

Chapter 1

Figure 1.7a	Ecological Society of America
Figure 1.7b,c	Cambridge University Press, Cambridge, *North American Terrestrial Vegetation*, Chapter 3, RK Peet
Figure 1.8	California Botanical Society, Berkeley

Chapter 2

Figure 2.1	American Meteorological Society, Boston

Chapter 3

Figure 3.1	Methuen and Co., New York, *Boundary Layer Climates*, TR Oke
Figure 3.3	Methuen and Co., New York, *Boundary Layer Climates*, TR Oke
Figure 3.5	*Atmospheric Environment*, Volume 21, JA King, FH Shair, DD Reible, "The influence of atmospheric stability on pollutant transport by slope winds." Copyright 1987, Pergamon Press plc
Figure 3.6	McGraw-Hill, Inc, New York, *Weather and Energy*, Schwoegler and McClintock, Copyright 1981
Figure 3.18	Air & Waste Management Association, Pittsburgh

Chapter 5

Figure 5.2	Longman Group UK, *Air Pollution and Acid Rain: The Biological Impact*, AR Wellburn
Figure 5.3	Air & Waste Management Association, Pittsburgh
Figure 5.4	Elsevier Applied Sciences, London
Figure 5.10	Kluwer Academic Publishers, Dordrecht, Holland

Chapter 6
    Figure 6.4   Cambridge University Press, Cambridge; *North American Terrestrial Vegetation*, Chapter 3, RK Peet

Chapter 7
    Figure 7.5   Air & Waste Management Association, Pittsburgh
    Figure 7.9   Air & Waste Management Association, Pittsburgh
    Figure 7.10   Air & Waste Management Association, Pittsburgh

Chapter 9
    Figure 9.1   American Geophysical Union, Washington, DC

Chapter 10
    Figure 10.2   World Resources Institute, Washington, DC

Chapter 11
    Figure 11.3   Edward Arnold, UK
    Table 11.1   Edward Arnold, UK

Chapter 12
    Figure 12.9   American Meteorological Society, Boston
    Table 12.3   American Meteorological Society, Boston

Preparation of this book has been funded by the US EPA as part of the joint US EPA–USDA Forest Service Forest Response Program. The book has not been subjected to policy review by the EPA or Forest Service, and should not be construed to represent the policies of either agency.

Richard Olson
Dan Binkley
Margi Böhm

CONTENTS

Section II: Regional Studies of Forest Growth and Condition

Section III: Summary and Projections

CONTRIBUTORS

Michael Arbaugh, U.S.D.A. Forest Service, Pacific Southwest Forest and Range Experiment Station, Riverside, CA 92507

Dan Binkley, Department of Forest and Wood Science, Colorado State University, Fort Collins, CO 80523

Margi Böhm, ManTech Environmental Technology, Inc., U.S. EPA Environmental Research Laboratory, Corvallis, OR 97333

Linda Brubaker, College of Forest Resources, University of Washington, Seattle, WA 98195

Andrzej Bytnerowicz, U.S.D.A. Forest Service, Pacific Southwest Forest and Range Experiment Station, Riverside, CA 92507

Steven Cline, ManTech Environmental Technology, Inc., U.S. EPA Environmental Research Laboratory, Corvallis, OR 97333

Ed Cook, Tree-ring Laboratory, Lamont-Doherty Geological Observatory, Palisades, NY 10964

Terry Droessler, ManTech Environmental Technology, Inc., U.S. EPA Environmental Research Laboratory, Corvallis, OR 97333

Dan Duriscoe, Eridanus Research Associates, Three Rivers, CA 93271

Christopher Earle, College of Forest Resources, University of Washington, Seattle, WA 98195

David Ford, Center for Quantitative Science, University of Washington, Seattle, WA 98195

Donald Graybill, Laboratory of Tree-Ring Research, University of Arizona, Tucson, AZ 85721

Nancy Grulke, U.S.D.A Forest Service, Pacific Northwest Research Station, Corvallis, OR 97331

Jeffrey Miller, 7045 NW Grandview Drive, Corvallis, OR 97333

Paul Miller, U.S.D.A. Forest Service, Pacific Southwest Forest and Range Experiment Station, Riverside, CA 92507

Richard Olson, ManTech Environmental Technology, Inc., U.S. EPA Environmental Research Laboratory, Corvallis, OR 97333

David Peterson, U.S.D.A. Forest Service, Pacific Southwest Forest and Range Experiment Station, Riverside, CA 92507

Christine Ribic, Center for Quantitative Science, University of Washington, Seattle, WA 98195

Martin Rose, Desert Research Institute, Reno, NV 89506

Gerardo Segura, College of Forest Resources, University of Washington, Seattle, WA 98195

Ken Stolte, Air Quality Division, U.S.D.I. National Park Service, Lakewood, Colorado 80225

Silvia Vega-Gonzalez, Center for Quantitative Science, University of Washington, Seattle, WA 98195

INTRODUCTION

The Response of Western Forests
to Air Pollution

R. K. Olson

This book addresses the relationships between air pollution in the western United States and trends in the growth and condition of Western coniferous forests. The West is defined in this case as the eleven conterminous states of California, Oregon, Washington, Idaho, Nevada, Arizona, New Mexico, Utah, Colorado, Wyoming, and Montana. Approximately one-third of the West is forested, primarily by coniferous forest types.

The major atmospheric pollutants to which forests in this region are exposed are sulfur and nitrogen compounds and ozone. Ozone is a secondary pollutant formed by photolytic reactions involving nitrogen oxides and hydrocarbons. Sulfur and nitrogen are primary pollutants which are emitted as gases from a variety of human and natural sources and, following atmospheric transport and chemical reactions, enter forests in gaseous, particulate, and dissolved forms. The types and amounts of pollutants entering Western forests depend on the spatial relationships between emissions sources, forests, and patterns of air movement.

The potential effects of atmospheric pollution on these forests include foliar injury, alteration of growth rates and patterns, soil acidification, shifts in species composition, and modification of the effects of natural stresses. Effects can occur at many different scales and biological levels of organization (e.g., cellular, leaf, branch, single tree, stand, ecosystem), and primary effects at one level often result in secondary effects at other levels.

Assessing the impact of air pollution on a forest is a difficult process. Time and monetary constraints preclude measuring all but a few of the possible indicators of effects. High natural variability in most indicators makes interpretation of results difficult as does the lack of clearly defined baselines against which to compare results. Very few indicators are pollutant-specific; they respond to many stresses and further confound results. Integrating results across scales or biological levels (e.g., what does an increase in chlorotic mottling mean for stand biomass increment?) often requires a mechanistic understanding which is beyond the current state of science. Developing an appropriate research approach is complex, and there is no "best" approach to this problem.

This book focuses on five regional studies of air pollution effects on Western coniferous forests (Chapters 8–12). These studies, conducted by four different groups of investigators, serve as examples of approaches for assessing air pollution effects. The study regions were selected to meet two main criteria: (1) regionally elevated levels of air pollution, and (2) significant areas of coniferous forests. Within each region, a subset of the forest types was chosen as the study focus. The resulting region/ forest type combinations are: (1) western Washington/old-growth Douglas-fir (Chapter 8); (2) Colorado Front Range/mixed conifers in unmanaged old-growth stands, and ponderosa pine in second-growth stands (Chapter 9); (3) central and southern Arizona and central New Mexico/pinyon-juniper, ponderosa pine, and mixed conifers (Chapter 10); (4) Sierra Nevada, California/ponderosa pine (Chapter 11); and (5) San Bernardino Mountains, California/mixed conifers (Chapter 12).

As indicators of forest condition, all five studies include measurements of trends in the radial growth of trees, and three of the studies quantify visible foliar damage. For each region, one step in the interpretation of the indicator measurements is the comparison of indicator values between areas with greater vs. lesser amounts of air pollution. Four of the studies also analyze tree-ring series. These series provide a record of growth trends during pre-pollution times which are used to develop baseline characterizations of natural variability in growth. Recent growth is then analyzed for deviations from these baselines.

While the same general approaches are used in each regional study, the details differ according to differences in regional situations. For example, techniques appropriate for evaluating the growth and condition of old-growth Douglas-fir may not be best suited for dealing with young stands of ponderosa pine. Open grown pinyon-juniper present a different situation than denser stands of lodgepole pine.

And finally, the science of evaluating air pollution effects on forests, especially the application of dendrochronological techniques to this task, is still developing. Investigators have different ideas on how to proceed. It is interesting and instructive to compare the different approaches to similar problems which are presented in this book.

Chapters 1–7 present background information which establishes a context for the regional studies, and aids in interpreting their results. Results from research projects other than the regional case studies are also surveyed, particularly from controlled exposure studies and other experimental work. Chapter 1 describes the forest types and physiography of the West. The climate of the West is described in Chapter 2, with special emphasis on differences between the case study regions, and on long-term patterns of moisture availability within the regions. Chapter 3 describes air quality in the West as part of a broader discussion of emissions sources and processes governing transport, transformation, and deposition of pollutants to forests.

Chapter 4 leads from an overview of mechanisms of soil acidification to an assessment of the sensitivity to acid deposition of soils in four Western regions. Chapter 5 then reviews current knowledge of the physiological effects of air pollution on Western tree species at scales from subcellular to whole trees. Chapter 6 reviews expected patterns of stand development as a baseline against which to compare actual stand growth and development, then discusses potential effects of pollution on these expected patterns. The introductory portion of the book concludes with a review in Chapter 7 of field methods for evaluating the effects of air pollution on forests. The methods used in the regional case studies are described along with methods used in other Western studies of forest condition.

Following the descriptions of the regional case studies in Chapters 8–12, the book concludes with Chapter 13—a summary of the main results from earlier chapters and a discussion of future trends in the condition of Western forests.

Section I

Background

1

Physiography and Forest Types

R. K. Olson

The eleven Western states occupy 308,278,680 ha or about 40% of the land area of the conterminous United States (USDA 1981). The West is characterized by extreme relief with massive north-south oriented mountain ranges separated by basins, valleys, and plateaus. Topographic modification of regional climate determines which areas can support forests. Pollutant loading to Western forests is also influenced by topographic patterns that modify local air flow patterns.

The current composition and structure of Western forests results in part from present and historical interactions of climate and topography. Evolution of the modern Western flora was driven by large changes in global and regional climate during the past 65 million years, and by concurrent changes in landforms (King 1977, Axelrod and Raven 1985).

This chapter presents a broad overview of Western physiography and forest types as a framework for later chapters. Discussion of patterns within forest types focuses on topographic and moisture relationships.

Physiography

The western United States is dominated by three parallel mountain ranges: the Coast Ranges, Sierra-Cascade Ranges and the Rocky Mountains (Figures 1.1, 1.2). Differences in relief are extreme, ranging from a maximum elevation of 4418 m (Mt. Whitney, CA) to 86 m below sea level (Death Valley, CA).

The West includes four main physiographic units with substantial areas of forests: the Pacific Coast Mountains and Valleys, the Rocky Mountains, the Intermontane Plateaus and Basins, and the Cascade-Sierra Ranges (Fenneman 1930, 1931; Figure 1.2). A fifth unit, the Great Plains, occupies large areas of eastern Montana, Wyoming, Colorado and New

0   100   200   300   400   500   600   KILOMETERS

Figure 1.1 Physiographic map of the western United States. From USGS (1970).

Figure 1.2 Physiographic provinces of the western United States. From Fenneman (1930).

Mexico, abutting the Rocky Mountains except in southern New Mexico where it adjoins the Basin and Range Province. Elevation at the western boundary averages 1680 m and ranges from 610 to 2130 m (Fenneman 1931). In all of the western sections of the Great Plains, drainages such as the Arkansas, Platte and Missouri Rivers create a pattern of valleys separated by uplands of slight to moderate relief. These river valleys along with isolated mountain ranges such as the Black Hills, Bear Paw Mountains and Little Rocky Mountains offer most of the limited area suitable for forests in this region of the West (Peet 1988).

Pacific Border Province

Virtually the entire Pacific Coast of the western United States is bordered by mountain ranges. The coastal plain, where present, is narrow. Over most of their length, these coastal ranges are separated from the Cascade-Sierra Ranges to the east by interior lowlands with elevations less than 300 m. The presence of large areas at elevations close to sea level is unique to this region of the West. The geomorphology of this province is complex, resulting in part from the extreme forces generated at the interface of the North American and Pacific plates. Only the Olympic Mountains and parts of the adjacent Puget Trough were glaciated. Glaciers can still be found at high elevations in the Olympic Mountains.

The interior lowlands include the Puget Trough and Willamette Valley that together extend 560 km from the Canadian border of northwest Washington to central Oregon. With a maximum width of 80 km at their northern end, the elevation of these lowlands is usually less than 150 m. The other main interior lowland, the Central Valley, stretches 650 km down the middle of California. The Central Valley is characterized by very little relief and elevations lower than 120 m.

The Olympic Mountains occupy the Olympic Peninsula of Washington. The area is characterized by high terrain except for a 15–30 km wide coastal plain on the western side of the Peninsula and several west-east oriented river valleys. The mountains generally range between 1375 and 1525 m in elevation, although isolated peaks reach 2400 m.

The Oregon Coast Range is a set of low mountains lying between the Olympic Range and the Klamath Mountains in southern Oregon. They fall between a narrow coastal plain in the west and the Puget Trough and Willamette Valley to the east, although south of 44°N they border directly on the Cascades. Peaks range from 520 m to 1060 m with elevations typically increasing to the south.

The Klamath Mountains occupy southwestern Oregon and northwestern California and border directly on the Cascades with no intervening lowlands. They are the highest of the coast ranges with crest levels from 600 m near the coast to 2150 m in the interior. Individual peaks reach over 2750 m in elevation.

The California Coast Ranges are small ranges trending northwest from the western edge of the Central Valley. Elevations average 600–1200 m with intervening valleys about 300 m lower than the crests. As in Oregon, the coastal plain is usually only a few kilometers wide.

The Transverse Ranges begin on the coast north of Santa Barbara with the Santa Ynez Mountains and proceed inland to the San Gabriel and San Bernardino Mountains northeast of Los Angeles. The latter two ranges are significantly higher than the more northern coast ranges with an average elevation of 1830 m in the San Gabriel Mountains and peaks above 3050 m in the San Bernardinos. They separate the Los Angeles basin from the Mohave Desert to the northeast.

The Peninsular Ranges form the southern end of the coast ranges and lie between the Transverse Ranges and the Mexican border. Major ranges include the San Jacinto, Santa Rosa, Santa Ana, and Laguna Mountains. Peaks generally rise to 2150–2750 m.

Rocky Mountains

The mountains known collectively as the Rocky Mountains extend more than 1500 km within the United States from northern New Mexico to northeastern Washington. Although most of the individual ranges are oriented north-south, the Rockies as a whole shift westward from about 105°W in Colorado to 120°W in Washington.

A distinct characteristic of the Rocky Mountains is the overall high elevation of the region. Many peaks exceed 4270 m in elevation and intervening valleys and basins typically dip to no lower than 2750–3050 m. As a result, peaks often rise less than 1000 m above the level of the surrounding terrain. Glacial features such as cirques and troughs are common in all regions of the Rockies, although alpine glaciers are only found in a few locations.

The Southern Rocky Mountains occupy central Colorado with extensions north into Wyoming and south into New Mexico. They consist primarily of a series of broad, north-south ranges overlapping but offset from east to west. More than 50 peaks exceed 4250 m elevation, and over 300 summits rise above 3950 m.

Abutting the Great Plains between the Arkansas River and the Cache la Poudre River in northern Colorado is the Colorado Front Range. It consists of a broad line of alpine ridges and peaks with well known summits including Pikes Peak (4300 m) and Long's Peak (4345 m). The northern end of the Colorado Front Range splits into eastern and western lobes that extend into Wyoming as the Laramie Range and Medicine Bow Mountains respectively. The Laramie Range declines in elevation to 2438 m, presenting a low ridge approximately 450 m above the surrounding plains. The Medicine Bow Mountains retain a higher aspect with peaks between 3050 m and 3660 m.

Three high elevation basins separate the Front Range from ranges to the west: North Park, Middle Park and South Park. The floors of these basins range from 2440 to 3050 m in elevation, and are typically characterized by gentle relief. The basins abut the Park Range of northern Colorado and southern Wyoming, the Sawatch Range in central Colorado, and the Sangre de Cristo Mountains extending from the southern edge of South Park to northern New Mexico and the southern terminus of the Rocky Mountains. Peaks in the three ranges rise from 3050 m to higher than 4270 m. Mt. Elbert, the highest point in the Rocky Mountains, is found in the Sawatch Range and reaches 4399 m in elevation.

Other major mountain groups in the Southern Rockies include the San Juan Mountains, a group of 3950–4270 m peaks of volcanic origin in southwestern Colorado, and the White River Plateau, the westernmost of the four lobes of the Southern Rockies that extend into the Wyoming Basin. Ranges connecting the Southern and Central Rockies are buried beneath the sediments of the 2130 m floor of the Wyoming Basin.

The Central Rocky Mountains lie northwest and west of the Southern Rockies. The two mountain provinces are similar in that most ranges are linear uplifts forming single orographic units. Ranges in the Central Rockies are more widely separated, with basins and valleys occupying a larger proportion of the total area and exerting a greater influence on the physiography of the region. Also, ranges adhere less strictly to a north-south orientation. The dominant peaks of all the main ranges in this region fall between 2750 m and 3950 m elevation, giving them a rise of 1525 m or less above the elevated floors of intervening basins and valleys.

The Big Horn, Wind River, Teton and Absaroka Ranges occupy northcentral and northwestern Wyoming. The Absaroka Ranges lie just east of Yellowstone National Park, a volcanic plateau averaging 2300 to 2600 m in elevation. The Wyomide Ranges extend down the Wyoming-Idaho border to the Wasatch Range in northern Utah. This 320 km long range drops abruptly on its western edge to the Great Basin. Perpendicular to the Wasatch on its eastern edge, the 250 km long Uinta Mountains are the largest east-west mountain range in the United States outside Alaska.

The Northern Rocky Mountains occupy western Montana, northern Idaho and northeastern Washington and are the largest of the three Rocky Mountain regions. Elevation is generally less than in the Southern or Central Rockies and mountains are less often oriented in distinct linear ranges. Valleys separating ranges are generally narrow and range between 300 and 1525 m in depth. Broad basins characteristic of the Central and Southern Rockies are absent.

Average elevations of crests in the Northern Rocky Mountains range from 1830 to 3200 m. The highest summits are found in the Lemhi and Lost River Ranges of east central Idaho with an average crest level of 3200 m and peaks exceeding 3660 m. Average heights decrease north and west of these ranges.

Intermontane Plateaus and Basins

The Rocky Mountains are separated from the Sierra-Cascade Mountains by three distinctive physiographic provinces: the Columbia Plateau, the Basin and Range Province, and the Colorado Plateau. Topographic differences are great both within and between these three units.

The Colorado Plateau is an uplifted area of about 340,000 km^2 surrounding the four corners intersection of Utah, Colorado, Arizona and New Mexico. Most of the Plateau is above 1525 m. The greatest elevations are reached in the High Plateaus of Utah along the western boundary of the Colorado Plateau (3000–3500 m). Other high points include the volcanic San Francisco Peaks, with maximum elevation 3870 m, and the Mogollon Plateau on the southwestern edge. The surface of the Colorado Plateau is less deformed than that of the Rocky Mountain regions. In place of mountain ranges and interstitial basins and valleys, the general pattern of relief constitutes numerous individual plateaus, basins and canyons. The largest canyon is the Grand Canyon on the Colorado River in northwestern Arizona. The Grand Canyon reaches a depth of 1500 m to 1800 m below the plateau. The degree of dissection varies within the province, being greatest in the central or Canyonlands section and more limited on the broad Mogollon and San Francisco Plateaus.

The Columbia Plateau, the northernmost of the Intermontane Provinces, is a massive lava plain in eastern Washington, eastern Oregon, and southern Idaho. Greatest relief and elevation occurs in the Blue Mountains of northeastern Oregon where antecedent mountain ranges of 2750 m rise above the surface of the lava plain, which is at 1830–2150 m in elevation.

The Blue Mountains are separated from the Northern Rockies by the Snake River Canyon that is 1700 m deep. The Owyhee Mountains of southwestern Idaho rise to 2400 m. Elsewhere, elevation is less and relief is variable. The Snake River Plain in Idaho exhibits relatively flat terrain as it slopes from 1830 m at its northeastern end to 1000 m in southcentral Idaho. North of the Blue Mountains, the Palouse Country consists of broadly rolling wheatlands at about 1070 m. The southwestern corner of the Columbia Plateau Province contains the Great Sandy Desert, an extremely flat and dry area with overall elevation of 1200 m.

The Basin and Range Province is the largest of the Intermontane Provinces. Most of Nevada, western Utah, southern Arizona and southwestern and southcentral New Mexico are included in this province as are smaller portions of Oregon, California and Idaho. The main characteristic of the Basin and Range Province is a landscape of subparallel mountain ranges, oriented north-south, and separated by desert basins of low relief. Regions within the province differ in the proportion of area occupied by mountains as well as in the presence or absence of external drainage.

Both the Great Basin, lying between the Rocky Mountains and the Sierra-Cascade Ranges, and the Mexican Highlands of southern Arizona and New Mexico have even proportioning of surface area between mountains and basins. Basin floors are generally between 1200 and 1525 m in elevation. Ranges in the Great Basin average 80–120 km long by 10–24 km wide with common elevations between 2150 m and 3050 m. The Mexican Highlands in southeastern Arizona includes the Santa Catalina, Santa Rita and Pinaleno Mountains that range between 2750 and 3350 m in elevation. In New Mexico, the province includes a series of mountains along its eastern edge that range between 2400 and 3660 m in height. The Sonoran and Mohave Deserts in southwestern Arizona and southeastern California have fewer and lower mountain ranges than the other regions.

Sierra-Cascade Ranges

The Sierra-Cascade Ranges stretch more than 1600 km from the Washington/Canadian border to southern California. They average 80 to 100 km in width and separate the Columbia Plateau in the north and the Basin and Range Province in the south from the interior valleys and coastal ranges of the Pacific Province. Although peak elevations are typically lower than in the Southern and Central Rockies, the Sierra-Cascade Ranges start from lower base elevations, thus offering equal or greater elevational differentials than in the Rockies. Glaciation was extensive throughout these ranges during the Pleistocene, and many small alpine glaciers remain in both the Cascades and Sierra Nevada.

The Sierra Nevada stretch along the eastern boundary of California for 650 km from north of Lake Tahoe to San Emigdio Mountain in southern California. Most of the range developed during uplift and westward tilting of a single granitic block. As a result, the range rises gradually from the Central Valley over distances of 80–130 km to the crest located on the eastern edge of the mountains. A relatively abrupt drop to the Great Basin follows. Elevations are highest in the southern Sierra Nevada with some peaks exceeding 4250 m. Mt. Whitney, at 4418 m, is the highest mountain in the lower 48 states. Crest line declines towards the north and ranges between 1830 and 2400 m in northern California.

The Southern Cascades in northern California are volcanic in origin, with a landscape of volcanic cones separated by low elevation lava beds. Mountains do not form distinct ranges. The Southern Cascades border the Central Valley to the southwest and the Klamath Mountains to the northwest. High peaks include Mt. Shasta (4317 m) and Lassen Peak (3187 m).

The Northern Cascades are a dissected upland with steep slopes and rugged crests. Peaks range from 1825 to 2600 m in elevation whereas intervening valleys are often 760–1060 m deep with steep sides. A few isolated peaks, primarily volcanic, project above the main level to elevations above 3050 m. From the Canadian border, this section of the Cascades extends south to approximately east of Seattle.

The Central Cascades are a transition between the Northern and Southern Cascades, both geographically and physiographically. Mountains form more distinct ranges than in the Southern Cascades and have a more volcanic nature than the Northern Cascades. Average peak heights are 2280 m or less, although individual volcanoes rise to much greater heights.

Forest Types and Distribution

The western United States covers a large area characterized by a diversity of topography, climates, and soils, leading to a corresponding diversity in species and forest types. Although Western forests as a whole share certain characteristics that set them apart from other forest regions (Gleason and Cronquist 1964), differences among Western forests serve as the basis for numerous classification schemes. This chapter emphasizes the Society of American Foresters Forest Type Group (Eyre 1980, Appendix I) in describing Western forests.

General Characteristics of Western Forests

Land, of which at least ten percent is stocked by forest trees of any size, or formerly having such cover and not currently developed for nontimber use is considered to be forested (USDA 1981). About 92,887,000 ha, or one-third of the West, are forested (Table 1.1), with coverage by state ranging from 11% in Nevada to 55% in Washington (Figure 1.3). Approximately 61% of Western forest land is federally owned.

Forest composition

Conifers dominate Western forests, as exemplified by the Pacific Northwest where conifers exceed hardwoods 1000:1 by volume (Küchler 1946, Franklin 1988). A large number of conifer species are important throughout the West (see Front Endpaper), occupying a wide range of habitats. Conifer dominance derives in part from their better adaptation to the summer drought—winter precipitation climate of much of the West. The coniferous habit allows greater photosynthesis outside the growing season when moisture is available (Waring and Franklin 1979). This and other physiological characteristics (Waring and Running 1978, Tyree and Dixon 1983) make conifers well adapted to the Western precipitation regime. Three conifer species—ponderosa pine (*Pinus ponderosa*), lodgepole pine (*Pinus contorta*), and Douglas-fir (*Pseudotsuga menziesii*) are notable for their wide distribution over most of the West (Figure 1.4). These species are common threads throughout Western forests and are the basis for three of the main SAF Forest Type Groups (Appendix I, Table 1.1).

Only 14% of forested land in the West is classified as a non-conifer forest type (USDA 1981, Table 1.1), and hardwoods are usually important only in specialized situations such as riparian habitats (Pace and Layser 1977), early successional stages (Lassoie et al. 1985, Smith 1985) or certain xeric, low elevation habitats (Griffin 1988). The only major hardwood with distribution throughout the West is quaking aspen (*Populus tremuloides*), pure stands of which are best developed at mid-elevations in the Southern Rocky Mountains (Morgan 1969).

Topography

Topography largely determines forest distribution. A comparison of Western physiography (Figure 1.1) with the forest type map in Appendix I shows that, with the exception of the Pacific Northwest, extensive Western forests occur only in the mountains (Gleason and Cronquist 1964). A generally dry climate, exacerbated in many areas by rain

Table 1.1 Forest land area in the western United States by forest type and state for 1977 (USDA 1981). Area given in hectares x 1000.

State	Douglas-fir	Ponderosa pine	Western white pine	Fir-spruce	Hemlock-Sitka spruce	Larch	Lodgepole pine	Redwood	Other Western softwoods	Western hardwoods	Non-stocked	Chaparral	Pinyon-Juniper	Other	Total: All Forests
Arizona	91	1653	0	68	0	0	0	0	2	95	49	659	4867	0	7484
California	1312	3156	44	2297	62	0	423	313	0	1704	452	3057	1091	2339	16249
Colorado	579	808	0	1759	0	0	900	0	31	1294	493	1214	1783	152	9013
Idaho	2775	823	117	1867	472	327	1501	0	246	226	203	73	163	0	8793
Montana	2282	1058	23	1065	97	517	2596	0	730	218	92	140	193	119	9120
Nevada	8	28	0	63	1	0	22	0	516	102	3	481	1885	0	3109
New Mexico	518	1712	0	300	0	0	0	0	23	178	83	173	4322	0	7309
Oregon	4094	2276	21	1523	459	55	900	5	0	993	525	26	978	208	12064
Utah	332	205	0	605	0	0	298	0	20	797	26	387	3625	0	6296
Washington	3116	947	24	1650	1788	240	370	0	0	908	325	0	0	14	9381
Wyoming	476	391	0	926	0	0	1464	0	231	187	82	54	234	15	4059
Total:	15582	13056	229	12124	2878	1139	8473	318	1798	6702	2334	6264	19142	2847	92887

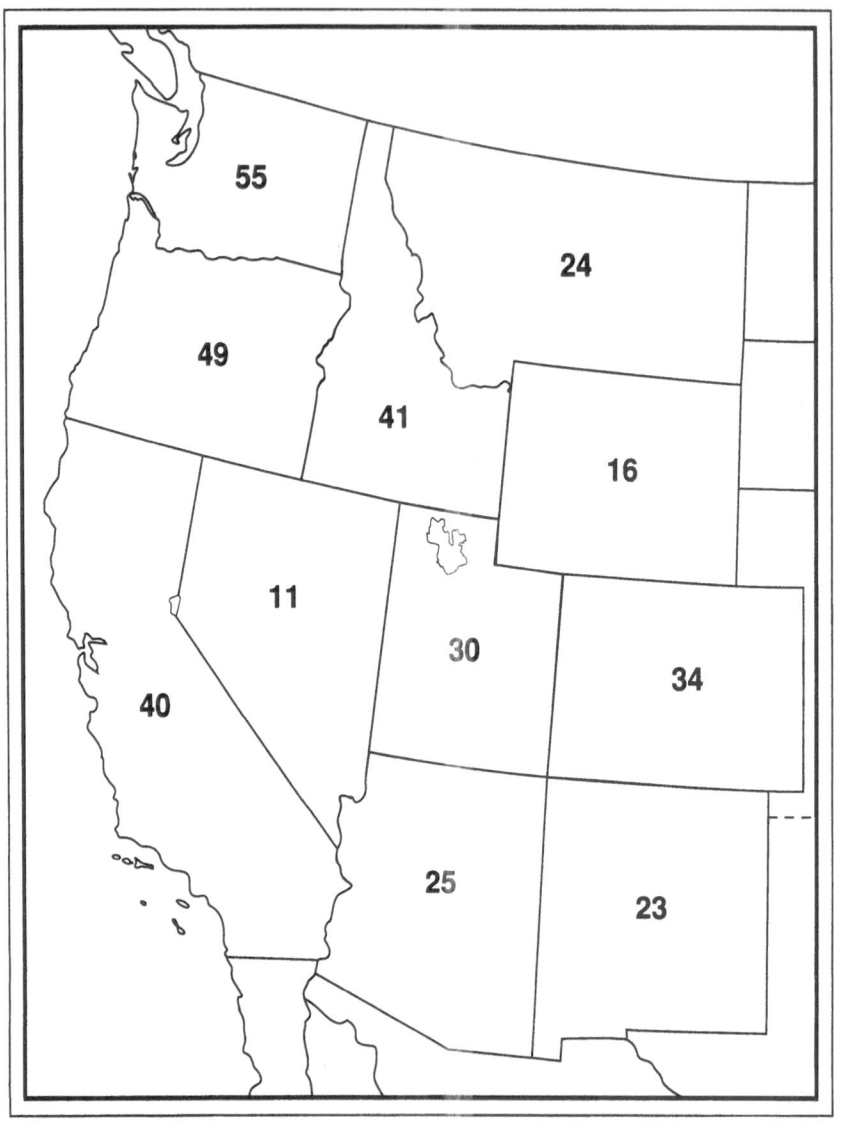

Figure 1.3 Percent area, by state, classified as forest land (USDA 1981).

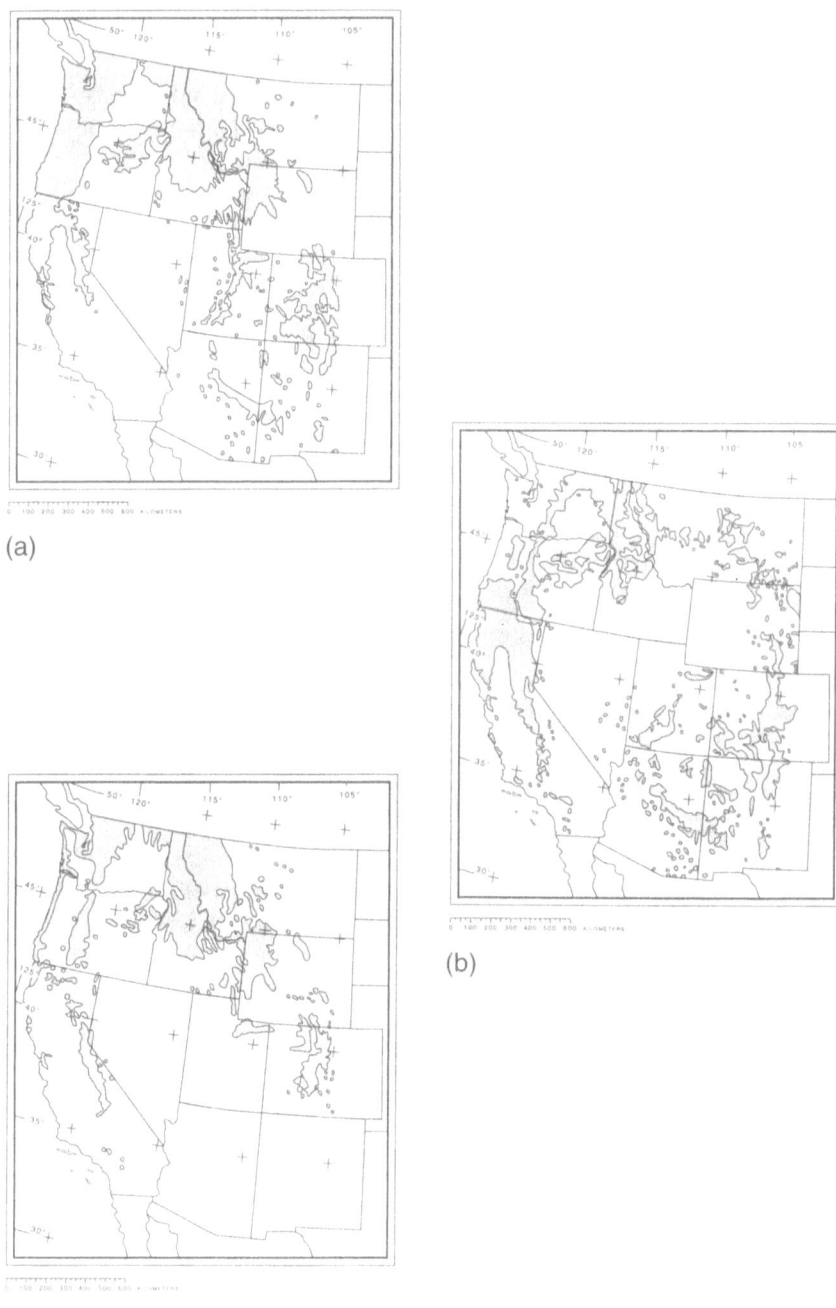

(a)

(b)

(c)

Figure 1.4   Ranges of (a) Douglas-fir, (b) ponderosa pine, and (c) lodgepole pine. From Little (1971).

shadows of large mountain ranges, restricts most forests to medium and high elevations where evapotranspiration demand is lower and precipitation is greater.

Elevations of both upper and lower treeline tend to rise with decreasing latitude. Upper treeline changes at about 110 m per degree latitude (Daubenmire 1954) so that conifer forests on the west slope of the Northern Cascades (48°N) are found in the elevational range from sea level to 1400–1750 m (Franklin and Dyrness 1973), while conifer forests in the San Bernardino Mountains (36°N) occur from 600 m to between 2600 and 3500 m. Upper treeline is generally lower in coastal mountains with maritime climates than in interior mountains (Franklin and Dyrness 1973, Smith 1985).

Within elevational zones, topographic and moisture gradients complicate forest distribution patterns (Franklin and Dyrness 1973, Vankat 1982, Peet 1988). Forest types tend to occur at higher elevations on xeric, south-facing slopes and at lower elevations on north slopes, mesic canyons and riparian zones.

Disturbance

Natural disturbances greatly influence forest structure and species composition in the West (Oliver 1981). At any given time, a large proportion of Western forests have been recently affected by fire, insect or disease outbreaks, or wind-throws. Douglas-fir, lodgepole pine, and quaking aspen are major Western species requiring disturbance in most areas of their ranges to maintain their type.

Fire is the most important natural disturbance in much of the West. In the Rocky Mountains, pre-settlement fire frequencies ranged from 5 to 12 years for low elevation ponderosa pine (Arno 1980, Gruell 1985, Peet 1988); from 20 to 40 years for low elevation Douglas-fir in Montana (Gruell 1985); and from 200 to 400 years for subalpine forests in the Central and Southern Rockies (Peet 1981, Romme and Knight 1981). Prior to 1875, mid-montane forests in the Sierra Nevada experienced fires on an average of every 8 years for pine sites and every 16 years for fir sites (Barbour 1988). Fire frequencies in the Cascade Mountains of Washington are low, 300–400 years (Hemstrom and Franklin 1982, Franklin 1988), but are often catastrophic when they do occur. However, infrequent fires are important in maintaining long-lived seral species such as Douglas-fir (Franklin 1988).

Current management practices have altered fire frequency in much of the West. Active fire suppression, or indirect suppression due to forest fragmentation, has decreased the frequency of fires (Peet 1988). As a result, more intense (Barbour 1988) and more extensive (Romme and

Despain 1989) fires may occur than under a normal fire regime. Alternatively, human-ignited fires may increase the overall fire frequency, as has occurred in some areas of the California evergreen chaparral (Keeley and Keeley 1988).

Insects and pathogens also have major influences on Western forest condition. Between 1979 and 1983, outbreaks of western spruce budworm were recorded on 2.8 million ha of forested land. Mountain pine beetle infestations occurred on a reported 1.7 million ha. Dwarf mistletoe occurred on 9.2 million ha of forested land (USDA 1988). Root diseases are current management concerns on 6 million ha of Western commercial forest land (USDA 1988). Effects of insects and pathogens range from reduced growth to increased mortality, although increases in stand-level growth from thinning effects are possible (Romme et al. 1986).

Human disturbances of Western forests have been and continue to be extensive. For example, as much as 90% of Pacific Northwest old-growth Douglas-fir present in 1800 has been eliminated, largely through logging (Henderson et al. 1989); by 1973, 75% of private timberlands in California had undergone some harvesting (USDA, unpub. data); much of the pinyon-juniper woodlands are used for grazing as are large portions of ponderosa pine, California hardwoods, and other Western forest types (USDA 1981); and only about 12% of Western forest lands are excluded from commodity production through incorporation in parks, wilderness areas, or other set-asides (Waddell et al. 1989). Harvest practices greatly alter forest structure, composition, and function. Clear-cutting creates even-aged stands and, along with associated road building, fragments the forest landscape (Harris 1984, Franklin and Forman 1987). Selective cutting potentially alters age distributions, species composition, and growth rates of remaining trees. Reforestation efforts usually involve only a few species from a limited number of seed sources, causing decreased species and genetic diversity (California Gene Resources Program 1982, Maser 1988, Wilcove 1988).

Mining, agriculture, and development of buildings and roads convert forests to non-forest uses and fragment remaining stands. In California between 1950 and 1980, 338,000 ha of forests were converted to agricultural or urban uses (State of California 1988). During the same period, 486,000 ha of timberland rated "highly productive" were lost in the three lower Pacific Coast states, with greatest losses due to road building and grazing clearings in Oregon, and urban expansion in the Puget Sound region of Washington (USDA 1981).

Western Forest Regions

Pacific Northwest

Forests of the Pacific Northwest occupy the area between the crests of the Northern and Central Cascades and the Pacific coast, and extend into the coast ranges of northern California. A climate of mild, wet winters and dry summers has fostered development of some of the most massive coniferous forests in the world, both in terms of size of individual trees (Table 1.2) and total biomass (Lassoie et al. 1985). Topographic and moisture gradients play primary roles in determining distributions of the forests (Figure 1.5). As in the rest of the West, disturbance also has a large influence on community composition. Widely distributed species such as Douglas-fir and red alder (*Alnus rubra*) depend on periodic disturbances to maintain their populations. Fire and wind are the main natural disturbances and timber harvesting is the most important human disturbance in forests of the Pacific Northwest.

Forest types

Pacific Northwest forests are comprised of four main SAF Forest Type Groups: Hemlock-Sitka spruce, Douglas-fir, Fir-spruce, and Hardwoods (Appendix I).

Table 1.2. Selected Pacific Northwest species and typical sizes attained on productive sites (from Franklin and Dyrness 1973).

Species	Age (yrs)	Diameter (cm)	Height (m)
Pacific silver fir	400+	90–110	45–55
noble fir	400+	100–150	45–70
Alaska yellow cedar	1000+	100–150	30–40
Engelmann spruce	500+	100+	45–50
Sitka spruce	800+	180–230	70–75
Douglas-fir	750+	150–220	70–80
western redcedar	1000+	150–300	60+
western hemlock	400+	90–120	50–65

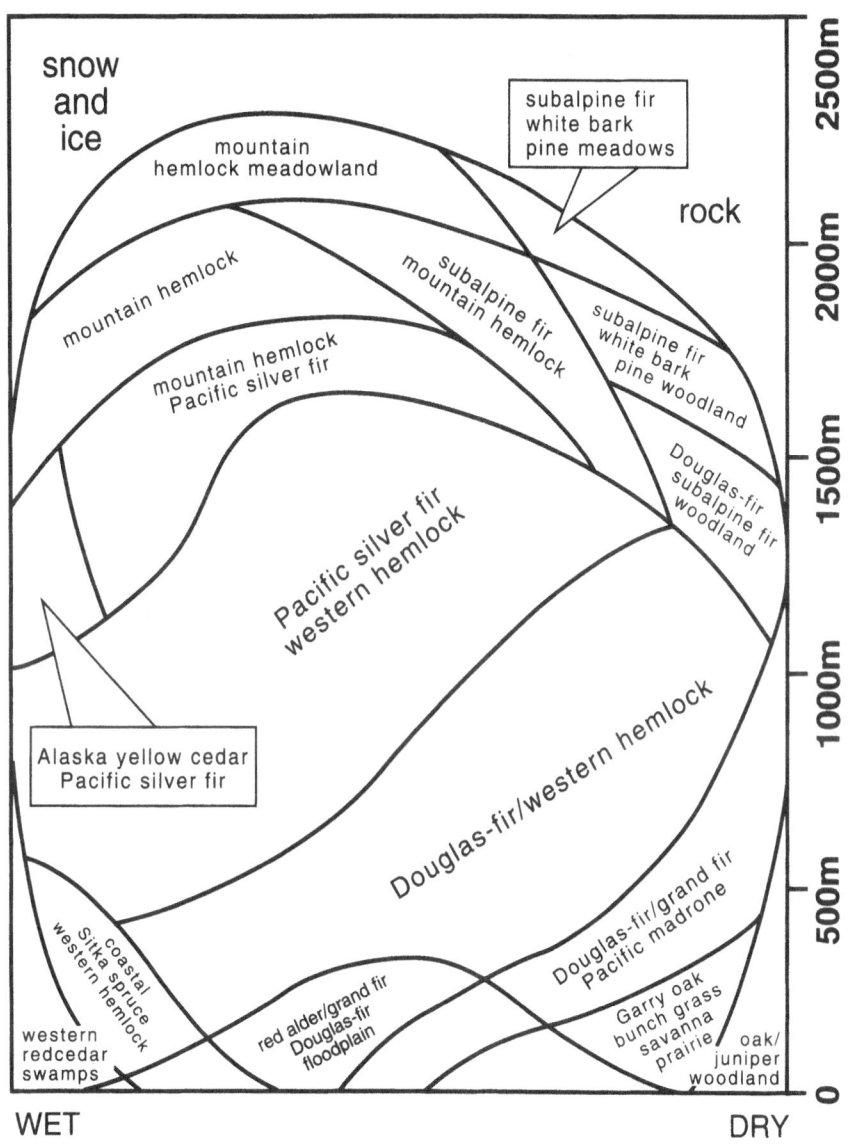

Figure 1.5 Generalized elevation and moisture relationships for Washington and Oregon forests west of the Cascades crest. Figure developed by Robert VanPelt, University of Washington.

Western hemlock-Sitka spruce (*Tsuga heterophylla-Picea sitchensis*) forests are best developed in a foggy zone along the Washington/Oregon coast within 80 km of the sea and at elevations up to 450 m (Eyre 1980, Figure 1.5). The species share dominance on fertile, well-drained soils on moist slopes, river edges and lake margins. In most cases, western hemlock is the theoretical climax species. Sitka spruce is found within virtually all of these forests since it is a long-lived species (Table 1.2) and actively colonizes disturbed sites (Franklin 1988). Sitka spruce tends to dominate wetter sites and because of its salt tolerance may form pure stands in the ocean spray zone (Eyre 1980). Especially large individuals remain in the Olympic Peninsula rain forest (Fonda and Bliss 1969). Associate species vary in importance with site conditions and include Douglas-fir, Pacific silver fir (*Abies amabilis*), western redcedar (*Thuja plicata*), Alaska yellow cedar (*Chamaecyparis nootkatensis*) and Port Orford cedar (*Chamaecyparis lawsoniana*). Lodgepole pine can share salt spray and boggy areas with Sitka spruce. Hardwoods serve as early successional species following disturbance, and include red alder, bigleaf maple (*Acer macrophyllum*), and vine maple (*Acer circinatum*).

Douglas-fir is the main forest type west of the Northern and Central Cascades crests at elevations from sea level to 700–1000 m (Franklin 1988). It reaches higher elevations farther south in the Klamath Mountains and California Coast Ranges to north of San Francisco (Eyre 1980). These forests are dominated by Douglas-fir, western hemlock, and western redcedar. Douglas-fir is generally seral with western hemlock the putative climax species. Douglas-fir dominates large areas, often as pure stands following disturbance. These old-growth, or ancient, forests often exceed 400 years in age and are characterized by huge Douglas-fir trees with understory of smaller western hemlock and western redcedar. Douglas-fir can be the climax on drier sites, while western redcedar achieves major importance only on very wet sites (Franklin 1988). The dominants are sometimes joined by grand fir (*Abies grandis*), Sitka spruce, and western white pine (*Pinus monticola*). Hardwoods are generally restricted to early successional roles following disturbance or to riparian or other extreme sites. Red alder, black cottonwood (*Populus trichocarpa*) and Oregon white oak (*Quercus garryana*) fill these roles throughout much of this zone. In southern Oregon, incense-cedar (*Calocedrus decurrens*), sugar pine (*Pinus lambertiana*) and ponderosa pine are associates.

Fir-spruce forests occupy intermediate to high elevations in the Cascades and Olympic Mountains (Figure 1.5). The coastal true fir-hemlock type lies at elevations between the Douglas-fir–western hemlock forests and subalpine forests (Zobel et al. 1976). The presence of permanent winter snowpacks differentiates the zone from lower elevations (Franklin 1988). Dominants include Pacific silver fir, western hemlock, noble fir (*Abies* *procera*), Douglas-fir, and western white pine. At the top of its elevation range, the coastal true fir-hemlock forest grades into the mountain hemlock (*Tsuga mertensiana*) forest type (Figure 1.5).

Mountain hemlock is the highest and coldest forest type in the Olympic Mountains and Cascades. Elevational boundaries of the mountain hemlock zone range from 1300–1700 m in northern Washington to 1700–2000 m in the southern Central Cascades (Franklin and Dyrness 1973). In addition to mountain hemlock, major species include lodgepole pine, western white pine, whitebark pine (*Pinus albicaulis*), Engelmann spruce (*Picea engelmannii*), subalpine fir (*Abies lasiocarpa*) and Alaska yellow cedar. South of the McKenzie River in southern Oregon noble fir is replaced by California red fir (*Abies magnifica*) as the Cascade and Sierra Nevada forest types intersect. At the upper elevation limit of this zone, closed forests of mountain hemlock and associates give way to a parkland of forests, meadows and alpine communities. Parklands are developed to their fullest extent in the Olympic Mountains and Northern Cascades.

The eastern slopes of the Cascades show a mixture of coastal and Rocky Mountain forests in response to the drier interior climate (Franklin and Dyrness 1973). Major differences from the west slope include western juniper (*Juniperus occidentalis*) woodlands in central and southern Oregon, and extensive ponderosa pine forests at slightly higher elevations throughout.

Rocky Mountains and Intermontane Regions

Rocky Mountain-type forests occur in the three Rocky Mountain physiographic provinces (Figure 1.2) as well as in the adjacent Colorado Plateau, Columbia Plateau, Basin and Range Provinces, and even the lower east slopes of the Cascade-Sierra Ranges (Appendix I). Although certain species and forest types serve as common threads to integrate the Rocky Mountain and Intermontane regions, important differences occur at scales ranging from local to regional. Several of the differences between Rocky Mountain regions are worth noting as a means of characterizing the whole.

Many tree species found in Pacific Northwest forests are also important in the Northern Rocky Mountains. Western hemlock and western redcedar (Figure 1.6), Pacific yew (*Taxus brevifolia*), mountain hemlock and subalpine larch (*Larix lyalli*) are among the species showing this distribution. This eastward extension of west coast species, referred to as the "Pacific peninsula" (Daubenmire 1943), follows a pattern of intrusions of Pacific air masses that moderate summer drought and temperature in this region (Mitchell 1976).

Southern Arizona and New Mexico are distinguished from the rest of the Rocky Mountain and Intermontane regions by a high incidence of species with affinities to the Madrean flora (Axelrod and Raven 1985, Peet 1988). Examples include Mexican pinyon (*Pinus cembroides*), Apache pine (*Pinus engelmannii*), Arizona pine (*Pinus ponderosa arizonica*), southwestern white pine (*Pinus strobiformis*), and Emory oak (*Quercus emoryi*). Climatic controls on the northern limits of Madrean species include decreasing summer rain and decreasing winter temperatures (Mitchell 1976, Neilson and Wullstein 1983).

Upper treeline increases from less than 2600 m in the Northern Rockies to above 3600 m in northern New Mexico (Peet 1988). Lower treeline also increases, and all forest types show a general upward shift in their elevation zones from north to south. Topography, edaphic factors, and species composition influence the basic vegetation patterns at particular locations (Figure 1.7). Disturbances then overlay these factors, often creating a mosaic of forest and vegetation types in a particular zone, and frequently preventing achievement of a climax or steady-state community (Peet 1981, Parker and Peet 1984).

Forest types

Forests of the Rocky Mountains are comprised of eight main SAF Forest Type Groups: Hardwoods, Pinyon- juniper woodlands, Ponderosa pine, Douglas-fir, Western white pine, Western larch, Lodgepole pine, and Engelmann spruce and subalpine fir.

Hardwood forests in the Rocky Mountain and Intermontane Regions are dominant only in specific environmental or successional situations. Riparian zones and moist canyon bottoms offer mesic habitats that support mixtures of willow (*Salix*) and cottonwood (*Populus*) (Eyre 1980). These forests are found primarily in Arizona and New Mexico, where oak species are also components at low elevations (Peet 1988), and in the Great Basin, Colorado Plateau, and Rocky Mountain foothills to the Canadian border (Eyre 1980). Open woodlands dominated by evergreen oaks form a climax at lower-middle elevations (1200–1800 m) in southern Arizona and New Mexico (Figure 1.7a). Dominants in these xeric forests include Emory oak, Arizona white oak (*Quercus arizonica*), Mexican blue oak (*Q. oblongifolia*) and silverleaf oak (*Q. hypoleucoides*) as well as assorted junipers (*Juniperus* spp.). As elevations increase, there is a transition to pine-oak forests including Chihuahua pine (*Pinus leiophylla*), Mexican pinyon, and Arizona pine (Peet 1988).

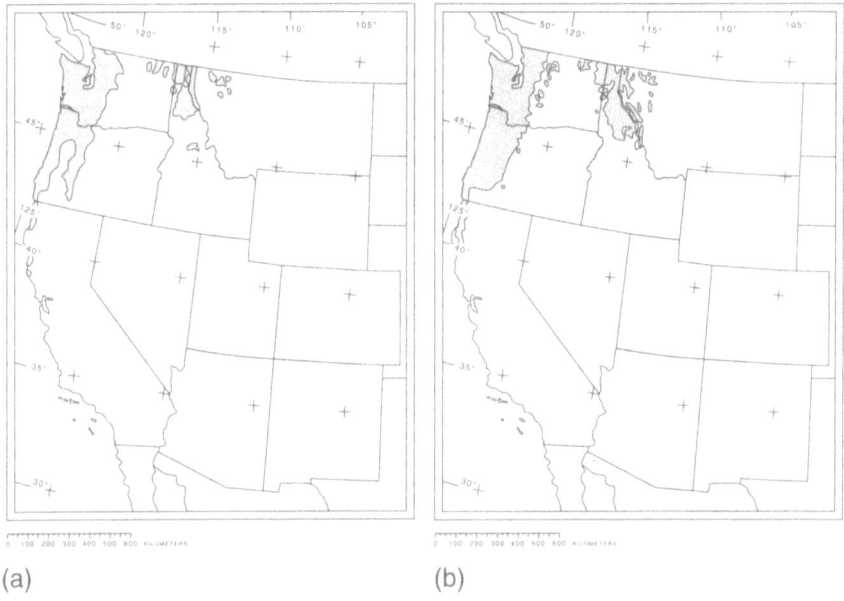

(a)   (b)

Figure 1.6 Ranges of (a) western hemlock and (b) western redcedar. From Little (1971).

Quaking aspen is the most widely distributed tree species in North America and the only major deciduous tree species in the Rockies. Typically seral, quaking aspen reproduces by root sprouts following fire, resulting in clonal stands as large as 20 ha (Barnes 1975). The type reaches its maximum development in the Colorado Rockies between 2000–3000 m on moist north slopes (Eyre 1980, Figure 1.7b). Aspen competes throughout much of its range with lodgepole pine. Barring disturbance, it is most frequently succeeded by Douglas-fir at lower elevations and by Engelmann spruce and subalpine fir at higher elevations (Peet 1988).

Pinyon-juniper woodlands are the climax forest type for large portions of the Basin and Range, Colorado Plateau, and Southern Rocky Mountain Provinces. These woodlands consist of a shifting mix of pine and juniper species that are adapted to low precipitation and hot summers. Stand structure is characterized by open, though often shrub-like, woodland with stems less than 9 m (Eyre 1980). Junipers tend to have the advantage on drier sites, pines on higher, wetter sites (Peet 1988). Dominant species in different regions (West 1988) are: Southern Rockies—one-seed juniper (*Juniperus monosperma*), pinyon pine (*Pinus edulis*); Northern Rockies—Utah juniper (*Juniperus osteosperma*), Rocky Mountain

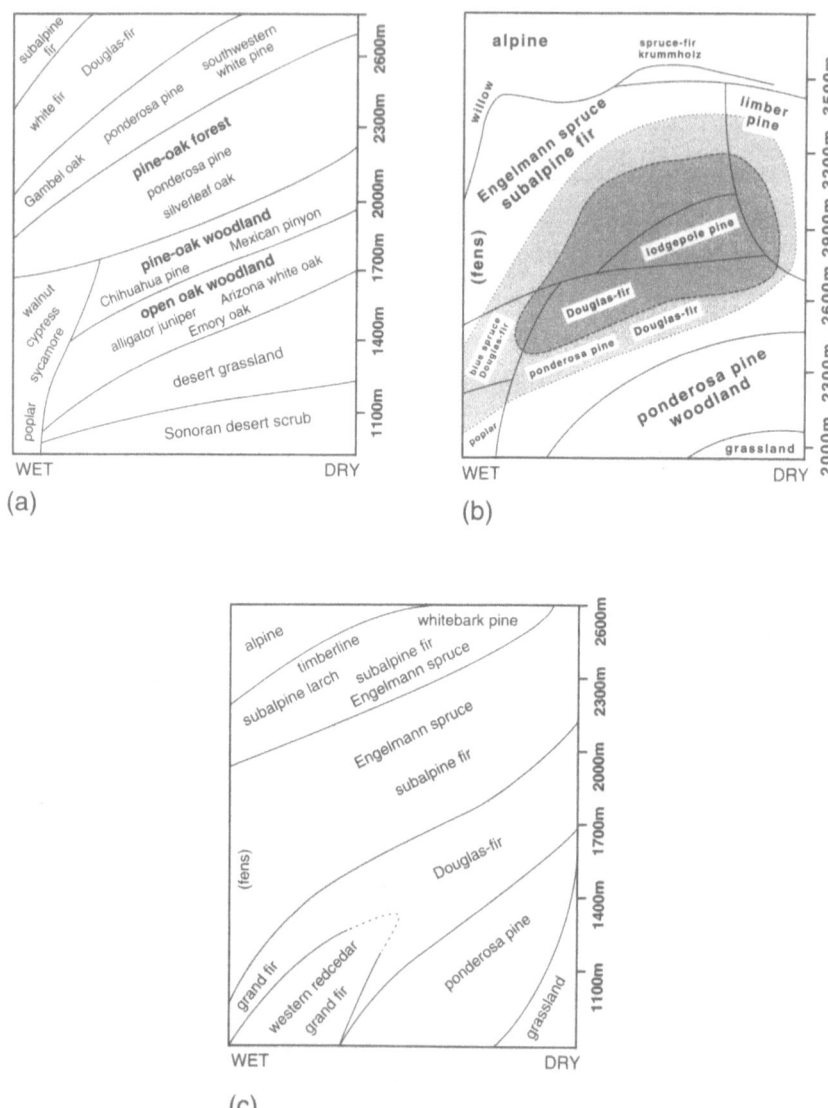

Figure 1.7 Elevation and moisture relationships for forests in (a) Santa Catalina Mountains, Arizona, (b) Front Range, Colorado, and (c) Bitterroot Mountains, Montana. In (b), dark stippling represents conditions where lodgepole pine is an important post-disturbance species, while the combined area of dark and light stippling shows conditions where quaking aspen is an important post-disturbance species. Figure (a) from Whittaker and Niering (1965); (b) and (c) from Peet (1988).

juniper (*Juniperus scopulorum*); Great Basin—Utah juniper, singleleaf pine (*Pinus monophylla*); Colorado Plateau—Utah juniper, one-seed juniper, pinyon pine.

Ponderosa pine is found throughout the Rocky Mountain and Intermontane regions. These forests occur at elevations between 300–1800 m in the Northern Rockies (Figure 1.7c) and between 1800–2600 m in the Southern Rockies (Eyre 1980). At lower elevations the species typically forms a climax of open stands with extensive grass or shrub understories. At higher elevations, closed stands are the norm, but are often seral to Douglas-fir (Peet 1988). In Arizona and New Mexico, ponderosa pine associates with Rocky Mountain juniper at low elevations; at higher elevations ponderosa pine is a component of a mixed conifer forest that includes Douglas-fir, white fir, blue spruce (*Picea pungens*), quaking aspen and southwestern white pine. Throughout its range, fire tends to maintain ponderosa pine forests while fire exclusion speeds conversion to other forest types.

Douglas-fir functions as a climax dominant in many locations and as a seral species in others. It grows in the Northern Rockies between 370–2440 m (Eyre 1980) and is typically in association with western hemlock, western redcedar, grand fir, western white pine, western larch (*Larix occidentalis*), lodgepole pine and quaking aspen at low elevations. At higher elevations, Engelmann spruce, subalpine fir, mountain hemlock, and whitebark pine are associates. Douglas-fir often follows lodgepole pine or ponderosa pine, and is succeeded by western hemlock or grand fir.

In the Southern Rockies, the elevational range of Douglas-fir is 1830–2590 m (Eyre 1980). Douglas-fir is often replaced by ponderosa pine at the lower boundary and by Engelmann spruce and subalpine fir at the upper boundary. In the Mexican Highlands (Figure 1.2), Douglas-fir is restricted to mesic sites (Brady and Bonham 1976, Niering and Lowe 1984) where it may associate with white fir (*Abies concolor*) or blue spruce. On less mesic sites, ponderosa pine is an associate.

Western white pine dominates sufficient area only in the Northern Rockies to be recognized as a forest type. The species is always seral and depends on fire or other disturbances to maintain the type (Eyre 1980). Western white pine is best developed on moist sites from 450 to 1200 m in elevation. It rarely forms pure stands and is commonly associated with western redcedar, western hemlock, grand fir, and western larch. At higher elevations, subalpine fir and Engelmann spruce mix with western white pine. The species suffers badly from attack by the mountain pine beetle and white pine blister rust.

Western larch forests are limited to the Northern Rockies, the adjacent
Blue Mountains, and the east slope of the Northern Cascades (Eyre 1980).
Western larch is a seral species most frequently succeeded by Douglas-
fir, grand fir and subalpine fir. Other associates include Engelmann
spruce, western redcedar, and western hemlock.

Lodgepole pine (var. *latifolia*) sometimes forms a stable climax (Peet
1981), but is most often a seral species whose wide distribution in the
Rocky Mountain region is due to the high frequency of disturbance,
especially fire (Peet 1988). The species is absent only from the Mexican
Highlands and Colorado Plateau (Little 1971, Figure 1.4c) where quaking
aspen often occupies habitats suitable for lodgepole. Lodgepole pine in
the Colorado Rockies is found in the elevational range of 2150 to 3350 m
(Eyre 1980).

Lodgepole pine is so broadly adaptable that it can be found in association somewhere with most of the major Rocky Mountain species (Pfister
and Daubenmire 1975). Among its most frequent associates are subalpine fir, Engelmann spruce, white fir, and Douglas-fir, all of which may
eventually replace it if fire does not occur. Climax stands of lodgepole
pine occupy infertile sites or sites with no local seed sources of more
shade tolerant species (Despain 1973, 1983). Adaptations of lodgepole
pine to fire may include serotiny, although the degree of serotiny varies
and depends upon factors including the role of fire in stand establishment (Lotan 1975, Muir and Lotan 1985).

Engelmann spruce and subalpine fir are the main components of the
subalpine forest throughout the Rockies. They generally occupy the zone
between Douglas-fir and timberline. As timberline rises to the south,
subalpine forests may extend to the top of individual peaks or may be
excluded, especially in the Mexican Highlands and the Great Basin. In
other cases, isolated peaks may host only one of the two species (Sawyer
and Kinraide 1980, Harper et al. 1978).

Engelmann spruce and subalpine fir form climax or long-lived seral
communities, either in mixed or pure stands (Eyre 1980, Knapp and
Smith 1982). In the Northern Rockies, subalpine fir is the main climax
species (Daubenmire and Daubenmire 1968, Pfister et al. 1977), while
both species are generally part of the climax in the Central and Southern
Rockies (Alexander 1974). Common associates include Douglas-fir and
lodgepole pine. In the Northern Rockies, subalpine spruce-fir forests
often include Pacific silver fir, mountain hemlock, and subalpine larch; in
the Southern Rockies associates include blue spruce, white fir and
quaking aspen. More xeric sites in the subalpine zone are often occupied
by five-needle pines. Whitebark pine is important in this role in the
Northern Rockies, while bristlecone pine (*Pinus aristata*) is found in the

Southern Rockies. Great Basin bristlecone pine (*Pinus longaeva*), found on peaks in the Great Basin, has the distinction of reaching the oldest age (more than 4900 years) of any tree species (Schulman 1954).

California

California has more forested land than any other Western state (Table 1.1). A high diversity of climates, land forms, and soils has led to a corresponding diversity in forest species and types. Endemic and rare communities such as coastal redwood, giant sequoia (*Sequoiadendron giganteum*) groves in the Sierra Nevada, and Engelmann oak (*Quercus engelmannii*) and California walnut (*Juglans californica*) woodlands in the Transverse and Peninsular Ranges (Barbour and Major 1988) add to the overall richness and importance of California forests.

Dry moisture regimes preclude forests over much of the Central Valley, south coast basins, and southeastern California. On the east slope of the Sierra Nevada, montane conifer forests and pinyon-juniper communities tend to grade directly into sagebrush steppe or arid scrublands (Küchler 1988). On the west side of the Sierra Nevada, Transverse and Peninsular Ranges and the interior slopes of the Coast Ranges, a variety of hardwood, mixed evergreen and chaparral types separate montane coniferous forests from grassland or coastal sagebrush communities (Figure 1.8).

Conifers are components of most of these hardwood forest types, but dominance is held by a number of evergreen or deciduous oaks. Blue oak (*Quercus douglasii*) is the main species in the blue oak woodland that extends around the inner edge of most of the Central Valley at elevations of 100–1200 m (Barbour 1988). Other species include digger pine (*Pinus sabiniana*), coast live oak (*Quercus agrifolia*), interior live oak (*Quercus wislizenii*), valley oak (*Quercus lobata*), and black oak (*Quercus kelloggii*). Groups of trees are often interspersed with grassland and chaparral in landscape patterns determined by topography, soils, and disturbances (Barbour 1988).

As elevation and moisture increase, interior live oak, coast live oak, black oak, and canyon live oak (*Quercus chrysolepis*) increase in importance and are dominants in different oak woodlands (Eyre 1980, Vankat 1982). In the Klamath Mountains and Coast Ranges, a mixed evergreen forest (Franklin 1988, Sawyer et al. 1988) combines oak woodland and montane conifer species. Canyon live oak, bigleaf maple, and California-laurel (*Umbellularia californica*) are found throughout the range of this forest type. Douglas-fir, Pacific madrone (*Arbutus menziesii*), coast live oak, tanoak (*Lithocarpus densiflorus*), canyon live oak, and golden chinquapin (*Castanopsis chrysophylla*) are dominants or important associates in many

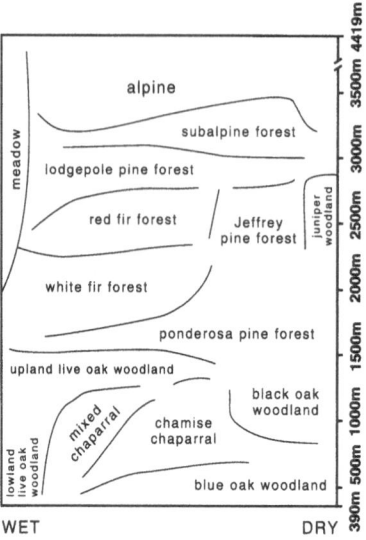

Figure 1.8 Elevation and moisture relationships for forests in Sequoia National Park, southern Sierra Nevada, California. From Vankat (1982).

areas. In the Sierra Nevada, ponderosa pine and white fir are conifer components of mixed evergreen forests, while in the Transverse and Peninsular Ranges, Coulter pine (*Pinus coulteri*) and big cone Douglas-fir (*Pseudotsuga macrocarpa*) occur with canyon live oak (Barbour 1988).

Forest types

Coniferous forests in the Sierra Nevada are comprised of five main SAF Forest Type Groups: Redwoods, Ponderosa pine, Fir, Lodgepole pine, and Subalpine.

Redwood forests are a southern extension of the Pacific Northwest hemlock-Sitka spruce forest and extend from extreme southwestern Oregon down the California coast to below San Francisco. This type is characterized by the dominance of redwood (*Sequoia sempervirens*) and generally occurs in a narrow band starting 2–3 km inland with a maximum width of 35 km (Zinke 1988). It is restricted to areas of summer fog, moist slopes and river bottoms at low to moderate elevations (Eyre 1980). Douglas-fir is usually present. Other associates include tanoak, western hemlock, Sitka spruce, knobcone pine (*Pinus attenuata*), incense-cedar, madrone, and red alder (Franklin 1988).

Ponderosa pine forests occur extensively in the Klamath Mountains, east slopes of the northern Coast Ranges, southern Cascades, Sierra Nevada, Transverse and Peninsular Ranges. North of the Sierra Nevada, ponderosa pine occurs at elevations from 300–1220 m, often with Douglas-fir, sugar pine, incense-cedar, Pacific madrone, and various oaks. In the Sierra Nevada, ponderosa pine dominates in a band from 300–1800 m in the north and 1200–2100 m in the south (Rundel et al. 1988). The species occurs in pure stands or in various mixtures with white fir, incense-cedar, black oak, and sugar pine. Canopies range from 30–60 m height and 50–80% cover (Barbour 1988). Southward, at higher elevations, and on the eastern slope, Jeffrey pine (*Pinus jeffreyi*) replaces ponderosa pine.

Fir dominated forests replace ponderosa pine and Jeffrey pine on mesic sites in their respective elevation zones in the Sierra Nevada (Figure 1.8). White fir dominates mesic sites at mid-elevations (1250–2200 m) (Vankat and Major 1978, Vankat 1982). Sugar pine, incense-cedar, giant sequoia, and at lower elevations, black oak and ponderosa pine, are associates (Rundel et al. 1988). California red fir replaces white fir as elevation increases, often forming pure stands that are considered the climax for the zone (Eyre 1980). Trees reach 30–45 m with an average 60% canopy cover.

Lodgepole pine (*Pinus contorta* var. *murrayana*) is important in the Klamath Mountains, southern Cascades, and along the full length of the Sierra Nevada. It is a high elevation species in the Sierra Nevada, with greatest importance above the California red fir and Jeffrey pine zones (Figure 1.8). Lodgepole pine dominates forests up to 2400 m in the northern Sierra Nevada and 3000 m in the southern Sierra Nevada (Barbour 1988). Unlike the Rocky Mountain variety, *Pinus contorta* var. *murrayana* cones are not serotinous (Parker 1986, Pfister and Daubenmire 1975), although fire does facilitate reproduction by preparing a mineral soil seedbed. Trees are generally less than 20 m tall (Barbour 1988). Best development occurs on arid sites with shallow soils or on wet, cold sites (Rundel et al. 1988). Quaking aspen or black cottonwood may replace lodgepole pine at wet sites.

Subalpine forests occupy the highest forested zones in the Southern Cascades and southward. The subalpine zone lies between 2300–2900 m in the northern Sierra Nevada, 3000-3400 m in the southern part of the range, and 2800–3500 m in the Transverse and Peninsular Ranges (Barbour 1988). About 85% of precipitation in this zone of the Sierra Nevada falls as snow, and growing seasons are short. Trees tend to grow in isolated clusters covering only 5–40% of the landscape. Canopy heights range from 30 m to low krummholz (Eyre 1980). Important tree species include whitebark pine, limber pine (*Pinus flexilis*), lodgepole pine, western white pine, mountain hemlock, and foxtail pine (*Pinus balfouriana*).

Summary and Conclusions

There are a number of characteristics of Western forests that increase the difficulty of evaluating the effects of air pollution. These are generally factors that increase the variability of the background "noise" from which pollution signals must be extracted, and include:

- High species diversity impedes generalizations about forest responses. Sensitivity to pollutants differs between species (Davis and Wilhour 1976) and between different genotypes (see Chapter 5). Experimental work such as controlled exposures of seedlings to pollutants is necessarily limited by time and funding to a few species and genotypes, making extrapolation to the "real world" difficult. For the same reasons, field surveys of forest condition are limited to a small number of sites and forest types, raising questions about completeness of coverage and representativeness of results.

- Complex topography interacts with broader patterns of climate (see Chapter 2) and soils (see Chapter 4) to create an almost infinite number of combinations of environmental factors that influence forest response to air pollution. For example, water availability, through influence on stomatal closure, affects tree response to ozone (see Chapter 5). Experiments simulate only a few of these combinations. Western topography also complicates prediction of pollutant exposure for specific forests. Pollutant transport models are not well developed for complex terrain, and differences in elevation can cause large changes in wet and dry deposition over short distances (see Chapter 3).

- Disturbance is common throughout the West, and creates changes in growth patterns that can mask pollutant effects. Catastrophic events such as major fires or clear-cutting start new sequences of stand development and tree growth (see Chapter 6). Measured growth must be evaluated in terms of deviations from these expected trends in order to detect growth effects of air pollution (see Chapter 7). Growth changes caused by insects, disease, wind damage, and management practices add "noise" to these background patterns, and complicate interpretation.

Physiographic and forest patterns in the West present a difficult challenge to air pollution research. The remaining chapters in Section I define this challenge more completely, while Section II describes some of the efforts to meet it.

References

Alexander RR (1974) *Silviculture of Subalpine Forests in the Central and Southern Rocky Mountains: The Status of Our Knowledge.* U.S. Department of Agriculture Forest Service Research Paper RM-121

Arno SF (1980) Forest fire history in the northern Rockies. *Journal of Forestry* 78:460–465

Axelrod DI, Raven PH (1985) Origins of the Cordilleran flora. *Journal of Biogeography* 12:21–47

Barbour MG (1988) California upland forests and woodlands. In: Barbour MG, Billings WD (eds) *North American Terrestrial Vegetation.* Cambridge University Press, New York, pp 131–164

Barbour MG, Major J (1988) Introduction. In: Barbour MG, Major J (eds) *Terrestrial Vegetation of California.* California Native Plant Society, Sacramento, pp 3–10

Barnes BV (1975) Phenotypic variation of trembling aspen in western North America. *Forest Science* 21:319–328

Brady W, Bonham CD (1976) Vegetational patterns on an altitudinal gradient, Huachuca Mountains, Arizona. *Southwestern Naturalist* 21:55–65

California Gene Resources Program (1982) *Douglas-fir Genetic Resources: An Assessment and Plan for California.* National Council on Gene Resources, Berkeley, CA

Daubenmire RF (1943) Vegetational zonation in the Rocky Mountains. *Botanical Review* 9:326-393

Daubenmire RF (1954) Alpine timberlines in the Americas and their interpretation. *Butler University Botanical Studies* 11:119–136

Daubenmire R, Daubenmire JB (1968) *Forest Vegetation of Eastern Washington and Northern Idaho.* Washington Agricultural Experiment Station Technical Bulletin 60

Davis DD, Wilhour RG (1976) *Susceptibility of Woody Plants to Sulfur Dioxide and Photochemical Oxidants.* EPA-600/3-76-102, U.S. Environmental Protection Agency, Washington, DC

Despain DG (1973) Vegetation of the Big Horn Mountains, Wyoming, in relation to substrate and climate. *Ecological Monographs* 43:329–355

Despain DG (1983) Nonpyrogenous climax lodgepole pine communities in Yellowstone National Park. *Ecology* 64:231–234

Eyre FH (ed) (1980) *Forest Cover Types of the United States and Canada.* Society of American Foresters, Washington, DC

Fenneman NM (1930) *Physical Divisions of the United States.* Map at 1:7,000,000. U.S. Department of the Interior, U.S. Geological Survey, Washington, DC

Fenneman NM (1931) *Physiography of the Western United States.* McGraw-Hill, New York

Fonda RW, Bliss LC (1969) Forest vegetation of the montane and subalpine zones, Olympic Mountains, Washington. *Ecological Monographs* 39:271–301

Franklin JF (1988) Pacific Northwest forests. In: Barbour MG, Billings WD (eds) *North American Terrestrial Vegetation.* Cambridge University Press, New York, pp 103–130

Franklin JF, Dyrness CT (1973) *Natural Vegetation of Oregon and Washington.* Oregon State University Press, Corvallis

Franklin JF, Forman RTT (1987) Creating landscape patterns by forest cutting: Ecological consequences and principles. *Landscape Ecology* 1:5–18

Gleason HA, Cronquist A (1964) *The Natural Geography of Plants.* Columbia University Press, New York

Griffin JR (1988) Oak woodland. In: Barbour MG, Major J (eds) *Terrestrial Vegetation of California.* California Native Plant Society, Sacramento, pp 383–415

Gruell GE (1985) Fire on the early western landscape: An annotated record of wildland fires 1776-1900. *Northwest Science* 59:97–107

Harris LD (1984) *The Fragmented Forest: Island Biogeography Theory and the Preservation of Biotic Diversity.* University of Chicago Press, Chicago

Harper KT, Freeman DC, Ostler WK, Klikoff LG (1978) The flora of Great Basin Mountain Ranges: Diversity, sources, and dispersal ecology. *Great Basin Naturalist Memoirs* 2:81–103

Hemstrom MA, Franklin JF (1982) Fire and other disturbances of the forests in Mount Rainier National Park. *Quaternary Research* 18:32–51

Henderson S, Olson RK, Noss RF (1989) Current and potential threats to biodiversity in forests of the lower Pacific Coast States. In: Olson RK, Lefohn AS (eds) *Effects of Air Pollution on Western Forests.* Transactions Series, No. 16, Air & Waste Management Association, Pittsburgh, PA, pp 325–336

Keeley JE, Keeley SC (1988) Chaparral. In: Barbour MG, Billings WD (eds) *North American Terrestrial Vegetation.* Cambridge University Press, New York, pp 165–207

King PB (1977) The Evolution of North America. Princeton University Press, Princeton

Knapp AK, Smith WK (1982) Factors influencing understory seedling establishment of Engelmann spruce (*Picea engelmannii*) and subalpine fir (*Abies lasiocarpa*) in southeast Wyoming. *Canadian Journal of Botany* 60:2753–2761

Küchler AW (1946) The broadleaf deciduous forests of the Pacific Northwest. *Annals of the Association of American Geographers* 36:122–147

Küchler AW (1988) The map of the natural vegetation of California (with map at 1:1,000,000). In: Barbour MG, Major J (eds) *Terrestrial Vegetation of California.* California Native Plant Society, Sacramento, pp 909–938

Lassoie JP, Hinckley TM, Grier CC (1985) Coniferous forests of the Pacific Northwest. In: Chabot BF, Mooney HA (eds) *Physiological Ecology of North American Plant Communities.* Chapman and Hall, New York, pp 127–161

Little EL Jr (1971) *Atlas of United States Trees, Vol.1, Conifers and Important Hardwoods.* U.S. Government Printing Office, Washington, DC

Lotan JE (1975) The role of cone serotiny in lodgepole pine forests. In: *Proceedings of the Symposium on Management of Lodgepole Pine Ecosystems, vol. 2.* Washington State University, Pullman

Maser C (1988) *The Redesigned Forest.* R&E Miles, San Pedro, CA

Mitchell VL (1976) The regionalization of climate in the western United States. *Journal of Applied Meteorology* 15:920–927

Morgan MD (1969) Ecology of aspen in Gunnison County, Colorado. *American Midland Naturalist* 82:204–228

Muir PS, Lotan JE (1985) Disturbance history and serotiny of *Pinus contorta* in western Montana. *Ecology* 66:1658–1668

Neilson RP, Wullstein LH (1983) Biogeography of two southwestern American oaks in relation to atmospheric dynamics. *Journal of Biogeography* 10:275–297

Niering WA, Lowe CH (1984) Vegetation of the Santa Catalina Mountains: Community types and dynamics. *Vegetatio* 58:3–28

Oliver CD (1981) Forest development in North America following major disturbances. *Forest Ecology and Management* 3:153–168

Pace CP, Layser CE (1977) *Classification of Riparian Habitat in the Southwest*. U.S. Department of Agriculture Forest Service General Technical Report RM-43

Parker AJ (1986) Persistence of lodgepole pine forests in the central Sierra Nevada. *Ecology* 67:1560–1567

Parker AJ, Peet RK (1984) Size and age structure of conifer forests. *Ecology* 65:1685–1689

Peet RK (1981) Forest vegetation of the Colorado Front Range: Composition and dynamics. *Vegetatio* 45:3–75

Peet RK (1988) Forests of the Rocky Mountains. In: Barbour MG, Billings WD (eds) *North American Terrestrial Vegetation*. Cambridge University Press, New York, pp 63–101

Pfister RD, Daubenmire R (1975) Ecology of lodgepole pine (*Pinus contorta* Dougl.). In: *Proceedings of the Symposium on Management of Lodgepole Pine Ecosystems, vol. 1*. Washington State University, Pullman, pp 27–46

Pfister RD, Kovalchik BL, Arno SE, Presby PC (1977) *Forest Habitat Types of Montana*. U.S. Department of Agriculture, Forest Service General Technical Report, INT-34, Intermountain Forest and Range Experimental Station, Odgen

Romme WH, Knight DH (1981) Fire frequency and subalpine forest succession along a topographic gradient in Wyoming. *Ecology* 62:319–326

Romme WH, Knight DH, Yavitt JB (1986) Mountain pine beetle outbreaks in the central Rocky Mountains: Regulators of primary productivity? *American Naturalist* 127:484–494

Romme WH, Despain DG (1989) Historical perspective on the Yellowstone fires of 1988. *BioScience* 39:695–699

Rundel PW, Parsons DJ, Gordon DT (1988) Montane and subalpine vegetation of the Sierra Nevada and Cascade Ranges. In: Barbour MG, Major J (eds) *Terrestrial Vegetation of California.* California Native Plant Society, Sacramento, pp 559–599

Sawyer DA, Kinraide TB (1980) The forest vegetation at higher altitudes in the Chiricahua Mountains, Arizona. *American Midland Naturalist* 104:224–241

Sawyer JO, Thornburgh DA, Griffin JR (1988) Mixed evergreen forest. In: Barbour MG, Major J (eds) *Terrestrial Vegetation of California.* California Native Plant Society, Sacramento, pp 359–381

Schulman E (1954) Longevity under adversity in conifers. *Science* 119:396–399

Smith WK (1985) Western montane forests. In: Chabot BF, Mooney HA (eds) *Physiological Ecology of North American Plant Communities.* Chapman and Hall, New York, pp 95–126

State of California (1988) California's forests and rangelands: Growing conflict over changing uses. California Department of Forestry and Fire Protection, Sacramento, CA

Tyree MT, Dixon MA (1983) Cavitation events in *Thuja occidentalis* L.: Ultrasonic acoustic emissions from the sapwood can be measured. *Plant Physiology* 72:1094–1099

U.S. Department of Agriculture (1981) *An Assessment of the Forest and Range Land Situation in the United States.* USDA Forest Service, Forest Resource Report No. 22. U.S. Government Printing Office, Washington, DC

U.S. Department of Agriculture (1988) *Forest Health through Silviculture and Integrated Pest Management: A Strategic Plan.* USDA Forest Service, Washington, DC

U.S. Geological Survey (1970) *The National Atlas of the United States.* U.S. Department of the Interior, Washington, DC

Vankat JL (1982) A gradient perspective on the vegetation of Sequoia National Park, California. *Madroño* 29:200–214

Vankat JL, Major J (1978) Vegetation changes in Sequoia National Park, California. *Journal of Biogeography* 5:377–402

Waddell KL, Oswald DD, Powell DS (1989) *Forest Statistics of the United States, 1987.* Resource Bulletin PNW-RB-168, Portland, Oregon. U.S. Department of Agriculture, Forest Service, Pacific Northwest Research Station, 106p

Waring RH, Franklin JF (1979) Evergreen coniferous forests of the Pacific Northwest. *Science* 204:1380–1386

Waring RH, Running SW (1978) Sapwood water storage: Its contribution to transpiration and effect upon water conductance through the stems of old-growth Douglas-fir. *Plant, Cell and Environment* 1:131–140

West NE (1988) Intermountain deserts, shrub steppes, and woodlands. In: Barbour MG, Billings WD (eds) *North American Terrestrial Vegetation.* Cambridge University Press, New York, pp 209–230

Whittaker RH, Niering WA (1965) Vegetation of the Santa Catalina Mountains, Arizona: A gradient analysis of the south slope. *Ecology* 46:429–452

Wilcove DS (1988) National forests: Policies for the future. *Protecting Biological Diversity, Vol. 2.* The Wilderness Society, Washington, DC

Zinke PJ (1988) The redwood forest and associated north coast forests. In: Barbour MG, Major J (eds) *Terrestrial Vegetation of California.* California Native Plant Society, Sacramento, pp 679–698

Zobel DB, McKee A, Hawk GM, Dyrness CT (1976) Relationships of environment to composition, structure, and diversity of forest communities of the central western Cascades of Oregon. *Ecological Monographs* 46:135–156

2

Climate

M. R. Rose, M. Böhm, and R. K. Olson

Introduction

Climate determines the types of forests which can occupy a particular region, and climate as modified by topography largely controls the spatial patterns of forest types within a region. These relationships stem from the influence of climate and weather on tree growth and physiology. Seasonal and annual variations in precipitation and temperature often result in corresponding variations in annual growth increments of individual trees (Chapter 7). Since analyses of trends in annual growth increments are central to the regional air pollution effects studies described in Chapters 8–12, an understanding of the general patterns and variability of climate in the West is necessary for interpreting the results of those studies.

In this chapter, the general patterns and causes of climate over the western United States are described. More detailed climatic descriptions are then presented for the five regions of the western United States discussed later in this volume: western Washington (Chapter 8), the Colorado Front Range (Chapter 9); Arizona and New Mexico (Chapter 10); the Sierra Nevada (Chapter 11); and southern California (Chapter 12). Mean seasonal temperature and precipitation are characterized for selected climatic divisions within the regions. The Palmer Drought Severity Index is used to discuss temporal variability in the climate of these regions.

General Atmospheric Characteristics

The general atmospheric circulation influencing climate across the West consists of a broad belt of westerlies extending from the subtropics to the polar region in which cyclones and anticyclones of the middle and high latitudes are embedded (Bryson and Hare 1974, Barry and Chorley 1982). The air within an anticyclone or high pressure system slowly subsides and moves outward in a clockwise direction. The subsiding air warms adiabatically causing mild weather with little cloudiness and precipitation. Cyclones or low pressure systems are the storms of middle latitudes. The air circulates counter clockwise with an inward rising motion causing stormy conditions with substantial precipitation (Barry and Chorley 1982).

Two aspects of the mean global circulation control the climate of the western United States. These are light variable winds in summer associated with the subtropical anticyclone (Pacific High) located around latitudes 30°– 40°N; and frontal storms associated with cyclones embedded in the southwesterlies between latitudes 40°– 60°N in the winter. The anticyclonic and cyclonic systems range from several hundred to several thousand kilometers in diameter and their strength varies seasonally. Anticyclones tend to persist longer than cyclones. Some pressure systems are migratory, while the subtropical Pacific High and subpolar Aleutian Low are considered semipermanent. The movement of migratory systems is closely related to meanderings of the belt of westerly winds aloft and to the jet stream embedded in the westerlies.

The general circulation and therefore the climate of the West varies seasonally. In the following discussion, seasons are simply defined as three-month blocks with winter defined as December through February. While this definition of seasons is not applicable to all Western forests, it suffices for a basic description of general circulation patterns in the West.

Winter

In winter, the continental mass of western North America cools faster than the Pacific Ocean. High pressure develops over land and semipermanent lows form over the ocean (Tang and Reiter 1984). The Pacific High weakens and moves southward. The Aleutian Low is well developed in the northern Pacific Ocean. The mean position of the polar front is displaced southward, and cold polar highs moving south eventually merge with semipermanent highs around 30°N latitude. The belt of westerlies is expanded and extends farther south than in summer.

Winter weather and precipitation in the western United States is associated with the position of the Aleutian low pressure system, and with large scale cyclonic storms embedded in the prevailing westerlies. The cyclonic systems normally follow a northerly path around the semipermanent subtropical high pressure ridge off the West Coast, intercepting land in the vicinity of the Oregon/Washington border. The maritime air is moist and warmer than the continent. Moisture condenses and is often precipitated as the maritime air moves inland over cooler land. Further condensation and precipitation occurs as the maritime air mass rises up the western slopes of the Cascades and Sierra Nevada. This results in a winter precipitation maximum across the Pacific Coast.

Cyclonic systems continue to move eastward en route to the Atlantic Ocean. Most of the moisture is deposited over mountain ranges west of the continental divide in Colorado; eastern slopes usually receive relatively small amounts of precipitation from these storms (Berry 1968).

The southwestern United States receives significant winter precipitation only when there is a westward displacement of the high pressure ridge in the Pacific and formation of a semipermanent low pressure trough over the western United States. Instead of taking a northerly route, cyclonic storms follow prevailing southerly flow of fast moving air currents along the West Coast before entering the continent, often as far south as San Francisco. Once this flow pattern is established, it tends to persist. Moisture in the maritime air mass is precipitated over coastal and inland mountain ranges of California, Nevada, Arizona, and Utah. The Mogollon Rim in Arizona, oriented west-northwest to east-southeast, does not affect paths of these storms and about equal amounts of precipitation occur in northwestern and southwestern New Mexico during such events (Tuan et al. 1973). The continental divide and northern and central mountains in New Mexico induce precipitation of any moisture remaining in the maritime air mass and winter dryness is experienced in New Mexico's central valley and on eastern mountain slopes.

There is a noticeable inverse relationship between conditions in the Southwest and those in Oregon and Washington; heavy winter precipitation in the Southwest is associated with dry conditions in the Northwest, and vice-versa. El Nino, a period of warmer than average surface water temperatures in the eastern Pacific (Cane 1983), is associated with this inverse relationship. El Nino years tend to have increased winter precipitation in the Southwest and decreased precipitation in the Northwest (Cayan and Peterson 1990, Changnon et al. 1990) as a result of alterations in upper wind patterns.

A ridge of continental polar or Arctic air may extend farther south than is usual during winter. The Rocky Mountains effectively prevent most continental polar or Arctic air masses from moving westward, although on occasion, cold continental air masses do move southward west of the Rockies, bringing extreme cold to the area. When the cold, dry air mass interacts with warmer, moist tropical maritime air, the less dense moist air masses are displaced vertically, causing precipitation as part of frontal activity convergence storms (Dorrah 1946, Pianka 1978).

Spring

In spring the jet stream over California weakens, the Pacific High and westerlies move northward, and the Aleutian Low weakens and splits (Barry and Chorley 1982). Precipitation associated with cyclonic disturbances declines as low pressure systems follow more northerly routes. Frontal activity is limited to extreme northern portions of the region. Continental high pressure systems are weak and tend to be replaced frequently.

Warm, moist air from the Gulf of Mexico frequently moves northward into the Colorado region inducing the heaviest and most general rainfall of the year over the eastern part of the state. Severe thunderstorms associated with the vertical movement of moist air masses are frequent along the eastern slopes of the Front Range.

Summer

A major and rapid reorganization of the atmospheric circulation introduces summer to the West. Continental land masses heat up faster than adjacent oceanic areas resulting in continental low and oceanic high pressure (Tang and Reiter 1984). The Pacific High continues to migrate towards the north and the Aleutian Low disintegrates; any cyclonic activity is forced northward by the Pacific High and storm tracks rarely reach as far south as Oregon during the summer. Contrasts in temperature between the Equator and the North Pole are smaller in summer than winter, pressure gradients are weaker, and the resulting air motion is slower. The westerlies decrease in strength and are restricted to a narrow band. Aloft, the circumpolar vortex is small and the westerlies, jet stream, and polar front are displaced far to the north. The subtropical Bermuda High over the Atlantic builds from the east, influencing the southwestern states of New Mexico, Colorado, and Arizona.

A high pressure ridge, oriented towards the northeast, extends along the Pacific Coast from about 38°– 50°N. Northerly winds cause upwelling of cold water along the northern Pacific Coast, forming the California cold current. The climate along the coast is noticeably influenced by the California cold current; temperatures are depressed and coastal fogs are frequent. Coastal ecosystems, in particular the redwood forests, are dependent on moisture trapped during these fog events to alleviate water stress during dry summer months (Chapter 1). Subsidence in the eastern Pacific, together with cool air along the coast, increases atmospheric stability and decreases rainfall. These atmospheric conditions allow the accumulation of pollutants in areas such as the Los Angeles basin.

Maritime tropical air masses move over the arid Southwest from the Gulf of Mexico, and the Gulf of California and Pacific Ocean, abruptly ending the dry spring conditions (Carleton 1986, 1987). Tropical moist air moves inland from the Gulf of Mexico during late June and reaches as far west as Arizona before retreating eastward during September. Frequent storms, which are often of short duration and of small total precipitation, result from invasion of moist air coupled with mechanical or dynamic lifting mechanisms such as convection, orographic uplift, or air mass convergence. Though most of the moist air originates from the Gulf of Mexico, some of the heaviest precipitation recorded historically is a result of incursions of tropical air from the Pacific Ocean (Bryson and Lowry 1955). These deep surges of tropical air occur most frequently in late August and September and are usually associated with hurricanes or tropical depressions off the west coast of Mexico (Sellers and Hill 1974).

The Mogollon Rim complex of high plateaus and mountains lies in the path of the flow of moist air during summer. Areas to the north in Arizona and New Mexico lie in the "rain shadow" of the Mogollon Rim and receive little summer precipitation.

Autumn

The end of summer is marked by southward progression and intensification of the polar front, and redevelopment of the Aleutian Low. Low pressure systems from the Pacific Ocean move farther south, impacting the Pacific Northwest and northern Rocky Mountains. In the transition period before winter, the northern United States starts to experience increased storminess and onset of precipitation. These cyclonic systems usually pass north of Arizona and New Mexico.

Regional Climates of the West

Definition of Climatic Regions

The prevailing climate of the West is modified by topography, especially by the position of mountain ranges with respect to prevailing westerly winds. Elevation influences air temperatures, precipitation amounts, and the proportion of precipitation that falls as rain or snow (Barry 1981). North-south oriented mountain ranges along the western coast of the region inhibit eastward movement of maritime air masses, further increasing climatic variability across the region, both vertically and horizontally (Barry and Chorley 1982). This variability makes it difficult to discuss regional climates of the West using usual climate descriptors such as average temperature and precipitation (e.g., Baker 1944).

To address this problem, Mitchell (1976) examined regional climates of the western United States using equivalent potential temperatures (Rogers and Yau 1989:23) to differentiate major air masses. Because it is conservative in both adiabatic and pseudoadiabatic processes, equivalent potential temperature is a useful parameter in situations where air masses undergo frequent changes in altitude and changes of phase of water.

A monthly series of equivalent potential temperature maps reveals two basic patterns during the year, a winter pattern prevalent during November to March, and a summer pattern from June to September, though best developed in July and August. Because the characteristics of air masses influence climate, the equivalent potential temperature maps can be used to define six major climatic regions (Mitchell 1976) across the West (Figure 2.1).

Region I (the Pacific Northwest) is characterized by frequent passage of air masses from the Pacific Ocean during winter and much of summer. Region II (Idaho, Montana, and Wyoming) is also affected by the frequent intrusion of Pacific air masses during winter, but not during summer. Region III (northeastern Montana) is dominated by interior air masses in summer, with intrusions of Arctic and Pacific air in winter. Region IV (coastal California) is influenced by Pacific air during summer and winter. Region V (eastern California, Nevada, and Utah) is dominated by interior air masses during the summer. Region VI (Arizona, New Mexico, and Colorado) has a summer rainy season brought by the influx of moist, tropical air from the Gulfs of Mexico and California (Hales 1974, Mitchell 1976).

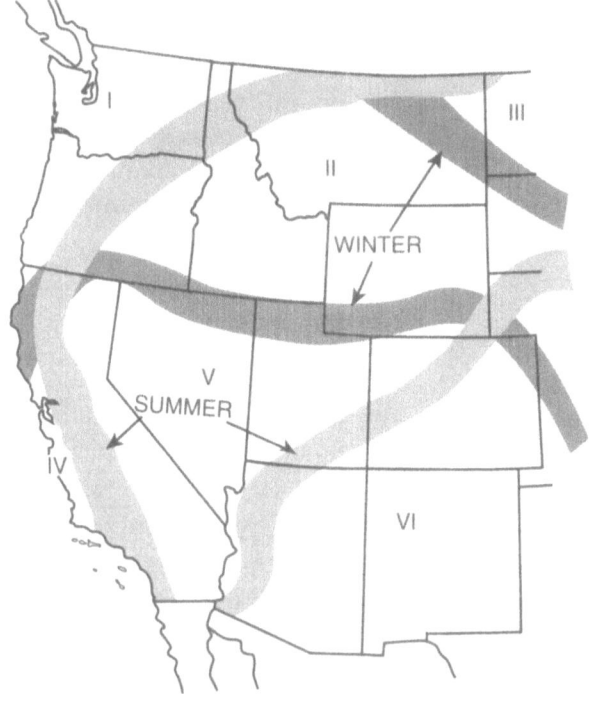

Figure 2.1 Major equivalent potential temperature boundaries and climatic regions for the western United States. From Mitchell (1976).

Topographic Influences on Climate

Most Western forests are located in mountainous terrain (Chapter 1). Topography generally influences climate by varying radiation input and modifying airflow. Topographical characteristics influence the angle at which incident radiation strikes the surface of the earth. Spatial variations in slope and azimuth angles determine radiant loading differences across the landscape; the slope directly facing the sun receives the most radiation. South-facing slopes can receive up to three times more net radiation than north facing slopes (Oke 1987), with corresponding differences in air and soil temperatures, and rates of evaporation. Hydrological activity also varies as a result of different rates of evaporation, lengths of snow cover retention, and probabilities of avalanching on different slopes. All of these factors have important consequences for forests.

Temperature decreases with elevation at an average rate of 5.96°C per 1000 m, although there is considerable variation with season and between different regions. In the mountains of the western United States, the temperature lapse rate averages 6.32°C per 1000 m in July and 5.24°C per 1000 m in January. In July, mean maximum temperatures have a lapse rate of 7.22°C per 1000 m and the mean minimum drops at 5.24°C per 1000 m. In January, the lapse rates for both maximum and minimum temperatures are 5.24°C per 1000 m (Baker 1944).

Topography also influences amount and, in many cases, physical form of precipitation. Precipitation amount generally increases with elevation because air masses cool as they rise, and if sufficient moisture is present, condensation occurs (Barry and Chorley 1982). For example, precipitation increases 316 mm per 1000 m gain in elevation along a transect in the Sierra Nevada of California (Stohlgren and Parsons 1987). The rate at which precipitation increases with elevation varies greatly throughout the West (Baker 1944), and not all high elevation sites receive large amounts of precipitation (e.g., Baker 1944, Lewis et al. 1984). Many high elevation areas are in a "rain shadow" and exhibit strikingly low precipitation amounts despite their high elevations.

Precipitation type differs between low and high elevations, especially during the winter months. Higher elevations in the West usually receive a greater proportion of total precipitation as snow and rime ice than lower elevation sites. For example, Seattle, Washington (elevation 38 m), receives about 940 mm of rain and 35 mm (water equivalent) snow per year, while a nearby site at 1821 m on Mt. Rainier in the Cascade Mountains receives approximately 2640 mm rain and 1362 mm (water equivalent) snow each year (Lassoie et al. 1985).

Climates of the Five Study Areas

Climatic Divisions

The five regional forest studies discussed in Chapters 8–12 are located within three of Mitchell's six climatic regions (Figure 2.1). The Douglas-fir forests of western Washington fall within Region I; the mixed conifer forests of the Sierra Nevada and San Bernardino Mountains within Region IV; and the conifer forests of Arizona and New Mexico and the Front Range of Colorado within Region VI.

Our more detailed description of the climates of the forest study areas is based largely on a summary of weather data from the 10 NOAA climatic divisions in which the study areas are located (Figure 2.2). Regional temperature, precipitation, and Palmer Drought Severity Index datasets were obtained from the US Historical Climatology Network established by the US Department of Energy and the National Oceanic and Atmospheric Administration. Details on meteorological station history, station selection, data adjustment, correction techniques, and other quality assurance activities have been documented (e.g., Boden 1987).

Descriptions of the physiography and forest types of the 10 climatic divisions can be found in Chapter 1, while a brief introduction follows:

The West Olympic-Coastal, East Olympic-Cascade Foothills, and Cascade Mountains West Climatic Divisions in western Washington (Figure 2.2a) include coastal lowlands, high mountains and foothills extending from the Pacific Ocean to the crest of the Cascade Mountains. Elevations range from 0 m to 4400 m. Forest research sites are located at elevations from 225 m to 1160 m (Chapter 8).

The Platte Drainage Climatic Division extends north from Colorado Springs to the Wyoming border, east of the continental divide (Figure 2.2b). Physiographic features range from semi-arid high plains to peaks up to 4300 m. The forest research sites are located in the far western portion of this division, in the Front Range and adjacent foothills at elevations from 1700 m to 3500 m (Chapter 9).

The Northeastern Plateau Climatic Division in Arizona (Figure 2.2c) covers the northeastern quarter of the state at elevations between 1400 and 2200 m. Study sites are located along the southwestern edge over the full range of elevations (Chapter 10). The Southeastern Climatic Division of Arizona consists of basin and range topography with forest research sites restricted to isolated peaks above 1800 m. The Southwestern Mountains Climatic Division in New Mexico contains study sites at elevations from 1750 m to 2700 m.

The Sacramento and San Joaquin Climatic Divisions (Figure 2.2d) contain the Sierra Nevada research sites (Chapter 11). From the floor of the Central Valley at 100 m, the land rises to between 1800 m and 4400 m at the crest of the Sierra Nevada. Research sites are located in the mixed conifer forests at elevations from 900 m to 2000 m.

The South Coast Drainage Climatic Division (Figure 2.2d) contains the San Bernardino Mountains and associated study sites (Chapter 12). Research sites are located at elevations from 1500 m to 2300 m.

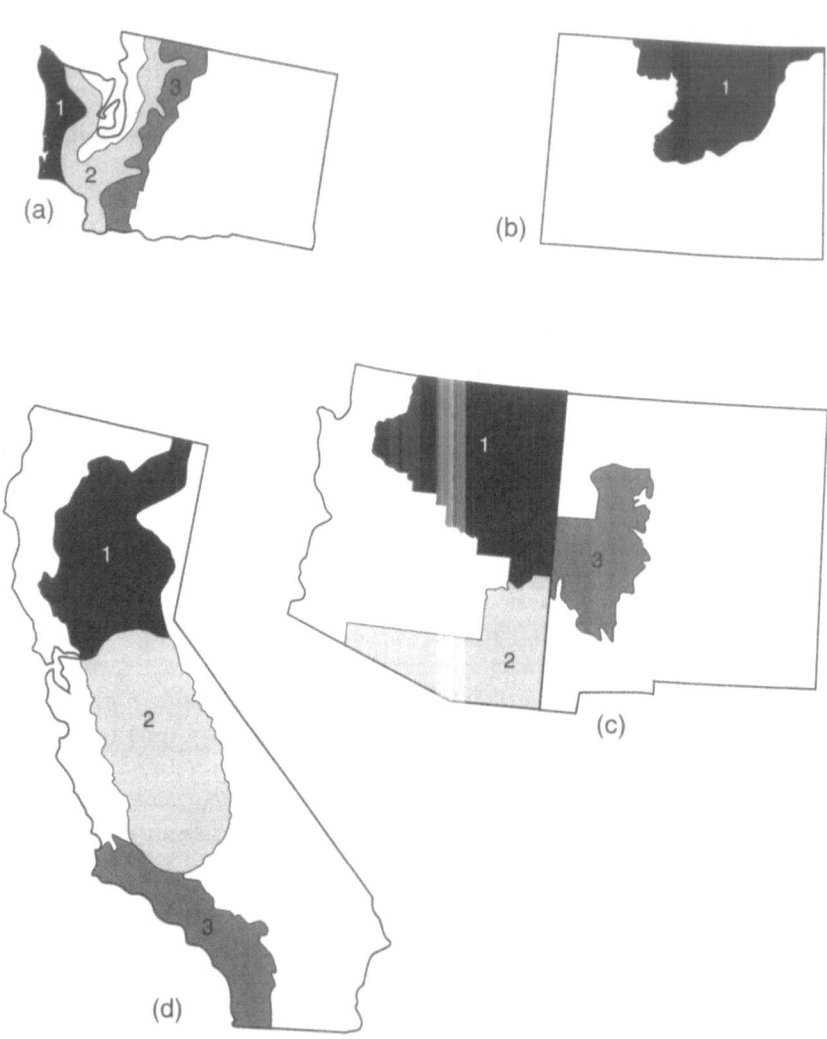

Figure 2.2 National Oceanic and Atmospheric Administration Climatic Divisions encompassing the regional air pollution effects study areas described in Chapters 8–12. (a) Western Washington: (1) West Olympic Coastal, (2) East Olympic–Cascade Foothills, (3) Cascade Mountains West; (b) Colorado: (1) Platte Drainage; (c) Arizona/New Mexico: (1) Northeast Arizona, (2) Southeast Arizona, (3) Southwestern Mountains, New Mexico; (d) California: (1) Sacramento Drainage, (2) San Joaquin Drainage, (3) South Coast Drainage. Based on Ruffner (1978).

Seasonal temperature and precipitation

Monthly temperature averages and precipitation totals were obtained by climatic division from the National Climate Data Center, Asheville, North Carolina (National Oceanic and Atmospheric Administration, Historic Climate Data Set). The data represent historical division averages, i.e. mean values of average temperatures and precipitation totals for all reporting stations within a NOAA climatic division. For all divisions except the South Coast Drainage Climatic Division, seasonal averages were calculated from data for the period 1896 through 1986. Seasonal averages were calculated for the South Coast Division for the period 1951–1980 (US National Climatic Center 1983).

Most weather stations are located at lower elevations while forests are restricted in most cases to the higher elevations (Chapter 1). Therefore, mean monthly temperatures for a division are generally higher than mean temperatures in forested areas of the division. Average precipitation totals for a division are generally lower than precipitation means for forested areas. However, division means do serve to characterize general trends and magnitude of seasonal precipitation and temperature. Elevation does not have a large effect on the distribution of precipitation among seasons (Baker 1944). Major differences between study regions are readily apparent, and analyses at the scale of climatic divisions provides a framework for the more specific climate–tree growth analyses described in Chapters 8–12.

Figure 2.3 shows mean seasonal temperatures for the ten climatic divisions which contain the five study areas. Seasons are defined as three month blocks with winter as December–February. The results are straight-forward. Mean temperatures generally increase as latitude and elevation decrease. Coastal divisions have more moderate winter temperatures than interior divisions.

Temperature data (National Park Foundation 1988) from Rocky Mountain National Park in the Colorado Front Range, and Sequoia National Park in the southern Sierra Nevada demonstrate how temperatures in forested areas of the climatic divisions may deviate from the division averages. Rocky Mountain National Park at elevations from 2300 m to 4300 m has a winter mean temperature of about -9°C, and a summer mean of about 13°C. Winter and summer mean temperatures at Sequoia National Park (elevation 390 m to 4419 m) are 1°C and 16°C respectively. These values can be compared with the higher division averages shown in Figure 2.3.

Figure 2.4 presents mean seasonal total precipitation for the ten climatic divisions. Differences in precipitation between regions are much greater than temperature differences, both in magnitude and seasonal pattern.

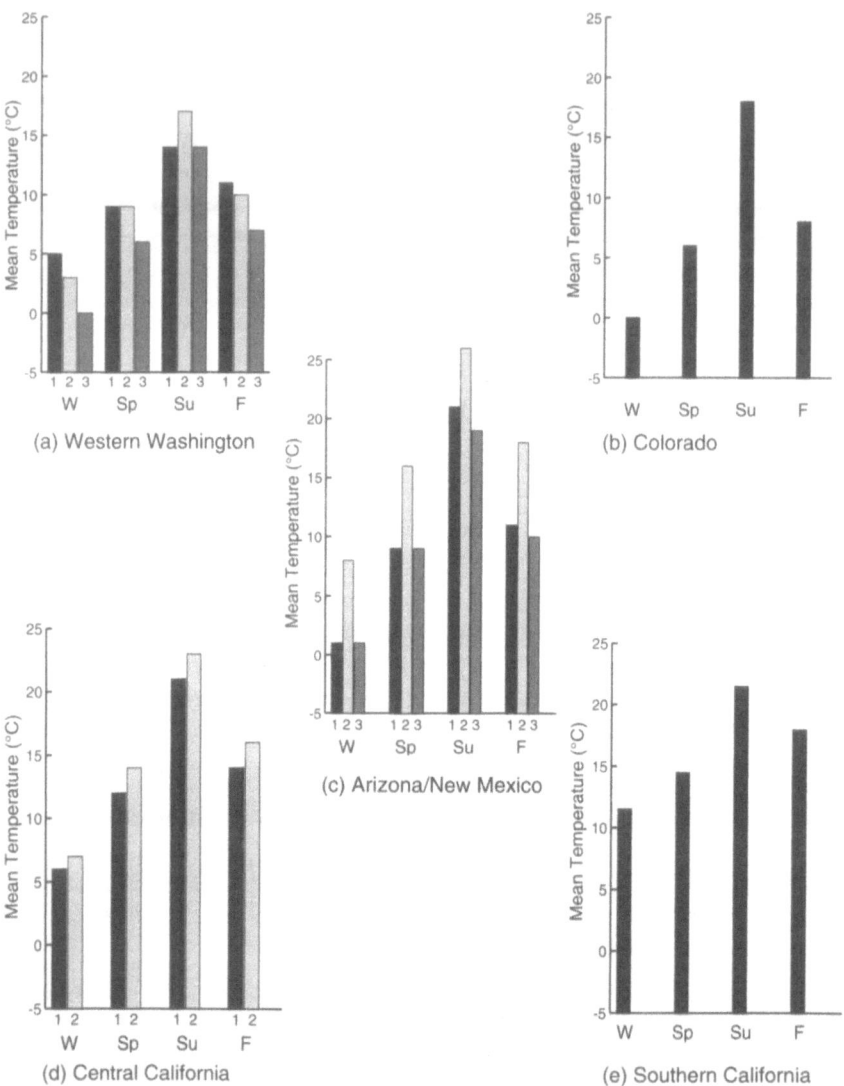

Figure 2.3 Mean seasonal temperatures for NOAA Climatic Divisions encompassing the regional air pollution effects study areas described in Chapters 8–12. a. Western Washington: (1) West Olympic Coastal, (2) East Olympic–Cascade Foothills, (3) Cascade Mountains West; b. Colorado, Platte Drainage; c. Arizona/New Mexico: (1) Northeast Arizona, (2) Southeast Arizona, (3) Southwestern Mountains, New Mexico; d. Central California: (1) Sacramento Drainage, (2) San Joaquin Drainage; e. Southern California, South Coast Drainage. Seasons are defined as three month blocks with winter consisting of December-February.

The climatic divisions described for western Washington (Figure 2.4a) receive approximately five times the total annual precipitation of the Colorado, Arizona, or southern California divisions (Figure 2.4b,c,e). The rain forests of the southwestern and western slopes of the Olympic Mountains receive the heaviest precipitation in the continental United States.

Seasonal patterns of precipitation also vary between regions. The West Coast climatic divisions have winter maximums for precipitation and pronounced dry summers; this contrasts with the Colorado and Arizona divisions which have summer maximums (Court 1974). It is interesting to note that even though summer is a precipitation minimum in western Washington and a maximum in the Colorado and Arizona divisions, the summer precipitation amounts are about the same for these regions (Figure 2.4).

This uniform summer dryness over the western United States increases the importance of snow-melt as a supplemental water source during late spring and early summer (e.g., Knight et al. 1985). Snow is the dominant form of precipitation during the winter for high elevation forests throughout the West (Baker 1944).

In general, division averages for precipitation underestimate the actual precipitation received by forests which occur at the higher, wetter elevations. For example, total mean annual precipitation at Sequoia National Park in the southern Sierra Nevada is greatest at 1550 m with annual amounts of about 1140 mm (Baker 1944). This compares to an annual mean for the San Joaquin Climatic Division of 520 mm (Figure 2.4d). Combined with lower temperatures, the greater precipitation at higher elevations results in an overall water balance conducive to forest growth.

Temporal trends and variability

Seasonal means for temperature and precipitation mask inter-annual variability and temporal trends. However, extreme years may be particularly important determinants of forest structure and development. Examples include the facilitation of ponderosa pine seedling establishment by especially wet years (Peet 1988) and the effects of drought-promoted fires (Christensen et al. 1989). Annual growth of forest trees is highly correlated with precipitation trends (Chapter 7).

Water availability is a better predictor for forest processes than precipitation per se. Availability is a function not just of precipitation trends and amounts, but also of temperature patterns and physical and biological site characteristics such as soil type and vegetative cover. A time series of

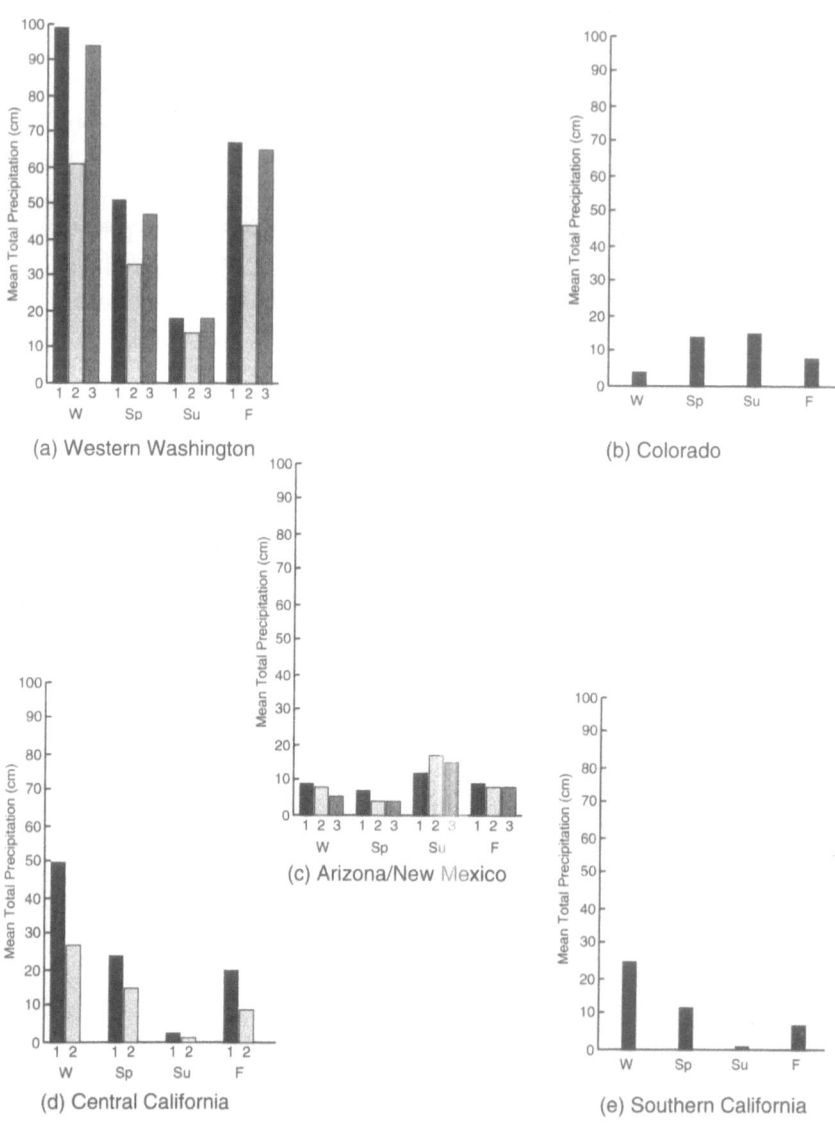

Figure 2.4 Mean seasonal total precipitation for NOAA Climatic Divisions encompassing the regional air pollution effects study areas described in Chapters 8–12. a. Western Washington: (1) West Olympic Coastal, (2) East Olympic–Cascade Foothills, (3) Cascade Mountains West; b. Colorado, Platte Drainage; c. Arizona/New Mexico: (1) Northeast Arizona, (2) Southeast Arizona, (3) Southwestern Mountains, New Mexico; d. Central California: (1) Sacramento Drainage, (2) San Joaquin Drainage; e. Southern California, South Coast Drainage. Seasons are defined as three month blocks with winter consisting of December-February.

some measure of water availability for a region is likely to be a better tool than precipitation or temperature values alone for evaluating climate in relation to changes in forest growth and condition over time.

Palmer (1965) developed a method for integrating data on temperature, precipitation and soils into an index of relative drought, the Palmer Drought Severity Index (PDSI). The PDSI is an index of meteorological drought that can also be used to assess the availability of water from the soil. A drought period is an interval of time, generally of the order of months or years in duration, during which the actual moisture supply is consistently lower than the climatically normal moisture supply. The severity of drought is a function of both the duration and magnitude of moisture deficiency (Palmer 1965).

The PDSI is one of many methods for estimating drought, but it is one of the few that can be calculated with long term monthly temperature and precipitation observations and very general information regarding soil moisture retention capabilities. Other hydrological accounting proce- dures more accurately portray conditions of meteorological drought, but they require information (such as pan evaporation, wind velocities at different elevations, and detailed temperature and precipitation observa- tions) that are not normally available over periods representative of the lifetime of a tree. Nor are such data usually collected in the mountainous regions occupied by most Western forests. Despite several basic assump- tions (Palmer 1965), the PDSI is especially useful because it is an integra- tive measure, i.e., temperature, precipitation, and soil moisture retention characteristics are employed in its calculation. By normalizing water availability against expected conditions for each region of interest, the PDSI allows comparisons through time of conditions in a single region and comparisons of conditions between regions.

Table 2.1 lists the index classes for the Palmer Drought Severity Index. The procedure for calculating PDSI values is described in Palmer (1965). For illustrating trends and regional differences for the West, a time series of the Index was plotted for four of the climatic divisions discussed in this chapter: East Olympic-Cascade Foothills, WA; Platte River Drainage, CO; Southeastern Arizona; and San Joaquin Drainage, CA (Figure 2.5a–d).

To develop these plots, PDSI data were obtained from the National Climate Data Center, Asheville, North Carolina. Data from 1895 through 1983 were used to compile the division averages. Index values for June are discussed as indicators of late spring and summer drought.

A number of observations can be made from Figure 2.5. Between-year variability in PDSI values is high in all divisions. Periods of three or more consecutive years with the same index class are rare, and year to year changes can be extreme. For example, the Platte Drainage Climatic

Table 2.1. Palmer Drought Severity Index class intervals.

PDSI Value	CLASS
$\geq$+4.00	Extremely wet
+3.00 to +3.99	Very wet
+2.00 to +2.99	Moderately wet
+1.00 to +1.99	Slightly wet
+.50 to +.99	Incipient wet spell
+.49 to -.49	Near normal
-.50 to -.99	Incipient drought
-1.00 to -1.99	Mild drought
-2.00 to -2.99	Moderate drought
-3.00 to -3.99	Severe drought
$\leq$ -4.00	Extreme drought

Division went from extreme drought in 1956 to very wet in 1957 (Figure 2.5b). However, within this variability, distinct periods or trends are apparent. Extended drought periods include western Washington, 1904–1941; northeastern Colorado, 1931–40 and 1952–56; southeastern Arizona, 1896–1904, 1943–48, and 1950–59; and central California, 1897–99 and 1959–62. The period 1903–30 was especially wet for northeastern Colorado, as were 1905–07 and 1915–20 in southeastern Arizona, and 1911–15 in central California. The lack of correlation between divisions in the timing of dry or wet intervals reflects the different aspects of the overall western circulation that control climate in these widely separated regions.

Variability in PDSI values is lower in the East Olympic–Cascade Foothills division than in the other three divisions. One measure of this is the number of years with June PDSI values ≥ 3.0 (very or extremely wet) or ≤ 3.0 (severe or extreme drought) during the period 1895–1983. The number of these years (with the number of severe or extreme drought years in parentheses) are: East Olympic–Cascade Foothills, 9(7); San Joaquin, 21(10); Southeastern Arizona, 30(13); and Platte Drainage, 38(8). The two divisions with the lowest annual precipitation have the greatest number of extreme years, although the number of years with severe or extreme drought are quite similar among the four divisions.

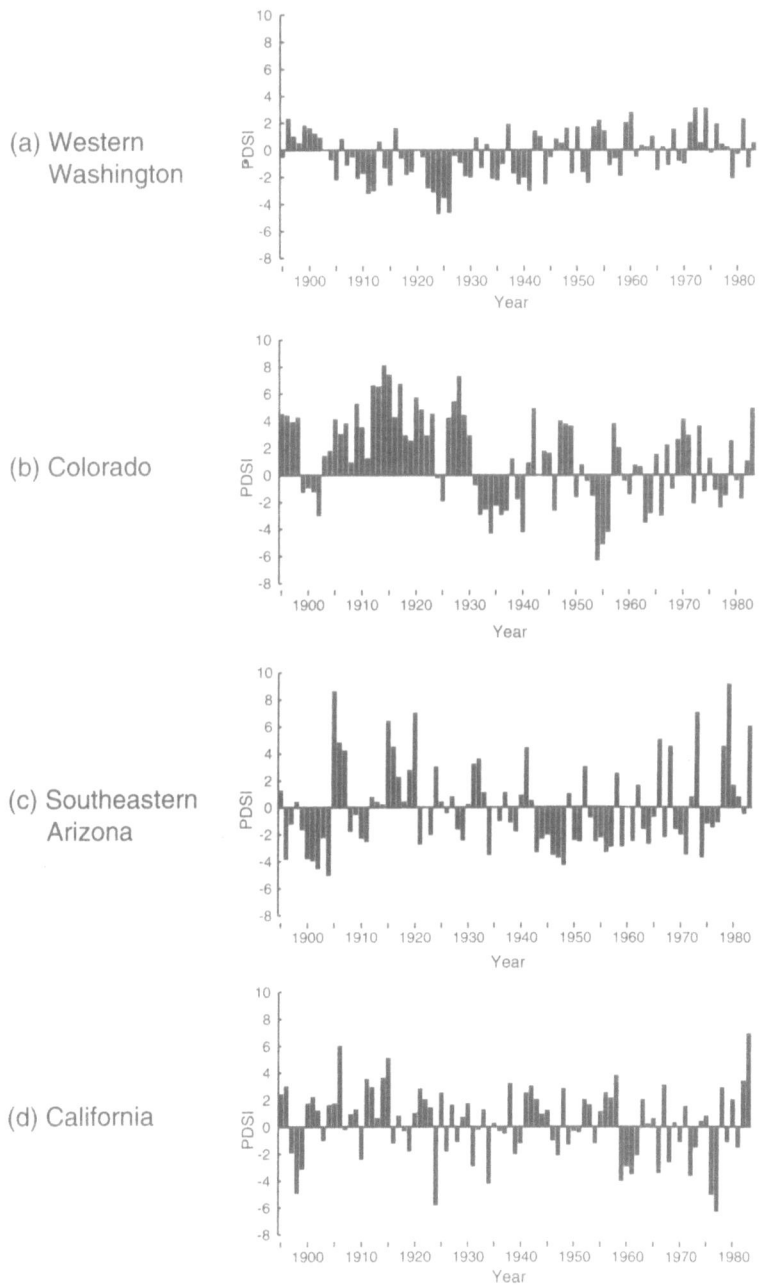

Figure 2.5 June values for the Palmer Drought Severity Index for four NOAA Climatic Divisions. (a) East Olympic–Cascade Foothills, Western Washington; (b) Platte Drainage, Colorado; (c) Southeastern Arizona; (d) San Joaquin Drainage, California.

Summary and Conclusions

In summer, a large subtropical high-pressure system, or anticyclone, materializes in the north-central Pacific Ocean east of Hawaii. Warm, dry air circulates clockwise around its center bringing fair weather to most of the West. Intrusion of tropical moist air from the Gulfs of Mexico and California bring summer precipitation to most of the Southwest. In winter, the anticyclone shifts southward and storms associated with low pressure systems, or cyclones, move eastward across the region. The frontal systems bring heavy rains to the West Coast, particularly of Washington, Oregon, and northern California. Heavy snows fall on the western slopes of the Cascades and Sierra Nevada.

The winter precipitation maximum is best developed over the Pacific Coast states, which are also characterized by summer drought. In Colorado some climatic divisions experience a significant amount of spring as well as summer precipitation, while farther to the southwest in Arizona and New Mexico the effects of the summer monsoon predominate.

The June Palmer Drought Severity Indices provide an integrated picture of the effects of temperature and precipitation patterns on moisture availability in late spring and early summer. The plots show that moisture availability varies constantly through time, but they also identify broad trends in drought and mesic conditions that have occurred during the past century. The effects of these trends on growth trends of trees need to be considered in any evaluation of the possible effects of air pollution on forest growth.

References

Baker FS (1944) Mountain climates of the western United States. *Ecological Monographs* 14(2):224–254

Barry RG (1981) *Mountain Weather and Climate.* Methuen, London, 313p

Barry RG, Chorley RJ (1982) *Atmosphere, Weather, and Climate, 4th Edition.* Methuen, London, 407p

Berry JW (1968) The climate of Colorado. In: *Climates of the States, Vol. II—Western States.* Water Information Center, Port Washington, NY

Boden TA (1987) *United States Historical Climatology Network (HCN) Serial Temperature and Precipitation Data.* CDIAC, Environmental Sciences Division, Oak Ridge National Laboratory, Oak Ridge, TN

Bryson RA, Hare FK (1974) The climates of North America. In: Bryson RA, Hare FK (eds) *Climates of North America, World Survey of Climatology, Volume 11.* Elsevier Scientific Publishing Company, Amsterdam, pp 1–47

Bryson RA, Lowry WP (1955) Synoptic climatology of the Arizona summer precipitation singularity. *Bulletin of the American Meteorological Society* 36:329–339

Cane MA (1983) Oceanographic events during El Nino. *Science* 222: 1189–1195

Carleton AM (1986) Synoptic-dynamic character of "bursts" and "breaks" in the southwest US summer precipitation singularity. *Journal of Climatology* 6:605–623

Carleton AM (1987) Summer circulation climate of the American Southwest, 1945–1984. *Annals of the Association of American Geographers* 77(4):619–634

Cayan DR, Peterson DH (1990) The influence of north Pacific atmospheric circulations on streamflow in the West. *Geophysical Monograph No. 55:375–397

Changnon D, McKee TB, Doesken NJ (1990) *Hydroclimatic Variability in the Rocky Mountain Region.* Climatology Report No. 90-3, Department of Atmospheric Science, Colorado State University, Ft. Collins

Christensen NL, Agee JK, Brussard PF, Hughes J, Knight DH, Minshall GW, Peek JM, Pyne SJ, Swanson FJ, Thomas JW, Wells S, Williams SE, Wright HA (1989) Interpreting the Yellowstone fires of 1988. *BioScience* 39(10):678–685

Court A (1974) The climate of the conterminous United States. In: Bryson RA, Hare FK (eds) *Climates of North America, World Survey of Climatology,Volume 11.* Elsevier Scientific Publishing Company, Amsterdam, pp 193–343

Dorrah VH Jr (1946) Certain hydrological and climatic characteristics of the Southwest. *University of New Mexico Publications in Engineering 1,* Albuquerque, NM

Hales JE Jr (1974) Southwestern United States summer monsoon source—Gulf of Mexico or Pacific Ocean? *Weather-wise* 27:148–155

Knight DH, Fahey TJ, Running SW (1985) Water and nutrient outflow from contrasting lodgepole pine forests in Wyoming. *Ecological Monographs* 55(1):29–48

Lassoie JP, Hinckley TM, Grier CC (1985) Coniferous forests of the Pacific Northwest. In: Chabot BF, Mooney HA (eds) *Physiological Ecology of North American Plant Communities.* Chapman and Hall, New York, pp 127–161

Lewis MW, Grant MC, Saunders JF (1984) Chemical patterns of bulk atmospheric deposition in the state of Colorado. *Water Resources Research* 20(11):1691–1704

Mitchell VL (1976) The regionalization of climate in the western United States. *Journal of Applied Meteorology* 15(9):920–927

National Park Foundation (1988) *The Complete Guide to America's National Parks 1988–1989 Edition.* Prentice Hall, New York, 572p

Oke TR (1987) *Boundary Layer Climates, 2nd Edition.* Methuen, London, 435 pp

Palmer WC (1965) *Meteorological Drought.* US Weather Bureau Research Paper 45, US Department of Commerce, Washington, DC

Peet RK (1988) Forests of the Rocky Mountains. In: Barbour MG, Billings WD (eds) *North American Terrestrial Vegetation.* Cambridge University Press, New York, pp 63–101

Pianka ER (1978) *Evolutionary Ecology.* Harper and Row Publishers, New York

Rogers RR, Yau MK (1989) *A Short Course in Cloud Physics, 3rd edition.* Pergamon Press, Oxford

Ruffner JA (ed) (1978) *Climates of the States, with Current Tables of Normals 1941–1970 and Means and Extremes to 1975. Volume 1: Alabama–Montana, Volume 2: Nebraska–Wyoming and Territories.* Gale Research Company, Detroit

Sellers WD, Hill RH (eds) (1974) *Arizona Climate, 1931–1972.* University of Arizona Press, Tucson

Stohlgren TJ, Parsons DJ (1987) Variation of wet deposition chemistry in Sequoia National Park, California. *Atmospheric Environment* 21(6):1369–1374

Tang M, Reiter ER (1984) Plateau monsoons of the Northern Hemisphere: A comparison between North America and Tibet. *Monthly Weather Review* 112:617–637

Tuan YF, Everand CE, Widdison JG, Bennet W (1973) *The Climate of New Mexico, Revised Edition.* State Planning Office, Santa Fe, New Mexico

US National Climatic Center (1983) *Climate Normals for the US (Base 1951–1980), First Edition.* Gale Research Company, Detroit

3

Air Quality and Deposition

M. Böhm

The pollution climate of an area is influenced by meteorology and emissions of air pollutants at local and regional scales. The physical and chemical state of the atmosphere determines pollutant transport, dilution, chemical transformation, and ultimately deposition. In many cases, meteorology is more important than atmospheric chemistry in controlling the location and the form in which the pollutants are deposited (Cape and Unsworth 1987). Estimating pollutant concentrations and loadings to forests in the West requires a detailed analysis of emissions, pollutant transport, dilution, chemical transformations, and deposition processes, together with estimates of the relative contribution by each depositional process to total deposition.

The first portion of this chapter examines atmospheric conditions that influence the transport and deposition of pollutants and applies this information to conditions experienced within Western forests. A brief discussion of the chemistry of sulfur, nitrogen, and ozone is included. The second portion of the chapter discusses air quality and wet deposition in and around Western forests. Emphasis is placed on nitrogen and sulfur oxides and their oxidation products. These pollutants contribute to acid rain as well as to other air pollution problems such as ozone and visibility reduction (Schwartz 1989). All aspects of air quality are presented; from emissions of pollutants to ambient concentrations of important chemical species in gaseous, particulate matter, and precipitation forms. The chapter concludes with a glossary of technical terms.

Pollutants in the Earth-Atmosphere System

Air pollutants from a variety of sources are emitted into the atmosphere and are eventually deposited on the earth. The amount of a pollutant absorbed by a forest depends on atmospheric processes that affect the delivery of the pollutant to the vegetation, and on biological processes that control uptake of the pollutant into trees. Most atmospheric processes that dilute, transport, transform, and aid in deposition of pollutants are influenced by physical and chemical interactions between the earth and the atmosphere. These interactions are mainly limited to the lowest 10 km of the atmosphere in a layer known as the troposphere (Figure 3.1). Over short time periods such as a day, earth-atmosphere interactions are restricted to the planetary boundary layer. The lowest portion of the planetary boundary layer, the mixed layer, is characterized by turbulence generated from thermal convection and frictional interactions between moving air masses and the ground. Mixing height is the vertical thickness of the mixed layer above the ground. Mixing height varies depending on the strength of surface-generated mechanical and thermal turbulence and the presence or absence of large scale weather systems or complex terrain. Typically the mixing height extends 1 to 2 km upwards from the ground. The physical characteristics of the mixed layer effectively control ground-level pollutant concentrations for a given emission density as well as determine the pattern of pollutant reaction, deposition, and transport (Smith and Hunt 1978).

The mixed layer consists of several sub-layers; the two of greatest interest in air pollution meteorology are the surface and interfacial layers. The surface layer is usually several hundred meters thick and is characterized by turbulence produced by surface roughness and convection. Forests are within the surface layer. The quiescent interfacial layer, where molecular transport of gases and Brownian motion of particles dominate over turbulent transport, is a few millimeters thick and is in immediate contact with surfaces such as foliage. The interfacial layer

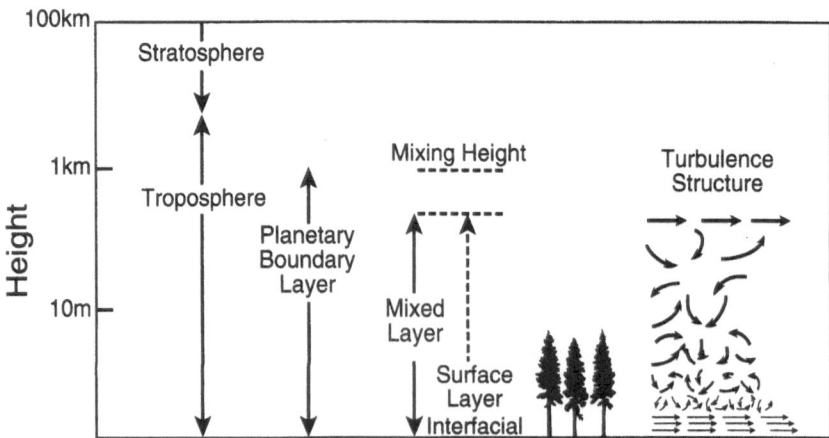

Figure 3.1 Vertical structure of the atmosphere (after Oke 1987).

restricts turbulent transfer of gases and particles to the surface of foliage. This resistance to mass transfer is related to the thickness of the interfacial layer and plays an important role in pollutant deposition mechanisms. The thickness of the interfacial layer depends on surface characteristics and wind speed. During high wind speeds, the layer is very thin and resistance to deposition is low. Plant physiologists often disregard interfacial layer resistances at wind speeds in excess of 2 m/s (Monteith 1973).

Air Pollution Meteorology

The dilution or accumulation and transport of air pollutants are influenced by the physical state of the mixed layer. Meteorology at time of emission and during the pollutant's lifetime aloft influences the size of the area affected and the amount of material deposited. Air pollution plumes usually diffuse when turbulent eddies are formed during unstable conditions. If the sizes of the turbulent eddies are smaller than the diameter of the plume, the plume is dispersed and diluted. Transport of the plume occurs when the eddies are larger than plume diameter. The mixing height sets an upper limit to the volume of air available for diffusion. Usually plume transport and dilution occur simultaneously, a process referred to as dispersion. Several factors influence the dispersion of pollutants, including emission height, atmospheric stability, convection, and terrain.

Emission height

Chimney effluents are buoyant and unstable because they often contain large amounts of steam and are usually emitted at high temperatures and velocities. The height to which a plume rises before reaching thermal equilibrium with its environment is the effective emission height (Figure 3.2). Plumes with low effective emission heights are dispersed slowly and usually cause local air pollution problems. Material in plumes emitted from tall stacks remains aloft for long periods and can be transported hundreds of kilometers by the stronger winds at higher altitudes. The pollutants are usually well dispersed prior to deposition. Consequently, stacks with large effective emission heights prevent local accumulation of pollutants (McElroy 1987).

How tall stacks moderate or enhance acid deposition to terrestrial ecosystems is not well understood. Plumes emitted above the mixed layer are transported for long distances. Long atmospheric residence times for the gases enable most chemical reactions to proceed to completion (Bubenick et al. 1983, McElroy 1987). While tall stacks facilitate

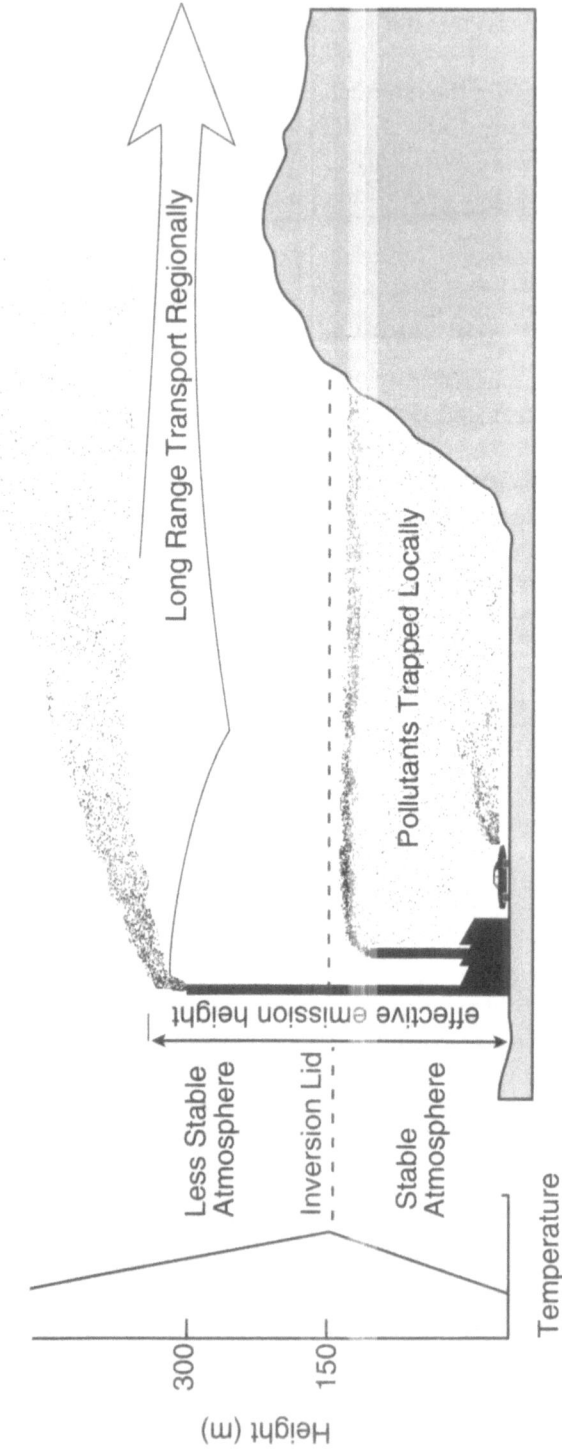

Figure 3.2 Influence of stack and emission heights on pollution transport and diffusion in the atmosphere. The temperature trace illustrates the height and strength of the surface inversion.

pollutant dispersion over large areas, thereby transforming local pollution problems into regional phenomena, there appears to be a gradient in deposition with highest deposition occurring near emission source areas.

The use of tall stacks capitalizes on the relative instability of air above a surface inversion. In order to take advantage of atmospheric instability in mountainous areas, stacks need to be tall enough so that emitted plumes clear surface inversions and topographical obstacles. In the western United States, point source emissions of sulfur and nitrogen oxides occur from stacks with mean and median heights of 26 m and 15 m. The tallest stack in the West in 1985 was 369 m high (National Emissions Data System Emissions Inventory 1985). Vehicles emit nitrogen oxides and hydrocarbons (ozone precursors) within a few meters of the earth's surface.

Atmospheric stability and convection

The vertical displacement of the plume is controlled by atmospheric stability and turbulence as well as by plume characteristics such as density and temperature. The best conditions for dispersion of pollutants occur during periods of strong atmospheric instability when a deep mixed layer is present, conditions characteristic of sunny days. On a sunny day the atmosphere may be dominated by large eddy structures that transport the plume up and down in an unsteady track known as looping (Figure 3.3a). Although plumes are dispersed rapidly under these conditions, undiluted plumes can come into contact with the ground at short distances from the stack (Oke 1987). Trees exposed to the undiluted plume are subjected to short, acute pollution exposures.

The worst conditions for dispersion occur during temperature inversions when the planetary boundary layer is stable. The inversion, with associated calm conditions, suppresses vertical mixing. Plumes emitted into the inversion layer are trapped and little dispersion of the pollutants ensues (Figure 3.3b). As the inversion breaks up, the air close to the surface of the earth becomes unstable, while at higher elevations the inversion remains intact. Remnants of the inversion prevent vertical expansion of the mixed layer, whereas instability at lower levels causes plumes to fumigate downward (Figure 3.3c). Forests may be exposed to relatively high pollutant concentrations.

Effluents emitted above the surface inversion are prevented from mixing downward by the stable surface layer. Moderately unstable air above the inversion allows the plume to loft upward, enhancing dispersion (Figure 3.3d). With this scenario, forests close to the pollution source are protected from high pollution exposures, although trees farther downwind may experience chronic pollution exposures.

Inversions that form by slowly sinking air over large areas are called subsidence inversions. They sometimes occur at the earth's surface, but more frequently, they are observed aloft (Ahrens 1988). The elevated layer of stable air effectively caps a less stable mixed layer below, preventing vertical dispersion of pollutants emitted into the mixed layer. Accumulation of pollutants beneath the subsidence inversion causes a murky mixed layer with a sharp upper boundary (Figure 3.3c). Since subsidence inversions are typically associated with large high pressure systems, such pollution episodes are regional and can persist for several days (Oke 1987, Ahrens 1988). Forests may be exposed to relatively high pollutant concentrations.

Plumes emitted into an elevated subsidence inversion may be trapped within the thin band of stable air sandwiched between two less stable air masses (Figure 3.3e). These plumes may be transported slowly for many kilometers with little or no dilution. Erratic wind behavior within the elevated subsidence inversion causes plumes to meander across the sky.

Plume behavior in the vicinity of large lakes or oceans is modified by land-sea breeze circulations. During the day, surface flow of air from the sea to the land induces fanning, fumigation, and looping of plumes. Land-sea breezes often form a closed circulation so a portion of the polluted air mass is carried aloft and concurrently transported out across the water (Figure 3.3f). Advection of this pollution back towards the land occurs during subsequent sea-breeze circulations (Oke 1987). Consequently, ocean air masses are occasionally contaminated with terrestrial emission products and coastal forests may be exposed to high pollutant concentrations, especially in the presence of coastal fog.

Valley inversions trap pollutants within the confines of a valley. As the inversion breaks up, pollutants are swept into the surrounding hills via anabatic (upslope) and valley winds allowing advection of high concentrations of pollutants into areas far from pollution sources. This sequence is common in the southern Sierra Nevada and the west slopes of the Cascades and may play an important role in introducing pollutants to forests in mountainous areas of the West.

The dispersion of pollutants is also influenced by thermal convection. Pollutants can be carried aloft during cloud formation and dissipation. Clouds created during weak convection, frontal, and orographic activities disperse pollutants vertically. During strong convectional activity associated with thunderstorms, pollutants are transported to great heights (Figure 3.4).

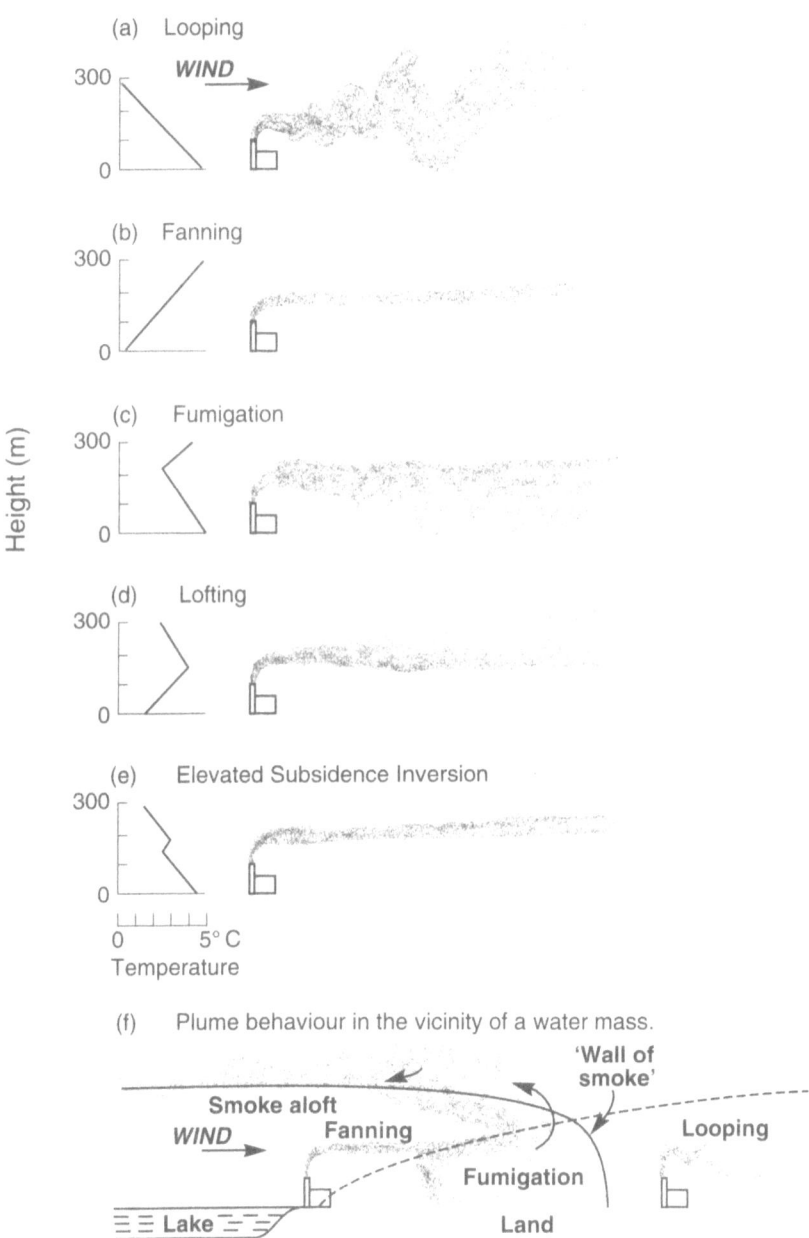

Figure 3.3(a-f) Dispersion of pollutants during different atmospheric stability conditions (after Oke 1987). The temperature traces describe the atmospheric stability conditions.

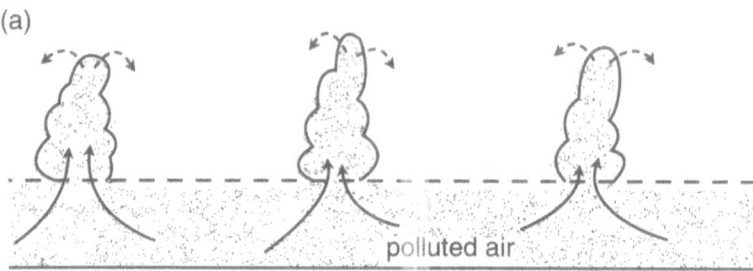

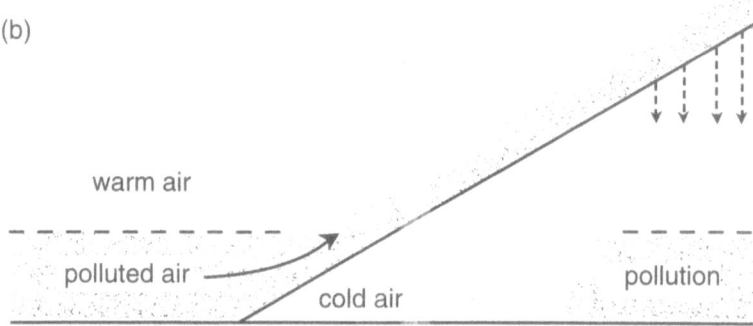

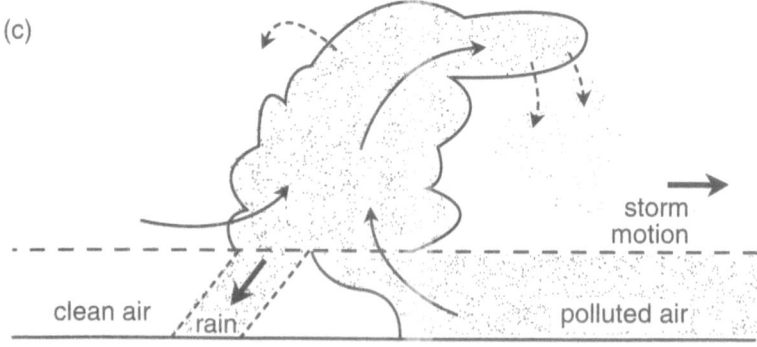

Figure 3.4 Transport of pollutants during cloud formation and frontal activity. (a) Vertical displacement during weak convection; (b) Warm frontal systems transporting pollutants vertically; (c) Transport to great heights during strong convection.

Terrain effects

Plume dispersion in complex terrain is influenced by mechanical and thermal turbulence. Mechanical effects include flow channeling due to valley or canyon walls and enhanced turbulence due to surface roughness. Thermal turbulence results from heating of the ground by the sun. Warming of mountain slopes during the day causes the air near the ground to be warmed by conductance and turbulence. If the air adjacent to the slope is at a different temperature than ambient air at the same altitude over the center of the valley, buoyant forces generate a circulation. During the day, these local circulations are anabatic (upslope) and valley flows whereas during the night, katabatic (downslope) and mountain flows develop. Anabatic and katabatic flows occur across the valley axis, whereas valley and mountain flows occur along the valley axis (Stull 1988).

Synoptic winds, associated with large-scale meteorological phenomena, often blow in a different direction than local flows. During stable conditions, pronounced directional shear between synoptic and local wind systems is common in mountain-valley situations. Mechanical turbulence together with entrainment and deepening of the boundary layer up the slope, increases plume dilution. Although cross wind dilution is limited by topographical channeling of plumes in mountainous terrain, estimates of dilution are greater than those predicted for flat terrain (Start et al. 1974, Whaley and Lee 1977, Reible and Shair 1981, King et al. 1987). A shallow, dynamically unstable layer of air can develop at the interface between the stable valley layer and the less stable synoptic (regional) air mass above. Turbulent exchanges between this intermediate unstable layer and the stable valley layer can retard upward growth of the surface inversion. If directional shear is reduced, due to weak synoptic winds, the growth of surface inversions is not inhibited, although a practical upper limit is the height of surrounding ridgetops (Fransioli and Weston 1989). However, at some locations dispersion of pollutants by local winds such as anabatic and valley flows is insensitive to changes in synoptic winds (e.g., Willson et al. 1983).

Terrain can also inhibit dispersion of plumes. Turbulent eddies in the lee of steep slopes or cliffs create aerodynamic traps, or downwash, where high concentrations of pollutants can be brought into contact with the earth's surface (Lott 1982, Oke 1987). Anabatic and valley flows transport pollutants emitted in valley basins to forests located on the slopes of hilly or mountainous terrain. Stability of the surface layer largely controls the depth and strength of upslope flows (King et al. 1987). During less stable conditions, anabatic and valley flows penetrate deep into mountains, with the depth of the penetrating layer increasing with elevation in a wedge-like fashion (Figure 3.5a). Under stable conditions, penetration of

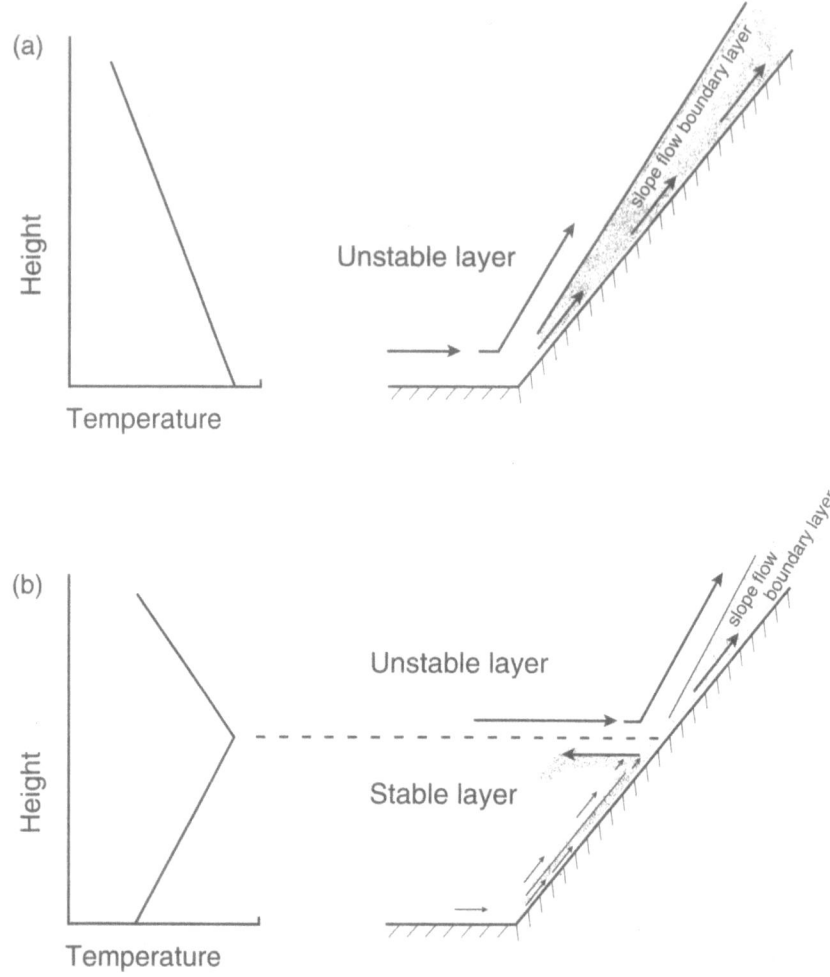

Figure 3.5 Effect of atmospheric stability on the transport of pollutants into mountainous areas during anabatic (upslope) flow (after King et al. 1987). (a) Anabatic flow during less stable conditions; (b) Anabatic flow during stable conditions.

upslope flows is restricted, and a shallow circulation develops that can return pollutants to the valley floor (Figure 3.5b). Circulations of this type occur in the Los Angeles Basin and cause forests at higher elevations to receive greater concentrations of polluted air under weakly stable meteorological conditions than under stable conditions (King et al. 1987). At night, katabatic and mountain flows return some of the polluted air to lower elevations.

Over distances of 10 to 50 km, most plumes lose their identity and contribute to a general contamination of the mixed layer (Oke 1987). Plumes with low effective emission heights lose their identities sooner than plumes from tall stacks. During periods of atmospheric instability, turbulence produces a rather homogeneous melange of pollutants throughout the mixed layer (Oke 1987). When the pollution source area is topographically confined and the inversion height is lower than surrounding ridges, the mixed layer becomes stagnant. Anabatic and katabatic flows tend to circulate the air with little net transport. Although individual plumes lose their identity, poor ventilation characteristics of closed valleys lead to little pollution flushing and an almost continuous increase of pollutant concentrations with time. Transport of pollutants out of valleys rarely occurs over mountain slopes, but rather through mountain passes (e.g., McElroy 1987). Passes with high crest elevations require strong upslope flows to achieve pollutant flushing, and consequently passes with lower crest elevations have greater flushing efficiencies (McElroy 1987).

Air pollution meteorology in the western United States

The potential for severe air pollution episodes in an area is determined by topographical and meteorological conditions conducive to the accumulation of atmospheric pollutants and is a function of vertical mixing, horizontal dispersion, and magnitude of emissions (Diab 1977). The climate in a specific area can be classified according to characteristics of major synoptic weather systems. Air quality can generally be related to several meteorological variables. Synoptic classification schemes can be used to investigate air pollution potentials for areas where detailed data are not available, such as much of the western United States (Yu and Pielke 1986). Pielke et al. (1984) provide an overview of air quality related aspects for five broad synoptic classes in the northern hemisphere. Generally, synoptic conditions associated with slow moving high pressure systems are most conducive to the accumulation of air pollutants.

Regional variation in air pollution potential across the United States has been estimated using the frequency of occurrence of high pressure systems (Figure 3.6, Schwoegler and McClintock 1981). Large areas in the West are subjected to non-ventilating weather as a result of intruding high pressure systems. The western United States is also characterized by complex terrain with small air basins (Chapter 1, Figure 1.1). The mountain ranges that surround these basins inhibit horizontal movement of air pollutants. In addition, the formation of valley inversions trap pollutants close to the surface of the earth. The potential for air pollution in the western United States is further enhanced by the social

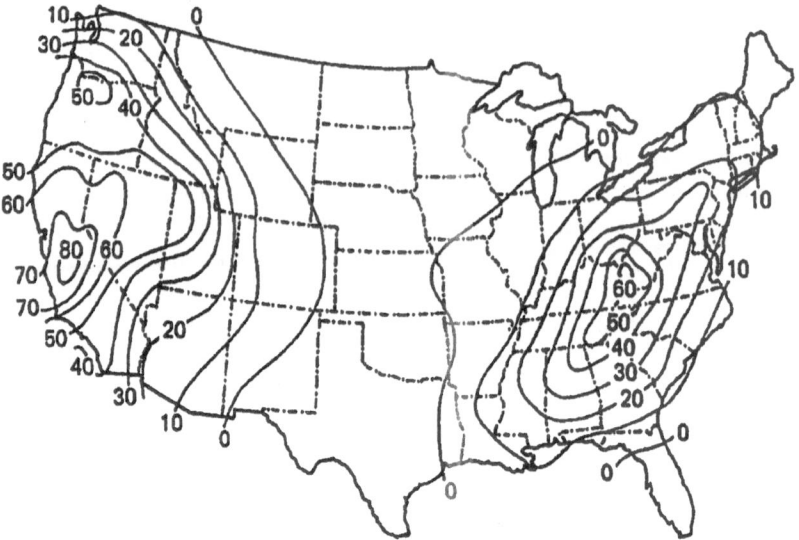

Figure 3.6 Zones subject to non-ventilating weather as experienced beneath intruding high pressure systems. The isoline surface represents the number of days per year with high potential for pollution accumulation (after Schwoegler and McClintock 1981).

structure of Western communities. Vehicle use continues to increase in the large, sprawling urban centers characteristic of the West, leading to increases in low level emissions of pollutants. Despite clear scientific evidence that traffic congestion leads to a three fold increase in vehicle emissions during peak traffic conditions (Anonymous 1987), there is little public support for mass transit or vehicle sharing programs.

Complex topography together with the greater influence of high pressure systems imply that the potential for air pollution accumulation across much of the western United States is great, perhaps even greater than that for the eastern United States. However, this does not mean that all the pollution emitted in the West remains trapped close to the surface or in valleys. Meteorological conditions conducive to long-range transport of pollutants do occur. For example, the Los Angeles urban plume has been implicated in degrading summer visibility over the southwestern United States, including the Grand Canyon (e.g., Hoffer et al. 1981), and the plume from the Mohave power plant in Arizona is detectable at ranges in excess of 100 km (Hegg et al. 1985). During summer and winter, the Mohave plume exhibited little vertical extent even at great distances downwind (Hegg et al. 1985), implying that the plume was emitted into a stable or neutral atmosphere (e.g., elevated subsidence inversion) that allowed long-range horizontal transport with little vertical diffusion (Figure 3.3e).

Pollutants in Forests

Meteorological conditions above and within forests affect dispersion of air pollutants within the forest boundary layer. The forest boundary layer consists of the surface layer between ground level and slightly above the top of the forest canopy. Within a forest, geometric factors such as crown coverage and tree height, and morphological characteristics, such as amount and height of branching and size and density of foliage, modify local meteorology. Forests deform vertical and horizontal wind fields and turbulence. Vertical and horizontal gradients in temperature, moisture, and solar radiation are also modified by the vegetation. Turbulent fluxes of momentum, heat, and water vapor, and ultimately the total energy balance of the terrain surface and of the canopy domain are modified. Consequently, stability of the forest boundary layer often differs from that of the overlying air mass, which in turn affects depositional behavior of pollutants (Cionco 1989).

Atmospheric transfer of mass typically occurs via turbulent diffusion along the mean concentration gradient of the pollutant. Formally, the flux density is related to height and concentration gradient by the diffusivity of the pollutant of interest. This turbulent diffusivity is height dependant and may be inferred from the local energy balance or wind profile. Unfortunately, studies in forests indicate that this standard approach to atmospheric transfer of mass is inappropriate for tall, rough surfaces such as forests (Denmead and Bradley 1985). Measurements of temperature and humidity gradients in coniferous forests (McBean 1968, Denmead and Bradley 1985, Finnigan 1985) imply that turbulent transport of mass in the canopy is a sporadic process unrelated to the shape of the mean concentration profile. Large coherent motions displace canopy air with air from aloft, together with its associated pollutants. Thus, it is important to have some understanding of wind, turbulence, temperature, and humidity gradients in forests when estimating pollutant concentrations and deposition to forests.

Wind speed

Variations in vertical profiles of wind speed result from differences in the structure and density of forest canopies. Wind speeds decrease exponentially from above the canopy to a level within the canopy where wind speeds are close to zero (Figure 3.7). This plane of zero displacement represents the apparent height at which bulk drag exerted by the vegetation on air is greatest. The zero plane is often displaced 2/3 of the height of the forest (Jarvis et al. 1976, Fritsche et al. 1989), but displacement height depends on bulk drag exerted by forest elements and on vertical distributions of leaf area. Less bulk drag is exerted on the air by forests with open canopies than by forests with dense canopy elements.

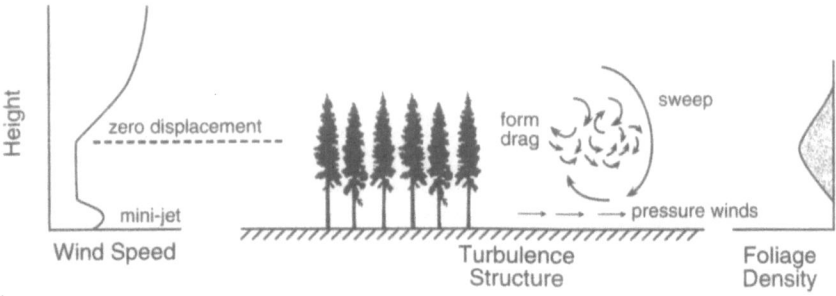

Figure 3.7 Schematic diagram of the influence of foliar density on wind speed and turbulence within the forest canopy.

A correspondingly greater penetration into less dense canopies occurs, resulting in higher mean wind speeds throughout the stand depth. Usually, the zone of maximum foliage density coincides with the zero plane for wind speed (Oke 1987).

Increased wind speeds in the trunk space often occur as a mini-jet of air (Figure 3.7). Strength of the mini-jet depends on the density of tree trunks (Amiro et al. 1989) and the presence or absence of an understory (Fritschen 1985).

Turbulence

Turbulence is several orders of magnitude more effective for mass transfer, e.g., particles and gases, into forest canopies than is molecular diffusion (Stull 1988). Large coherent eddies or gusts are the primary agents of transfer close to rough surfaces such as forests (Figure 3.7). The behavior of these eddies is independent of surface geometry and responds instead to gross roughness of the surface (Finnigan 1985). Transfer of particles and gases to forest canopies occurs via (1) upward ejection or burst of relatively slowly moving air from near the surface; (2) downward sweep or gust of relatively fast moving air from above the forest boundary layer; (3) outward transfer or upward diffusion of fast moving air; and (4) inward transfer or downward transfer of slow moving air. Net downward transfer of mass occurs primarily during sweep and gust and secondly during ejection or burst processes. Upward transportation occurs during outward and inward transfer. Sweep and gust processes tend to dominate over other transfer processes, although this relationship is not always obvious (Shaw 1985). Forests characterized by relatively few trees with dense canopies experience

strong downward transfers of momentum and kinetic energy within the
upper canopy (Amiro and Davis 1988, Baldocchi and Hutchinson 1988,
Baldocchi and Meyers 1988a,b; Amiro et al. 1989). Fluctuations in wind
velocity in lower portions of deep multistory forests may also be
influenced by local fluctuations of horizontal pressure gradients (e.g.,
Holland 1989).

Turbulent eddies in the canopy have smaller dimensions than those in
the lower trunkspace (Baldocchi and Meyers 1988b). Within the forest
canopy, turbulent eddies loose energy due to normal processes of eddy
cascading. Additional losses in energy may be due to drag by canopy
elements. Turbulent eddies generated by canopy elements dissipate
rapidly and, therefore, contribute little to the energy of turbulence close
to the ground (Amiro et al. 1989). Turbulence in the trunkspace originates mainly from sweep or gust processes that transfer large turbulent
eddies from above the forest canopy towards the ground (Baldocchi and
Meyers 1988b).

Edge effect

When airflow encounters the leading edge of a forest (Figure 3.8), most
of the air is deflected up and over the forest wall for a distance of about
10 to 20 times the height of the trees, before downdrafts re-introduce the
air into the forest canopy (Cionco 1985). However, blow-through can also
occur. Under such conditions, a small jet-like structure develops below
the crowns of the trees at the leading edge of the forest (Cionco 1985).
The penetration of this jet appears to be limited to a distance of 2 to 3
times the height of trees, although wind penetration is greater through
recently created edges when the understory has not had an opportunity
to regrow (Fritschen 1985).

The presence of large clearings in forests can also alter the wind patterns.
Wind data taken in a pine and aspen forest 60 m from the edge of a 0.5
km^2 clearing indicate that the strongest and most frequent winds below
the canopy were directed towards the clearing center during daytime
hours (Figure 3.8a). At night when the atmosphere was stable, the flows
below the canopy tended to be away from the clearing center (Figure
3.8b). Wind patterns above the forest canopy were less organized with
respect to the clearing than winds below canopy level (Leahey and
Hansen 1987).

Temperature and humidity

During the day, temperature and humidity profiles within the forest
boundary layer exhibit maxima at the zone of maximum foliage density
as a result of radiative absorption and transpiration of water vapor (Oke

(a)

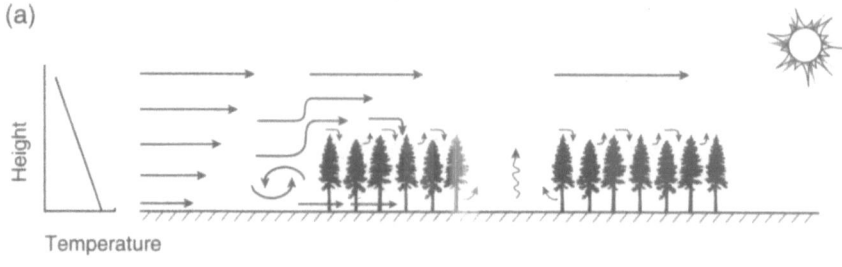

(b)

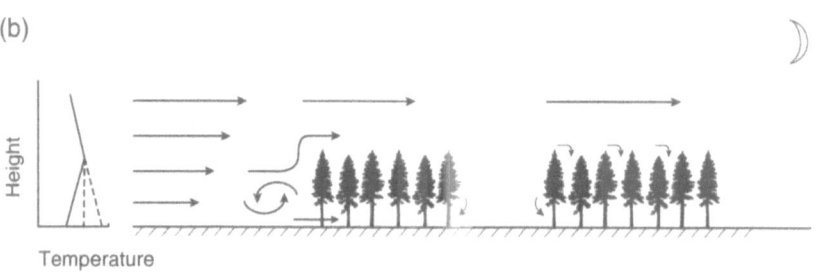

Figure 3.8 Schematic representation of the influence of forest edges and gaps on the turbulent structure of the windfield in the forest boundary layer.

1987). A temperature inversion is created because the canopy is warmer than the forest floor. Strong temperature inversions also occur above the forest canopy (Cionco 1985, Druilhet et al. 1989). Thus, the unstable atmosphere of the open terrain need not prevail within or directly above the forest canopy during the day, especially in summer.

Radiation losses from the forest canopy to the sky at night are greater from upper portions of the canopy than from within the canopy. Temperature profiles are the reverse of those measured during the day, and humidity is low in the crown layer and higher above and below the crown. Thus, at night the forest boundary layer is neutral or slightly unstable (Cionco 1985, Oke 1987) and relatively mild conditions prevail within the stand.

The temperature and humidity of the forest boundary layer vary season-ally, with little diurnal variation and vertical stratification during winter and larger diurnal fluctuations with pronounced vertical stratification during summer months (e.g., Helms 1970). The moisture and tempera-ture profiles within a forest are also sensitive to climatic phenomena such as summer drought. Even drought-resistant trees need to conserve water during periods of extended drought. Decreased stomatal conduc-tance reduces the exchange of water vapor with the environment and results in less evaporative cooling of the needles. Consequently, surface temperature within the canopy rises and the strength of the surface inversion increases.

Air pollution potential of forests

The degree of vertical penetration and distribution of pollutants within forests is influenced by forest meteorology. Pollutants in the air above a forest are readily mixed into the upper canopy via turbulent eddies. These turbulent eddies loose energy in the canopy and the degree of mixing decreases with distance from the top of the canopy, possibly reaching a minimum at the plane of zero displacement. Large turbulent eddies, generated by sweep or gust processes, introduce air and its associated pollutants from above the forest into the trunk-space.

In the absence of large turbulent eddies sweeping pollutants downwards from above, diffusion of gases and particles probably dominate transport of pollutants to the forest floor, certainly during the day when tempera-ture inversions occur in the lower forest boundary layer. A gradient in pollutant concentrations probably exists under these conditions, with highest concentrations near the forest top and lowest concentrations towards the forest floor.

Modifications of the wind field by forests imply spatial variations in pollutant exposure. For example, at a forest edge, wind penetration can directly introduce pollutants to the trunk space for a distance of up to 3 times the height of the trees. Deflection of air by the leading edge of the forest can result in a gap as large as 20 times the height of the trees before downdrafts re-introduce the polluted air into the forest canopy. The presence of large clearings in forests also modify wind patterns and consequently air pollution exposure (Figure 3.8). These issues have typically been investigated for dry deposition of pollutants. Recent studies have shown significantly larger dry deposition of pollutants at the forest edge or at the edge of large clearings (e.g., Beier and Gundersen 1989).

Chemical Transformation of Gaseous Pollutants

The atmosphere is an efficient oxidizing medium. Primary pollutants, such as gaseous sulfur dioxide, nitrogen oxides, and volatile organic compounds, undergo a series of reactions to produce secondary pollutants that are usually characterized by higher chemical oxidation states than their precursors. Oxidation is often accompanied by an increase in polarity and hence water solubility (Schroeder and Lane 1988). The rate at which reactions occur is determined by complex interactions between catalysts and environmental conditions. For slow chemical reactions, turbulence succeeds in mixing the primary constituents before the chemical reactions occur or become completed. Time scales of chemical reactions are typically of the same order of magnitude as the turbulent mixing time so that reaction rates are controlled by the ability of turbulence to bring reacting species together (Schumann 1989).

The form of secondary pollutants (gas, particle, or droplet) determines their modes of deposition and may influence their effect on trees. Secondary pollutants are usually aerosols, although ozone and a variety of nitrogen-based species are in the gaseous phase. An aerosol is defined as a relatively stable suspension of solid or liquid particles in a gas. The size of the solid particles reflects their origin. Secondary pollutants usually have aerodynamic diameters smaller than 2.5 μm whereas naturally produced particles, such as dust, are larger than 2.5 μm (Finlayson-Pitts and Pitts 1986). Most of the sulfate mass is found in the aerodynamic diameter range of 0.1 to 1.0 μm (Garland 1978). Many aerosols are hygroscopic; they have an affinity for water vapor and act as cloud condensation nuclei.

Sulfur chemistry

Once emitted, gaseous sulfur dioxide is oxidized in the atmosphere to sulfuric acid aerosols or sulfates by gas phase or homogeneous reactions, or during heterogeneous reactions in the liquid phase or on surfaces of solids. Homogeneous oxidation of sulfur dioxide can occur as direct photo-oxidation (Figure 3.9a), oxidation by species formed in photochemical reactions, especially the hydroxyl radical (Figure 3.9b), oxidation by the stabilized Criegee biradical RCHOO (Figure 3.9c), or oxidation by nitrogen dioxide and oxygen (Figure 3.9d). Liquid-phase transformation is controlled by adsorption of sulfur dioxide into cloud or rain droplets. The adsorption of sulfur dioxide is a relatively rapid process that depends on the size of the droplet, the efficiency of adsorption, and the chemical nature of the surface (Finlayson-Pitts and Pitts 1986). Once in the aqueous phase, sulfur dioxide is transformed to sulfate in the presence of strong oxidants, (such as hydrogen peroxide and ozone, Figure 3.9e) and catalytic metals, (such as manganese and iron, Figure

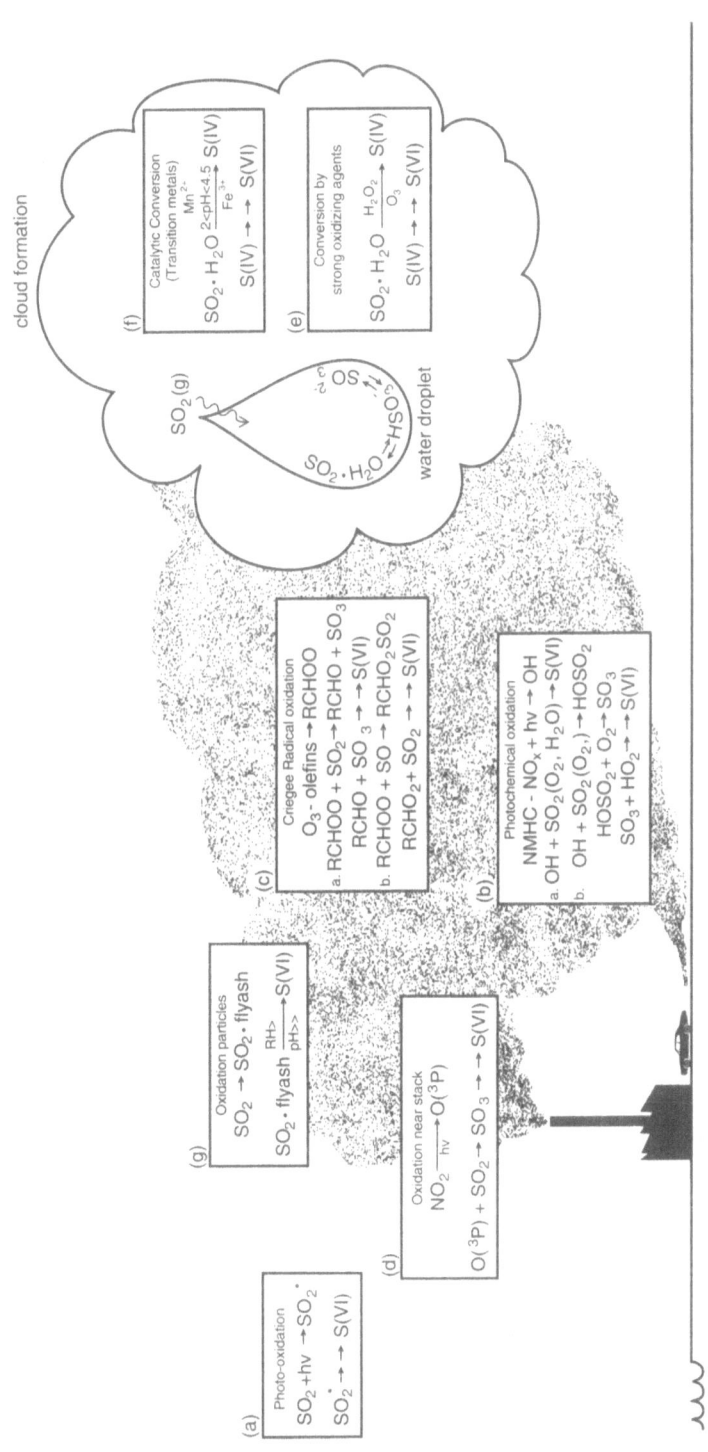

Figure 3.9 *Schematic diagram showing the various chemical processes whereby sulfur dioxide transforms to sulfate in the atmosphere. S(IV) refers to dissolved sulfur dioxide in solution (i.e. $SO_2 \bullet H_2O + HSO_3^- + SO_3^{2-}$) whereas S(VI) refers to sulfuric acid and sulfate (i.e. $H_2SO_4 + SO_4^{-2}$). $O(^3P)$ refers to ground state oxygen atoms.*

3.9f). Sulfur dioxide can also be oxidized by oxygen in the liquid phase without metal catalysis, although measured rates of oxidation vary considerably; if the highest rates measured do occur in the atmosphere, this reaction may be a significant source of acid formation in droplets with no catalytic metals and pH > 5 (Finlayson-Pitts and Pitts 1986). Adsorption of sulfur dioxide onto the surfaces of particles (e.g., flyash) depends on the nature of the surface. Once adsorbed, transformation to sulfate depends on the presence of co-pollutants such as nitrogen dioxide and, in the case of carbonaceous surfaces, on relative humidity (Figure 3.9g). The relative contributions of homogeneous and heterogenous oxidation in transforming sulfur dioxide to particle sulfate are not known. It is likely that both mechanisms are at times dominant, depending on meteorological conditions, such as the presence of fog, and on the particular mix of atmospheric pollutants. For example, the oxidation of sulfur dioxide is enhanced by the presence of ozone and fog as illustrated by high sulfate concentrations in the Los Angeles Basin following foggy nights (Jacob et al. 1987).

The rates of conversion of sulfur dioxide to sulfate appear to be higher at noon compared to nighttime and in summer compared to winter. Sunlight intensity, the presence of oxidants, relative humidity, and the presence of cloud influence observed conversion rates. For example, rates of oxidation of sulfur dioxide in plumes from electric power plants are usually less than 10% per hour during the day (e.g., Newman 1981), although much higher rates have been observed when plumes pass through a cloud or fog bank (Eatough et al. 1984). Oxidation rates as high as 30% per hour have been recorded (Breeding et al. 1976). Nighttime oxidation rates of sulfur dioxide are generally less than 5% per hour (Finlayson-Pitts and Pitts 1986), although oxidation rates as high as 11% per hour have been measured in Los Angeles (Cass and Shair 1984).

These rates can be used to estimate the residence time of sulfur dioxide in the atmosphere, i.e., the time taken for 100% conversion to sulfate. Depending on atmospheric conditions and effective emission height, the longer a pollutant remains in the atmosphere, the farther it is transported and the more it is diluted. Some indication of the degree of transport and dispersion of sulfur dioxide is useful when estimating regional exposure of forests to this pollutant. In summer, the residence time of sulfur dioxide is about a day in the presence of cloud or high relative humidity, or five days in a dry atmosphere. In winter, the residence time for sulfur dioxide can be as long as 20 days (Newman 1981, Richards et al. 1981, Finlayson-Pitts and Pitts 1986).

In contrast to sulfur dioxide, the residence time of sulfate, the oxidation product of sulfur dioxide, is usually determined by deposition characteristics rather than by chemical conversion rates. Sulfate is hygroscopic and acts as cloud condensation nuclei in the presence of moisture. Thus the atmospheric residence time of sulfate is very short in a precipitation condition. In the absence of high relative humidities, sulfate can remain suspended in the atmosphere for extended periods of time depending on atmospheric conditions as discussed earlier (e.g., Garland 1978). Longer residence times for sulfate in a dry atmosphere increases the potential for regional dispersion, facilitating deposition of sulfate to forests located far from the source area.

Nitrogen chemistry

The chemistry of nitrogen compounds is complex due to the large number of chemical compounds involved and their many chemical and photochemical reactions. The conversion of nitrogen oxides to nitrate occurs via a series of reactions during which the participating nitrogen oxides switch back and forth between various stages of oxidation and eventually end up as nitrates (Record 1981). The deposition of nitrogen species and the formation of aerosol nitrate is sensitive to dinitrogen pentoxide (N_2O_5) hydrolysis, dry deposition rates, aerosol scavenging processes, and temporary storage of nitrogen oxides in the form of aerosol nitrates and peroxyacetyl nitrate (PAN). For example, the half-life of nitrogen oxides emitted in the Los Angeles Basin in summer is about 24 hours and the major removal process is dry deposition of nitric acid (HNO_3), PAN, and nitrogen dioxide. Much of the nitrogen left in the air column at the end of 24-hours is considered to be associated with nitrogen dioxide and PAN (Russell et al. 1985).

Oxidation rates of nitrogen oxides in power plant plumes range between 0.2 and 12% per hour, with the major products being PAN and nitric acid, and to a lesser extent particle nitrate (e.g., Hegg and Hobbs 1979). The rate depends on the ratio of non-methane hydrocarbon to nitrogen oxides, with higher rates during the day than during the night and under conditions of high relative humidity. Conversion rates of up to 24% per hour have been recorded in urban plumes (Chang et al. 1979, Spicer 1982). A detailed discussion of the chemistry of nitrogen is given in Finlayson-Pitts and Pitts (1986).

Ozone formation

Ozone is formed by the photolysis of nitrogen oxides in the presence of non-methane hydrocarbons such as volatile organic compounds. On a global scale, natural sources of nitrogen oxides and volatile organic compounds are as large as anthropogenic sources (Finlayson-Pitts and Pitts 1986). Since anthropogenic emissions are generally concentrated in relatively small areas, tropospheric ozone is considered an anthropogenic pollutant that is prevalent in and around urban centers. Although nitrogen dioxide is the primary precursor of ozone, nitric oxide, formed during the photolysis reaction, reacts rapidly with ozone to form nitrogen dioxide and oxygen. Thus, nitric oxide and ozone do not co-exist in the atmosphere and ambient ozone concentrations are a function of the balance between formation and scavenging (Finlayson-Pitts and Pitts 1986).

In an urban environment or source area for nitrogen oxides and volatile organic compounds, ozone formation and removal follow reproducible patterns. In the early morning, concentrations of nitric oxide rise and peak at the time of maximum vehicular traffic. Nitric oxide reacts rapidly with ozone to form nitrogen dioxide. Ozone levels, which are relatively low in the early morning, increase rapidly around noon when nitric oxide concentrations drop following complete oxidation to nitrogen dioxide. Ozone concentrations decrease during the late afternoon as sun intensity decreases and nitric oxide levels increase following the second daily traffic maximum. Thus, the ozone concentrations in urban areas are low during the night and peak during the middle of the day (Finlayson-Pitts and Pitts 1986). However, this pattern can be distorted by changes in meteorological conditions such as inversion formation and break-up.

Downwind of urban centers, high ozone concentrations occur during the late afternoon. The formation of surface inversions can isolate ozone and other pollutants from the ground, allowing transport of polluted air with little or no deposition or scavenging. Profiles of ozone at sites intercepting the elevated polluted air mass exhibit less diurnal variation and high ozone concentrations can occur at night (e.g., Singh et al. 1978). During break-up of surface inversions, the polluted layer of air is mixed down to the ground yielding diurnal profiles with a dramatic increase in ozone concentrations shortly after sunrise and exposing forests to relatively high ozone concentrations.

Depositional Processes

Vegetation, especially coniferous forests, is an important sink for air pollutants. Leaves indiscriminantly absorb pollutant gases during normal gas-exchange procedures. Foliage provides a surface of interaction for particle matter and gases. If chemically altered by the plant, the pollutants may be detoxified or even utilized as nutrients. The deposition of primary and secondary pollutants to forests depends on physical and chemical characteristics of both pollutants and receptor surfaces, as well as on the dynamics of the lower atmosphere.

Dry deposition

Dry deposition is the transfer of gases and particles from the atmosphere to the ground via atmospheric turbulence and diffusion without the intervention of precipitation. Dry deposition occurs when turbulent transport or sedimentation introduces gases and particles to the laminar interfacial layer surrounding deposition surfaces. The gases and particles are transported through this layer via convection, diffusion, or inertial processes until chemical or physical capture of the pollutants by the surface occurs (Voldner et al. 1986). The rate of dry deposition of gases and particles is controlled by atmospheric conditions, such as the degree of turbulence in the surface layer and the thickness of the interfacial layer; chemical properties of the pollutant; aerosol dynamics; receptor surface characteristics and chemistry; and plant anatomy and physiology.

Dry deposition of particles depends on particle size. Sedimentation, or gravitational settling, is the dominant dry deposition process for large particles (aerodynamic diameter > 3 µm). The effect of gravity on particles smaller than 3 µm is reduced and diffusion and inertial processes become as effective or more effective removal mechanisms than sedimentation. Deposition of these gases and particles is influenced by turbulent diffusion and the depth of the quasi-laminar interfacial layer of air that adheres to all surfaces. Turbulence in the atmosphere can transport gases and particles to, or close to, the receptor surface, depending on the depth of the interfacial layer. Once the gases and particles have been introduced into the interfacial layer, molecular (gases) and Brownian (small particles) diffusion are the dominant transfer mechanisms. Particles with aerodynamic sizes > 1 µm can have sufficient inertia to enhance movement through the interfacial layer to the receptor surface. Particles with aerodynamic diameters < 0.3 µm are primarily transported by Brownian diffusion (Sehmel 1980, Hosker and Lindberg 1982).

Particles coming into contact with the receptor surface can bounce off if the momentum of the particle is not absorbed on impact. The degree of bounce off is determined by the mass, size and surface characteristics of the particle, and by presence or absence of a soft surface layer to cushion the impact. Bounce off is important when the obstacle and thickness of the laminar layer are small. For particles with an aerodynamic diameter of less than 5 µm, bounce off does not occur unless the incident velocity is high and the surface is hard and smooth. Efficiency of deposition depends on the micro-roughness of the receptor surface, particularly the presence of hairs. Once particles are at rest on a surface, surface tension and other forces hold them and the drag of the wind is reduced by the laminar layer so the particles are not easily disturbed. If the surface of the plant is sticky or wet, deposition of particles is very effective (Chamberlain 1975).

Quantification of dry deposition at a regional scale cannot be made directly. Usually, estimation of dry deposition involves sophisticated models that attempt to deal with physical and chemical complexity, as well as with the variety of possible interactions between the atmosphere, pollutant, and surface characteristics. Estimation of deposition rates, or deposition velocity V_d, to heterogeneous surfaces and complex topography typical of Western forests is not yet possible (Cape and Unsworth 1987). For this reason, V_d is not addressed in this chapter and the interested reader is referred to Garland (1978), Sehmel (1980), and Voldner et al. (1986).

Wet deposition

Wet deposition involves the removal of pollutants during precipitation events. Acid rain or acid precipitation refers to wet deposition episodes whose chemistry is significantly altered by anthropogenic activities. The natural pH of precipitation typically ranges between 4.5 and 5.6. A precipitation event is usually considered "acid" if the pH is less than 5.0 (Charlson and Rodhe 1982, Lefohn and Krupa 1988).

Pollutant removal by rain and snow occurs during rainout (within-cloud scavenging) and washout (below-cloud scavenging). Rainout occurs when water vapor condenses on hygroscopic nuclei to form cloud droplets or when gases or particles diffuse into the cloud droplets. The most important pollutant scavenging process for particles within the cloud is via nucleation (Garland 1978, Schroeder and Lane 1988). Deposition occurs when the weight of the hydrometeors exceeds the force of updrafts within the cloud. Washout involves the removal of particles and gases by rain drops or snow flakes en route to the ground by impaction, interception, and diffusion. Rainout of condensation nuclei in a moderately polluted atmosphere is the main wet deposition mechanism for the removal of sulfate. Washout of large sulfate particles by raindrops may contribute to the initial concentrations of sulfate in precipitation, but this fraction of the aerosol will be exhausted by the first few millimeters of rain (Garland 1978).

Wet deposition is the total mass of deposited ion over the period of interest. Estimates of wet deposition are complicated by great variability in the occurrence of precipitation events, the instability of hydrogen ion and nitrogen species within most precipitation samples (Galloway and Likens 1976), and problems with measuring precipitation amount, especially for snow. Since the solubility of most pollutants differs in ice, snow, and rain (Scott 1981, Finlayson-Pitts and Pitts 1986), the form of precipitation influences the concentrations of ions deposited. This is an important consideration when extrapolating from data collected at low elevation sites, which usually receive less snow than forests at high elevations.

Estimates of wet deposition rates of acidic species such as sulfate and nitrate are more accurate than those for dry deposition. Although scavenging and deposition of small particles by wet and cloud deposition is more rapid and efficient than removal under dry conditions (Garland 1978), precipitation events usually occur intermittently. Therefore, the total contribution by dry deposition often exceeds that from wet deposition, sometimes by several orders of magnitude (e.g., Garland 1978).

Cloud deposition

Cloud droplets are highly effective at scavenging particles and gases from the air. Hygroscopic particles act as condensation nuclei causing physical changes in aerosol size, thereby facilitating deposition of small particles such as sulfate (Jacob et al. 1984). Cloud droplets are important sinks for soluble gases such as nitric acid, ammonia, and sulfur dioxide. Furthermore, cloud droplet capture by the forest canopy has been recognized as an important hydrological, and therefore chemical, input to some Western forests (Azevedo and Morgan 1974, Muir and Böhm 1989).

The rate of deposition of cloud droplets is controlled by wind speed and turbulence, canopy and leaf geometries, and liquid water content and size distribution of the droplets. Cloud droplets are introduced into the forest canopy by turbulent transfer and sedimentation. Droplet capture occurs by impaction or sedimentation (Fowler 1984, Waldman et al. 1985). Although the liquid water contents of cloud droplets are lower than those for rain and snow, solute concentrations can be several orders of magnitude higher in cloud droplets than in rain and snow. Thus, total deposition of ions during cloud events may approach or exceed that during wet deposition. For example, in the Sierra Nevada, annual estimates of hydrogen ion, ammonium, nitrate, and sulfate deposited during cloud events were of the same order of magnitude as annual wet deposition at the same site (Collett et al. 1989).

Deposition in Forests

The structure of forest stands exerts considerable influence on exchanges of heat, mass, and momentum between the atmosphere and vegetation. A number of experiments on pollutant deposition to and dispersion within forests have been conducted (Fox 1985). Unfortunately, the data paint a confused picture that fails to determine whether existing models accurately predict pollutant removal within the canopy. Canopy parameters have an overwhelming effect on the magnitude of removal (Fox 1985), and airflow within the canopy is complex as described earlier in this chapter.

Deposition of gases, particles, and cloud droplets

Gases, particles and cloud droplets are introduced to the canopy via turbulent diffusion, mainly sweep and gust processes, and via sedimentation for aerosols with aerodynamic diameters greater than 5 μm. Turbulence varies considerably throughout the forest, especially natural forests of non-uniform age. Turbulence at the forest edge and a mini-jet in the trunk space can transport polluted air into the forest (Figure 3.8). Towards the interior of the forest, turbulent structures are a function of low wind speeds and stability of the atmosphere above the forest canopy.

Under stable atmospheric conditions with low wind speeds in air layers above the canopy, the polluted air mass probably does not penetrate the canopy much below the zero displacement plane for wind (Figure 3.7). The reduction in wind speed with distance from the top of the canopy together with reduction in turbulent transfer and the formation of an inversion between the ground and height of maximum foliar density, inhibits turbulent diffusion of gases and small particles into the lower portion of the canopy and forest (e.g., Fritschen and Edmonds 1976). Under such conditions, most deposition of gases and particles probably occurs above the height of maximum foliar density, although sedimentation of larger particles and cloud droplets to the forest floor may occur. During windy conditions above the forest boundary layer, the polluted air mass penetrates deep into the canopy via vertical turbulent diffusion, such as sweep and gust processes. Under these conditions, concentrations of pollutants in the lower canopy are probably a few percent lower than values above the canopy. Thus, most elements of the forest are similarly exposed to pollutants. Coniferous forests with fine needles are very efficient at collecting wind-driven cloud droplets (Unsworth and Wilshaw 1989). Intercepted cloud droplets wet foliar surfaces but do not necessarily flush them clean as does rain. Dissolution of previously deposited material can lead to more concentrated solutions in contact with the foliage than in the incident cloud water (e.g., Waldman et al.

1985). Furthermore, surface moisture can significantly increase deposition of particles by reducing bounce-off and resuspension (Chamberlain 1967) as well as provide a sink for soluble gases (Brimblecombe 1978).

Diffusion through the stomata into the leaf is the main pathway for the uptake of pollutant gases (Guderian 1985, Cape and Unsworth 1987). Gases deposited on the cuticle partake in chemical transformation and remain on the leaf surface (Cape and Unsworth 1987, Kerstiens and Lendzian 1989). Rates of dry deposition of most gases are therefore controlled by stomatal response and environmental factors that influence stomatal conductance. Large areas of the western United States experience summer drought implying that species that close their stomata in response to drought stress are partially protected against high ozone concentrations that occur during mid to late summer. Furthermore, the metabolic rates for young needles (65–95% of final expansion) are higher than for older needles (Guderian 1985), and therefore gaseous uptake is expected to be higher in outer parts of the canopy dominated by young needles.

Dry deposition via diffusion of submicron particles is negligible compared to wet deposition via rain and snow or by cloud deposition (Garland 1978). However, submicron sulfate may enter stomata and affect internal cells before affecting leaf surface cells (Gmur et al. 1983, Chevone et al. 1986).

Wet deposition

Unlike deposition of gases, plant metabolism does not actively control the uptake of pollutants from rain and snow, although forest and individual plant morphology does determine the degree of interaction between falling raindrops or snowflakes and foliage. When precipitation comes into contact with the foliage, the behavior of water droplets at the leaf surface is determined by the physical and chemical nature of the cuticle and by the form of epicuticular wax (Cape and Unsworth 1987). Once in contact with the foliage, the water droplet may act as a source or sink for inorganic or organic compounds from both the leaf and the atmosphere. Soluble gases may dissolve from the atmosphere into the liquid film surrounding the leaf. Particles deposited during dry weather can be washed off and introduced to the soil.

High elevation forests typically receive more precipitation than forests at low elevations. In the West, high elevation forests receive a large percentage of their precipitation as snow. The concentrations of ionic species may vary considerably between high and low elevation sites, partly because of a concentrating effect and partly because many pollutants have different solubilities in ice, rain, and snow (Finlayson-Pitts and Pitts

1986). A preliminary study on the influence of elevation on precipitation chemistry in the Colorado Rockies indicated that sulfate concentrations in snow at the high elevation site were generally lower than those at the low elevation site (Warren et al. 1990). In addition, sulfate concentrations in snow at the high elevation site were between 70 and 90% of sulfate concentrations in rain at both elevations.

Rimed snow and rime ice occur by the impaction of cloud droplets on snow flakes and to vegetative surfaces. The chemistry of these frozen condensates has not been well quantified, but the riming process appears to concentrate atmospheric pollutants. For example, inclusion of rime deposits on snow crystals provided up to 86% of deposition of trace constituents measured in a snowpack in Colorado (Borys et al. 1988); deposition of hydrogen ions, nitrate, and sulfate in rime ice were two to three times that in rain and snow during the winter in the central Sierra Nevada (Berg and Dunn 1988); and concentrations of nitrate, sulfate, and ammonium ions were two to four times higher in rime than in snow in the Washington Cascades (Duncan 1990). Rime ice is common in the Sierra Nevada, where on average about 20% of winter days experience rime ice with as many as 70% of winter days having observable riming (Berg 1988). Rimed snow is common in the central Washington Cascades (Duncan 1990). About 10% of water equivalent of snowpack in the Colorado Rockies is attributable to rime deposits (Hindman et al. 1983).

Forest canopies are coupled closely with the atmosphere, and precipitation intercepted by the canopy can evaporate rapidly (Unsworth and Wilshaw 1989). As polluted rain evaporates, a concentrating of chemical species occurs in the film of water surrounding foliage. Ion exchange between the leaf interior and surface liquid may also occur, neutralizing acidity with potassium, calcium, and magnesium. If evaporation of intercepted water occurs more rapidly than ion exchange, acid concentrations in the surface film may reach damaging levels. For example, leaf necrosis has been reported at pH lower than 3 (Unsworth and Wilshaw 1989).

The chemistry of rain and snow reaching the forest floor is influenced by the chemical and hydrological characteristics of incident precipitation, washout of dry deposited materials from canopy surfaces, and absorption or release of substances by the canopy during the precipitation event (Lovett et al. 1989). The most important canopy transformations of incident precipitation are direct assimilation of nutrients by the foliage, hydrogen ion buffering in the canopy and metal cation leaching (Bredemeier 1989). Importantly, hydrogen ion buffering in the canopy removes free acidity from the throughfall, but total hydrogen ion loading to the soil is not decreased.

Total deposition to a forest

The deposition of gases, particles, and cloud droplets is generally higher at the upwind edge than inside a forest (e.g., Godt and Mayer 1988). Precipitation amount increases with elevation, although deposition of pollutants in rain and snow may not necessarily increase to the same extent (Warren et al. 1990). Cloud deposition probably increases with elevation since cloud interception is usually higher at higher elevations; however, fog events at low elevations are also important along the Pacific Coast (Muir and Böhm 1989).

There is no question that exposed portions of forests receive greater pollutant loadings than unexposed portions. Increased precipitation and frequency of occurrence of cloud events implies greater pollutant loadings at high elevations than at lower elevations. The implication becomes less clear when the role of meteorology and topography in dispersing and transporting pollutants from industrial and urban complexes is considered. For example, during restrictive meteorological conditions, low elevation forests surrounding urban and industrialized valleys may be exposed to higher pollutant concentrations than high elevation sites. The overall contribution of these intermittent exposures to total deposition at low relative to high elevation forests in the West is unclear.

Differential pollutant loading probably occurs within the forest. Exposure to the polluted atmosphere is higher at the top of the canopy than within the forest canopy. A gradient in dry and cloud deposition of pollutants from the top of the canopy to within the canopy is expected (e.g., Fritsche et al. 1989). Ozone uptake should also be higher at the top than within the canopy since metabolic rates of young needles are typically greater than those for older needles (e.g., Guderian 1985). Interestingly, gradients in damage from pollutants have been observed which decrease from the top to lower parts of the canopy (Fritsche et al. 1989), possibly as a consequence of different pollutant loadings.

Dense forests should experience different deposition amounts than open forests given the same pollution scenarios. Wiman and Ågren (1985) used a modelling environment to investigate the influences of forest structure on total deposition of particles. Their results indicate that increasing leaf area (leaf area index = 10) causes a pronounced increase in deposition close to the forest edge. However, deep inside the forest, the high foliage density acts strongly on wind speed, deposition mechanisms, and turbulent transfers, giving only a moderate increase in deposition rate. Since leaf area indices are usually less than 6, the modelling results apply to extremely dense forests. Nevertheless, the results are informative and indicate that increasing leaf area has the effect of *increasing deposition at the edges of forests with little effect on deposition deep inside forests*.

Pollutant Exposure of Western Forests

Emissions

Levels of local and regional pollutants in the western United States depend less on energy demands driven by weather and climate patterns than do pollutant levels in Europe and the eastern United States. Power utilities account for less than a quarter of annual emissions of sulfur and nitrogen oxides in the West (Roth et al. 1985).

Emissions data are not measured values that are inherently accurate. Instead, the emission factor is multiplied by the activity level of the source for a given time period to yield an estimate of emissions from that source. Consequently, emission inventories possess varying degrees of uncertainty. Much of the following discussion is based on the National Acid Precipitation Assessment Program 1985 Emission Inventory, which is considered by many to be the most complete and accurate large-scale inventory currently available for the western United States (Sellars and Norris 1989). Annual emission estimates for large point sources are reliable, but annual area source estimates are less certain.

Finally, relationships between emissions and precipitation chemistry are not direct. No clear understanding of the historical trends in pH of precipitation is possible without consideration of area-specific trends in emissions of particulate matter and other changes that affect atmospheric chemistry (Goklany and Hoffnagle 1984).

Sulfur oxides

Sulfur compounds are emitted into the atmosphere during combustion of fossil fuels, smelting of ores, manufacturing of steel, and refining of petroleum. These emissions are in the form of gaseous sulfur oxides, mainly sulfur dioxide (Cullis and Hirschler 1980). Natural sources of sulfur compounds include soil micro-organisms, vegetation, oceanic biological activity, and geothermal activity. These natural emissions are in the form of reduced sulfur compounds such as hydrogen sulfide, dimethyl sulfide, and particle sulfate (Cullis and Hirschler 1980). On a national scale, anthropogenic sources of sulfur dioxide are estimated at 21 Tg/year (1 Tg = 10^{12} g); natural sources of sulfur are estimated at about 4% of these (Placet and Streets 1987).

Anthropogenic sources of sulfur dioxide are scattered across the eleven Western states. In 1985, the estimated total sulfur dioxide emissions for the West was 2.15 Tg, 10% of the national estimate (Placet and Streets 1987). Individual states contributed from 0.04 to 0.59 Tg/year (Table 3.1, Placet and Streets 1987). Emissions by county varied between 0 and 0.3

Table 3.1 State level estimates of annual average sulfur dioxide emissions from all anthropogenic sources in 1975, 1980, 1985, and 1987/1988 (Tg per year). Estimates normalized by area of the state are presented in parentheses for 1985 (10^{-6} Tg year $^{-1}$ km $^{-2}$). 1 Tg = 1 million metric tons.

State	1975	1980[a]	1985[b]	1987/1988[c]
Arizona	1.37	0.71	0.59 (2.00)	na
California	0.47	0.43	0.30 (0.74)	na
Colorado	0.10	0.13	0.13 (0.48)	na
Idaho	0.05	0.04	0.05 (0.23)	na
Montana	0.20	0.14	0.12 (0.32)	na
Nevada	0.22	0.13	0.11 (0.39)	na
New Mexico	0.37	0.28	0.27 (0.86)	na
Oregon	0.04	0.04	0.04 (0.16)	0.04
Utah	0.09	0.09	0.08 (0.38)	0.06
Washington	0.35	0.28	0.25 (1.45)	0.12
Wyoming	0.14	0.19	0.21 (0.84)	na
Total U.S.	25.80	23.6	21.5 (15.7)	na

a: Total U.S. emissions for 1980 presented here are about 0.9 Tg lower than those in the NAPAP 1980 Emissions Inventory (Wagner et al. 1986) due to the possible omission of emissions from combustion of nonpurchased fuels in the industrial sector.

b: Totals presented for 1985 are the sum of the 1985 power plant emissions (calculated using the same methodology as Knudson [1986]) and the 1984 values from Knudson (1986). They do not reflect reductions from non-utility sources during 1984–1985.

c: Estimates obtained from state Departments of Air Quality.

na: not available at time of printing.

Tg/year (Figure 3.10). Major point sources were primarily located in Arizona, New Mexico, Colorado, Wyoming, Montana, and Washington (Figure 3.11). The metal industry accounted for most of the sulfur dioxide emissions in the West, followed by electricity generation (Figure 3.12). With the exception of Arizona, sulfur dioxide emission densities for Western states were less than 1.5 Mg/km^2 (Table 3.1, Chinkin et al.

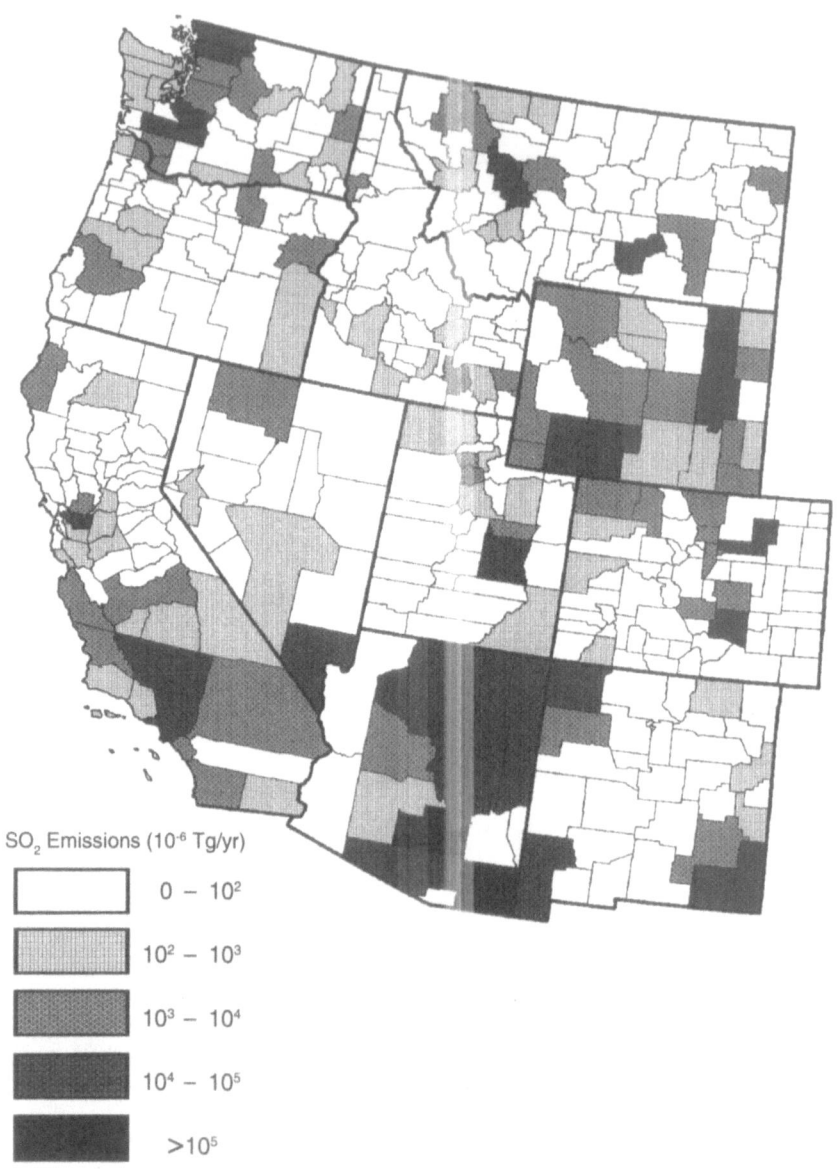

Figure 3.10 Annual average estimates (10^{-6} Tg) of total sulfur dioxide emissions during 1985 by county (National Emissions Data System Emissions Inventory 1985). Map compiled by Terralyn Vandetta, Department of Computer Science, Oregon State University, Corvallis, OR. 1 Tg = 1 million metric tons.

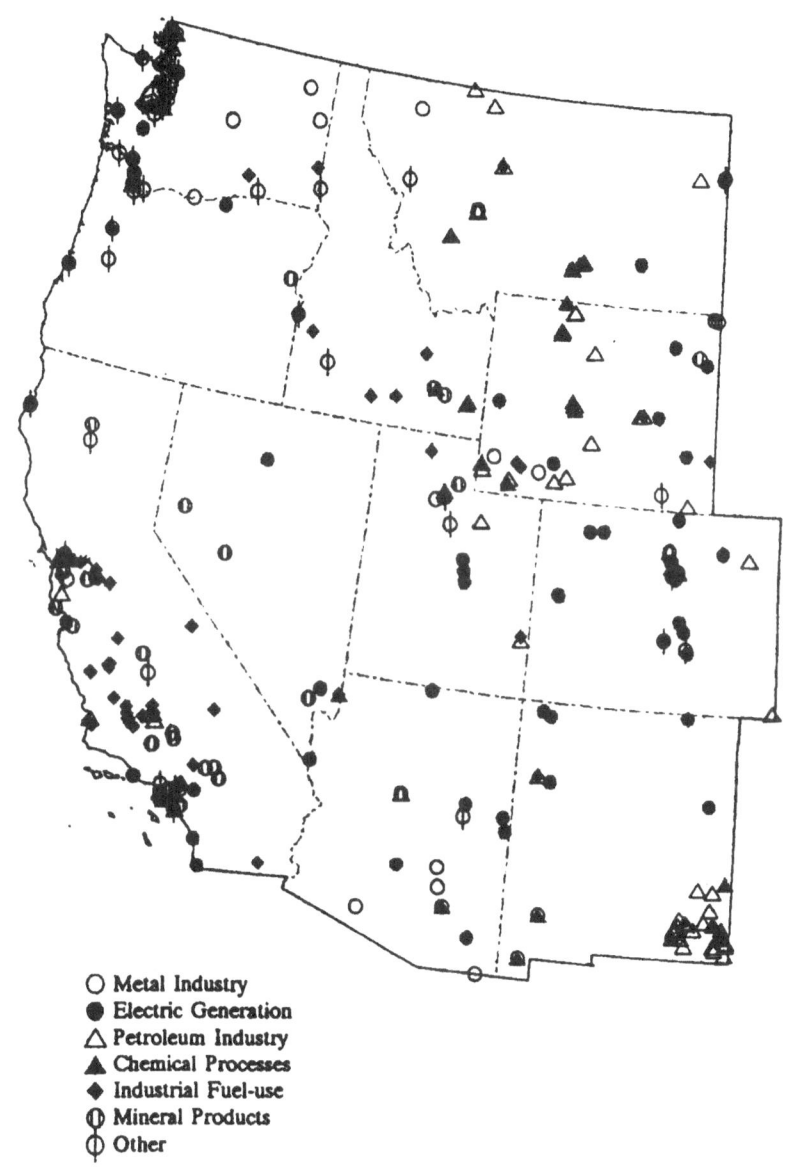

O   Metal Industry
●   Electric Generation
△   Petroleum Industry
▲   Chemical Processes
◆   Industrial Fuel-use
Φ   Mineral Products
Φ   Other

Figure 3.11 Locations of major point sources of sulfur dioxide (annual estimated emissions ≥ 0.1 Tg/year) in the West during 1985 (National Emissions Data System Emissions Inventory 1985). Map compiled by Terralyn Vandetta, Department of Computer Science, Oregon State University, Corvallis, OR. 1 Tg = 1 million metric tons.

(a)

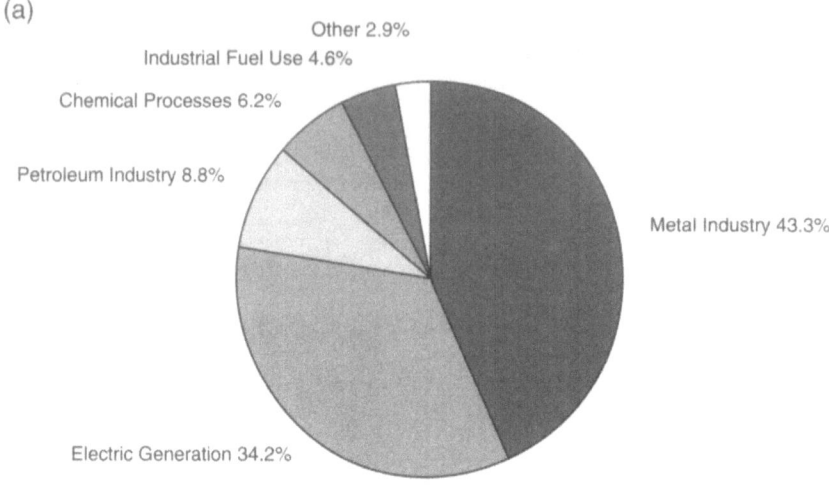

(b)

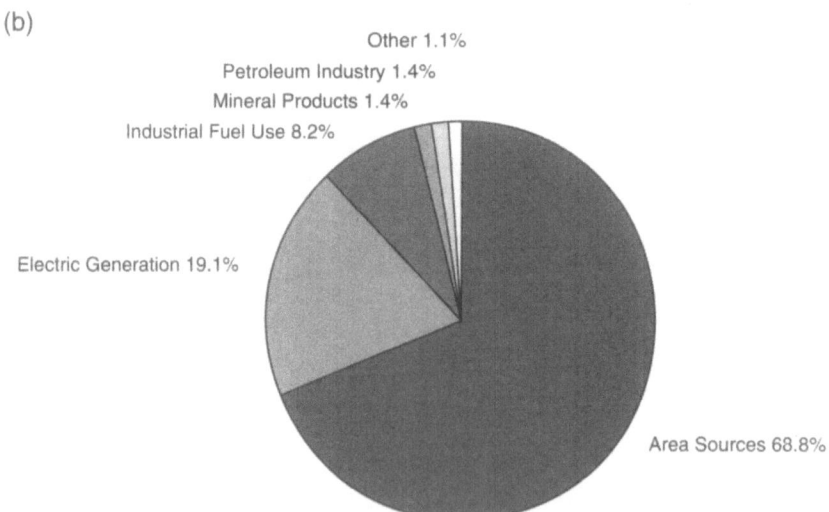

Figure 3.12 Percent contribution to total (a) sulfur dioxide and (b) nitrogen dioxide emissions in the West during 1985 by source category (National Emissions Data System Emissions Inventory 1985).

1987). Meteorological and topographical conditions in many source areas may prevent dispersion of plumes and localized pockets of high ambient sulfur dioxide levels probably exist.

Geothermal activity in North America is concentrated in Alaska, the western contiguous United States, and Mexico and therefore natural sulfur dioxide emissions may play an important, but sporadic, role in the West (Placet and Streets 1987). For example, the Mount St. Helens eruptions between 1980 and 1982 contributed an estimated 0.34 Tg of sulfur dioxide to the atmosphere, more than the annual contribution from man-made sources in Washington and Oregon.

Sulfur dioxide emissions in the West have decreased slightly during the last decade (Placet and Streets 1987). An intensive trend detection analysis and time series characterization of monthly emissions by state for the period 1975–84 also indicated decreases across the West, except for Wyoming where increases in emissions occurred, and Idaho and Colorado, where no change in emissions could be detected. The greatest reductions nation-wide were found in Arizona (-62%) and California (-49%) (Lins 1987). Sulfur dioxide emissions increased between 1980–86 by an estimated 25% in Idaho and by 10.5% in Wyoming, remained unchanged in Oregon and Colorado, and decreased slightly elsewhere (Placet and Streets 1987). These results are partially reflected in the annual emissions at state level given in Table 3.1.

Since 1986, total sulfur dioxide emissions in Oregon, Utah, and Washington have decreased (Table 3.1). Although state-wide emissions generally decreased, increases occurred at a county level. This implies that a few large point sources closed down or improved emission abatement, but at the same time, industrial growth continued. It also means that after 1985 larger areas in the West became exposed to sulfur dioxide than prior to 1985, although the total level of emissions was lower. Emission estimates post 1985 were not available for Arizona, California, Colorado, Idaho, Montana, Nevada, New Mexico, and Wyoming.

After 1990, increasing emissions from stationary fuel combustion sources should outweigh the continuing decline in smelter emissions, resulting in a gradual increase in sulfur dioxide emissions by 2000. These projections are approximate since emissions from several large smelters in northwestern Mexico are expected to increase through the 1990s, raising sulfur dioxide levels in Arizona and New Mexico (Young et al. 1988).

Nitrogen oxides

Nitrogen oxides are emitted during combustion processes. Transportation, primarily highway vehicles, is the major source of anthropogenic nitrogen oxides in the West (Hidy and Young 1986). Consequently, large emissions occur in and around cities, with fewer sources in rural and remote areas, except around large point sources. Natural sources of nitrogen oxides include lightning, biogenic processes in soils, stratospheric injection, and photolysis in the oceans. The national estimate of anthropogenic emissions of nitrogen oxides was 19 Tg during 1985 (Placet and Streets 1987). Inventories of natural emissions of nitrogen oxides are incomplete, although natural emissions account for an estimated 3–23% of total emissions of nitrogen oxides in North America.

An estimated 3.01 Tg of nitrogen oxides were emitted in the West during 1985, 15.8% of the national total (Placet and Streets 1987). The highest emissions (estimated at 1.11 Tg/year) were for California (second only to Texas on a national scale), with Oregon and Idaho (0.21 and 0.08 Tg/year resp.) having the lowest emissions in the West (Table 3.2, Placet and Streets 1987). The largest emission densities occurred in large urban complexes such as Los Angeles, San Francisco, Puget Sound (Seattle-Tacoma), Salt Lake City, Las Vegas, Phoenix, Albuquerque, and Denver (Figure 3.13, Chinkin et al. 1987).

Regionally, nitrogen oxides emissions in the West decreased slightly between 1980 and 1985 (Placet and Streets 1987). At a state level, Arizona, Colorado, and Wyoming show larger estimates for 1985 than for 1980 as a result of increased power plant emissions. Montana and Nevada experienced increased emissions from power plants and highway vehicles. California, Oregon, and Washington show decreases in nitrogen oxides emissions (Table 3.2). A gradual increase in nitrogen oxides emissions is projected through the 1990s (Young et al. 1988).

Volatile organic compounds

Volatile organic compounds and oxides of nitrogen are important in the formation of photochemical smog and ozone. While sulfur and nitrogen oxides are primary precursors of acid precipitation, volatile organic compounds are important constituents in many chemical reactions in the atmosphere (Gschwandtner et al. 1989).

Transportation activity, particularly combustion of gasoline in automobiles and trucks, is the largest anthropogenic source of volatile organic compounds in the United States. Other man-made sources include organic solvents and paints used in industry, residential, and commercial activities; storage of gasoline, crude oil, and other petroleum products; and wood burning. A variety of hydrocarbons are emitted from natural sources, primarily vegetation (Altshuller 1983, Duce et al. 1983). Approximately 90% of estimated natural emissions in the United States are attributable to forests. The emissions from deciduous species account for approximately 30%, while conifers contribute approximately 60% of

Table 3.2 State level estimates of annual average nitrogen oxides emissions from all anthropogenic sources in 1975, 1980, 1985, and 1987/1988 (Tg per year). 1 Tg = 1 million metric tons.

State	1975	1980[a]	1985[b]	1987/1988[c]
Arizona	0.18	0.23	0.25	na
California	1.18	1.21	1.11	na
Colorado	0.24	0.26	0.28	na
Idaho	0.08	0.08	0.08	na
Montana	0.10	0.12	0.13	na
Nevada	0.08	0.09	0.10	na
New Mexico	0.20	0.22	0.22	na
Oregon	0.20	0.22	0.21	0.19
Utah	0.11	0.13	0.13	na
Washington	0.25	0.27	0.25	0.27
Wyoming	0.17	0.21	0.25	na
Total U.S.	18.5	19.4	19.1	na

[a]Total U.S. emissions for 1980 presented here are about 1.3 Tg lower than those in the NAPAP 1980 Emissions Inventory (Wagner et al. 1986) due to the possible omission of emissions from combustion of nonpurchased fuels in the industrial sector and exclusion of emissions from forest fires and similar burning.

[b]Totals presented for 1985 are the sum of the 1985 power plant emissions (calculated using the same methodology as Knudson, 1986) and the 1984 estimates for vehicular emissions. They do not reflect reductions from non-utility sources during 1984–1985.

[c]Estimates obtained from state Departments of Air Quality.

na: not available at time of printing.

total estimated natural emissions. Crops are estimated to produce only 3% of the estimated total natural volatile organic compound emissions. About 50% of volatile organic compound emissions from deciduous forests are isoprenes. Conifers mainly emit terpenoid compounds, 25% of which are α-pinene. The rate and amount of natural VOC emission

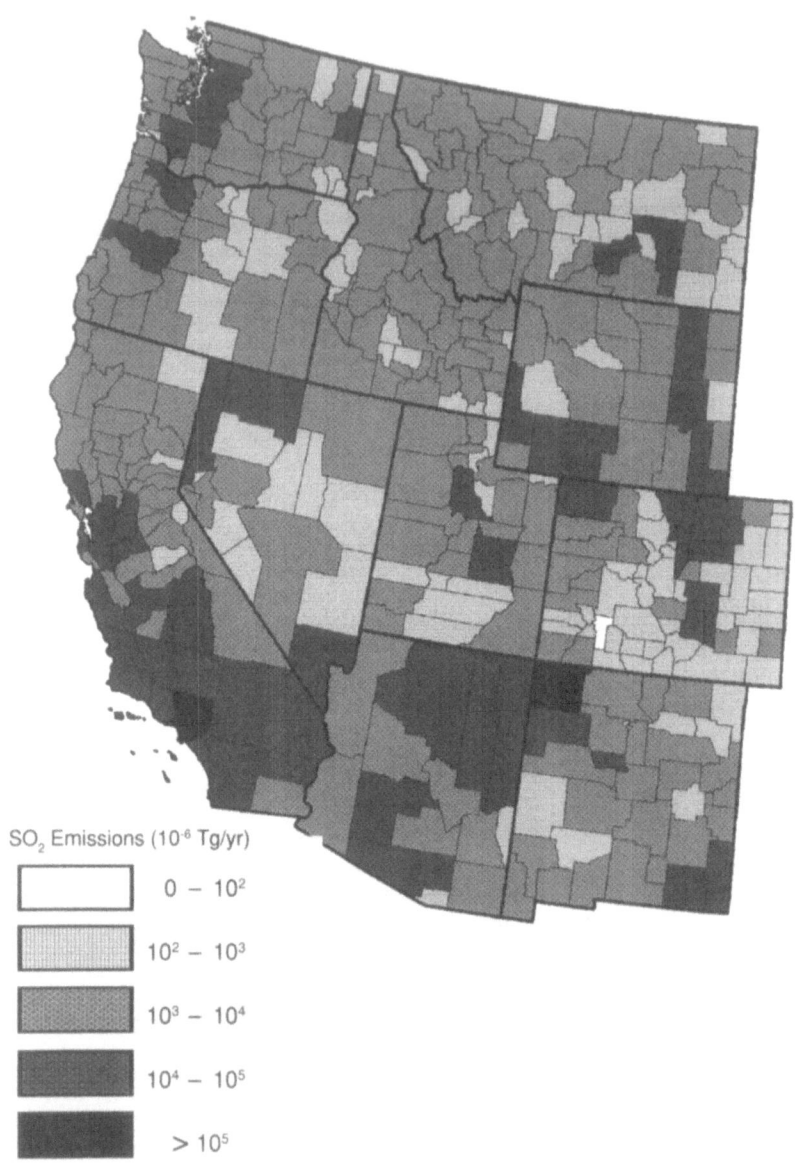

Figure 3.13 Annual average estimates (10^{-6} Tg) of total nitrogen dioxide emissions during 1985 by county (National Emissions Data System Emissions Inventory 1985). Map compiled by Terralyn Vandetta, Department of Computer Science, Oregon State University, Corvallis, OR. 1 Tg = 1 million metric tons.

depends on species and environmental conditions. Lamb et al. (1987) estimated that the national total natural non-methane hydrocarbon emissions were 30.7 Tg/year. The 1980 emissions estimate of anthropogenic volatile organic compounds was 21 Tg/year (Placet and Streets 1987).

In 1980, the total emissions of anthropogenically generated volatile organic compounds in the West were estimated at 4.2 Tg/year, 20% of the national estimate (Placet and Streets 1987). State-wide emission patterns were similar to those for nitrogen oxides. California recorded the greatest emissions in the United States with 2.01 Tg volatile organic compounds during 1980 (Table 3.3, Placet and Streets 1987).

Forest fires

Burning of logging slash and naturally occurring forest fires produce particles, carbon monoxide, nitrogen oxides, and gaseous hydrocarbons (Westberg et al. 1981). The ratio of hydrocarbon to nitrogen oxides is often optimal for photochemical oxidant formation, and the contribution of forest fires to ozone production in wilderness and rural areas may be important. Ozone accumulates close to the location of a burn and substantial increases in the concentrations of ozone have been detected downwind of burn areas and near tops of the plumes (Stith et al. 1981, Westberg et al. 1981). An evaluation of ozone concentrations during three prescribed burns (N=45) indicate that 76% of measurements showed increases from 1 to 44 ppb (1–52%) over ambient, with maximum increases towards the tops of a plume generated by burning of conifer slash (Stith et al. 1981).

Air Pollution Exposure of and Deposition to Western Forests

Assessing the response of forests to pollutants requires information on the amount and chemistry of deposited material. Unfortunately, there is a paucity of meteorological and air quality data throughout the West, and most meteorological data within a forest originate from one tower or location. Considerable variation of environmental conditions may be observed over relatively small areas within forests, and consequently point measurements do not extrapolate well. Most air quality sites are located close to urban areas and thus, available air quality data probably do not reflect the conditions experienced in most Western forests. Finally, numeric simulation has not developed to the point of accurate prediction of deposition in complex terrain.

Table 3.3 State level estimates of annual average emissions of volatile organic compounds from all anthropogenic sources during 1980 and 1985 (Tg per year). 1 Tg = 1 million metric tons.

State	1980[a]	1985[b]
Arizona	0.23	0.28
California	2.01	2.28
Colorado	0.32	0.29
Idaho	0.20	0.11
Montana	0.21	0.13
Nevada	0.09	0.10
New Mexico	0.17	0.16
Oregon	0.31	0.33
Utah	0.15	0.15
Washington	0.42	0.43
Wyoming	0.09	0.11
Total U.S.	21.00	21.67

[a]Placet and Streets 1987

[b]Gschwandtner et al. 1989

The following section discusses air pollution exposure and deposition at sites located within 20 kilometers of coniferous forests in the West. Data are available to estimate deposition of pollutants in rain and snow. Data on the ambient concentrations of gases and particles, together with that on cloud water chemistry are presented as an estimation of the chemical nature of the atmosphere in contact with forests. Estimates of total deposition are not possible since dry and cloud deposition cannot be reliably estimated from available data.

Ozone

Ozone is the only regionally dispersed pollutant known to injure foliage and lead to tree mortality at ambient levels (Miller 1973, Woodman 1987). Ozone effects can occur at concentrations above 60 ppb (Pell 1974); however, effects generally develop after short exposures to concentrations greater than 80 ppb (Taylor 1973). Few data on ozone concentrations in Western forests are available since most ozone monitoring sites are located close to urban complexes. Furthermore, ozone monitoring sites in the West cluster in certain areas, leaving large gaps in information for forests in Idaho, Montana, Wyoming, Nevada, and New Mexico (Böhm and Vandetta 1990).

The response of vegetation to ozone is influenced by diurnal and seasonal patterns in exposure (Hogsett et al. 1985), as well as by the magnitude of the ozone concentrations (Tingey and Taylor 1982, Guderian 1985). Table 3.4 presents growing season (May through October) means, percentiles, and percent occurrence of hourly ozone concentrations above 60, 80, 100 and 120 ppb for all ozone sites near Western forests since 1980. Figure 3.14 shows examples of the dominant patterns in daily ozone concentrations across Western forests (Böhm and Vandetta 1990, Böhm et al. 1991).

Sites located far from urban or point source areas experience patterns with little variation in hourly ozone concentrations (Figure 3.14a,d,f). The lowest ozone concentrations in the West were recorded on the Olympic Peninsula in Washington (growing season mean = 5–16 ppb). In Arizona, sites at Apache-Sitgreaves National Forest (growing season mean 42 ppb), and Saguaro National Monument (growing season mean 39 ppb) recorded 24-hour means between 22 and 51 ppb during most of the growing season. The western slopes of the Washington Cascades (growing season mean 31–35 ppb), northern California (Lassen and Redwood National Parks, growing season means 22 and 38 ppb), and Grand Canyon National Park (growing season mean 29 ppb) experienced daily means between 22 and 36 ppb during more than half of summer days. Yellowstone National Park, Wyoming, had a growing season mean of 35 ppb and 24-hour means of 22 to 41 ppb during 75% of summer days (Böhm and Vandetta 1990).

Sites on the fringe of urbanized centers or valleys experience patterns with some variation in hourly ozone concentrations. Higher ozone levels usually occur during the late afternoon (e.g., Figure 3.14b). Yosemite National Park (growing season mean of 47 ppb) and Sequoia National Park (62 ppb) receive pollutants from highly urbanized areas and had 24-hour means ranging from 36 to 85 ppb on 75% of summer days. Lake

Table 3.4 Growing season (May through October) summary statistics for ozone monitoring sites in or near forests for the period 1980 through 1988. Percentiles and means were generated using the entire data set (1980 –1988; May–October). % hours ≥ x are normalized to represent the average occurrence of ozone levels ≥ x during May through October. % data capture = # valid hours/(4416) * 100, where 4416 is the total number of hours during the period May through October. Generated from Böhm and Vandetta 1990. Site abbreviations: NP=National Park, NM=National Monument

Site	Latitude °N	Longitude °W	Elev (m)	% Data Capt	Percentiles (ppb)								% hours ≥ x			
					5	10	25	50	Mean ± std	75	90	95	60	80	100	120
Albuquerque, NM	35.1	106.6	1585	89	1	5	15	29	29.8 ± 19	43	55	61	6	1	0	0
Apache-Sitgreaves, AZ	33.7	109.0	2462	94	25	30	35	40	42.3 ± 12	50	60	65	12	1	0	0
Aptos, CA	37.0	121.9	78	100	0	10	10	20	25.1 ± 15	30	40	50	3	0	0	0
Arches NP, UT	38.8	109.6	1567	32	28	31	36	43	42.8 ± 09	49	54	58	4	0	0	0
Ash Mtn, CA (AIRS)	36.5	118.8	526	50	20	30	50	60	64.1 ± 26	80	100	110	64	36	12	2
Ash Mtn, CA (NPS)	36.5	118.8	610	57	20	30	47	61	62.9 ± 24	80	93	100	59	29	8	1
Azusa, CA	34.1	117.9	185	93	0	0	0	20	43.3 ± 56	70	130	160	28	22	17	12
Banning, CA	33.9	116.9	722	98	10	10	20	40	49.6 ± 35	70	100	120	35	19	11	6
Bishop, CA	37.4	118.4	1260	84	10	10	20	30	31.5 ± 16	40	50	60	7	0	0	0
Boulder County, CO	40.0	105.3	1635	95	8	14	24	35	36.0 ± 18	47	60	69	11	2	0	0
Bountiful, UT	40.9	111.9	1335	87	8	14	25	38	38.3 ± 20	49	62	72	12	3	1	0
Burbank, CA	34.2	118.3	170	95	0	0	0	20	36.3 ± 45	50	100	130	25	18	12	8
Camp Mather, CA	37.9	119.8	1432	33	22	26	36	46	47.5 ± 16	59	70	76	24	3	0	0
Carmel Valley, CA	36.5	121.7	131	86	10	10	20	30	28.4 ± 14	40	50	50	4	1	0	0
Cedar River, WA	46.4	122.0	210	82	11	14	19	28	31.4 ± 17	39	53	64	7	2	0	0
Clackamas County, OR	45.3	122.6	174	94	4	8	14	23	25.3 ± 16	33	45	55	4	1	0	0
Cochise County, AZ	31.6	110.3	1401	56	13	17	26	37	37.4 ± 15	49	58	63	8	0	0	0
Colorado Springs, CO	38.8	104.8	1842	88	0	2	10	25	26.3 ± 18	40	51	57	4	0	0	0
Colorado NM, CO	39.1	108.7	1750	30	30	32	37	42	44.1 ± 14	48	54	57	2	0	0	0
Columbia County, OR	45.8	122.8	6	88	1	4	11	20	21.3 ± 14	29	39	46	2	0	0	0

Table 3.4 (continued)

Site	Latitude °N	Longitude °W	Elev (m)	% Data Capt	Percentiles (ppb)								% hours ≥ x			
					5	10	25	50	Mean ± std	75	90	95	60	80	100	120
Crook County, OR	44.2	119.7	1372	90	20	25	30	35	36.5 ± 09	40	50	55	2	0	0	0
Denver, CO	39.8	105.0	1591	96	0	2	7	19	21.9 ± 18	33	47	55	3	1	0	0
Douglas County, NV	39.0	120.0	1951	60	6	11	20	35	35.5 ± 19	49	62	69	12	1	0	0
Eugene, OR	44.0	123.1	187	77	1	3	9	18	21.4 ± 17	30	42	52	3	1	0	0
Flagstaff, AZ	35.2	111.6	2117	77	17	24	34	44	43.7 ± 15	53	62	67	13	1	0	0
Fresno County, CA	37.1	119.3	1723	85	20	20	30	40	44.9 ± 17	60	70	80	26	5	0	0
Grand Canyon NP, AZ	36.1	112.1	2073	56	17	20	23	27	29.4 ± 09	33	43	46	0	0	0	0
Great Sand Dunes, CO	37.7	105.5	2487	54	24	27	33	39	38.4 ± 09	44	49	52	1	0	0	0
King County, WA	47.6	122.0	22	90	0	0	0	10	14.9 ± 17	20	40	50	3	1	0	0
Lake Gregory, CA	34.2	117.3	1397	93	10	20	40	60	72.5 ± 49	100	140	170	55	37	26	18
Larimer County, CO	40.6	105.1	1522	90	1	4	14	27	27.9 ± 18	40	52	59	5	0	0	0
Lassen NP, CA	40.5	121.6	1788	36	17	21	28	36	37.8 ± 14	46	58	64	9	0	0	0
Logan, UT	41.7	111.8	1382	45	8	12	20	32	32.5 ± 15	45	52	58	4	0	0	0
Kaweah, CA (AIRS)	36.6	118.8	1901	35	10	20	40	60	59.7 ± 26	80	90	100	57	32	8	1
Kaweah, CA (NPS)	36.6	118.8	1890	58	21	30	41	56	56.3 ± 21	71	83	90	44	15	2	0
Mammoth Lakes, CA	37.6	119.0	2395	92	20	30	40	50	46.6 ± 16	60	70	70	30	5	0	0
Marion County, OR	44.8	122.9	102	94	1	1	7	18	20.3 ± 16	30	41	50	3	1	0	0
Medford, OR	42.3	122.8	503	93	1	3	9	22	24.8 ± 18	37	50	59	5	1	0	0
Monterey, CA	36.6	121.9	23	86	10	10	20	30	27.3 ± 12	30	40	50	1	0	0	0
Ogden, UT	41.2	112.0	1314	97	0	1	9	30	29.8 ± 22	46	58	65	8	1	0	0
Ojai, CA	34.4	119.0	233	87	10	10	20	40	42.3 ± 26	60	80	90	30	12	3	1
Olympic NP, WA (DOE)	47.9	123.4	100	85	0	0	10	20	16.3 ± 11	20	30	40	0	0	0	0
Olympic NP, WA (NPS)	48.1	123.4	125	26	0	1	1	2	4.8 ± 07	3	17	23	0	0	0	0
Pack Forest, WA	46.8	122.3	24	80	10	10	20	30	30.0 ± 18	40	50	70	8	3	1	0

Table 3.4 (continued)

Site	Latitude °N	Longitude °W	Elev (m)	% Data Capt	5	10	25	50	Mean ± std	75	90	95	60	80	100	120
									Percentiles (ppb)					% hours ≥ x		
Pasadena, CA	34.1	118.1	255	89	0	0	10	20	47.8 ± 58	70	130	170	30	24	18	14
Pierce County, WA	47.2	122.3	14	85	0	0	0	10	15.1 ± 16	20	40	40	2	0	0	0
Pima County, AZ	32.3	111.0	695	86	1	2	10	28	29.8 ± 22	46	60	68	10	1	0	0
Pinnacles NM, CA	36.5	121.2	335	66	10	16	26	41	42.8 ± 22	58	72	80	22	5	1	0
Port Angeles, WA	48.1	123.4	30	71	0	0	1	2	8.4 ± 10	10	20	30	0	0	0	0
Prescott, AZ	34.6	112.5	1673	69	5	9	16	30	29.9 ± 16	43	52	55	2	2	0	0
Provo, UT	40.3	111.7	1402	72	2	5	14	29	32.1 ± 22	49	62	68	12	0	0	0
Redwood NP, CA	41.6	124.1	233	49	8	10	15	22	22.0 ± 09	28	34	39	0	0	0	0
Reno, NV	39.5	119.8	1280	92	0	10	10	30	28.4 ± 20	40	50	60	10	1	0	0
Rocky Mt. NP, CO	40.3	105.5	2743	49	25	31	38	46	46.0 ± 12	54	60	65	10	1	0	0
Saguaro NM, AZ	32.2	110.7	933	66	19	22	30	38	38.8 ± 14	47	57	63	8	1	0	0
Salt Lake City, UT	40.8	111.9	1305	87	2	4	11	28	30.4 ± 22	45	59	70	10	3	1	0
San Bernardino, CA	34.1	117.3	320	80	0	0	0	30	50.2 ± 57	80	140	170	35	28	21	15
Santa Monica Mt. CA	34.1	118.4	191	55	0	2	10	30	39.6 ± 35	59	86	110	25	13	7	4
Santa Barbara, CA	34.5	120.0	25	96	0	10	20	30	32.2 ± 19	40	60	60	11	2	0	0
Santa Barbara Co., CA	34.4	119.8	12	96	0	10	20	30	31.5 ± 20	40	60	70	13	2	1	0
Scotts Valley, CA	37.1	122.0	171	79	0	0	10	20	22.4 ± 18	30	50	50	5	1	0	0
Snohomish Co., WA	48.1	122.0	120	83	0	0	0	10	17.0 ± 15	30	40	40	2	0	0	0
South L. Tahoe, CA	38.9	120.0	1907	88	10	20	20	40	37.8 ± 17	50	60	60	18	1	0	0
Spokane, WA	47.7	117.4	584	74	0	0	10	20	20.9 ± 16	30	40	50	2	0	0	0
Stampede Pass, WA	47.3	121.3	1217	80	20	20	30	30	35.2 ± 14	40	50	60	7	0	0	0
Ventura County, CA	34.7	119.1	1600	83	0	10	20	40	36.2 ± 22	50	60	70	18	5	1	0
Wawona Valley, CA	37.5	119.7	1280	66	9	15	27	42	44.0 ± 23	61	76	83	26	7	1	0
Yellowstone NP, WY	44.6	110.4	2484	58	15	19	27	36	35.4 ± 12	44	51	55	2	0	0	0
Yreka, CA	41.7	122.6	809	80	0	0	10	20	25.9 ± 18	40	50	60	6	0	0	0

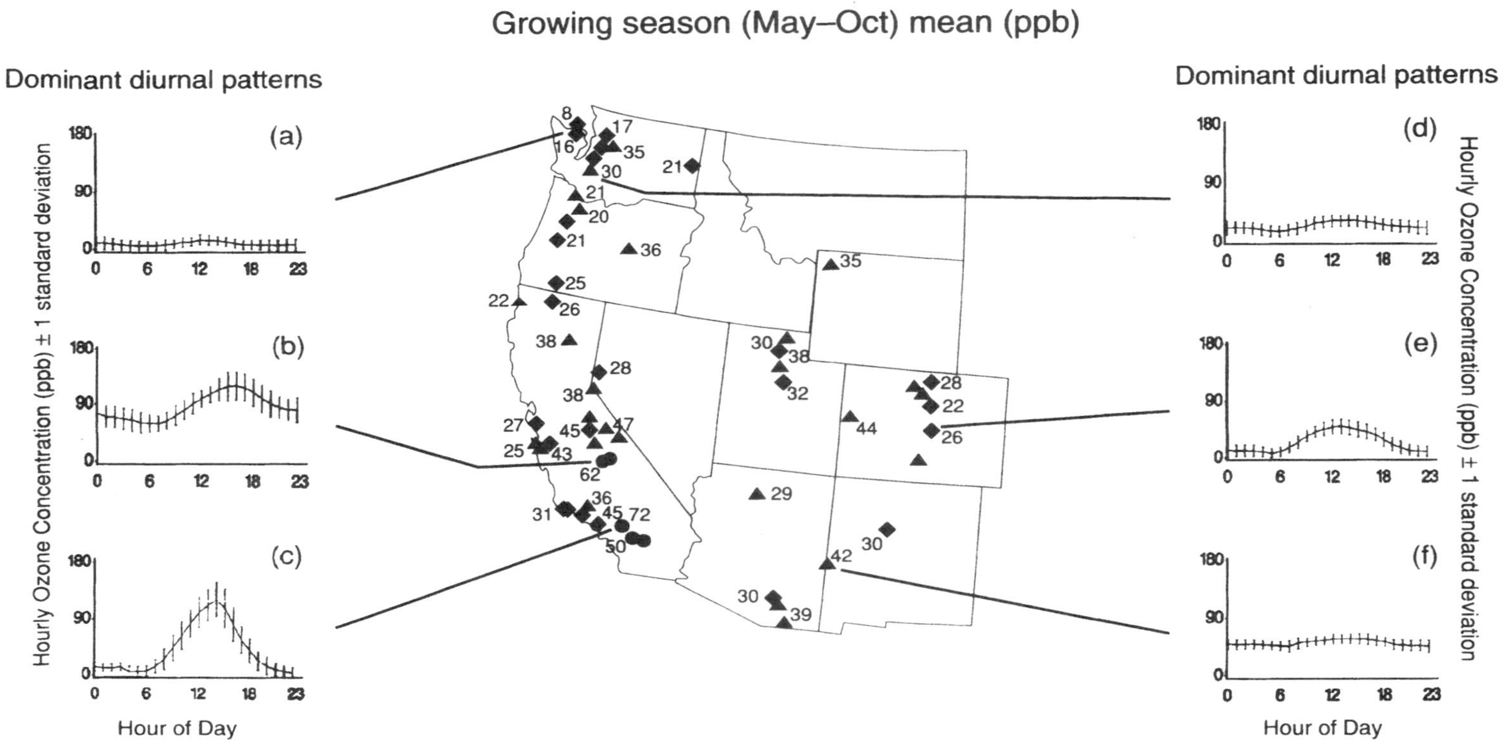

Figure 3.14 Examples of the dominant diurnal patterns in hourly ozone concentrations at sites located (1) far from urban sources (a, d, f), (2) rural but under urban influence (b), and (3) within urban areas (c, e) (after Böhm and Vandetta 1990, Böhm et al. 1991). Numbers on map are growing season (May–October) mean ozone concentrations (ppb).

Gregory, on the eastern fringe of the Los Angeles Basin, had a growing
season mean of 73 ppb. Diurnal patterns with means ranging from 85 to
100 ppb occurred during 49% of summer days.

Urban sites have patterns of diurnal ozone concentrations with marked
scavenging of ozone at night (Figure 3.14c,e). In Washington, growing
season means at urban sites near forests ranged between 12 and 21 ppb,
between 28 and 37 ppb in Utah, between 20 and 35 ppb in Colorado, and
between 32 and 58 ppb in the Los Angeles Basin.

The lowest hourly ozone concentrations were recorded in remote forests
where levels did not exceed 60 ppb (e.g., Olympic National Park,
Washington). Forests around Albuquerque, in Apache-Sitgreaves
National Forest, Arizona, and in the central Sierra Nevada, usually
record hourly ozone concentrations between 60 and 80 ppb during 10–
30% of measured hours in summer (May though October) with few
concentrations greater than 80 ppb. Forests located on the rims of valleys
with large urban areas experienced ozone concentrations greater than
100 ppb. The frequency of occurrence of such high ozone levels appears
to be related to the size of the city and the air pollution potential of the
area. Sites on the west slopes of the Washington Cascades, east slopes of
the Front Range in Colorado, southern Sierra Nevada, and slopes of
mountains surrounding urban settlements in Utah recorded ozone
concentrations in excess of 100 ppb during about 5–10% of the time
during mid-summer (Jun–Aug), with the forests of the southern Sierra
Nevada experiencing concentrations as high as 140 ppb. The San
Bernardino National Forest east of Los Angeles was exposed to ozone
levels greater than 100 ppb during all seasons. During winter, 1–2% of
measured hours had ozone levels in excess of 100 ppb. In summer, high
ozone levels occurred during 10% of measured hours (Böhm 1989).
Similar patterns were found during the 1970s (Miller et al. 1986).

Sulfur dioxide

Thresholds reported for direct injury to plants from 1-hour exposures to
sulfur dioxide ranged between 500 and 2500 ppb for sensitive species
(Heck and Brandt 1977, McLaughlin 1981). Long-term thresholds have
proven more difficult to estimate (McLaughlin 1981). There is some
evidence that annual average concentrations as low as 10–20 ppb,
together with occasional peaks of 40–80 ppb can reduce tree growth (Last
and Rennie 1982).

Sulfur dioxide concentrations since 1980 have been generally low across
the West (Böhm 1989, Böhm and Vandetta 1990). At most sites sulfur
dioxide concentrations rarely exceeded 40 ppb. Sites located close to
large point sources were the exception. During the early 1980s, several

sites in Arizona, Idaho, and Montana were frequently exposed to relatively high sulfur dioxide concentrations with between 10 and 20% of measured hours having concentrations that exceeded 80 ppb during each month of the year (Böhm 1989).

Higher sulfur dioxide concentrations generally occurred during the winter months when poor ventilation prevented dispersion of pollutants. Some sites showed sub-maxima during spring and summer, possibly related to rapid mixing in an unstable lower atmosphere capped by subsidence inversions (Böhm 1989).

Nitrogen oxides

Nitrogen oxides are expected to react primarily inside the leaf tissue and exchange slowly with the plant (Hosker and Lindberg 1982). The effects of nitrogen oxides on plants are not frequently studied and were not even considered by Skelly et al. (1987). Since nitrogen oxides are readily oxidized, they have short residence times in the atmosphere and any biological damage is more likely from secondary pollutants such as peroxyacetyl nitrate (PAN) and nitrate.

Sites near major cities recorded few nitrogen dioxide concentrations less than 10 ppb, whereas concentrations between 10 and 80 ppb occurred during 150 hours/month or 20% of the time. The remainder of the sites usually experienced nitrogen dioxide concentrations between 10–20 ppb. Higher concentrations (> 80 ppb) of nitrogen dioxide occurred more frequently in winter than in summer (Böhm 1989).

Particulate matter

Particle mass and chemistry have been monitored across the western United States during several regional programs of short duration (e.g., Pacific Northwest Regional Aerosol Mass Apportionment Study, South Coast Air Quality Study, Visibility Impairment due to Sulfur Transformation in the Atmosphere, Western Regional Air Quality Studies, Western Fine Particle Study). Two programs have been operational for several years since the early 1980s, viz., Subregional Cooperative Electric Utility Study and the National Park Service Particulate Monitoring Network. This section concentrates on results of the National Park Service Network (Cahill et al. 1985).

In June 1982, the National Park Service began to monitor fine (aerodynamic diameter less than 2.5 μm) and coarse (aerodynamic diameter between 2.5 and 15 μm) particulate matter at 28 sites in rural and remote areas across the West (Figure 3.15). Two samples of three day duration are collected each week (Cahill et al. 1985).

Fine particles play an important role in visibility impairment since their aerodynamic diameters are within the range for most efficient scattering of light (0.2 to 1.0 μm). Usually fine particles are secondary pollutants formed during the oxidation of gases such as sulfur dioxide and nitrogen oxides.

Three major groups of particles contribute to the total fine mass in the West. These are ammonium sulfate, soil, and soot, with sulfur being the most dominant element. The remaining fine mass is associated with organic material, hydrocarbons, and some nitrates. Average concentrations of fine particles range from 3 to 6 μg/m³. Sites in a band from northern California and southern Oregon to northern Arizona and southern Utah recorded the lowest levels. Highest fine particle concentrations were found at sites in northern Washington and Montana, western Wyoming and the southwestern region of southern California and central and southern Arizona and New Mexico (Figure 3.16). These areas have the highest emissions of sulfur dioxide in the West (Figure 3.10). Sulfur contributes around 20% of fine mass in the Northwest, except around Mt. Rainier and the industrial region of Puget Sound where the contribution is higher. The percentage contribution by sulfur to fine mass increases to the east and south, with between 40 and 50% contribution in Arizona and New Mexico.

Fine particle concentrations vary with season (Figure 3.17). Fine sulfur concentrations are lower during winter (average fine sulfur concentrations range between 50 and 350 ng/m³) than in summer (average fine sulfur concentrations range between 190 and 600 ng/m³). Regional patterns in fine sulfur concentrations are also seasonally variable. In winter, lowest concentrations occur in the western portion of the region with increased levels to the east and southeast (Figure 3.17a). In spring there is an increase in concentrations but the west to east and southeast progression appears to be similar to that during winter (Figure 3.17b). Sulfur levels in the southwest, however, increase disproportionately relative to northern sites during spring. By summer, fine sulfur concentrations in the southern portion of the region are much higher than those to the north (Figure 3.17c). Interestingly, low fine sulfur concentrations always occur in northern California and southern Oregon with highest concentrations in southern California, Arizona, New Mexico, Utah, and Colorado. These general results mimic the emissions map for sulfur dioxide (Figure 3.10). The Mt. Rainier site in Washington has a distinct peak in fine sulfur concentrations during summer that is not regionally consistent. Puget Sound has several large point sources of sulfur dioxide (Figure 3.10) which, together with latitude (and associated actinometric factors) and local meteorology, may cause the observed regional abnormality.

Figure 3.15 Location of sites in the National Park Service's Particulate Monitoring Network (after Cahill et al. 1985).

Fine particle potassium is associated with soil elements and smoke from field and forest burning. Concentrations of non-soil fine particle potassium in the Pacific Northwest increase dramatically during July through October, a period traditionally associated with field burning and large forest fires. In the Southwest, non-soil fine particle potassium concentrations are highest in May through July. Elsewhere, levels are generally low with slight increases during summer months.

Cloud

Many forests in the western United States experience drought stress during the growing season. Fog and cloud events alleviate this stress by providing moisture to the system; in some cases, the amount of moisture trapped from clouds contributes significantly to the annual water budget

(a)

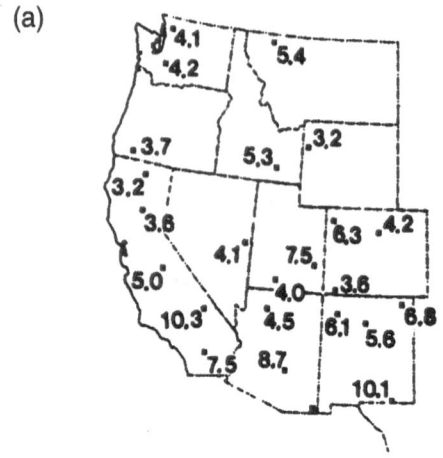

(b)

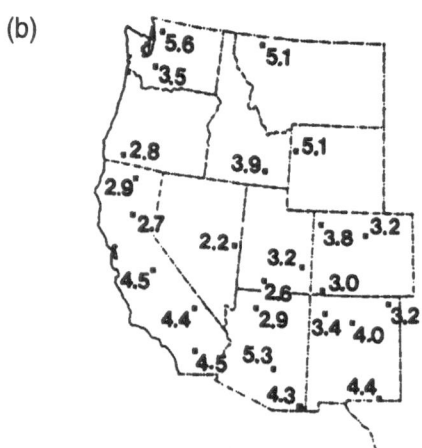

Figure 3.16 Mean annual estimates of total (a) coarse and (b) fine particle mass
($\mu g/m^3$). After Cahill et al. (1985).

(Oberlander 1956, Azevedo and Morgan 1974, Harr 1982). In addition,
concentrations of ions in cloudwater are generally higher than those of
rain or snow. Fog and clouds may provide an important contribution to
the total input of chemicals to montane ecosystems in the western United
States.

Stratus and stratocumulus cloud interception by Western forests was
estimated by Muir and Böhm (1989) from Warren et al. (1986) (Figure
3.18). Little is documented on regional occurrences of orographic cloud

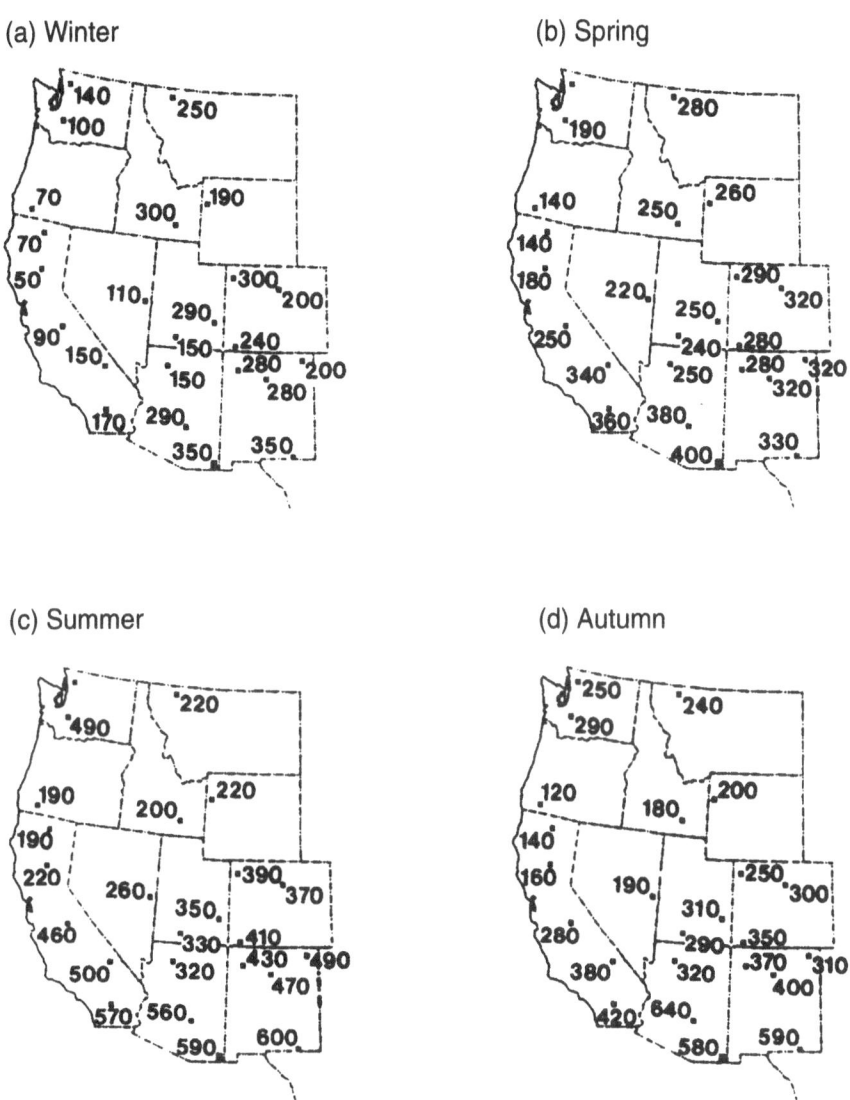

Figure 3.17 Seasonal estimates of fine particle sulfur across the West (ng/m³) after Cahill et al. (1985).

and therefore the following discussion probably underestimates total cloud interception in the West. The minimum number of days/year during which stratus and stratocumulus cloud interception could occur ranged from approximately 100 days in the Cascade Mountains of Washington and Oregon to 30 days in the Sierra Nevada; from 43 days in the northern Rockies to 25 days in the southern Rocky Mountains. The stratus and stratocumulus cloud-base heights varied with season, and ranged between approximately 750–900 m above sea level in the Cascade

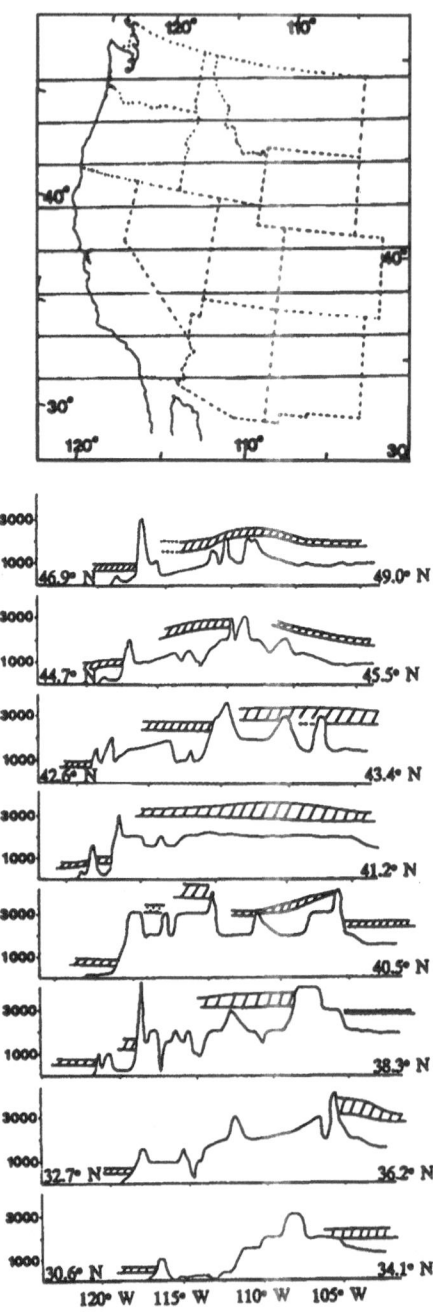

Figure 3.18 West-east topographical transects across the western United States showing elevational variations in the range of mean cloud base for above ground stratus and stratocumulus clouds. Map illustrates location of transects (after Muir and Böhm 1989).

Mountains of Washington and Oregon to between 1200–1500 m in the Sierra Nevada; from between 2400–2700 m in the northern Rockies to between 2800–3700 m in the southern Rocky Mountains. In winter, 25% more stratus and stratocumulus clouds occurred along the northern Pacific Coast than along the southern Pacific Coast, and the converse is true for summer. In spring and fall the entire West Coast was exposed to similar amounts of stratus and stratocumulus. The forests along the western slopes of the Cascades were exposed to more stratus and stratocumulus clouds than those along the western slopes of the Sierra Nevada (twice as much in winter and three times in spring and fall). Forests to the west of the Cascades and Coast Range, California, were exposed to 9–14% more stratus and stratocumulus clouds in winter, 16% more in spring and fall, and 21–32% more in summer than forests to the east of these primary orographic barriers. Finally, the occurrence of stratus and stratocumulus clouds was similar across the intermountain area (Muir and Böhm 1989).

The chemical composition of clouds impacting Western forests has been measured in Alaska and Oregon (Bormann et al. 1989), in the Washington Cascades (Basabe et al. 1989a), in the redwood forests of northern California (Bicknell 1989), along the western slopes of the southern Sierra Nevada (Collett et al. 1990) and at Mt. Werner in the northwestern Colorado Rockies (Miller et al. 1989a). Sulfate, chloride, and sodium dominated the cloudwater samples from Alaska, Oregon, and northern California, whereas sulfate, nitrate, hydrogen ion, and ammonium ion dominated at the other sites (Table 3.5).

The range in cloudwater pH at sites in Alaska was 3.8 to 5.6, similar to the range of 4.0 to 5.6 that was recorded in the Coast Range of Oregon (Bormann et al. 1989, Bormann pers comm). pH values between 3.1 and 5.9 were recorded at all the sites in the Washington Cascades (Basabe et al. 1989a). The pH of cloudwater collected at Lower Kaweah in Sequoia National Park ranged from 3.9 to 6.5; cloudwater collected at Turtleback Dome in Yosemite had pH values between 3.8 and 5.2. During periods of simultaneous collection, the Yosemite samples were more acidic than those from Sequoia. The pH differential appeared to be related to relatively small differences in concentrations of nitrate, sulfate, and ammonium at the two sites. In the absence of large ammonia inputs, sample pH values in the Sierra Nevada may fall below 3.0 and occur regularly at values less than 4.0 (Collett et al. 1990). The pH of samples measured at Redwood National Park ranged between 3.6 and 5.2 and between 3.0 and 5.2 at Mt. Werner. The pH of montane clouds did not drop below 3.5 very often; such events were associated with low liquid water contents and were usually of short duration.

Table 3.5 Summary of cloudwater chemistry (unweighted averages in μeq/L, standard deviation, and number of samples) in selected forests of the western United States. [1] Bormann et al. 1989, [2] Collett et al. 1990, [3] Basabe et al. 1989a, [4] Miller et al. 1989, [5] Bicknell 1989; * refers to 22 events consisting of a total of 63 samples, ** refers to 15 events consisting of 43 samples.

	Douglas Is.[1] Alaska			Mary's Peak[1] Oregon			Lower Kaweah[2] California			Turtleback Dome[2] California			Hurricane Ridge[3] Washington		
	Av.	sd	N	Av.	sd	N	Av.	sd	N	Av.	sd	N	Av.	sd	N
H^+	30.0	40.0	20	20.0	20.0	14	19.2	23.7	22*	28.5	24.9	15**	26.4	23.6	12
NH_4^+	7.8	9.9	23	12.7	11.1	14	348.6	453.1	22	62.2	89.3	15	16.1	6.9	12
SO_4^{2-}	45.8	58.1	23	35.6	38.1	14	90.6	110.9	22	37.2	46.3	15	29.4	29.0	12
NO_3^-	5.6	8.4	23	9.8	11.4	14	232.3	315.9	22	47.2	57.0	15	15.2	14.9	12
Cl^-	47.7	117.3	23	44.3	66.3	14	27.6	45.1	22	8.3	5.9	15	10.4	8.5	12

	Burley Mt.[3] Washington			Stampede Pass[3] Washington			Cascade Pass[3] Washington			Mt. Werner[4] Colorado			Redwood National Park[5] California		
	Av.	sd	N	Av.	sd	N	Av.	sd	N	Av.	sd	N	Av.	sd	N
H^+	155.8	88.8	47	133.6	90.3	176	47.6	33.4	11	216	-	48	74	73	22
NH_4^+	84.6	65.8	47	62.1	47.9	176	51.3	44.6	11	57	-	51	92	56	14
SO_4^{2-}	154.2	93.9	46	162.4	118.4	176	72.1	69.5	11	162	-	54	156	95	16
NO_3^-	99.3	105.5	46	84.2	75.0	176	37.2	50.0	11	97	-	54	46	29	16
Cl^-	43.2	35.3	46	36.7	81.9	175	12.6	10.5	11	11	-	54	275	264	15

Cloudwater had higher concentrations of all ions than were measured in rainwater. Ammonium ion, sulfate, and nitrate concentrations were 4.3–4.8 times and chloride, sodium, and magnesium were 0.7–1.5 times more concentrated in cloudwater than in rainwater at the Alaskan and Oregon sites (Bormann et al. 1989). In the southern Sierra Nevada, the average concentrations of ammonium and nitrate in cloudwater were more than 10 times higher than those in precipitation, and cloudwater concentrations of sulfate were more than 3 times those observed in precipitation (Hoffmann et al. 1989). The concentrations of hydrogen ion, ammonium ion, sulfate, and nitrate in cloudwater measured at Mt. Werner were 4–8 times those measured in precipitation (Miller et al. 1989a).

Rain and snow

The chemical species responsible for depressing the pH of rain and snow are secondary pollutants, derived primarily from the oxidation of sulfur and nitrogen oxides. The National Atmospheric Deposition Program/ National Trends Network operates a small number of precipitation monitoring sites within or near forests in the western United States (Figure 3.19, Table 3.6). The concentrations and deposition of major ions in Western rain and snow are generally 15–25% of Eastern values. Volume weighted mean pH ranges between 4.8 and 5.5 (Figure 3.20b). Mean weekly concentrations range between 5.1–26.2 μmol/L for sulfate (oxidation state VI), 2.5–66.1 μmol/L for nitrate, and 3.6–33.2 μmol/L for hydrogen ions; and the 75th percentile for concentrations ranges between 6–24 μmol/L for sulfate, 3–60 μmol/L for nitrate, and 4–24 μmol/ L for hydrogen ions (Böhm 1989). Mean annual wet deposition estimates for sulfate range between 1 and 9 kg/ha (Figure 3.20c), between 1 and 5 kg/ha for nitrate (Figure 3.20d), and between 0.03 and 0.1 kg/ha for hydrogen ions.

The chemical composition of precipitation across the West exhibits a spatial pattern related to the proximity of sites to anthropogenic and natural sources of sulfur and nitrogen (Laird et al. 1986, Böhm 1989). The concentrations of ions recorded at sites removed from anthropogenic sulfur emission sources (e.g., Hoh River, WA; H.J. Andrews Forest, OR; Yosemite NP, CA; Headquarters, ID; Glacier NP, MT) were lower, with weekly mean sulfate concentrations ranging between 5 and 7 μmol/L, than concentrations at sites in close proximity to or downwind of paper/ pulp mills (e.g., Alsea Forest, OR, 8.6 μmol/L) and fossil fuel refineries and power plants (e.g., San Gabriel Mountains-Tanbark Flat, CA, 26 μmol/L; and the Front Range, CO where weekly mean sulfate concentrations range between 15 and 20 μmol/L). Patterns in sulfate deposition were more complex, possibly a consequence of the marine influence at

Figure 3.19 Location of National Atmospheric Deposition Program / National Trends Network precipitation monitoring sites in or near Western forests. Sites that passed NADP criteria for inclusion in two or more Annual Summaries during the period 1985–1988 are shown.

remote sites such as Hoh River, Alsea Forest, and H.J. Andrews Forest in conjunction with the high precipitation amounts recorded in these areas (Figure 3.20a). The greatest sulfate depositions occur along the Pacific Northwest coast and near major emission areas in Arizona and New Mexico (Figure 3.20c).

Nitrate concentrations in rain and snow at sites close to urban/industrial complexes were much higher than those at more distant sites (Böhm 1989). Hoh River and Alsea Forest experience very low nitrate concentrations with weekly mean nitrate concentrations ranging between 2.5 and 4 μmol/L respectively; slightly higher values, 6–9 μmol/L, were recorded at H.J. Andrews Forest, Headquarters, and Glacier National Park which are located near towns or small cities; and much higher concentrations, 16–66 μmol/L, were found at sites downwind of large urban complexes such as California's Central Valley (Yosemite and Sequoia NP), Los

Angeles Basin (Tanbark Flat), and urban areas in Colorado and New Mexico (Manitou, CO; Alamosa, CO; and Bandelier National Monument, NM). Deposition of nitrate followed similar spatial patterns to concentration, with areas in close proximity to urban/industrial centers recording higher loadings (Figure 3.20d).

A general increase in sulfate concentrations occurs during the summer, possibly related to lower volume precipitation events, and a decrease occurs during winter, although at more remote sites this pattern was not easily discernable. The range in concentration was generally lower during winter than during summer, possibly related to seasonal patterns in occurrence of snow versus rain, and the origin of air masses (Böhm 1989). Similar results have been noted at Sequoia NP, CA (Stohlgren and Parsons 1987), and in Colorado (Nagamoto et al. 1983, Harte et al. 1985). Extreme concentrations such as 350 μmol/L sulfate, 917 μmol/L nitrate, and 645 μmol/L hydrogen ions occur in the forests surrounding Los Angeles. Such concentrated precipitation events are infrequent, are often associated with small precipitation amounts, and usually occur in spring and summer.

Air Quality in the Five Study Areas

Chapters 8–12 evaluate relationships between forest growth and air pollution in five areas of the western United States. The following section provides a summary of air quality and deposition in each area: western Washington, Front Range of Colorado, Arizona, the Sierra Nevada, and southern California.

Western Washington

Forests surrounding the Puget Sound, especially east and southeast of Seattle, are more likely to be impacted by pollution than forests elsewhere in the Pacific Northwest (Figure 3.21):

- Ozone levels in the Olympic Peninsula were very low with growing season means of 16 ppb and hourly concentrations less than 60 ppb. Ozone levels along the western slopes of the Washington Cascades ranged from growing season means of 17 ppb in the north to between 31 and 35 ppb east of Seattle. Hourly concentrations rarely exceeded 80 ppb, although values as high as 196 ppb have been recorded (Basabe et al. 1989b). At monitoring sites in the Cascades foothills, diurnal cycles in ozone had means of between 22 and 51 ppb during 70% of summer days (Böhm and Vandetta 1990).

Table 3.6 Annual summary statistics for wet deposition sites in or near Western forests for the period 1985 through 1988 (National Atmospheric Deposition Program (IR-7)/National Trends Network (1990). Sites that passed NADP criteria for inclusion in two or more Annual Summaries are presented. Data refer to means of annual statistics for the period 1985–1988. See Figure 3.19 for site locations. Site abbreviations: NP=National Park, RS=Ranger Station, EF=Experimental Forest, ES=Experimental Station, NM=National Monument

Site	Lat °N	Long °W	Elev (m)	Ppt[a] (mm)	pH x[b]	pH 75[c]	pH 90[d]	Sulfate (SO$_4^{2-}$) dep[e]	Sulfate x	Sulfate 75	Sulfate 90	Nitrate (NO$_3^-$) dep	Nitrate x	Nitrate 75	Nitrate 90	Ammonium (NH$_4^+$) dep	Ammonium x	Ammonium 75	Ammonium 90
Cascades NP, WA	48.5	121.4	120	1888	5.1	4.9	4.7	6.0	3.37	6.45	10.55	4.9	4.30	10.75	18.93	0.3	0.92	2.40	5.91
La Grande, WA	46.8	122.3	617	871	5.1	4.8	4.5	4.4	5.34	11.59	22.66	2.2	4.14	10.27	21.13	0.3	1.85	4.62	15.34
Olympic NP, WA	47.9	123.9	176	2858	5.4	5.3	5.1	8.3	3.07	4.74	7.65	1.2	0.65	1.53	5.08	0.3	0.55	1.11	1.11
Bull Run, OR	45.4	122.2	267	1531	5.2	5.0	4.7	6.0	4.13	8.95	15.48	4.4	4.62	14.84	26.40	0.7	2.59	9.79	18.11
Alsea Guard RS, OR	44.4	123.6	84	1427	5.3	5.3	5.1	5.6	4.13	7.11	10.38	1.0	1.13	2.53	5.38	0.2	0.74	1.11	3.33
H.J. Andrews EF, OR	44.2	122.3	436	1883	5.4	5.2	5.1	3.8	2.13	4.71	8.59	2.0	1.73	4.76	12.34	0.3	0.97	1.66	7.21
Starkey EF, OR	45.2	118.5	1253	473	5.3	5.1	4.9	1.0	2.19	4.24	8.09	1.1	3.87	8.39	17.78	0.1	1.80	2.77	12.75
Montague, CA	41.8	122.5	797	285	5.5	5.5	5.2	0.6	2.22	5.31	9.30	1.0	5.54	11.99	24.57	0.2	4.62	11.09	19.59
Hopland, CA	39.0	123.1	253	695	5.4	5.3	5.1	1.9	2.78	5.20	9.26	1.0	4.07	7.86	21.45	0.3	2.36	6.10	20.79
Sequoia NP, CA	36.6	118.8	1902	755	5.4	5.3	4.9	2.6	3.54	10.05	21.29	4.3	9.19	41.37	61.60	1.6	11.37	36.59	79.83
Chuchupate RS, CA	34.8	119.0	1614	278	5.1	5.1	4.7	1.3	4.75	10.20	27.83	2.3	13.07	27.96	73.02	0.4	7.58	16.82	37.51
Tanbark Flat, CA	34.2	117.8	853	548	5.0	4.6	4.4	3.2	6.12	19.08	43.38	5.0	15.32	57.46	103.64	0.8	8.59	33.40	65.70
Headquarters, ID	46.6	115.8	969	898	5.3	5.2	5.0	2.2	2.60	4.37	7.08	2.3	4.14	8.60	12.58	0.3	2.03	3.70	7.76
Bryce Canyon NP, UT	37.6	112.2	2477	364	5.1	5.0	4.7	2.5	7.15	13.95	26.75	2.5	11.24	22.15	34.09	0.3	3.88	8.50	15.15
Grand Canyon NP, AZ	36.1	112.2	2152	358	5.2	5.0	4.7	2.6	7.70	20.51	30.26	2.6	11.88	29.03	49.63	0.2	3.51	9.06	18.48
Glacier NP, MT	48.5	114.0	968	710	5.2	5.1	5.0	2.7	3.92	6.97	12.01	2.2	5.00	12.15	17.63	0.4	2.77	7.39	13.12
Havre ES, MT	48.5	109.8	815	364	5.3	5.3	5.0	2.0	5.99	13.01	17.38	1.9	9.52	17.02	23.31	0.5	8.04	13.86	34.37
Clancy, MT	46.5	112.1	1489	358	5.3	5.2	4.9	1.9	5.41	8.74	14.39	1.5	6.77	12.06	24.64	0.3	4.57	8.04	19.40
Custer NM, MT	45.6	107.4	957	321	5.3	5.2	4.9	2.4	7.89	13.40	25.45	2.1	10.44	20.65	32.14	0.4	7.35	14.00	23.56

Table 3.6 (continued)

Site	Lat °N	Long °W	Elev (m)	Ppt[a] (mm)	pH x[b]	pH 75[c]	pH 90[d]	Sulfate (SO_4^{2-}) dep[e]	x	75	90	Nitrate (NO_3^-) dep	x	75	90	Ammonium(NH_4^+) dep	x	75	90
Sinks Canyon, WY	42.7	108.8	2164	413	5.2	5.1	4.8	2.9	7.42	13.48	22.41	2.6	9.60	17.86	32.82	0.4	5.27	10.26	23.42
South Pass City, WY	42.5	108.8	2511	355	5.0	4.8	4.6	2.6	7.39	13.06	20.20	2.3	10.40	18.55	34.20	0.3	3.88	5.82	10.26
Newcastle, WY	43.9	104.2	1466	301	5.3	5.3	5.0	2.9	9.89	15.30	23.42	2.8	14.95	26.61	43.07	0.4	8.50	12.75	32.71
Rocky Mt. NP, CO	40.4	105.6	2490	371	5.1	4.9	4.7	3.0	8.47	16.38	23.77	3.5	15.22	30.97	40.27	0.7	9.42	18.85	31.60
Sugar Loaf, CO	40.0	105.5	2524	528	5.0	4.8	4.5	4.0	7.86	15.30	23.99	5.1	15.73	26.86	47.50	0.9	9.70	15.80	29.66
Manitou, CO	39.1	105.1	2362	338	4.8	4.7	4.5	3.4	10.58	18.29	26.48	4.2	20.49	35.59	49.36	0.5	7.39	15.71	24.76
Mesa Verde NP, CO	37.2	108.5	2172	518	4.8	4.7	4.4	5.4	11.03	20.38	33.05	4.3	13.79	30.00	48.75	0.5	4.99	11.23	22.45
Capulin Mt. NM, NM	36.8	104.0	2205	469	5.0	4.9	4.5	4.5	9.94	15.82	25.24	3.6	12.26	26.21	34.20	0.6	6.65	13.58	24.67
Cuba, NM	36.0	107.0	2124	362	4.9	4.7	4.5	3.3	9.63	16.06	30.45	2.9	13.19	24.15	42.42	0.3	5.40	11.23	20.24
Bandelier NM, NM	35.8	106.3	1998	508	5.0	4.8	4.6	4.7	9.60	15.38	25.27	4.0	13.19	24.84	39.96	0.5	6.10	10.39	18.99
Mayhill, NM	32.9	105.5	2009	624	5.1	5.0	4.7	6.5	10.93	18.69	46.71	3.8	10.00	18.17	24.57	0.6	5.36	10.16	22.91

[a] Annual average precipitation (mm).
[b] Annual average volume weighted mean (μmol/L for ions).
[c] Annual average 75th percentile (μmol/L for ions).
[d] Annual average 90th percentile (μmol/L for ions).
[e] Annual average deposition (kg/ha).

(a) Mean annual precipitation (cm)

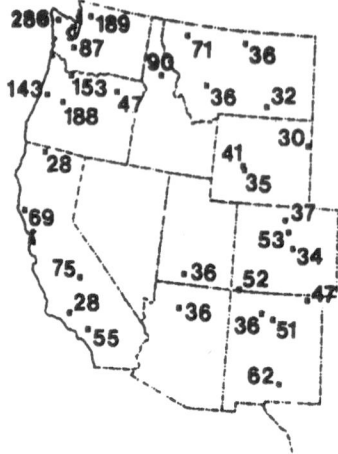

(b) Annual volume weighted average pH

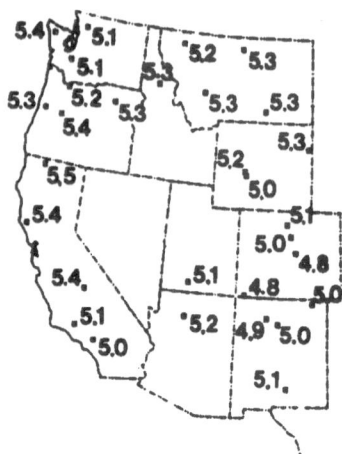

(c) Mean annual sulfate deposition (kg/ha)

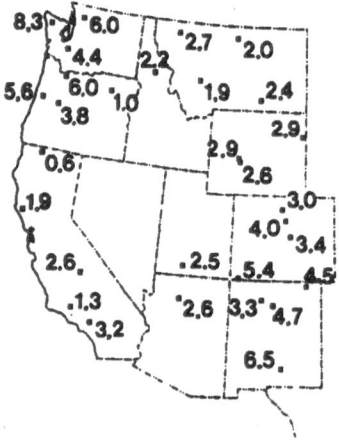

(d) Mean annual nitrate deposition (kg/ha)

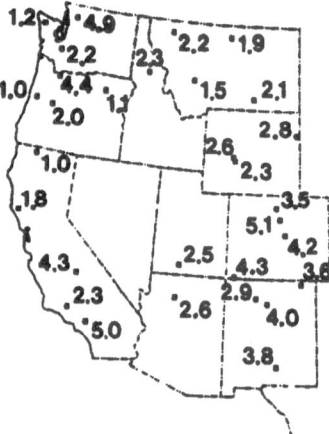

Figure 3.20 Mean annual estimates of precipitation amount and wet deposition in or near Western forests for the period 1985–1988 (National Atmospheric Deposition Program (IR-7)/National Trends Network (1990)). Sites included in two or more National Atmospheric Deposition Program Annual Summaries are represented. (a) Mean annual precipitation amount (cm); (b) Mean annual volume weighted pH; (c) Mean annual sulfate deposition (kg/ha); (d) Mean annual nitrate deposition (kg/ha). See Figure 3.19 for site names, and Table 3.6 for additional site data.

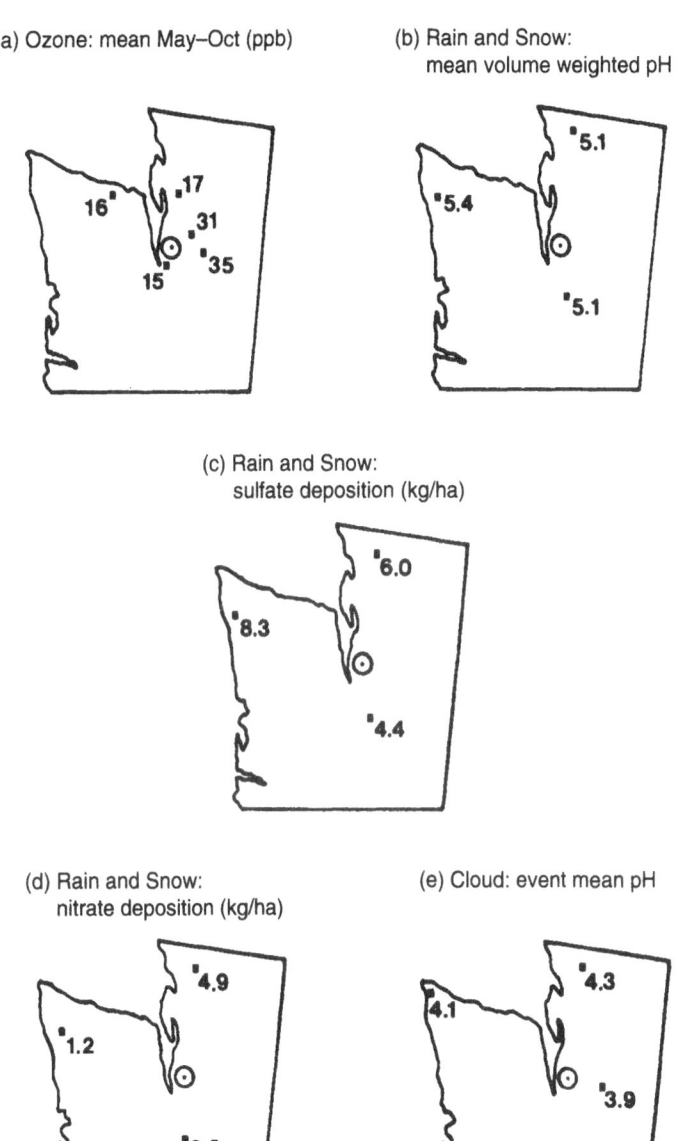

(a) Ozone: mean May–Oct (ppb)

(b) Rain and Snow:
mean volume weighted pH

(c) Rain and Snow:
sulfate deposition (kg/ha)

(d) Rain and Snow:
nitrate deposition (kg/ha)

(e) Cloud: event mean pH

Figure 3.21 Summary of air quality and wet deposition in forests of western Washington. (a) Growing season mean (May-Oct, 1980–1988) for ozone (ppb); (b) Mean annual volume weighted pH (1985–1988) for rain and snow; (c) Mean annual sulfate deposition (1985–1988) for rain and snow (kg/ha); (d) Mean annual nitrate deposition (1985–1988) for rain and snow (kg/ha); (e) Mean cloudwater pH (summer 1988). ⊙ =Seattle.

• Cloudwater collected at high elevation to the west of Seattle and along the western slopes of the northern Washington Cascades was less acidic with lower ionic strengths than that collected at high elevation along the western slopes of the central and southern Washington Cascades (Basabe et al. 1989a). The mean pH of cloud events impacting the western slopes of the Washington Cascades ranged from 4.3 in the north to 3.9 and 3.8 in the central portions and southern portions (Basabe et al. 1989a). The samples were usually dominated by sulfate and nitrate, hydrogen ions, and ammonium. Higher concentrations of acidic ions in cloudwater measured in the central Cascades occurred during northwesterly winds than during westerly or southwesterly winds, consistent with the general location of anthropogenic sources of pollution in the Puget Sound Basin. Cloud occurrence and interception at one site did not usually coincide with clouds at another site. These results appear to be consistent with the convective nature of summer clouds in the Washington Cascades (Basabe et al. 1989a).

• Higher concentrations of nutrient ions in clouds than in rainwater suggest that the deposition of cloudwater to vegetation may be an important mechanism for nutrient input in areas with frequent cloud cover. These inputs have been largely ignored in nutrient budgets for the West, particularly the old-growth forests of the Pacific Northwest.

• Volume weighted mean pH for precipitation in the Washington Cascades was around 5.0 with individual events recording pH values as low as 4.0. Annual wet deposition of sulfate was higher in the Olympic Peninsula (8.3 kg/ha) than in the northern (6.0 kg/ha) and southern Cascades (4.4 kg/ha). Annual wet deposition of nitrate was higher in the Cascades (2.2–4.9 kg/ha) than on the coast (1.2 kg/ha) (National Atmospheric Deposition Program (IR-7)/ National Trends Network 1990).

Sierra Nevada

Forests of the southern and central Sierra Nevada were exposed to high ozone concentrations, but precipitation acidity was low (Figure 3.22). In particular:

• Ozone concentrations were higher in the south than in the central and northern Sierra Nevada. Discontinuities in this gradient occurred due to intrusion of polluted air up canyons near urban complexes. Growing season means (May through October) varied from 60 ppb in the south to 40 ppb in the central portion and 26 ppb in the northern portion of the mountain range. Ozone values

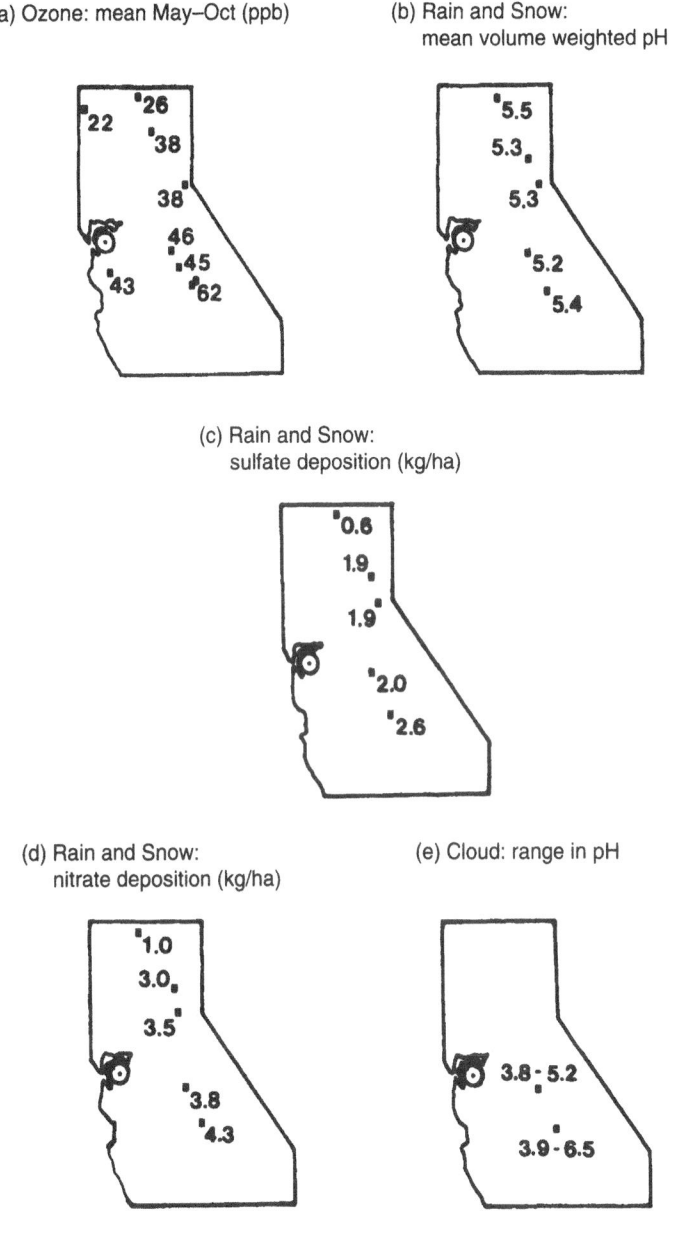

(a) Ozone: mean May–Oct (ppb)

(b) Rain and Snow: mean volume weighted pH

(c) Rain and Snow: sulfate deposition (kg/ha)

(d) Rain and Snow: nitrate deposition (kg/ha)

(e) Cloud: range in pH

Figure 3.22 Summary of air quality and wet deposition in forests of the Sierra Nevada. (a) Growing season mean (May-Oct, 1980–1988) for ozone (ppb); (b) Mean annual volume weighted pH (1985–1988) for rain and snow; (c) Mean annual sulfate deposition (1985–1988) for rain and snow (kg/ha); (d) Mean annual nitrate deposition (1985–1988) for rain and snow (kg/ha); (e) Range in cloudwater pH (1987–1988). ⊙ = San Francisco

were highest during the summer period June through September. Year-to-year trends in ozone concentrations showed a relationship with weather patterns. Years with hotter and drier summers recorded higher ozone concentrations than did cooler, wetter years. Diurnal patterns in ozone concentrations exhibited little variation; night-time hours in summer recorded levels of around 60 ppb. At Sequoia National Park, the diurnal variation in hourly ozone concentrations during 60% of summer days could be classified with two general patterns exhibiting little diurnal variation and with 24-hour means of 64 and 85 ppb (Böhm and Vandetta 1990). Daily maximum ozone concentrations usually occurred between 1400 and 2000 hours. Hourly ozone concentrations as high as 140 ppb were recorded at Sequoia National Park.

- Clouds intercepting the central and southern Sierra Nevada had pH values between 3.9 and 6.5 at Sequoia National Park and between 3.8 and 5.2 at Yosemite National Park (Collett et al. 1990). Regional patterns in cloud interception correlate closely with elevation because most intercepted clouds are associated with frontal systems approaching from the north or northwest. Cloud bases are often above 1500 m. Sites located along slopes of canyons or valleys with a north-south orientation experience relatively few cloud events, presumably due to terrain effects that inhibit most frontal cloud systems from impacting these sites. However, lower elevation sites are probably exposed to highly polluted radiation or Tulle fogs advected into the Sierra following destabilization of valley and radiation inversions (Hoffmann et al. 1989).

- Annual volume weighted pH of precipitation ranged between 5.2 and 5.5. Individual events had pH values as low as 4.5. Annual wet deposition was estimated at between 0.6 and 6.0 kg/ha for sulfate and between 1.0 and 7.0 kg/ha for nitrate (National Atmospheric Deposition Program (IR-7)/National Trends Network 1990, Blanchard et al. 1989). There appears to be a gradient in both sulfate and nitrate deposition from north (low) to south (high).

- A gradient of fine particle sulfur was measured in the Sierra Nevada from the north, with summer average concentrations of 220 ng/m^3, to south, with summer average concentrations of 500 ng/ m^3 (Cahill et al. 1986). Elevation rather than horizontal distance defined the occurrence and in some cases the magnitude of fine particle sulfur episodes. Loadings decreased with elevation in Sequoia National Park (Ashbaugh et al. 1989). Highest loadings of fine particle sulfur occurred during the summer months. The presence of arsenic indicated that the dominant sulfate, nitrate, and hydrogen ion fluxes came from sources to the south and east of Sequoia National Park, often as far away as Arizona. Nickel levels were enhanced at Sequoia National Park relative to other sites indicating that the southern portion of the Sierra Nevada experienced greater impacts from fuel combustion sources than did central and northern portions of the range (Cahill et al. 1986).

Southern California

Forests surrounding the Los Angeles Basin were exposed to very high ozone and particle concentrations as well as acidic cloud events. In particular:

- Ozone concentrations, both magnitude and frequency of occurrence of extreme values, were greater in the eastern and southeastern portions (e.g., Lake Gregory, growing season mean of 73 ppb) than in the western portion (e.g., Santa Monica Mountains National Recreation Area, growing season mean of 40 ppb) of the Los Angeles Basin (Table 3.4). Ozone values were highest during the summer period June through September, although hourly concentrations greater than 80 ppb were recorded all year round. Growing season means for ozone have decreased in the Los Angeles Basin since the 1970s (Walker 1985, Miller et al. 1989b). However, hourly concentrations during mornings and evenings in summer have not changed much over the last decade. Maximum ozone concentrations recorded during the 1980s were still high enough to cause injury to vegetation. Diurnal patterns in ozone concentrations at sites in the mountains surrounding the Los Angeles Basin exhibited noticeable fluctuations, but nighttime concentrations remained elevated. In contrast to sites in the southern Sierra Nevada, sites around the Los Angeles Basin were characterized by a large number of diurnal curves with means from as high as 100 ppb to as low as 26 ppb (Böhm and Vandetta 1990). A spatial gradient in ozone concentrations within the San Bernardino National Forest to the east of the Los Angeles Basin has been measured. Higher ozone concentrations were found in western and northern portions of the forest than to the east and south. The 24-hour averages in the northwestern part of the San Bernardino National Forest ranged between 90 and 140 ppb with average maximum values between 200 and 240 ppb (Miller et al. 1986).

- Cloud events in Riverside, CA, had pH values between 2.33 and 5.68. The large range in pH was attributed to ammonia emissions in the Chino portion of the valley (Munger et al. 1990). Forests on the lower slopes of the eastern portion of the Los Angeles basin may be exposed to similar pH values during cloud events.

- The volume weighted pH of precipitation in the mountains surrounding the Los Angeles Basin ranged between 4.8 and 5.1. Annual wet deposition estimates were 2.3 to 4.3 kg/ha for sulfate and 4.0 to 6.6 kg/ha for nitrate (National Atmospheric Deposition Program (IR-7)/National Trends Network 1990).

- The atmosphere was dominated by nitrogen compounds. Nitrate concentrations as high as 600 ng/m^3 and ammonium ion concentrations as high as 300 ng/m^3 were recorded in the San Gabriel Mountains (Bytnerowicz et al. 1987, Munger et al. 1990).

Arizona

The chemistry of the air impacting forests in Arizona is driven by emissions from large point sources (Figure 3.11). In particular (Figure 3.23):

- Ozone concentrations at non-urban sites across the state ranged from growing season means of 29 ppb at Grand Canyon National Park to 42 ppb at Apache-Sitgreaves National Forest and 37 ppb at Saguaro National Monument. Hourly concentrations rarely exceeded 80 ppb, indeed no values above 100 ppb were recorded at these sites. The diurnal cycles exhibited little fluctuation and patterns with means of 36, 51 and 22 ppb accounted for 70% of summer days (Böhm and Vandetta 1990). Forests near urban centers such as Phoenix and Tucson could be exposed to ozone concentrations between 100 and 120 ppb.

- Annual volume weighted pH of precipitation was more acidic in the south, pH=4.7, than in the northern portion of the state, pH=5.2. In the south, pH values as low as 3.8 have been recorded. Annual wet deposition of sulfate was also higher in the south (5.0 kg/ha) than in the north (2.6 kg/ha), although nitrate depositions were similar at both sites (National Atmospheric Deposition Program (IR-7)/National Trends Network 1990). Higher deposition in the south was substantiated by Blanchard and Stromberg (1987) who measured sulfate and nitrate deposition at sites in the southeastern portion of Arizona over a 13 month period during 1984 and 1985. Their data indicate sulfate depositions of 8.9 kg/ha and nitrate

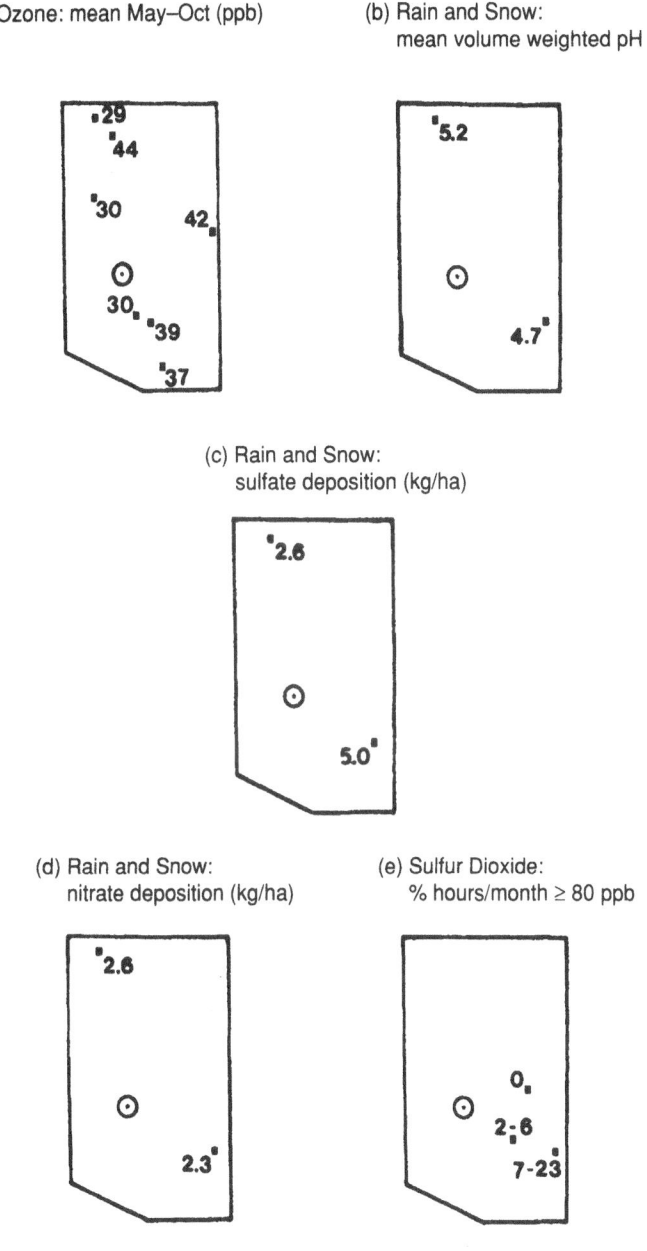

(a) Ozone: mean May–Oct (ppb)

(b) Rain and Snow:
    mean volume weighted pH

(c) Rain and Snow:
    sulfate deposition (kg/ha)

(d) Rain and Snow:
    nitrate deposition (kg/ha)

(e) Sulfur Dioxide:
    % hours/month ≥ 80 ppb

Figure 3.23 Summary of air quality and wet deposition in forests of Arizona. (a) Growing season mean (May-Oct, 1980–1988) for ozone (ppb); (b) Mean annual volume weighted pH (1985–1988) for rain and snow; (c) Mean annual sulfate deposition (1985–1988) for rain and snow (kg/ha); (d) Mean annual nitrate deposition (1985–1988) for rain and snow (kg/ha); (e) Annual range in percent hours per month with sulfur dioxide concentrations ≥ 80 ppb.
⊙ = Phoenix

depositions of 3.5 kg/ha. These high deposition rates correspond
with the regional locations of major sulfur dioxide emission sources
(Figure 3.11).

- Sulfur dioxide concentrations were high in the southeastern portion
  of the state where up to 23% of hours in winter had sulfur dioxide
  levels greater than or equal to 80 ppb.

Front Range, Colorado

Forests to the west and northwest of the Colorado Springs-Denver-
Boulder-Fort Collins urban corridor may experience elevated levels of air
pollution, although few monitoring data are available for the region
(Figure 3.24):

- The growing season mean for ozone concentrations in the Rocky
  Mountain National Park was 46 ppb. Diurnal cycles in hourly
  ozone concentrations showed little fluctuation, and patterns with
  24-hour means of 36 and 51 ppb accounted for 70% of summer days
  (Böhm and Vandetta 1990).

- Cloudwater pH at Mt. Werner ranged between 3.0 and 5.2. In
  general, clouds associated with winter storms from the north and
  northeast had relatively low hydrogen ion, ammonium, sulfate, and
  nitrate concentrations. The highest concentrations of these ions
  were observed in clouds associated with weather systems from the
  southwest during August and September (Miller et al. 1989a).

- Annual volume weighted pH for precipitation ranged from 5.1 to
  4.8. Events with pH values as low as 4.1 in the north and 4.2 in the
  south have been recorded. Annual deposition estimates for the
  Front Range were around 3.0–4.0 kg/ha for sulfate and 3.0–5.0 kg/
  ha for nitrate. The annual volume weighted pH of precipitation
  near Hayden and Craig, west of the continental divide, was 4.8
  with a minimum of 4.0. Sulfate and nitrate deposition were 7.6 and
  5.9 kg/ha respectively. Sulfate deposition near Hayden and Craig
  was almost twice as high as that along the Front Range (National
  Atmospheric Deposition Program (IR-7)/National Trends Network
  1990).

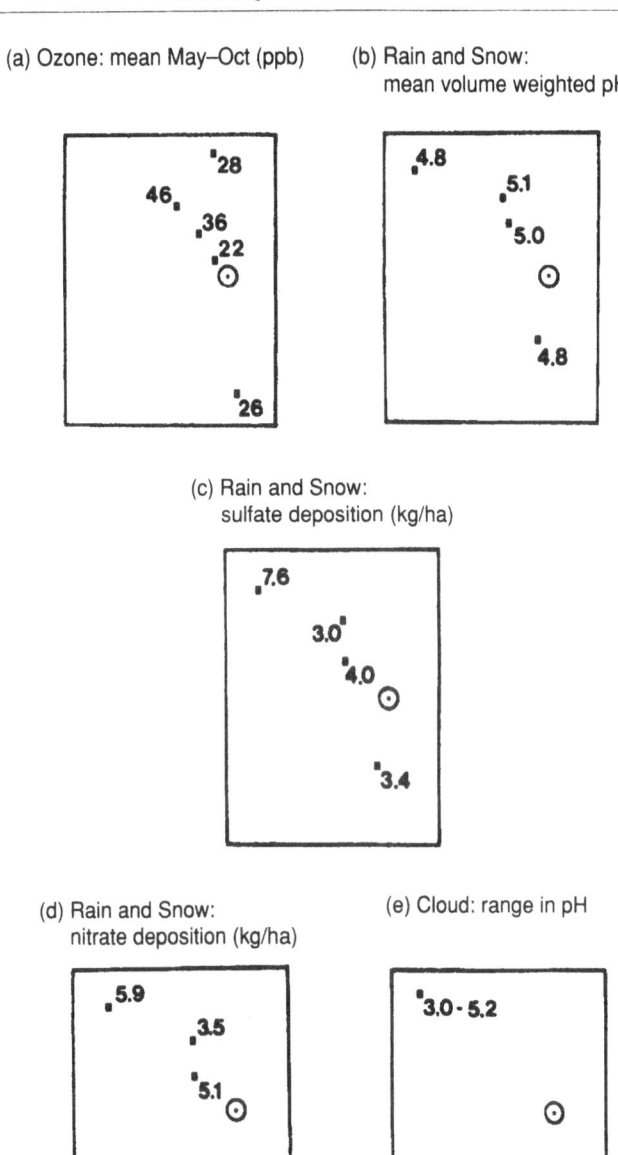

(a) Ozone: mean May–Oct (ppb)

(b) Rain and Snow: mean volume weighted pH

(c) Rain and Snow: sulfate deposition (kg/ha)

(d) Rain and Snow: nitrate deposition (kg/ha)

(e) Cloud: range in pH

Figure 3.24 Summary of air quality and wet deposition in forests of the Front Range, and other sites in Colorado. (a) Growing season mean (May–Oct, 1980–1988) for ozone (ppb); (b) Mean annual volume weighted pH (1985–1988) for rain and snow; (c) Mean annual sulfate deposition (1985–1988) for rain and snow (kg/ha); (d) Mean annual nitrate deposition (1985–1988) for rain and snow (kg/ha); (e) Range in cloudwater pH (summer 1987). ⊙ = Denver

Conclusions

The potential for severe air pollution episodes in an area is determined by meteorological and topographical conditions conducive to the accumulation of atmospheric pollutants. From a meteorological perspective, non-ventilating weather commonly occurs over large areas of the western United States. The topographical features of the West further inhibit horizontal dispersion of pollutants emitted in valleys. It appears that the potential for air pollution accumulation across the western United States is greater than in the eastern United States. Most areas in the West have been spared from regional pollution problems characteristic of the eastern United States because population densities and emissions of pollutants across most of the West are a fraction of those in the East. Areas in the West with high population densities, such as southern and central California, experience pollution episodes of equal if not greater severity than most areas in the eastern United States. As population continues to grow in most Western states, the frequency of occurrence and perhaps the severity of air pollution episodes are expected to increase. Furthermore, predictions of sulfur dioxide and nitrogen oxide emissions in the West imply gradual increases in the future. Comparisons of emission estimates for 1985 and 1987/88 indicate increases at the county level despite decreases in total emissions at the state level. This implies that a few large point sources have either closed down or activated greater abatement strategies but at the same time larger areas of the West are becoming exposed to higher pollution emissions than prior to 1985. In other words, regional air pollution emissions are expanding in the West and air pollution is no longer a local phenomenon. The challenge to Westerners, both civilian and those in public office, is to guide population and industrial growth as well as societal behavior in such a way to prevent regional air pollution episodes as experienced in the East and in Europe. With the threat of changing climate, the challenge is even greater for maintaining a quality of life cherished by visitors and often taken for granted by those who live here.

Ozone is the pollutant of most concern in the West, mainly because societal behavior has created a lifestyle characterized by large sprawling cities with a unique dependence on personal motor vehicles. Ozone levels in the San Bernardino Mountains and southern Sierra Nevada are high enough to cause visible injury and in some cases growth reductions on important forest species such as ponderosa pine. Ozone concentrations at levels greater than 80 ppb occur relatively frequently in the Washington Cascades, Wasatch Mountains, and Colorado Rockies during the summer months. Cloudwater acidities in southern California regularly drop to pHs below 3.5. In Washington and northwestern Colorado, several cloud events had pH values between 3.0 and 3.5. The chemistry of cloudwater impacting Western forests may be detrimental to forest ecosystems, although little is known about the regional extent of cloudwater acidity and specific responses of Western forest species to cloudwater chemistry. The chemistry of rain and snow in Western forests is of less concern since concentrations tend to be low. However, deposition of pollutants to high elevations may not be as insignificant as the concentration data imply. Very little is known about pollutant loadings to high elevation ecosystems in the West.

The emissions, air quality, and wet deposition data presented in this chapter imply that forests close to large urban and industrial complexes are more likely to receive higher air pollution exposures than forests farther from source areas of pollution. In particular, forests along the western slopes of the San Bernardino Mountains and Sierra Nevada in California, the Cascades in Washington and Oregon, the Wasatch Mountains in Utah and the Mazatzal Mountains in Arizona, together with the eastern slopes of the Colorado Rocky Mountains probably receive higher air pollution exposures than other forests in the West. However, large areas of the West lack data, and this conclusion may be refuted as more information on air quality, deposition, and forest condition become available for the West as a whole.

Detailed information on air quality and deposition in Western forests is generally only available after the 1970s. Forest stands in the West, on the other hand, range from decades to centuries in age. Since smelters emitting large quantities of sulfur dioxide and other toxic species were dotted across the western United States by 1900 (Quinn 1989), it is incorrect to consider air pollution as a recent (e.g., post 1940 or 1960 etc.) phenomenon in the West. Many Western forests have matured in a chemically changing environment and a detailed inventory of the location, type, and operation of early industry in the West together with more data on air pollution levels in forests is needed to complete the "emissions/air pollution/forest effects" triangle in the West.

Glossary of terms

(after Finlayson-Pitts and Pitts 1986, Oke 1987, Weast et al. 1987, Stull 1988).

Absorption—the process of assimilating gases or particles by uptake through pores or interstices.

Acute pollution exposure—exposure of vegetation to very high concentrations of a pollutant.

Adsorption—assimilation of gases or particles by the surface of a solid or liquid.

Advect—to transport quantities such as mass and energy via predominantly horizontal movement of air in the atmosphere.

Aerodynamic diameter—atmospheric particles are usually irregularly shaped. Since particle volume, mass, and settling velocity depends on size, an estimate of particle diameter is needed. The most commonly used estimate is aerodynamic diameter which translates to the diameter of a sphere of unit density ($1 \ g/cm^3$) which has the same terminal falling speed in air as the particle under consideration.

Air pollution meteorology—an applied form of micrometeorology studying boundary layer processes that influence the dispersion of pollutants.

Air quality—ambient levels of pollutants in the atmosphere at any given time.

Air pollution potential—the potential for frequent occurrences of severe air pollution episodes.

Anabatic winds—upslope winds due to surface heating that are gentle (less than 1 m/s) and that tend to hug the valley walls as they rise.

Anthropogenic—of man, applied in this chapter to pollutants emitted during human activities.

Area sources—non-stationary sources of air pollution such as motor vehicles, airplanes, and ships.

Atmospheric residence time—the length of time a pollutant remains suspended in the atmosphere between emission and deposition or chemical transformation.

Brownian motion—random motion of small particles suspended in a liquid or a gas, caused by collision with molecules of the surrounding medium.

Buoyancy—the upward force exerted on a parcel of air by virtue of density differences between the parcel and the surrounding air.

Chronic pollution exposure—exposure of vegetation to moderately low concentrations of a pollutant.

Cloud condensation nuclei—particles of micron and submicron size which have an affinity for water and serve as centers for condensation.

Cloud interception—physical interception of cloud droplets by vegetation or topographical obstacles.

Cloud deposition—deposition of pollutants in cloud water during interception of clouds by vegetation.

Convection—transfer of heat and mass between regions of unequal density by fluid motions resulting from nonuniform heating, usually applied to vertical motion in the atmosphere.

Diffusion—gradual mixing of molecules by random motion across a concentration gradient.

Diffusivity—the rate of diffusion of a property.

Dry deposition—transfer of gases and particles from the atmosphere to the ground without the intervention of precipitation.

Effective emission height—the height to which a plume rises before reaching thermal equilibrium with its environment.

Emission density—amount of pollutant emitted per unit area and time.

Flux density—transfer of a quantity such as energy or mass through a unit surface area per unit time.

Forest meteorology—meteorological characteristics of the forest boundary layer.

Forest boundary layer—portion of the surface layer from the ground to slightly above the forest canopy.

Heterogeneous reaction—reaction between chemical species of different physical forms, e.g., a gas reacting with a liquid.

High pressure system—a large scale atmospheric circulation system in which (in the northern hemisphere) the winds rotate clockwise and the air has higher atmospheric pressure than the surrounding air.

Homogeneous reaction—reaction between chemical species of the same physical phase, e.g., a gas reacting with a gas.

Hydrometeor—a precipitation body such as rain, snow, sleet, or hail, derived from the condensation of water in the atmosphere.

Hygroscopic—having properties that accelerate condensation of water vapor or readily allow absorption of moisture from the atmosphere.

Impaction—process of deposition of aerosols to receptor surfaces when aerosols transferred through the laminar interfacial layer by inertia or sedimentation fail to follow streamlines of airflow around obstacles.

Inertia—the resistance offered by a body to a change of its state of rest or motion.

Interception—process of deposition of aerosols to receptor surfaces when aerosols are transferred through the laminar interfacial layer by turbulent diffusion.

Interfacial layer—the layer of air immediately next to a fixed surface in which laminar airflow prevails and where molecular transport of quantities such as energy and mass dominates over turbulent transport.

Inversion height—elevation at which the upper limit of an inversion occurs.

Inversion—a departure from the usual decrease or increase with height of an atmospheric property, normally refers to an increase in temperature with height above the ground in contrast to the usual decrease in temperature with height.

Ion exchange—movement of ions across a concentration gradient through a semi-permeable membrane.

Katabatic wind—any wind blowing down an incline, often due to cold air drainage. These winds are shallow (2–20 m) and have velocities of around 1–5 m/s.

Land breeze—land surfaces have a lower heat capacity than do lakes and oceans. Rapid cooling of the land surface during the night leads to a pressure gradient between the land (high) and the water (low). Cold air from land flows towards the water at low levels, warms, rises, and returns aloft to land where it eventually descends to close the circulation.

Long-range transport—transport of primary and secondary pollutants in the atmosphere to great distances from the source area prior to deposition.

Mechanical turbulence—turbulence generated by mechanical friction acting on ambient flow or by interactions between air masses of different characteristics such as density and wind speed.

Mixed layer—layer of the lower atmosphere dominated by turbulence. The turbulence is usually driven by thermal convection caused by daytime heating of the earth's surface. Pollutants are presumed to mix throughout the layer.

Mixing height—the thickness of the mixed layer measured from the surface upward.

Molecular diffusion—gradual transport of molecules by diffusion.

Mountain wind—cold winds blowing down the valley onto the plains at night, depths range from 10 to 400 m and velocities from 1–8 m/s.

Neutral atmosphere—occurs when there is no capability for buoyant convection in the atmosphere, i.e., a displaced air parcel remains at the location to which it is displaced.

Non-ventilating weather—meteorological conditions that cause stagnation of air masses for extended periods of time. No net transfer of mass occurs from the air mass to surrounding air masses during the period of stagnation. Typically associated with pollutant build-up in valleys.

Nucleation—processes by which water droplets form on cloud condensation nuclei.

Oxidation—chemical process that increases the proportion of oxygen, acid, or radicals in a compound; a reaction in which an element's valence is increased as a result of losing electrons, e.g., sulfur (valence IV) as sulfur dioxide is oxidized to sulfur (valence VI) as sulfate.

Plane of zero displacement—height below the forest canopy at which quantities such as wind speed, temperature, and humidity reach a minimum.

Planetary boundary layer—layer of the atmosphere from the ground to the level where frictional influences are absent.

Point sources—stationary sources of air pollution such as coal-fired power plants, smelters, and other industrial activities.

Pollutant concentrations—mass per unit volume of pollutants in the ambient atmosphere during a given time period.

Pollutant exposure—ambient concentrations or loadings of pollutants in direct contact with the receptor organism just prior to uptake of the pollutants by the organism. Pollutant exposure equals pollutant uptake if there is no resistance to the uptake of deposited material by the receptor organism.

Pollutant loadings—mass per unit area of pollutants deposited on a receptor organism during a given time period. Pollutant loading equals pollutant uptake if there is no resistance to the uptake of the deposited material by the receptor organism.

Pollution episode—period of time during which pollutant concentrations exceed desirable levels.

Pressure gradient—gradient in surface pressure between two points in space, usually refers to horizontal gradients.

Primary pollutants—pollutants initially emitted during anthropogenic activities.

Rainout—removal of pollutants from the atmosphere during in-cloud scavenging and subsequent deposition on the earth's surface.

Runoff—surface flow of water deposited during a precipitation event.

Scavenging—removal of pollutants from the atmosphere during processes of wet, dry, and cloud deposition.

Sea breeze—land surfaces have a lower heat capacity than lakes and oceans. Rapid heating of the land surface during the day leads to a pressure gradient between the land (low) and the water (high). Cool maritime air flows towards land at low levels, warms, rises, and returns aloft to sea where it eventually descends to close the circulation.

Secondary pollutants—chemical transformation products of primary pollutants.

Sedimentation—process of dry deposition whereby large particles move down through the atmosphere under the influence of gravity.

Stable atmosphere—occurs when there is negative capability for buoyant convection in the atmosphere, i.e., displaced air parcels return to their starting point.

Surface layer—layer of the lower atmosphere characterized by turbulence generated by mechanical and thermal convection.

Surface film—film of water several molecules thick that covers surfaces such as foliage following condensation or deposition of moisture to the surface.

Surface roughness—index describing the characteristics of the surface elements. Surface roughness elements cause complex air flows around and within forests and play an important role in development of turbulent eddies.

Sweep and Gust—large coherent eddies that cause the downward transfer of relatively fast moving air from above the forest boundary layer to within the canopy and trunkspace.

Synoptic—atmospheric conditions over a large area at a given time.

Thermal turbulence—turbulence created by heating or cooling of the surface of the earth.

Throughfall—precipitation collected on the forest floor that has or has not interacted with the forest canopy during its passage to the ground.

Troposphere—the lowest 10–20 km of the atmosphere that is usually characterized by decreasing temperature with height and appreciable concentrations of water vapor and vertical motions.

Tropospheric ozone—ozone formed in the troposphere by both natural and anthropogenic activities. Usually considered a pollutant in contrast to stratospheric ozone that protects the earth from ultra-violet radiation.

Turbulence—a state of fluid flow in which the instantaneous velocities exhibit irregular and apparently random fluctuations.

Unstable atmosphere—occurs when there is capability for buoyant convection in the atmosphere, i.e., displaced air parcels rise from the ground or sink from cloud tops until they reach thermal equilibrium with their environment.

Valley wind—gentle flow of warm air up the valley axis.

Washout—removal of pollutants from the atmosphere during scavenging by rain drops, snow flakes, and hail stones as they fall from the base of the cloud to the ground.

Wet deposition—removal of pollutants from the atmosphere during precipitation events.

References

Ahrens DC (1988) *Meteorology Today: An Introduction to Weather, Climate, and the Environment, third edition.* West Publishing, 581 p

Altshuller AP (1983) Natural volatile organic substances and their effect on air quality in the United States. *Atmospheric Environment* 17:2131–2166

Amiro BD, Davis PA (1988) Statistics of atmospheric turbulence within a natural black spruce forest canopy. *Boundary Layer Meteorology* 44:267–283

Amiro BD, Ewing LL, Johnston FL (1989) Turbulence measurements within boreal forest canopies. In: *Proceedings of the Sixth Joint Conference on Applications of Air Pollution Meteorology*, Sponsored by American Meteorological Society and the Air and Waste Management Association, 30 January–03 February, Anaheim, CA, pp 85–88

Anonymous (1987) Southern California Air Quality Study (SCAQS) begins: Major effort to examine smog causes, chemistry. *Journal of the Air Pollution Control Association* 37(7):842–844

Ashbaugh LL, Watson JG, Chow J (1989) Estimating fluxes from California's dry deposition monitoring data. In: *Proceedings of the 82nd Annual Meeting and Exhibition of the Air and Waste Management Association*, Anaheim, CA, June 25–30, 1989, AWMA, Pittsburgh, PA

Azevedo HD, Morgan DL (1974) Fog precipitation in coastal California forests. *Ecology* 55:1135–1141

Baldocchi DD, Hutchinson BA (1988) Turbulence in an almond orchard: Spatial variations in spectra and coherence. *Boundary Layer Meteorology* 42:293–311

Baldocchi DD, Meyers TP (1988a) Turbulence structure in a deciduous forest. *Boundary Layer Meteorology* 43:345–364

Baldocchi DD, Meyers TP (1988b) A spectral and lag-correlation analysis of turbulence in a deciduous forest canopy. *Boundary Layer Meteorology* 45:31–58

Basabe FA, Edmonds RL, Chang WL, Larson TV (1989a) Fog and cloud water chemistry in western Washington. In: Olson RK, Lefohn AS (eds) (1989) *Effects of Air Pollution on Western Forests.* Transactions Series, No. 16, Air and Waste Management Association, Pittsburgh, PA, pp 33–49

Basabe FA, Edmonds RL, Larson TV (1989b) Regional ozone in western Washington and southwest British Columbia. *Proceedings of the 82nd Annual Meeting and Exhibition of the Air and Waste Management Association*, Anaheim, CA, June 25–30, 1989, AWMA, Pittsburgh, PA

Beier C, Gundersen P (1989) Atmospheric deposition to the edge of a spruce forest in Denmark. *Environmental Pollution* 60:257–271

Berg NH (1988) Mountain-top riming at sites in California and Nevada, U.S.A. *Arctic and Alpine Research* 20(4):429–447

Berg NH, Dunn PH (1988) Chemistry of rime ice at four sites in California. In: *Proceedings of the International Mountain Watershed Symposium*, Crystal Bay, NV, June 7–10, 1988.

Bicknell SH (1989) Fog and rain chemistry in northern California. In: Olson RK, Lefohn AS (eds) (1989) *Effects of Air Pollution on Western Forests*. Transactions Series, No. 16, Air and Waste Management Association, Pittsburgh, PA, pp 147–167

Blanchard CL, Stromberg MR (1987) Acidic precipitation in southeastern Arizona: Sulfate, nitrate and trace-metal deposition. *Atmospheric Environment* 21(11):2375–2381

Blanchard CL, Tonnessen KA, Ashbaugh LL (1989) Acidic deposition in California forests: Precipitation-chemistry measurements from the California Acid Deposition Monitoring Program. *Proceedings of the 82nd Annual Meeting and Exhibition of the Air and Waste Management Association*, Anaheim, CA, 25–30 June 1989

Böhm M (1989) A regional characterization of air quality and deposition in the coniferous forests of the western United States. In Olson RK, Lefohn AS (eds) *Effects of Air Pollution on Western Forests*. Transactions Series, No. 16, Air and Waste Management Association, Pittsburgh, PA, pp 221–244

Böhm M, Vandetta T (1990) *Atlas of Air Quality and Deposition in or near Forests of the Western United States*, EPA/600/3-90/081. Environmental Research Laboratory, Office of Research and Development, Environmental Protection Agency, Corvallis, OR 97333, 470 p

Böhm M, McCune B, Vandetta T (1991) Diurnal curves of tropospheric ozone in the western United States. *Atmospheric Environment* 25A(8): 1577–1590

Bormann BT, Tarrant RF, McClellan MH, Savage T (1989) Chemistry of rainwater and cloud water at remote sites in Alaska and Oregon. *Journal of Environmental Quality* 18:149–152

Borys RD, Hindman EE, DeMott PJ (1988) The chemical fractionation of atmospheric aerosol as a result of snow crystal formation and growth. *Journal of Atmospheric Chemistry* 7:213–239

Bredemeier M (1989) Forest canopy transformation of atmospheric deposition. *Water, Air, and Soil Pollution* 40:121–138

Breeding RJ, Klonis HB, Lodge JP Jr., Pate JB, Sheesley DC, Englert TR, Sears DR (1976) Measurements of atmospheric pollutants in the St. Louis area. *Atmospheric Environment* 10:181–194

Brimblecombe P (1978) Dew as a sink for SO_2. *Tellus* 30:151–157

Bubenick DV, Record FA, Kindya RJ (1983) Acid rain—An overview of the problem. *Environmental Progress* 2(1):15–32

Bytnerowicz A, Miller PR, Olszyk DM, Dawson PJ, Fox CA (1987) Gaseous and particulate air pollution in the San Gabriel Mountains of southern California. *Atmospheric Environment* 21(8):1805–1814

Cahill TA, Eldred RA, Feeney PJ (1985) *Particulate Monitoring and Data Analysis for the National Park Service 1982–1985.* Final Report to the National Park Service, Contract No. USDICX-0001-3-0056, 130 p

Cahill TA, Annegarn HJ, Ewell D, Feeney PJ (1986) *Particulate Monitoring for Acid Deposition Research at Sequoia National Park, California.* Final Report to the California Air Resources Board, Contract No. A4-124-32, Sacramento, CA

Cape JN, Unsworth MH (1987) Deposition, uptake, and residence of pollutants. In: Schulte-Hostede S, Darrall NM, Blank LW, Wellburn AR (eds) *Air Pollution and Plant Metabolism.* Elsevier Applied Science, NY, 381 p

Cass GR, Shair FH (1984) Sulfate accumulation in a sea breeze/land breeze circulation system. *Journal of Geophysical Research* 89(D1):1429–1438

Chamberlain AC (1967) Transport of Lycopodium spores and other small particles to rough surfaces. *Proceedings of the Royal Society* 296:45–70

Chamberlain AC (1975) The movement of particles in plant communities. In: Monteith JL (ed) *Vegetation and the Atmosphere,Volume 1: Principles*, Academic Press, London, pp 155–203

Chang TY, Norbeck JM, Weinstock B (1979) An estimate of the NO_x removal rate in an urban atmosphere. *Environmental Science and Technology* 13:1534–1537

Charlson RJ, Rodhe H (1982) Factors controlling the acidity of natural rainwater. *Nature* 295:683–685

Chevone BI, Herzfeld DE, Krupa SV, Chappelka AH (1986) Direct effects of atmospheric sulfate deposition on vegetation. *Journal of the Air Pollution Control Association* 36:813–815

Chinkin LR, Latimer DA, Mahoney LA (1987) *Western States Acid Deposition Project, Volume 2: A Review of Emission Inventories Needed to Regulate Acid Deposition in the Western United States.* Final Report SYSAPP-87/072, Western Governor's Association, Denver, CO, 148 p

Cionco RM (1985) Modeling windfields and surface layer wind profiles over complex terrain and within vegetative canopies. In Hutchison BA, Hicks BB (eds) *The Forest-Atmosphere Interaction.* D. Reidel Publishing Company, Boston, pp 501–520

Cionco RM (1989) Forest winds coupled to surface layer windfields. In: *Proceedings of the Sixth Joint Conference on Applications of Air Pollution Meteorology.* Sponsored by American Meteorological Society and the Air and Waste Management Association, 30 January–03 February, Anaheim, CA, pp 196–199

Collett JL, Daube BC, Munger JW, Hoffmann MR (1989) Cloudwater chemistry in Sequoia National Park. *Atmospheric Environment* 23(5):999–1007

Collett JL, Daube BC, Hoffmann MR (1990) The chemical composition of intercepted cloudwater in the Sierra Nevada. *Atmospheric Environment* 24A:959–972

Cullis CF, Hirschler MM (1980) Atmospheric sulfur: Natural and man-made sources. *Atmospheric Environment* 14:1263–1278

Denmead OT, Bradley EF (1985) Flux-gradient relationships in a forest canopy. In Hutchison BA, Hicks BB (eds) *The Forest-Atmosphere Interaction.* D. Reidel Publishing Company, Boston, pp 421–442

Diab RD (1977) Estimates of air pollution potential over southern Africa. *South African Journal Science* 73:270–274

Druilhet A, Schayes G, Britto R, Lyra R (1989) Energy budget and turbulence above and within a pine forest. In: Proceedings of the Sixth Joint Conference on Applications of Air Pollution Meteorology, Sponsored by American Meteorological Society and the Air and Waste Management Association, 30 January–03 February, Anaheim, CA, pp 96–99

Duce RA, Mohnen VA, Zimmerman PR, Grosjean D, Cautreels W, Chatfield R, Jaenicke R, Ogren JA, Pellizzari ED, Wallace GT (1983). Organic material in the global troposphere. *Rev. Geophys. Space Physics* 21:921–952

Duncan C (1990) The chemistry of rime and snow collected at a site in the Central Washington Cascades. Submitted to *Environmental Science and Technology*

Eatough DJ, Arthur RJ, Eatough NL, Hill MW, Mangelson NF, Richter BE, Hansen LD, Cooper JA (1984) Rapid conversion of $SO_2(g)$ to sulfate in a fog bank. *Environmental Science and Technology* 18:855–859

Finlayson-Pitts BJ, Pitts JN Jr (1986) *Atmospheric Chemistry: Fundamentals and Experimental Techniques.* John Wiley & Sons Inc., New York, 1098 p

Finnigan JJ (1985) Turbulent transport in flexible plant canopies. In: Hutchison BA, Hicks BB (eds) *The Forest-Atmosphere Interaction.* D. Reidel Publishing Company, Boston, pp 443–480

Fowler D (1984) Transfer to terrestrial surfaces. *Royal Society Philosophical Transactions Series B* 305:281–297

Fox DG (1985) Forestry. In: Houghton DD (ed) *Handbook of Applied Meteorology.* John Wiley and Sons, New York, pp 605–666

Fransioli PM, Weston RF (1989) Role of shear in valley flow dynamics. In: *Proceedings of the Sixth Joint Conference on Applications of Air Pollution Meteorology*, sponsored by American Meteorological Society and the Air and Waste Management Association, 30 January–03 February, Anaheim, CA, pp 195–197

Fritsche U, Gernert M, Schindler C (1989) Vertical profiles of air pollutants in a spruce forest—Analysis of adherent water, throughfall and deposits on surrogate surfaces. *Atmospheric Environment* 23(8):1807–1814

Fritschen LJ (1985) Characterization of boundary conditions affecting forest environmental phenomena. In Hutchison BA, Hicks BB (eds) *The Forest-Atmosphere Interaction.* D. Reidel Publishing Company, Boston, pp 3–23

Fritschen LJ, Edmonds JR (1976) Dispersion of fluorescent particles into and within a Douglas-fir forest. In *Atmospheric and Surface Exchange of Particulates and Gaseous Pollutants*, 1974 ERDA Symposium Series CONF-740921

Galloway JN, Likens GE (1976) Calibration of collection procedures for the determination of precipitation chemistry. *Journal of Water, Air, and Soil Pollution* 6:241–258

Garland JA (1978) Dry and wet removal of sulfur from the atmosphere. *Atmospheric Environment* 12:349–362

Gmur NF, Evans LS, Cunningham EA (1983) Effects of ammonium sulfate aerosols on vegetation-II. Mode of entry and responses of vegetation. *Atmospheric Environment* 17(4):715–721

Godt J, Mayer R (1988) Deposition rates of airborne substances to forest canopies in relation to surface structure. In: Unsworth MH, Fowler D (eds) *Acid Deposition at High Elevation Sites.* Kluwer Academic Publishers: Series C Mathematical and Physical Sciences, London, 670 p

Goklany IM, Hoffnagle GF (1984) Trends in emissions of PM, SO_x, and NO_x and VOC: NO_x ratios and their implications for trends in pH near industrialized areas. *Journal of the Air Pollution Control Association* 34(8):844–846

Gschwandtner G, Husar RB, Mobley JD (1989) National and regional trends in VOC and NO_x emissions from 1900 to 1987. *Proceedings: 82nd Annual Meeting and Exhibition of the Air and Waste Management Association*, paper no. 89–35.10, Anaheim, CA, June 25–30, 1989, AWMA, Pittsburgh, PA

Guderian R (ed) (1985) *Air Pollution by Photochemical Oxidants: Formation, Transport, Control, and Effects on Plants.* Springer-Verlag Ecological Series Volume 52, Berlin, 346 p

Harr DR (1982) Fog drip in the Bull Run municipal watershed Oregon. *Water Resources Bulletin* 18(5):785–789

Harte J, Lockett GP, Schneider RA, Blanchard C, Micheals H (1985) Acid precipitation and surface water vulnerability on the western slope of the High Colorado Rockies. *Water, Air, and Soil Pollution* 25:313–320

Heck WW, Brandt CS (1977) Effects on vegetation: Native, crops, forests. In: Stern AC (ed) (1977) *Air Pollution Volume II: Effects of Air Pollution*, Third Edition, Academic Press, New York, pp 158–229

Hegg DA, Hobbs PV (1979) Some observations of particulate nitrate concentrations in coal-fired power plant plumes. *Atmospheric Environment* 13:1715–1716

Hegg DA, Hobbs PV, Lyons JH (1985) Field studies of a power plant plume in the arid southwestern United States. *Atmospheric Environment* 19(7):1147–1167

Helms JA (1970) Summer net photosynthesis of ponderosa pine in its natural environment. *Photosynthetica* 4(3):243–253

Hidy GM, Young JR (1986) *Acid Deposition and the West: A Scientific Assessment*, ERT Document No. P-D572-503, March

Hindman EE, Borys RD, DeMott PJ (1983) Hydrometeorological significance of rime ice deposits in the Colorado Rockies. *Bulletin of American Water Resources Association* 19(4):619–624

Hoffer TE, Miller DJ, Farber RJ (1981) A case study of visibility as related to regional transport. *Atmospheric Environment* 15:1935–1942

Hoffmann MR, Collett JL, Daube BC (1989) *Characterization of Cloud Chemistry and Frequency of Canopy Exposure to Clouds in the Sierra Nevada*. Final Report to the California Air Resources Board, Contract No. A6-185-32, CARB, Sacramento, CA

Hogsett WE, Tingey DT, Holman SR (1985) A programmable exposure control system for determination of the effects of pollutant exposure regimes on plant growth. *Atmospheric Environment* 19:1135–1145

Holland JZ (1989) On pressure-driven wind in deep forests. *Journal of Applied Meteorology* 28(12):1349–1355

Hosker RP, Lindberg SE (1982) Review: Atmospheric deposition and plant assimilation of gases and particles. *Atmospheric Environment* 16(3):889–910

Jacob DJ, Waldman JM, Munger JW, Hoffmann MR (1984) A field investigation of physical and chemical mechanisms affecting pollutant concentrations in fog droplets. *Tellus* 36B:272–285

Jacob DJ, Shair FH, Waldman JM, Munger JW, Hoffmann MR (1987) Transport and oxidation of SO_2 in a stagnant foggy valley. *Atmospheric Environment* 21(6):1305–1314

Jarvis PG, James GB, Landsberg JJ (1976) Coniferous forest. In: Monteith JL (ed) *Vegetation and the Atmosphere, Vol. 2: Case Studies*, Academic Press, London, pp 171–240

Kerstiens G, Lendzian KJ (1989) Interactions between ozone and plant cuticles I. Ozone deposition and permeability. *New Phytology* 112:13–19

King JA, Shair FH, Reible DD (1987) The influence of atmospheric stability on pollutant transport by slope winds. *Atmospheric Environment* 21(1):53–59

Knudson DA (1986) *Estimated Monthly Emissions of Sulfur Dioxide and Oxides of Nitrogen for the 48 Contiguous States, 1975–1984, Vols. 1 and 2.* Argonne National Laboratory, Report ANL/EES-TM-318

Laird LB, Taylor HE, Kennedy VC (1986) Snow chemistry of the Cascade-Sierra Nevada Mountains. *Environmental Science and Technology* 20(3):275–290

Lamb B, Guenther A, Gay D, Westberg H (1987) A national inventory of biogenic hydrocarbon emissions. *Atmospheric Environment* 21(8):1695–1705

Last F, Rennie P (1982) Forest changes expert meeting No. 3. Stockholm Conference on Acidification of the Environment, June 20–30, 1982 (Swedish Ministry of the Environment)

Leahey DM, Hansen MC (1987) Observations of winds above and below a forest canopy located near a clearing. *Atmospheric Environment* 21(5):1227–1229

Lefohn AS, Krupa SV (1988) The relationship between hydrogen and sulfate ions in precipitation—a numerical analysis of rain and snowfall chemistry. *Environmental Pollution* 49(4):289–311

Lins HF (1987) Trend analysis of monthly sulfur dioxide emissions in the conterminous United States 1975–1984. *Atmospheric Environment* 21(11):2297–2309

Lott RA (1982) Terrain-induced downwash effects on ground level SO_2 concentrations. *Atmospheric Environment* 16(4):635–642

Lovett GM, Reiners WA, Olson RK (1989) Factors controlling throughfall chemistry in a balsam fir canopy: A modeling approach. *Biogeochemistry* 8:239–264

McBean GA (1968) An investigation of turbulence within the forest. *Journal of Applied Meteorology* 7:410–416

McElroy JL (1987) Estimation of pollutant transport and concentration distributions over complex terrain of southern California using airborne lidar. *Journal of the Air Pollution Control Association* 37:1046–1051

McLaughlin SB (1981) SO_2, vegetation effects, and the air quality standard: Limits of interpretation and application. A Specialty Conference on the Proposed SO_x and Particulate Standard, Air Pollution Control Association, Atlanta, GA

Miller DF, Borys RD, Graw R (1989a) Chemistry of summer cloud, precipitation, and air at a Rocky Mountain-top location. In: Olson RK, Lefohn AS (eds) (1989) *Effects of Air Pollution on Western Forests*. Transactions Series, No. 16, Air and Waste Management Association, Pittsburgh, PA, pp 105–115

Miller PR (1973) Oxidant-induced community change in a mixed conifer forest. In: Naegele JA (ed) *Air Pollution Damage to Vegetation*. Advances in Chemistry Series 122, American Chemical Society, Washington, DC, pp 101–137

Miller PR, Taylor OC, Poe MP (1986) Spatial variation of summer ozone concentrations in the San Bernardino Mountains. *Proceedings: 79th Annual Meeting of the Air Pollution Control Association*, Minneapolis, MN, June 22–27, 1986

Miller PR, McBride JR, Schilling SL, Gomez AP (1989b) Trend of ozone damage to conifer forests between 1974 and 1988 in the San Bernardino Mountains of southern California. In: Olson RK, Lefohn AS (eds) (1989) *Effects of Air Pollution on Western Forests*. Transactions Series, No. 16, Air and Waste Management Association, Pittsburgh, PA, pp 309–323

Monteith JL (1973) *Principles of Environmental Physics*. Edward Arnold, London, 241 p

Muir PS, Böhm M (1989) Cloud chemistry and occurrence in the western United States: A synopsis of current information. In: Olson RK, Lefohn AS (eds) *Effects of Air Pollution on Western Forests*. Transactions Series, No. 16, Air and Waste Management Association, Pittsburgh, PA, pp 73–101

Munger JW, Collett J, Daube B, Hoffmann MR (1990) Fogwater chemistry at Riverside, California. *Atmospheric Environment* 24B(2):185–205

Nagamoto CT, Parungo F, Reinking R, Pueschel R, Gerish T (1983) Acid clouds and precipitation in eastern Colorado. *Atmospheric Environment* 17(6):1073–1082

National Atmospheric Deposition Program (1989) *NADP/NTN Annual Data Summary: Precipitation Chemistry in the United States 1988*. NADP/NTN Coordination Office, Natural Resource Ecology Laboratory, Colorado State University, Fort Collins, CO 80523

National Atmospheric Deposition Program (IR-7)/National Trends Network (1990) Data Tape 17 August, 1990. NADP/NTN Coordination Office, Natural Resource Ecology Laboratory, Colorado State University, Fort Collins, CO 80523

National Emissions Data System Emissions Inventory (1985) Data Tape. Available from National Air Data Branch, EPA Research Triangle Park, NC 27709

Newman L (1981) Atmospheric oxidation of sulfur dioxide: A review as viewed from power plant and smelter plume studies. *Atmospheric Environment* 15(10/11):2231–2239

Oberlander GT (1956) Summer fog precipitation on the San Francisco peninsula. *Ecology* 37:851–852

Oke TR (1987) *Boundary Layer Climates*, Second Edition. Methuen & Co Ltd., London, 435 p

Pell EJ (1974) The impact of ozone on the bioenergetics of plant systems. In: Dugger M (ed) *Air Pollution Effects on Plant Growth, Volume 3*. American Chemical Society, ASC Symposium Series, Washington, DC, pp 106–114

Pielke RA, Yu C-H, Arritt RW, Segal M (1984) Mesoscale air quality under stagnant conditions. Air Pollution Effects on Parks and Wilderness Areas Conference, 20–31 May, Mesa Verde National Park, CO

Placet M, Streets DG (1987) *NAPAP Interim Assessment: The Causes and Effects of Acid Precipitation, Volume II: Emissions and Control*. NAPAP Headquarters, Washington, DC, 243 p

Quinn M-L (1989) Early smelter sites: A neglected chapter in the history and geography of acid rain in the United States. *Atmospheric Environment* 23(6):1281–1292

Record FA (1981) *Acid Rain Information Book*. New Jersey, Noyes Data Corporation, 228 p

Reible DD, Shair FH (1981) Plume dispersion and bifurcation in directional shear flows associated with complex terrain. *Atmospheric Environment* 15(7):1165–1172

Richards LW, Anderson JA, Blumenthal DL, Brandt AA, McDonald JA, Waters N, Macias ES, Bhardwaja PS (1981) The chemistry, aerosol physics, and optical properties of a western coal-fired power plant plume. *Atmospheric Environment* 15(10/11):2111–2134

Roth P, Blanchard C, Harte J, Micheals H, El-Ashry MT (1985) *The American West's Acid Rain Test*. Research Report #1, World Resources Institute, 50 p

Russell AG, McRae GJ, Cass GR (1985) The dynamics of nitric acid production and the fate of nitrogen oxides. *Atmospheric Environment* 19(6):893–903

Schroeder WH, Lane DA (1988) The fate of toxic airborne pollutants. *Environmental Science and Technology* 22(3):240–246

Schumann U (1989) Large-eddy simulation of turbulent diffusion with chemical reactions in the convective boundary layer. *Atmospheric Environment* 23(8):1713–1727

Schwartz SE (1989) Acid deposition: Unraveling a regional phenomenon. *Science* 243(4892):753–763

Schwoegler B, McClintock M (1981) *Weather and Energy*. McGraw-Hill, Inc., 230 p

Scott BC (1981) Sulfate washout ratios in winter storms. *Journal of Applied Meteorology* 20:619–625

Sehmel GA (1980) Particle and gas dry deposition: A review. *Atmospheric Environment* 14:983–1011

Sellars FM, Norris WB (1989) Uncertainties in the NAPAP 1985 emission inventory. *Proceedings: 82nd Annual Meeting and Exhibition of the Air and Waste Management Association*, Anaheim, CA, June 25–30, 1989. AWMA, Pittsburgh, PA

Shaw RH (1985) Gust penetration into plant canopies. *Atmospheric Environment* 19(5):827–830

Singh HB, Ludwig FL, Johnson WB (1978) Tropospheric ozone: Concentrations and variabilities in clean, remote atmospheres. *Atmospheric Environment* 12:2185–2196

Skelly JM, Davis DD, Merrill W, Cameron EA, Brown HD, Drummond DB, Dochinger LS (eds) (1987) *Diagnosing Injury to Eastern Forests*. National Acid Precipitation Assessment Program, Forest Response Program/Vegetation Survey Research Cooperative, Research Triangle Park, NC, 122 p

Smith FB, Hunt RD (1978) Meteorological aspects of the transport of pollution over long distances. *Atmospheric Environment* 12:461–478

Spicer CW (1982) Nitrogen oxide reactions in the urban plume of Boston. *Science* 215:1095–1097

Start GE, Dickson CR, Ricks NR (1974) Effluent dilutions over mountainous terrain and within mountain canyons. Symposium on Atmospheric Diffusion and Air Pollution, American Meteorological Society, Boston, MA

Stith JL, Radke LF, Hobbs PV (1981) Particle emissions and the production of ozone and nitrogen oxides from burning of forest slash. *Atmospheric Environment* 15:73–82

Stohlgren TJ, Parsons DJ (1987) Variation of wet deposition chemistry in Sequoia National Park, California. *Atmospheric Environment* 21(6):1369–1374

Stull RB (1988) *An Introduction to Boundary Layer Meteorology*. Kluwer Academic Publishers, Dordrecht, 666 p

Taylor OC (ed) (1973) *Oxidant Air Pollution Effects on a Western Coniferous Forest Ecosystem*. Task B Report, University of California Air Pollution Research Center, Riverside, CA

Tingey DT, Taylor Jr GE (1982) Variation in plant response to ozone: A conceptual model of physiological events. In: Unsworth MH, Ormrod DP (eds) (1982) *Effects of Gaseous Air Pollution in Agriculture and Horticulture*. Butterworth Scientific, London, pp 113–138

Unsworth MH, Wilshaw JC (1989) Wet, occult, and dry deposition of pollutants on forests. *Agricultural and Forest Meteorology* 47:221–238

Voldner EC, Barrie LA, Sirois A (1986) A literature review of dry deposition of oxides of sulfur and nitrogen with emphasis on long-range transport modelling in North America. *Atmospheric Environment* 20:2101–2123

Wagner JK, Walters RA, Maiocco LJ, Neal DR (1986) *Development of the 1980 NAPAP Emissions Inventory*. GCA Corp., Report GCA-TR-86-12-G

Waldman JM, Munger JW, Jacob DJ, Hoffmann MR (1985) Chemical characterization of stratus cloudwater and its role as a vector for pollutant deposition in a Los Angeles pine forest. *Tellus* 37B:91–108

Walker HM (1985) Ten-year ozone trends in California and Texas. *Journal of the Air Pollution Control Association* 35(9):903–912

Warren SG, Hahn CJ, London J, Chervin RM, Jenne RL (1986) *Global Distribution of Total Cloud Cover and Cloud Type Amounts Over Land*. United States Department of Energy, DOE/ER/60085–H1, Available from NTIS

Warren WG, Böhm M, Link D (1992) A statistical methodology for exploring elevational differences in precipitation chemistry. *Atmospheric Environment* 26A:159–169

Weast RC, Astle MJ, Beyer WH (eds) (1987) CRC *Handbook of Chemistry and Physics*. CRC Press Inc, Boca Raton, FL

Westberg H, Sexton K, Flyckt D (1981) Hydrocarbon production and photochemical ozone formation in forest burn plumes. *Journal of the Air Pollution Control Association* 31(6):661–664

Whaley H, Lee GK (1977) Plume dispersion in a mountainous river valley during spring. *Journal of the Air Pollution Control Association* 27:1001–1005

Willson R, Shair F, Reynolds B, Greene W (1983) Characterization of the transport and dispersion of pollutants in a narrow mountain valley by means of an atmospheric tracer. *Atmospheric Environment* 17(9):1633–1647

Wiman BLB, Ågren GI (1985) Aerosol depletion and deposition in forests—a model analysis. *Atmospheric Environment* 19(2):335–347

Woodman JN (1987) Pollution induced injury to North American forests: Facts and suspicions. *Tree Physiology* 3:1–15

Young JR, Ellis EC, Hidy GM (1988) Deposition of air-borne acidifiers in the western environment. *Journal of Environmental Quality* 17(1):1–26

Yu CH, Pielke RA (1986) Mesoscale air quality under stagnant synoptic cold season conditions in the Lake Powell area. *Atmospheric Environment* 20(9):1751–1762

4

Sensitivity of Forest Soils in the Western U.S. to Acidic Deposition

D. Binkley

Introduction

Air pollutants, including acidic deposition, may affect forests through direct impacts on leaves, or through indirect effects on soil chemistry, microbiology, and tree roots. A great deal of interest and research has been focused on the effects of acidic deposition on forest soils; for reviews see Berdén et al. (1987) for Scandanavia, Schulze et al. (1989) for Germany, and Binkley et al. (1989a) for the southeastern United States. Much less attention has been focused on the response of forest soils to acidic deposition in the western United States, due to the lack of evidence for widespread forest decline and to the relatively low rates of acidic deposition in the West.

In this chapter, I examine the sensitivity of forest soils in four regions of the West to acidic deposition by synthesizing information on geology, soil types, and H^+ budgets. The assessment and conclusions are limited by the nature of the landscape and the information available on soils and forests in the West. Mountainous topography is associated with great variations in geology and soils over relatively short distances; generalized statements about regions may overlook substantial exceptions at local scales. Regional deposition estimates are available to represent the West (see Chapter 3), but the rates of H^+ generation (and consumption) from internal ecosystem processes have been characterized for only a few forests. Despite these limitations, the assessment provides a picture of the current state-of-knowledge, and perhaps more importantly it identifies the limits to current knowledge on the potential effects of acidic precipitation on forest soils in the West. This assessment is an expansion of the work by Binkley (1989).

Acidic Deposition and Forest Soils

Potential problems with acidified soils include impaired plant nutrition, toxicity of high concentrations of aluminum in soil solution, and acidification of aquatic ecosystems receiving water that drains from acidified soils. Some experts are skeptical about the potential of acidic deposition to acidify forest soils, because forest soils are often acidic under unpolluted conditions (Tabatabai 1985). However, many of the studies that have directly assessed long-term changes in forest soil chemistry have found rates of acidification that ranged as high as 1 pH unit within 50 years (Brand et al. 1986, Johnston et al. 1986, Nilsson 1986, Tamm and Hällbacken 1986, Falkengren-Grerup 1987, Johnson et al. 1987, Anderson 1987, Binkley et al. 1989b, Van Miegroet et al. 1989, Binkley and Valentine 1991, Binkley and Sollins 1990). Most of these studies attributed the soil acidification to the effects of species or stand development over time, without a direct assessment of the role of acidic deposition. The implication is that natural forest processes, such as production of carbonic acid from root respiration and the accumulation of nutrient cations in biomass, acidify some forest soils on a time scale of decades. If rates of acidic deposition are of a similar or greater magnitude, then deposition should be expected to accelerate acidification of sensitive soils. No long-term data have been published for the chemistry of forest soils in the western United States.

Soil Acidification

Soil acidity is most commonly characterized by measurements of pH, but this is only one aspect of soil acidity (see Binkley et al. 1989a,b for more details). The concentration of H^+ maintained in solution depends on the equilibrium between the solution and the exchange complex of the soil. For many forest soils, the quantity of H^+ in soil solution at a single time is less than 1 $kmol_c$/ha, whereas the exchange complex may contain much more than 1000 $kmol_c$/ha (Binkley and Richter 1987). Changes in soil pH derive from changes in the exchange complex that influence the equilibrium with the solution. Four types of changes in the soil solution and exchange complex can alter the pH of soil solution (Table 4.1). An increase in the total quantity of anions present in the soil solution "allows" more cations to be present in the solution, and in most soils some of these extra cations will be H^+ (see Richter et al. 1988). A moderate concentration of anions might be about 250 μmol_c/L; a decrease down to 25 μmol_c/L would lower the concentration of all cations, and might raise pH (= lower concentrations of H^+) by 0.3 units (Table 4.2). The second factor is the quantity of exchangeable acidity (i.e., H^+ and Al associated with organic acids and hydrated aluminum ions). Decreasing the exchangeable acidity from 100 mmol$_c$/kg to 50 mmol$_c$/kg might allow pH to rise by about 0.1 units (Table 4.2). Third, the proportion of the exchange complex occupied by acidic cations is also important. If the quantity of acidic cations were held constant at 100 mmol$_c$/kg, but the size of (= acid saturation, or the converse of base saturation) the exchange complex were reduced from 200 mmol$_c$/kg to 125 mmol$_c$/kg, the drop in pH might be about 0.3 units due to the increase in the proportion of sites occupied by acidic cations. The last factor is the tendency of the exchange complex to retain H^+ (and aluminum) relative to other cations. This tendency is the acid strength of the exchange complex (= pK in simple chemical solutions). An increase in acid strength, represented by

Table 4.1 The pH of soil solutions may differ over time or between sites due to differences in four factors that determine pH.

Soil Solution

1. Quantity of anions in solution

Soil Exchange Complex

2. Quantity of exchangable H^+ and Al

3. Proportion of acid cations on exchange complex

4. Acid strength of exchange complex

Table 4.2 Representative change in the pH of a solution from a typical soil if single factors were changed.

Soil Solution pH = 5.0	Expected pH if:
1. 250 µmol$_c$/L of anions	1. 25 µmol$_c$/L of anions, pH = 5.3
Soil Exchange Complex	
2. 100 mmol$_c$/kg of exchangeable H^+ and Al	2. 50 mmol$_c$/kg of exchangeable H^+ and Al, pH = 5.1
3. Exchange complex = 200 mmol$_c$/kg	3. Exchange complex= 125 mmol$_c$/kg pH = 4.7
4. Acid strength pK = 5.2	4. Acid strength pK = 4.2, pH = 4.2

a 1 unit decline in pK of the exchange complex, would lower pH by 1 unit (Table 4.2). In poorly aerated soils, changes in redox potential can also drive short-term, readily reversible changes in soil pH.

Of the four factors that determine soil solution pH, acidic deposition is generally viewed to have a potential impact on the acid saturation of the exchange complex, and on the ionic strength of the solution passing through the soil. The acidification of soils depends both on the quantity of H^+ deposited from the atmosphere (and other sources of H^+), and the processes that may neutralize the H^+. In forest soils, the major neutralizing mechanisms are cation exchange reactions, mineral weathering, biologic assimilation of nitrate and sulfate, and sulfate adsorption.

The adsorption of sulfate by iron and aluminum oxides can displace OH^-, which then consumes H^+ to form water and buffer the soil against acidification. Later desorption may regenerate the acidity. For these reasons, the geochemical adsorption of sulfate (and the reversibility of the reaction) are important components of soil acidification. Sulfate immobilization by microbes also involves consumption of acidity. Unfortunately, both of these processes are very difficult to assess under ambient conditions, and interpretations relative to changes in rates of sulfate deposition are problematic. The quantity of information on sulfate adsorption in Western conifer forests is sparse. Harrison et al. (1989) present data on sulfate adsorption for the three Washington sites involved in the Integrated Forest Study of the Electric Power Research Institute and Oak Ridge National Laboratory. All sites adsorbed added sulfate from solution, and part of the adsorbed sulfate entered pools that were not readily extractable with water, implying that the adsorption was at least partially irreversible. The Emerald Lakes study site in the Sierra Nevada also appeared to have some sulfate adsorption capability (Lund et al. 1987). J. Clayton (USDA Forest Service, personal communication) found that most soils examined in the Wind River mountains in Wyoming adsorbed sulfate, with maximum adsorptions of < 1.5 mmol/ kg. Too little information is available to allow regional inferences.

Forest Sensitivity to Soil Acidification

In general, trees are not very sensitive to the concentrations of H^+ in solution. The relationships between tree health and soil pH derive from other soil chemical processes that relate to soil pH.

Aluminum toxicity to trees

Many tree species (especially hardwoods) are sensitive to the concentrations of free Al^{3+} in the soil solution (cf. Hutchinson et al. 1986, Raynal et al. 1990); the concentration of Al^{3+} in soil solution may increase by 1000 fold as pH declines from 5.0 to 4.0 (Bohn et al. 1985, Wolt 1987, Binkley et al. 1989a). Whether such a drastic increase in Al^{3+} actually occurs during soil acidification may be largely determined by the nature of soluble organic compounds. In many soils, organic compounds chelate (bind) the Al^{3+}, greatly reducing potential toxicity to trees.

No cases of aluminum toxicity to trees have been demonstrated under normal conditions in forests in the West, but this might reflect an absence of sufficient information rather than an absence of a problem. Although a variety of excellent seedling studies have examined threshold levels of aluminum concentrations that affect seedling development for species from the eastern United States (Hutchinson et al. 1986, Thornton et al. 1986a,b, Kelly et al. 1987, Raynal et al. 1990), similar studies are lacking for the West. Some research from Europe has suggested that the critical level of Ca:Al (molar ratio) is about 1:1 (cf. Rost-Siebert 1983); I know of no soil solution data for the West that approach this level.

Forest nutrition

Acidic deposition interacts with forest nutrition in four major ways. First, nitrogen and sulfur are essential plant nutrients, and where the supply rates of these nutrients limits forest growth, the deposition of nitric and sulfuric acids would probably stimulate growth. Indeed, assimilation of nitrate or sulfate involves neutralization of H^+ (Binkley and Richter 1987). Second, the increased ionic strength of soil solutions could lead to accelerated leaching losses of essential base cations. Third, soil pH affects phosphorus availability. Finally, elevated solution concentrations of aluminum could be toxic to roots or microbes, or could interfere with the uptake and use of calcium and magnesium.

The current availability of nitrogen probably limits the growth of most forests in the West (Powers 1983, Miller et al. 1986, Weetman 1988, Binkley et al. 1990), and increases in the rate of nitric acid deposition would probably increase growth. Sulfur limitations are less common, but do appear important in regulating the growth response of Douglas-fir to nitrogen fertilization in some cases (Blake et al. 1988).

As noted above, a major mechanism by which acidic deposition may acidify soils is through leaching of base cations, leading to lower base saturation of the exchange complex. An exchange complex that is low in base saturation is dominated by acid cations (Al and H^+), and an acid-dominated exchange complex maintains a low pH in soil solution. For these reasons, soil acidification may be associated with a declining supply of nutrient cations such as calcium, magnesium, and potassium. Johnson et al. (1985) estimated that high rates of deposition of sulfate in Tennessee have increased leaching losses of base cations by two-to-three fold. Some declining stands of Norway spruce (*Picea abies* L.) show strong magnesium deficiencies (Zöttl and Hüttl 1986, Oren et al. 1988, Schulze 1989, Schulze et al. 1989), but the relative importance of acidic deposition and other ecosystem factors (such as management history and parent materials) are unknown.

Little is known of the degree to which the availability of nutrient cations limits forest growth in the West. At least some lodgepole pine forests appear to be limited by K^+ availability (Weetman 1988), and presumably these forests might be at risk from any reduction in K^+ supply resulting from acidic deposition. In the cases where the exchangeable pools of calcium, magnesium and potassium have been reduced in the presence of alders, there is no indication (based on foliage analysis) that the supply rates of these nutrients have become limiting.

Soil pH affects the equilibrium of phosphate compounds in soil solutions. A decline in pH from 5.0 to 4.0 might lower the concentration of P in soil solution by 10 fold (Lindsay and Vlek 1977). Other soil processes, such as the production of organic enzymes and chelates that enhance the availability of P, may offset or exacerbate the reductions in P solubility with acidification. Little work has been done to examine changes in P availability in response to acidification of forest soils.

Another potential effect of soil acidification on forest nutrition involves indirect effects on the microbial populations responsible for nutrient cycling. In a review, however, Myrold (1987) concluded that although soil acidification may alter microbial populations and activities, overall effects on microbial activity in nutrient cycling should be minor except at very high (>>1 kmol H^+/ha annually) deposition rates.

Forest Soils of the Western United States

The forest soils of the western United States have developed on a wide range of parent materials under a wide range of influences of topography and biota over time spans ranging from decades to millenia. Few generalizations apply across the entire West, although many forest soils in the region share in common cold winters, and many experience periods of summer drought. Large parts of the region have very steep slopes with shallow, poorly developed soils. The range of forest soil types within each region of the West is as large as the range from one region to another. Each region has a wide range of variation in topography, parent material and climate (which is strongly related to elevation), and these features overide any generalities that might be expected within regions. The best way to generalize about the types of forest soils found in the West is to discuss relationships with topography, parent material and climate.

Topographic variations in the West influence soils in two major ways. Steep topography tends to be associated with relatively young, poorly developed soil types because erosional processes are fast enough to keep up with soil development. Topography is also important because of effects on soil moisture and parent material. Soils formed at the bottom of slopes tend to have greater moisture than mid- and upper-slope soils (unless the range in elevation is large enough to span substantial differences in precipitation). Greater water availability promotes the soil-forming activities of plants and microbes. Colluvial inputs of minerals from upslope can also be important. Soils forming on the shoulders of slopes tend to be drier and shallow, with more poorly developed horizons.

Soils on steep slopes tend to be Inceptisols and Entisols, which are two orders of soil classification for soils demonstrating very little profile development. Depending on parent material, these types of soils may be very resistant to acidification (especially for carbonate rocks), or relatively susceptible (especially for granitic rocks). Soil development is more advanced at the bottom of many slopes, ranging from Histosols (organic soils) in wetlands in valley bottoms, to Alfisols (soils high in base saturation, with high clay content in the B horizon) and occasionally Spodosols (typically in wet, cool environments in the West, characterized by an accumulation of iron and aluminum oxides in the B horizon) or Mollisols (in drier environments on base-rich soils) in the better-drained toeslopes.

Parent material is especially important in determining rates of mineral weathering, which in turn affect sensitivity of soils to acidification. Rapidly weathering minerals include carbonates, such as limestone and dolomite, and fine-grained minerals rich in calcium and magnesium, such as basalts. Weathering is much slower for coarse-grained minerals that are rich in aluminum and silica, such as granites. Sandstones, which are comprised mostly of silica sand, weather relatively rapidly but the lack of consumption of H^+ during weathering affords little resistance to soil acidification.

Climate affects soil development in a variety of physical, chemical and biological ways. Erosion depends largely on water movement, and in high rainfall environments, parent materials may be moved physically across a landscape (downhill, of course). Cycles of freezing and thawing, wetting and drying, can have substantial effects on the breakdown of rocks and minerals. Soil moisture affects almost all aspects of soil development: leaching of the byproducts of weathering, translocation of clay micelles from upper soil layers to lower depths. Soils in wet, leaching environments generally acidify more quickly than soils in drier environments. Finally, climate strongly affects the activity of plants (responsible for generating organic matter) and microbes (responsible for processing organic matter), and soils under different vegetation typically differ in major ways. The accumulation of soil organic matter can lower soil pH, because soil organic matter is comprised of complex organic acids. At the same time, soil organic matter tends to buffer soil pH, providing resistance to acidification from other sources.

Around the West, combinations of geology, climate, and vegetation combine to produce general patterns of distribution of soil types in the regions examined by the case studies of forest condition later in the book (Figure 4.1). Typical soils in the Puget Sound region developed since the last glaciation (10,000 to 15,000 years ago), and fall within the Inceptisol order. The Cascade Mountains are dominated by Inceptisols, and on older parts of the landscape, by Ultisols (old, acid soils low in base saturation and high in clay content of the B horizon). The Front Range of Colorado is characterized by Alfisols on stable parts of the landscape, and by Inceptisols on steeper slopes. The drier environments in southern Arizona have led to the development of Mollisols in most forests, although Alfisols are also common. In southern California, the complex geology limits generalizations about soil types, and Alfisols, Mollisols, Inceptisols and Entisols are all common.

No map of soil sensitivity to acidification has been developed for the West, but a map of lake sensitivity (based on mean annual total alkalinity) provides some insights (Figure 4.2). Alkalinity represents the ability of a solution to buffer inputs of acidity. Lakes with less than 50 $\mu mol_c/L$ of alkalinity may be very sensitive to acidification, whereas lakes with more than 200 $\mu mol_c/L$ of alkalinity should be very resistant to acidification. Much of the alkalinity in lakes derives from the biogeochemical processes operating in the surrounding terrestrial ecosystems, so low alkalinity lakes tend to occur in watersheds where soil processes provide little buffer capacity to acidification. As a general trend, lakes that are sensitive to acidification tend to occupy watersheds that may also be sensitive to acidification.

Across the West, low alkalinity (= high sensitivity to acidification) tends to be associated with high elevation sites, which also correspond largely to glaciated sites (Omernik and Griffith 1986). In Colorado, most of the lakes in the igneous intrusive and metamorphic Front Range and Park Range have alkalinities below 100 $\mu mol_c/L$, with many below 50 $\mu mol_c/L$. The mountains dominated by volcanic parent materials (such as the

San Juans) generally show much greater alkalinities. Low alkalinities also characterize many of the lakes in the Wind River, Big Horn, and Absaroka Mountains in Wyoming, as well as the Uinta Mountains in Utah and the Bitteroot Mountains in Montana and Wyoming. Low alkalinity lakes are common across much of the Cascade Mountains in the Pacific Northwest. The greatest proportion of low alkalinity lakes in the West occurs in the Sierra Nevada, where many lakes have less than 50 $\mu mol_c/L$ of alkalinity.

Pacific Northwest

The geology of western Oregon and Washington involves complex interactions of folding, volcanism, erosion, and deposition (McKee 1972, Heilman et al. 1979). The Coast Range along the Pacific Ocean is comprised of marine and estuarine sedimentary rocks uplifted by folding processes, with strong influence by volanism in some portions. The Willamette Valley to the east of the Coast Range is filled with alluvium of volcanic and sedimentary rocks. A series of glaciations in the Puget Sound region left behind a complex stratigraphy of deposits, derived largely from volcanic and granitic rocks. Farther to the east, the Cascade Mountains developed from repeated periods of deposition of volcanic strata, intrusion of granitic magmas, and uplifts followed by extensive erosion. Major portions of the Cascade region are covered by basalt flows (about 10 million years old) and by wind-blown deposits of ash (within the past 100,000 years).

Forest soils within western Oregon and Washington fall into several Orders (Mitchell 1979). Entisols are found on recent alluvial sites, primarily floodplains, landslides, and mudflow deposits. Inceptisols are extensive in the region, occuring on young or unstable landscapes on readily weathered parent materials, and on older landscapes with more resistant parent materials. Alfisols occur on very old surfaces, usually where precipitation has been too low to allow further leaching and acidification to Ultisols. Ultisols are found on very old surfaces along the Coast, where high precipitation has led to extensive weathering and leaching.

Long-term soil acidification has been examined in three chronosequences in the Pacific Northwest. At the southern extreme of the region, Dickson and Crocker (1954) examined the development of soils on mud flows of various ages around Mt. Shasta. They found essentially no sign of soil acidification from young deposits (27 years) to old deposits (estimated > 1,200 years); in fact, slight alkalization (rise in pH of 0.1 to 0.2 units) was apparent. At Mount Rainier in western Washington, Bollen et al. (1967) found that soil pH declined from about 7.2 on new avalanche materials

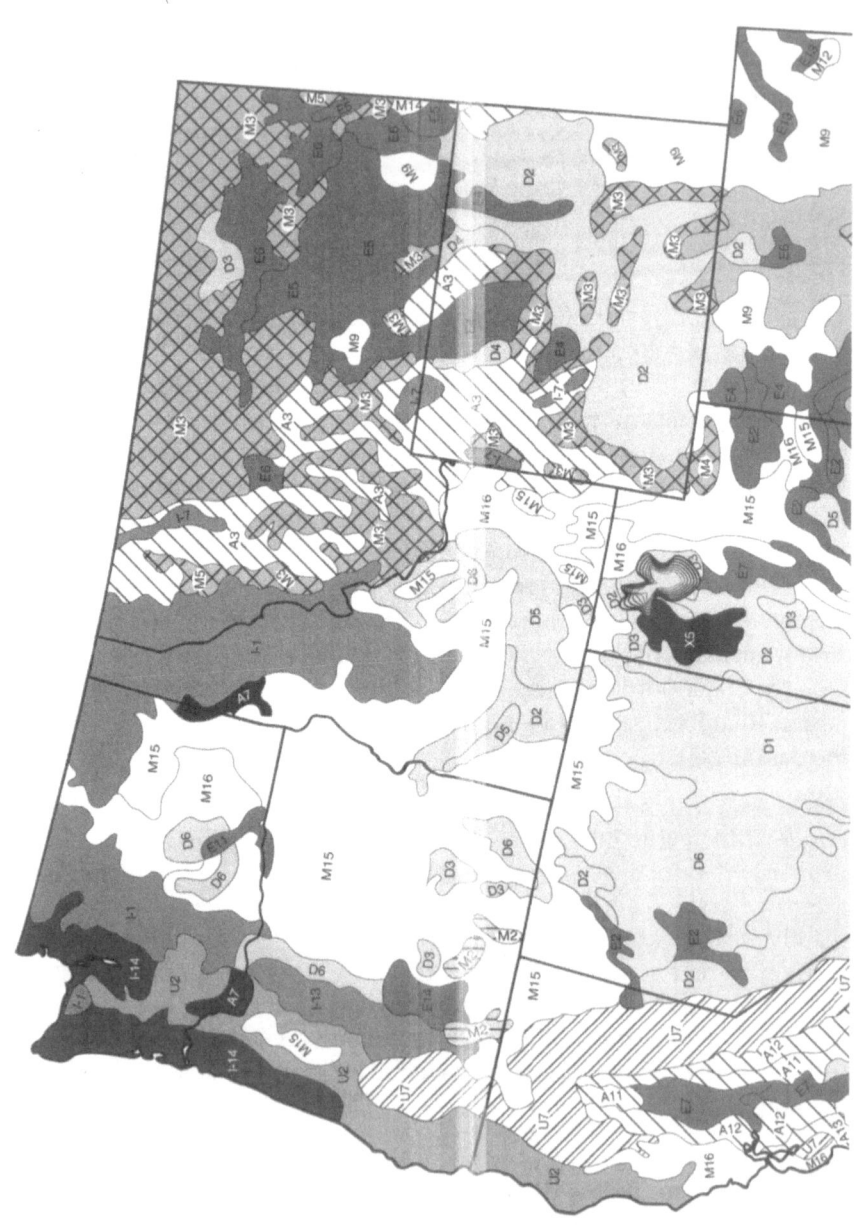

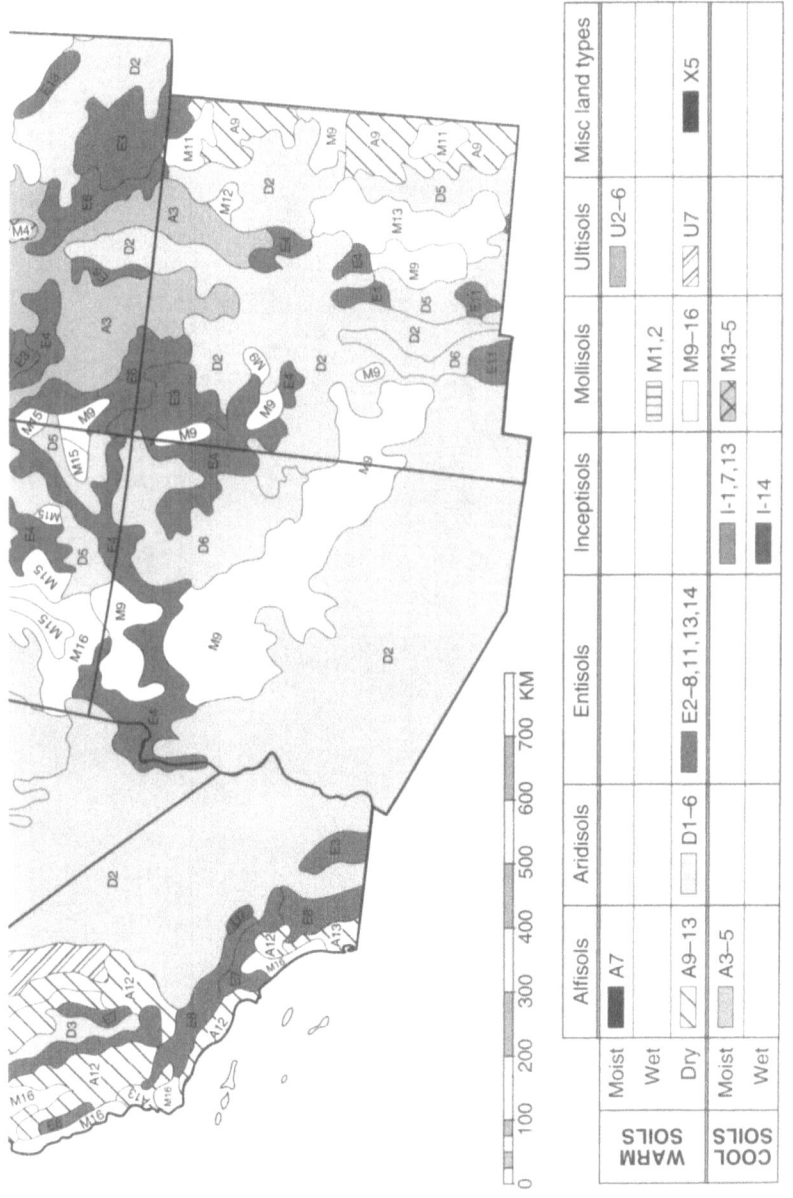

Figure 4.1 Generalized soils map of the western United States, from the National Atlas (USGS 1970).

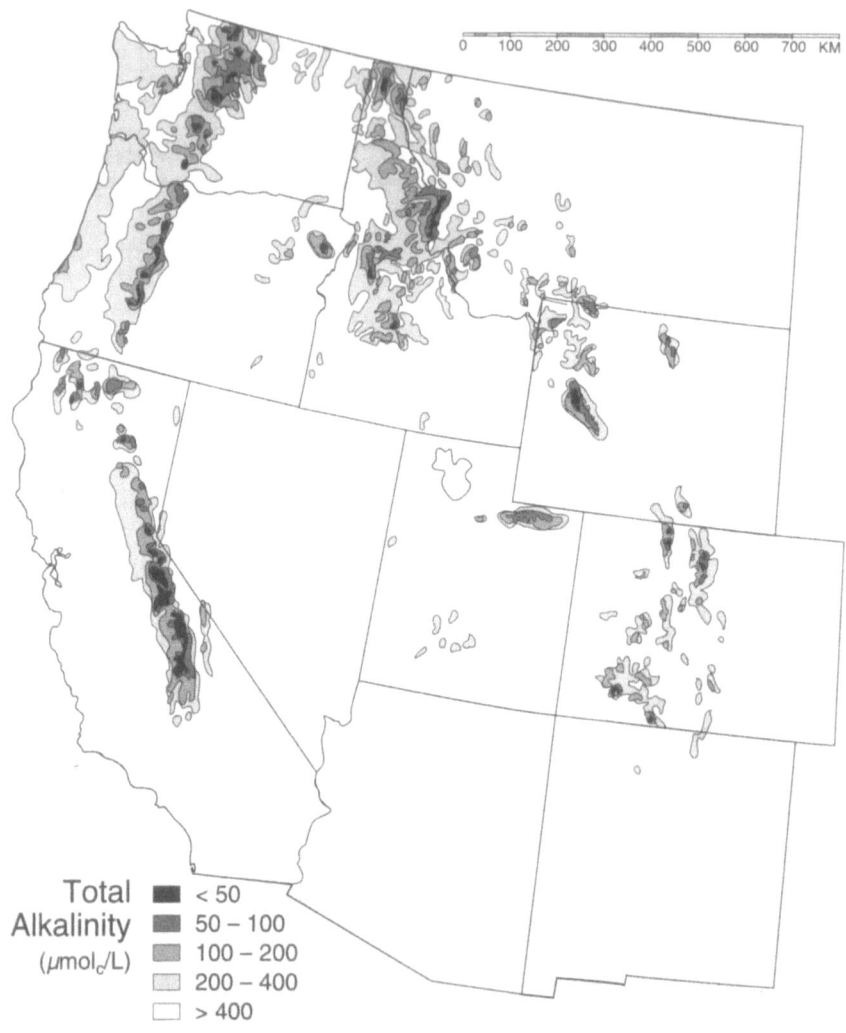

Figure 4.2 *Alkalinity map of lakes in the western United States. Values below
50 μmol$_c$/L may indicate watersheds that are poorly buffered relative to inputs of
acids (from Omernik and Griffith 1986).*

to 5.7 on similar materials that had been in place for 66 years. At the northern end of the region, Singleton and Lavkulich (1987) examined soil genesis in a chronosequence of soils on an aggrading beach on the west coast of Vancouver Island. Within 350 years, the soil profile had developed into a Typic Haplorthod. By 550 years, the upper soil pH dropped from about 5.8 to 4.5–5.0, soil organic matter had increased from near 0 to 2.0%, and oxalate-extractable aluminum had increased from 0.1% to 0.3%. Exchangeable base cations (alkali+alkaline earth cations) showed no major changes, so the decline in pH probably resulted from the accumulation of acidic organic matter and exchangeable aluminum.

The sensitivity of soils in the Pacific Northwest has also received little attention on shorter timescales of stand development and secondary succession. The best information related to soil acidification comes from ecosystems where nitrogen-fixing alders greatly accelerated ecosystem nitrogen cycling and nitrate leaching (summarized by Van Miegroet et al. 1989). These studies indicate that when acidification occurs under the influence of alder, it results in part from the input of H^+ from nitrification (coupled with nitrate leaching) of perhaps 1 to 3 kmol H^+/ha annually, and from the accumulation of strongly acidic organic matter. Where acidification has not occurred with alders, there seems to have been minimal nitrate production and leaching. In combination, these studies indicate that some soils in the Pacific Northwest would be acidified (with decreases in soil pH, exchangeable base cations, and solution alkalinity, coupled with increases in aluminum concentrations in soil solutions) if the rates of acidic deposition resulted in net inputs of H^+ on the order of 1 to 3 kmol H^+/ha annually. The current rate of acidic deposition in the Pacific Northwest is only about 0.05 to 0.30 kmol H^+/ha annually (Chapter 3), so the present effects of acidic deposition on soil chemistry should be much less pronounced than the effects of alder.

Detailed H^+ budget information is available for five conifer forests in Oregon and Washington, and three of these were paired with adjacent alder or alder/conifer ecosystems. In a 450-year-old Douglas-fir forest in the H.J. Andrews Experimental Forest in the Cascade Mountains of Oregon, precipitation contributed only about 0.1 kmol H^+/ha annually (Sollins et al. 1980). The accumulation of nutrient cations in biomass generated about 0.6 kmol H^+/ha, and the production and dissociation of carbonic acid in the soil generated a remarkable 5.4 kmol H^+/ha annually. Assuming no net ecosystem acidification, they calculated that the rate of H^+ consumption in weathering in this Typic Dystrochrept was about 7.1 $kmol_c$/ha annually. This rate is very high, and may indicate that some acidification of the exchange complex was indeed occurring.

Near the coast of Oregon, Binkley and Sollins (1990) examined components of H^+ budgets for adjacent 55-year-old mixed conifer and alder/ conifer ecosystems on a Typic Dystrandept. Precipitation contributed less than 0.1 kmol H^+/ha annually, compared with a rate of cation accumulation in biomass of about 0.7 $kmol_c$/ha annually in both stands. The formation and dissociation of carbonic acid contributed about 0.9 $kmol_c$/ha of H^+ in both ecosystems, whereas the high rate of nitrate leaching contributed an additional 2.0 $kmol_c$/ha of H^+ to the alder/ conifer ecosystem. Binkley and Sollins (1990) also examined components of H^+ budgets for a pair of 55-year-old Douglas-fir and alder/Douglas-fir ecosystems on a poorer soil (Andic Haplumbrept). Precipitation again contributed less than 0.1 kmol/ha of H^+, and cation accumulation in biomass (above ground biomass plus forest floor) contributed 0.3 and 0.6 $kmol_c$/ha annually in the Douglas-fir and the mixed ecosystems, respectively. Nitrate leaching was negligible in both ecosystems, but the formation and dissociation of carbonic acid was greater in the mixed ecosystem (1.0 $kmol_c$/ha annually) than in the Douglas-fir ecosystem (0.4 $kmol_c$/ha).

D. Cole and associates at the University of Washington developed the information necessary for calculating H^+ budgets for three forests as part of the Integrated Forest Study of the Electric Power Research Institute and Oak Ridge National Laboratory. Deposition of H^+ at these sites is relatively low, about 0.3 kmol/ha annually (Binkley 1992). The accumulation of cations in biomass in a Douglas-fir forest and a silver fir (*Abies amabilis* Dougl. ex Forbes) forest generated about 0.7 $kmol_c$/ha of H^+, compared with 1.9 $kmol_c$/ha for a red alder forest. Net losses of nitrate from the alder forest accounted for an additional load of 2.5 $kmol_c$/ha of H^+. The net loading of H^+ experienced from all sources in the Douglas-fir and silver fir forests was about 1.0 kmol/ha, contrasted with 4.6 $kmol_c$/ ha for the red alder forest. Reuss (1989) examined seasonal variations in the concentrations of major ions in soil solutions under red alder, and found increases in cation concentrations as nitrate concentrations increased. He found that patterns among the cations followed predicted patterns (based on valence), with concentrations of trivalent aluminum increasing as the 3/2 power of the concentrations of divalent calcium and magnesium.

These H^+ budgets for Pacific Northwest forests indicate that within-ecosystem biogeochemical processes contribute greater inputs of H^+ than are received from atmospheric deposition. The acidification that occurred under the influence of alder appears to have resulted in large part from the strongly acidic nature of organic matter accumulated under alder, and from the 1 to 3 $kmol_c$/ha annual input of H^+ from nitrification coupled with nitrate leaching. These studies, along with others discussed by Van Miegroet et al. (1989) indicate that the biogeochemical effects of alders can lead to substantial changes in soil chemistry. If rates of acidic deposition in this region rose to the level found in parts of the Eastern United States and Europe, similar changes in soil chemistry might occur in the absence of alder.

Central Rocky Mountains

Forests cover most of the Front Range of Colorado between the elevations of 2,000 to 3500 m, on parent materials of granite, gneiss and schist (Chronic 1980). Soils on gentle slopes at lower elevations with ponderosa pine (*Pinus ponderosa*) tend to be Mollisols, grading into Alfisols on similar sites with aspen (*Populus tremuloides*) and lodgepole pine (*Pinus contorta*) at higher elevations. In the cool environment of forests of spruce (*Picea engelmannii* and *P. pungens*) and fir (*Abies lasiocarpa*), soils may be Mollisols, Alfisols, or Spodosols. Steeper slopes are typically covered by Inceptisols and Entisols, or simply by unforested talus slopes at higher elevations.

The only intensively characterized forest ecosystem in the Front Range within Colorado is the Loch Vale Watershed in Rocky Mountain National Park. The watershed ranges from 3100 m to 4000 m in elevation, with bedrock composed of biotite gneiss and granite (Walthall 1985, Mast 1989). Soils cover only about 6% of the watershed, and the forested (spruce and fir) soils are predominantly Cryoboralfs with very low pH (< 4) and low base saturation (< 35%) in surface soils. Soil solutions from the forested part of the watershed average about pH 5.0, with alkalinity between 50 and 100 μmol_c/L (Arthur 1990). Mast (1989) estimated the rate of denudation of cations from the watershed to be about 0.6 $kmol_c$/ha, which should be a good estimate of the rate of consumption of H^+ in weathering. About 40% of the weathering release of cations comes from traces of calcite (in hydrothermally altered fractions of the bedrock). Arthur (1990) compiled H^+ budgets for portions of the Loch Vale Watershed. In the forested part of the watershed, the total sources of H^+ summed to about 0.34 $kmol_c$/ha annually, and H^+ sinks summed to about 0.32 $kmol_c$/ha annually. Although the consumption of H^+ in mineral weathering was low, it appeared sufficient to buffer soil acidity at its current level.

The Medicine Bow Mountains lie directly north of the Front Range across the border in Wyoming. Soils in this area include Lithic and Typic Cryochrepts, and Typic Cryoboralfs, developed in glacial till dominated by quartzite and fluvial conglomerate. A variety of lodgepole pine forests have been intensively studied by researchers at the University of Wyoming and Colorado State University (Fahey 1983, Fahey et al. 1985, Knight and Fahey 1985, Fahey and Knight 1986, Pearson et al. 1987,

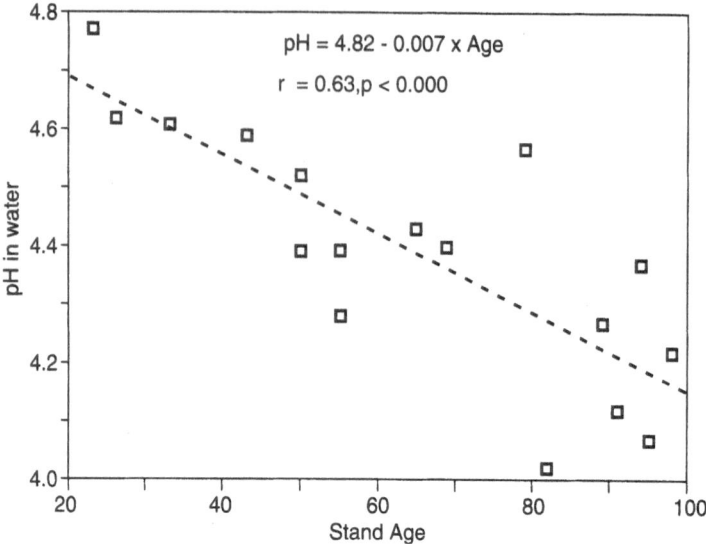

Figure 4.3 Soil pH decline (0–0.15 m depth) with stand age in natural stands of lodgepole pine in the Medicine Bow Mountains. The rate of decline averaged 0.7 units over 100 yr in this low pollution environment (Binkley, unpublished data).

Fahey and Yavitt 1988). The soils tend to be moderately to strongly acidic (pH 4.0 to 6.0) with moderate concentrations of base cations (50 to 80 $mmol_c$/kg). Soil pH relates strongly to stand age (Figure 4.3), probably resulting from the accumulation of base cations in biomass (Binkley unpublished data). Soil solutions average about pH 5.7 to 6.3, with alkalinities greater than 100 μmol_c/L. Concentrations of all ions in soil solutions show high values with the onset of snowmelt in spring, dropping by up to an order of magnitude into the summer. Knight et al. (1988) artificially acidified the snowpack in a lodgepole pine forest (by 10-fold and 50-fold); they concluded that the lack of substantial changes in soil solution chemistry indicated strong buffering capacity in the soil. Drever et al. (1987) examined the response of soils from three lodgepole pine forests to additions of sulfuric acid in the laboratory. The quantities added were less than 20% of the exchangeable base cations; H^+ was quickly removed from solution, and replaced primarily by calcium, with no dissolution of aluminum evident.

At a higher elevation in the Medicine Bow Mountains, the USDA Forest Service is conducting an intensive study on acidification of an alpine watershed. The Glacier Lakes Experimental Watershed ranges from 3300 m to 3660 m in a glaciated basin with fractured quartzite bedrock. Forests occupy a small portion of the watershed, and are dominated by Engelmann spruce and subalpine fir. The forested soils are Cryoboralfs, with pH < 5 and base saturation of < 30% in surface horizons (D. Fox, R. Hopper, and P. Walthall, personal communication). Soil solutions in the forested part of the watershed tend to have pH near 6.0, and alkalinity > 100 $\mu mol_c/L$ (F. Vertucci, personal communication). The rate of mineral weathering was estimated to be about 0.4 $kmol_c/ha$, based on the rate of cation denudation (Rochette et al. 1988). If the pH of precipitation dropped to 4.3, Rochette et al. (1988) estimated that West Glacier Lake would become permanently acidic after about 11 years, due to exhaustion of the soil's buffer capacity.

Farther west in the Wind River Mountains, Vertucci (1990) resampled 27 lakes that had been sampled in 1935. After 52 years, no evidence of lake acidification was apparent, which suggests no major acidification of soils in the watersheds. J. Clayton (USDA Forest Service, personal communication) examined the sensitivity of Wind River soils to acidification by leaching columns of soils with water and with acidic solutions. He found that weathering of abundant primary minerals was able to resupply the base cations leached by the acidic solutions, with little change in soil chemistry.

Litaor (1987) pointed out that many alpine soils in this region receive deposition of eolian dust that contains a significant quantity of alkalinity. Even trace amounts of calcium carbonate or other alkaline compounds could play large roles in mediating the effects of current rates of acidic deposition. However, quantitative estimates of the amount of alkalinity deposited in such dust are lacking.

Soils in the central Rocky Mountains do not appear to be at immediate risk of acidification under present levels of atmospheric deposition, but many soils in the region would likely be sensitive if deposition rates approached current levels for the East.

Central and Southern Arizona

The geology of central and southern Arizona has developed through extensive periods of erosion, folding, metamorphism, volcanism, and faulting. In the central highland region, stretching from Safford to Springerville across to Prescott (Chronic 1983), the parent materials for soil development in high-elevation areas include granite, gneiss, welded tuffs, basaltic lava, and sedimentary limestones. Where slopes are moderate (and especially in toe-slope positions), soil development has progressed to the stage of Mollisols or Alfisols (Figure 4.1). Steeper slopes generally have poorly developed Entisols and Inceptisols. Soil development is generally greater in more-weatherable parent materials, with limestone being most easily weathered, and granite and gneiss being least weatherable. Forests are restricted to high elevations. Soils within the ponderosa pine region are often moderately well developed, falling into the Mollisol or Alfisol orders. Higher elevation forests on steeper slopes are typically found on more poorly developed soils such as Entisols and Inceptisols.

The Basin and Range Province in southern Arizona is dominated by metamorphic core complex ranges. These ranges developed from intrusion of granitic magmas pushing into metamorphic layers. Volcanism has also been important, with some mountains veneered by basalt lavas or cemented tuffs. Soil development follows the same patterns as in the central highlands region.

For central and southern Arizona, I found no published H^+ budgets. I expect that soils in this region should be very resistant to acidification by acidic deposition, due to low precipitation, high potential evapotranspiration, and likely limitations of nitrogen supply on forest growth. Low precipitation combined with high potential evapotranspiration results in relatively small quanities of water leaching beyond the rooting zone in forests in this area. Low rates of leaching minimize acidification from formation of carbonic acid, allowing soils to maintain moderate to high levels of exchangeable base cations and buffer soil pH. Little work has been done on nutrient limitations in forests in this region, but nitrogen limitation is probably widespread (W. Covington, personal communication). Little work has been done on nitrate leaching in the region, but intact forests in New Mexico showed only trace concentrations of nitrate in soil solutions (Vitousek et al. 1979). Healthy forests are likely to retain any deposited nitric acid, effectively neutralizing the acidity. Studies on soil disturbance, however, demonstrate substantial potential for leaching losses of nitrate (Vitousek et al. 1979), so nitric acid deposited on disturbed sites may not be retained efficiently and could lead to accelerated leaching and acidification.

California

The Sierra Nevada are composed of granite (and granite-like) bedrock, and the soils in the region often developed in glacial tills, colluvium, and alluvium. The majority of forest soils are Inceptisols, with significant areas of Entisols and Spodosols at higher elevations (Weintraub 1986, Huntington and Akeson 1987).

The San Gabriel and San Bernardino Mountains in southern California are part of the Transverse Ranges that trend west-to-east around Los Angeles. These mountains are comprised of igneous and metamorphic rocks, with thick accumulations of sediment in the valleys (Sharp 1972). Soils in these mountains tend to be steep (half over 50% slope) and shallow (< 50 cm to fractured bedrock; USDA Forest Service no date). Most forested soils are Entisols (Xerothents and Xeropsamments on slopes, Xerofluvents on alluvial materials) or Alfisols (mostly Haploxeralfs).

A wide range of studies of soil sensitivity to acidification has been conducted in California, funded primarily by the California Air Resources Board, the electric utilities, and by the National Park Service. Several of these have included alpine systems as well as forests.

In one of the first studies, McColl (1981) performed leaching experiments on 17 forested soils primarily from granitic parent materials in Shasta, Fresno, and Siskiyou counties. Surface soils from each pedon were leached for two hours with solutions ranging in pH from 5.5 to 2.0 in a mechanical vacuum extractor. The leachates were collected and analyzed for a variety of properties. He found that ability of the soils to buffer added H^+ related well to the concentration of base cations on the exchange complex. Soils with less than about 70 $mmol_c$/kg of base cations appeared to be very sensitive to acidification. In general, the sensitivity of soils tended to increase with elevation.

Weintraub (1986) assessed the sensitivity of two alpine watersheds (Eastern Brook Lake and Emerald Lake) in the Sierra Nevada to acidification. Her approach involved mineralogic assays, estimates of weathering rates based on changes in mineralogic composition, and laboratory acidification experiments. She concluded that these watersheds would be very sensitive to acidification, due to the shallow depth of soils (where present) and low weathering rates. Assuming 10,000 years of weathering in the present landscapes, the estimates of weathering rates represented an annual consumption of only about 0.1 kmol H^+/ha, which is probably similar to the current rates of H^+ deposition in the Sierra. The response of soil columns to acid additions in the laboratory demonstrated that the rate of weathering might increase if H^+ loading increased, but this increase was judged to be small relative to observations from similar experiments on other soils. In another study of Emerald Lake, Lund et al. (1987) showed that the intermittent, shallow soils of this watershed are very low in exchangeable base cations (< 15 $mmol_c$/kg).

Wyels (1986) developed titration curves for 24 Sierran soils, and used these in conjunction with a soil sensitivity scheme to assign the soils to three sensitivity classes. Six categories of properties were: current pH, saturation of the exchange complex with base cations, cation exchange capacity, organic carbon, soil depth, and parent material. Some properties, such as pH, were weighted more heavily in her rating system than others, such as soil depth. This scheme led to classifying six soils as very sensitive to acidic deposition, 15 soils as sensitive, and three soils as unsensitive. Of the soils she rated as very sensitive, three had less than 70 mmol$_c$/kg of base cations.

Huntington and Akeson (1987) developed a soil resource inventory for the central part of Sequoia National Park. Although no estimates of sensitivity to acidification were made, chemical analyses of some representative pedons provided data on pH and exchangeable cations. The variation among soils within the same family was great, prompting caution in basing predictions of soil sensitivity on taxonomic classes. For example, the top 25 cm of one coarse-loamy, mixed, mesic Xeric Haplohumult had over 150 mmol$_c$/kg of base cations and a pH of 6.7, whereas another of the same family with a pH of 6.1 had less than 70 mmol$_c$/kg of these cations. No trend in the concentration of base cations was apparent across soil orders (Entisols, Inceptisols, Alfisols, and Mollisols); only soils with a cryic temperature regime (high elevation) had consistently low concentrations of these exchangeable cations.

The most thorough assessment of soil sensitivity to acidification in California comes from the work of Reilly and Zasoski (1989), and includes sensitivity classification of soil map units in an 800,000 ha region of the Sierra Nevada. They characterized the chemistry of 43 modal soil profiles from across the region, and applied a simple computer simulation model to evaluate sensitivity to acidic deposition. They ranked the soils according to the simulated base saturation after 50 years of deposition at a rate of 0.3 kmol H$^+$/ha annually. Sensitive soils were defined as dropping below 15% base saturation within the 50 year simulations. About 25% of the region was classed as least sensitive, with deep soils (> 1 m) on level or gently sloping landscapes with few rock outcrops. Soils classed as moderately sensitive comprised 18% of the region, and were characterized by shallow depths to bedrock on moderate-to-steep slopes. Highly sensitive soils covered 56% of the region, occuring primarily on steep slopes (> 30%). Their work included digitization and production of 45 maps at a scale of 1:62,500 that delineated soils by sensitivity criteria, and which may be of use in other projects in the future.

Fenn and Dunn (1989) measured decomposition rates of needle litter from ponderosa pine and Jeffrey pine (*Pinus jeffreyi*) across a pollution gradient in the San Bernardino Mountains. The rate of carbon dioxide evolution from litter collected in the more-polluted sites was about 40% faster (p < 0.01) than that of litter from less polluted sites. They speculated that the greater rate of decomposition from polluted sites was due to higher nitrogen concentrations in the litter which resulted from premature abscission of ozone damaged needles. Greater deposition of nitrogen was listed as a possible contributing factor, but no direct evidence was available.

Liu (1988) performed a thorough study using 12 pedons from four watersheds (including Eastern Brook Lake and Emerald Lake). His analyses included the basic soil chemistry parameters, titration curves, analyses of the cations removed during the H^+ titrations, and point of zero net charge (PZNC, the pH at which anion exchange capacity equals cation exchange capacity). He concluded that the difference between the current soil pH and the PZNC could be used to indicate the primary process that currently regulates pH. Specifically, if (pH - PZNC) was greater than 2.2 but less than 3.1, cation exchange reactions appeared to provide most of the buffering of added H^+. If the difference was between 1.3 and 2.2, both aluminum dissolution and cation exchange were the major buffering processes. Weathering of aluminosilicate minerals was indicated if the difference was below 0.9. Liu also developed a simple acidification model, and estimated that pedons from the Eastern Brook Lake watershed were very well buffered with respect to 100 years of acidic deposition at a relatively high rate of 1.8 kmol H^+/ha annually. However, the buffering in the most sensitive soils arose from dissolution of aluminum minerals. This process results in decreased solution alkalinity, and increased export of aluminum to aquatic systems. All of the soils studied by Liu had less than 20 mmol$_c$/kg of exchangeable base cations.

One study of soil acidification is available for low-elevation soils. Amundson and Tremback (1989) examined the development of three tree species on sand dunes in San Francisco. The pH of unstabilized sand dunes was about 7, compared with 7.5 under *Eucalyptus globulus*, 5.5 under radiata pine (*Pinus radiata*), and 5.3 under native live oak (*Quercus agrifolia*).

Weathering rates have also been estimated for two watersheds in the White Mountains; one dominated by dolomite weathered at a rate of 8.8 kmol$_c$/ha annually, compared with a rate of 1.3 kmol$_c$/ha annually for another dominated by Adamellite (Marchand 1971).

Estimates of ecosystem H^+ budgets are not available for any of the intensively studied sites in California, so it is difficult to evaluate the likely impacts of current (or future) rates of acidic deposition relative to natural ecosystem processes. Two watersheds in the Sierra Nevada have been examined with the ILWAS model (Gherini et al. 1985), and some implications for likely changes in soil acidity are evident.

The ILWAS model was calibrated for the Blue Lake Watershed in a joint project funded by Pacific Gas and Electric Company (Gilbert et al. 1989). The small watershed is underlain by metasedimentary and granodioritic bedrock, and soil covers only about 20% of the watershed. The vegetation is dominated by ponderosa and Jeffrey pines. The lake is typical of moderate-to-high elevation lakes in the Sierra Nevada, with moderately poorly buffered water (30–70 μmol_c/L ANC) of near-neutral pH. After calibrating the ILWAS model with intensively collected field data, five acidification scenarios were simulated for 21 years: base case (current deposition), doubled rate of sulfuric acid or nitric acid deposition, and halved rate of deposition for each acid. Although the model calculated a rate of weathering, it is not given in the report. The current rate of sulfate deposition led to no change in pH in 21 years, whereas doubling the rate decreased lake water pH by 0.05 units. Most of the deposited H^+ was buffered in the soil by the exchange complex, and much of the sulfate that reached the lake was consumed in redox reactions. The increased rate of nitric acid deposition actually led to an increase in lake alkalinity and pH, due to assimilation or reduction of the nitrate. Decreasing the deposition rate of sulfate or nitrate had no effect on lake pH. Although these simulations did not assess soil acidification directly, a lack of change in the poorly buffered lake suggests that no significant acidification of the soil would have occurred.

The application of the ILWAS model to Eastern Brook Lake in the central Sierra Nevada did not include intensive calibration with field data, so the simulation projections are tentative (Chen et al. 1988). Based on the chemistry of minerals and soil solutions, Weintraub (1986) estimated the weathering rate in Eastern Brook Lake watershed to be about 0.1 $kmol_c$/ ha annually. Eastern Brook Lake has greater alkalinity (about 70 to 250 μmol_c/L ANC) than Blue Lake, and ten year simulations with a five-fold increase in sulfate deposition had only a negligible effect on soil pH (a drop of 0.04 units).

Collectively, these California studies indicate that some Sierran soils, perhaps even a majority, may be sensitive to acidification if rates of deposition increased to levels found in parts of the eastern United States and Europe.

Summary and Conclusions

Forest soils of the western United States exhibit a great range of characteristics at both local (topographic) and regional scales. Some soils in the region have acidified rapidly through time or under the influence of alders, whereas others have shown no acidification even after 1000 years of natural soil development. In general, coarse-textured soils at high elevations would be expected to be sensitive to acidification, but even in these cases it is possible that minor contributions of alkalinity (from eolian dust or weathering of calcite inclusions) could offset acidifying processes. Finer textured soils at lower elevations might be expected to be very resistant to acidification, yet substantial acidification has been observed under the influence of red alder on a time scale of decades. Based on current knowledge, perhaps half or more of the forested soils of the Sierra Nevada would be classified as highly sensitive to acidic deposition. Current rates of acidic deposition are not well characterized for the West (see Chapter 3); across most of the region, rates are probably low enough to pose little threat to forest soils. Where deposition rates are highest (such as in southern California) or are increasing, it is likely that a large portion of the forested soils would show substantial changes under the influence of acidic deposition.

References

Amundson RG, Tremback B (1989) Soil development on stabilized dunes in Golden Gate Park, San Francisco. *Soil Science Society of America Journal* 53:1798–1806

Anderson M (1987) The effects of forest plantations on some lowland soils. I. A second sampling of nutrient stocks. *Forestry* 60:69–85

Arthur M (1990) *The Effects of Vegetation on Watershed Biogeochemistry at Loch Vale Watershed, Rocky Mountain National Park, Colorado.* PhD dissertation, Cornell University

Berdén , Nilsson SI, Rosén K, Germund T (1987) *Soil Acidification: Extent, Causes, and Consequences.* National Swedish Environment Protection Board Report 3292, Solna

Binkley D (1989) Sensitivity of forest soils in the western U.S. to acidic deposition. In: Olson R, Lefohn A (eds)*Effects of Air Pollution on Western Forests.* Transactions Series, No. 16, Air and Waste Management Association, Pittsburgh, pp 561–573

Binkley D (1992) H^+ budgets. In: Johnson D, Lindberg S (eds)*Atmospheric Deposition and Forest Nutrient Cycling: A Synthesis of the Integrated Forest Study* . Springer-Verlag, New York, pp 450–466

Binkley D, Richter D (1987) Nutrient cycles and H^+ budgets of forest ecosystems. *Advances in Ecological Research* 16:1–51

Binkley D, Sollins P (1990) Factors determining differences in soil pH in adjacent conifer and alder-conifer stands. *Soil Science Society of America Journal* 54:1427–1433

Binkley D, Valentine D (1991) Fifty-year effects on biogeochemistry in replicated plantations of green ash, white pine, and Norway spruce. *Forest Ecology and Management* 40:13–25

Binkley D, Smith FW, Long JN (1990) Nutrient limitation of leaf area with stand age in lodgepole pine forests. *Bulletin of the Ecological Society of America* 71(Suppl):92

Binkley D, Driscoll C, Allen HL, Schoeneberger P, McAvoy D (1989a) *Acidic Deposition and Forest Soils: Context and Case Studies in the Southeastern United States.* Ecological Studies #72, Springer-Verlag, New York, 150p

Binkley D, Valentine D, Wells C, Valentine U (1989b) An empirical analysis of factors contributing to 20-year decline in soil pH in an old-field plantation of loblolly pine. *Biogeochemistry* 8:39–54

Blake J, Webster S, Gessel S (1988) Soil sulfate-sulfur and growth responses of nitrogen-fertilized Douglas-fir to sulfur. *Soil Science Society of America Journal* 52:1141–1147

Bohn H, McNeal B, O'Connor G (1985) *Soil Chemistry.* Wiley, New York

Bollen WB, Lu KC, Trappe JM, Tarrant RF, Franklin JF (1967) *Primary Microbiological Succession on a Landslide of Alpine Origin at Mount Rainier.* USDA Forest Service Research Note PNW-50, Portland, OR

Brand D, Kehoe P, Connors M (1986) Coniferous afforestation leads to soil acidification in central Ontario. *Canadian Journal of Forest Research* 16:1389–1391

Chen CW, Gomez LE (1988) *Application of the ILWAS Model to Eastern Brook Lake Watershed in the Sierra Nevada Mountains.* Prepared for Southern California Edison Co., and Electric Power Research Institute, by Systech Engineering, Lafayette, CA

Chronic H (1980) *Roadside Geology of Colorado.* Mountain Press, Missoula

Chronic H (1983) *Roadside Geology of Arizona.* Mountain Press, Missoula

Dickson BA, Crocker RL (1954) A chronosequence of soils and vegetation near Mount Shasta, California. III. Some properties of the mineral soils. *Journal of Soil Science* 5:173–191

Drever J, Joyce G, Reiners W, Knight D (1987) *Laboratory Tests of Acid Treatment to Soils of the Medicine Bow Mountains, Wyoming.* Report #2 on contract 28-K5-360 to the USDA Forest Service Rocky Mountain Forest and Range Experiment Station, Ft. Collins, CO

Fahey T (1983) Nutrient dynamics of aboveground detritus in lodgepole pine (*Pinus contorta* ssp. *latifolia*) ecosystems, southeastern Wyoming. *Ecological Monographs* 53:51–72

Fahey T, Knight D (1986) Lodgepole pine ecosystems. *BioScience* 36:610–617

Fahey T, Yavitt J (1988) Soil solution chemistry in lodgepole pine ecosystems, southeastern Wyoming. *Biogeochemistry* 6:91–118

Fahey T, Yavitt J, Pearson J, Knight D (1985) The nitrogen cycle in lodgepole pine forests, southeastern Wyoming. *Biogeochemistry* 1:257–275

Falkengren-Grerup U (1987) Long-term changes in pH of forest soils in Southern Sweden. *Environmental Pollution* 43:79–90

Fenn M, Dunn P (1989) Litter decomposition across an air-pollution gradient in the San Bernardino Mountains. *Soil Science Society of America Journal* 53:1560–1567

Gherini S, Chen C, Mok L, Goldstein R, Hudson R, Davis G (1985) The ILWAS model: Formulation and application. *Water, Air and Soil Pollution* 26:425–459

Gilbert D, Sagraves T, Lang M, Munson R, Gherini S (1989) *R&D Lake Acidification Assessment Project: Blue Lake Acidification Study.* Pacific Gas and Electric Company, Department of Research and Development, San Ramon, CA

Harrison R, Johnson D, Todd D (1989) Forest soil sulfur pools and sulfate adsorption and desorption capacity following elevated inputs of sulfur. In: Olson R, Lefohn A (eds) *Effects of Air Pollution on Western Forests.* Transactions Series, No. 16, Air & Waste Management Association, Pittsburgh, pp 529–546

Heilman P, Anderson H, Baumgartner D (eds) (1979) *Forest Soils of the Douglas-fir Region.* Washington State University, Pullman

Huntington G, Akeson M (1987) *Pedologic Investigations in Support of Acid Rain Studies, Sequoia National Park, California; Soil Resource Inventory of Sequoia National Park, Central Part.* National Park Service

Hutchinson TC, Bozic L, Munoz-Vega G (1986) Responses of five species of conifer seedlings to aluminum stress. *Water, Air, and Soil Pollution* 31:283–294

Johnson DW, Richter D, Lovett G, Lindberg S (1985) The effects of atmospheric deposition on potassium, calcium, and magnesium cycling in two deciduous forests. *Canadian Journal of Forest Research* 15:773–782

Johnson DW, Henderson GS, Todd DE (1988) Changes in nutrient distribution in forests and soils of Walker Branch Watershed, Tennessee, over an eleven-year period. *Biogeochemistry* 5:275–293

Johnston AE, Goulding KWT, Poulton PR (1986) Soil acidification during more than 100 years under permanent grassland and woodland at Rothamstead. *Soil Use and Management* 2:3–10

Kelly JK, Joslin JD, Thornton FC, Schaedle M, Raynal D (1987) A comparison of the response of red spruce seedlings to Al in soil and in solution culture. *Agronomy Abstracts* 1987:260

Knight D, Fahey T (1985) Water and nutrient outflow from contrasting lodgepole pine forests in Wyoming. *Ecological Monographs* 55:29–48

Knight D, Reiners W, Joyce G, Drever J (1988) *Effects of Snowpack Acidification on Nutrient and Aluminum Outflow from Lodgepole Pine Forest Soils, Medicine Bow Mountains, Wyoming.* Report #3 on contract 28-K5-360 to the USDA Forest Service Rocky Mountain Forest and Range Experiment Station, Ft. Collins, CO

Lindsay W, Vlek P (1977) Phosphate minerals. In: Dixon J, Weed S (eds) *Minerals in Soil Environments.* Soil Science Society of America, Madison, Wisconsin, pp 639–672

Litaor MI (1987) The influence of eolian dust on the genesis of alpine soils in the Front Range, Colorado. *Soil Science Society of America Journal* 51:142–147

Liu WC (1988) *The Sensitivity of Selected Soils from the Sierra Nevada to Acidic Deposition.* PhD dissertation, University of California, Riverside, 102p

Lund L, Brown A, Lueking M, Nodvin S, Page A, Sposito G (1987) *Soil Processes at Emerald Lake Watershed.* Final Report, submitted to California Air Resources Board, Contract #A3-105-32, 114p

Marchand D (1971) Rates and modes of denudation, White Mountains, eastern California. *American Journal of Science* 270:109–135

Mast MA (1989) *A Laboratory and Field Study of Chemical Weathering with Special Reference to Acid Deposition.* PhD dissertation, University of Wyoming, Laramie, 176p

McColl J (1981) *Effects of Acid Rain on Plants and Soils in California.* Final
    Report submitted to California Air Resources Board, Contract
    #A8-136-31, 111p

McKee B (1972) *Cascadia: The Geologic Evolution of the Pacific Northwest.*
    McGraw-Hill, New York

Miller R, Barker P, Peterson C, Webster S (1986) Using nitrogen fertilizers
    in management of Coast Douglas-fir. I. Regional trends of
    response. In: Oliver C, Hanley D, Johnson J (eds) *Douglas-fir:
    Stand Management for the Future.* Contribution #55, College of
    Forest Resources, University of Washington, Seattle, pp 290–303

Mitchel R (1979) Soil formation, classification and morphology. In:
    Heilman P, Anderson H, Baumgartner D (eds) *Forest Soils of the
    Douglas-fir Region.* Washington State University, Pullman,
    pp 157–172

Myrold D (1987) *Acidic Deposition and Forest Soil Biology.* NCASI
    Technical Bulletin #527, National Council of the Paper Industry
    for Air and Stream Improvement, 260 Madison Ave., New York

Nilsson SI (1986) Critical deposition limits for forest soils. In: Nilsson J
    (ed) *Critical Loads for Nitrogen and Sulfur: Report from a Nordic
    Working Group.* Nordisk Ministerrad Miljo Rapport 11, pp 37–69

Omernik J, Griffith G (1986) *Total Alkalinity of Surface Waters: A Map of the
    Western Region.* USEPA EPA-600/D-85-219, Corvallis, Oregon

Oren R, Schulze E-D, Werk K, Meyer J (1988) Performance of two *Picea
    abies* (L.) Karst. stands at different stages of decline. VII. Nutrient
    relations and growth. *Oecologia* 77:163–173

Pearson J, Knight D, Fahey T (1987) Biomass and nutrient accumulation
    during stand development in Wyoming lodgepole pine forests.
    Ecology 68:1966–1973

Powers RF (1983) Forest fertilization in California. In: Ballard R, Gessell S
    (eds) *I.U.F.R.O. Symposium on Forest Site and Continuous
    Productivity* . USDA Forest Service General Technical Report
    PNW-163, Portland, OR, pp 388–397

Raynal D, Joslin J, Thornton FC, Schaedle M, Henderson G (1990)
    Sensitivity of tree seedlings to aluminum: III. Red spruce and
    loblolly pine. *Journal of Environmental Quality* 19:180–187

Reilly TA, Zasoski R (1989) *Draft Survey of Soil Map Unit Sensitivity to
    Acid Deposition in the Sierra Nevada, California.* Draft report to the
    California Air Resources Board, contract #A732-037, Sacramento

Reuss JO (1989) Soil-solution equilibria in lysimeter leachates under red alder. In: Olson R, Lefohn A (eds)*Effects of Air Pollution on Western Forests*. Transactions Series, No. 16, Air and Waste Management Association, Pittsburgh, pp 547–560

Richter D, Comer P, King K, Sawin H, Wright D (1988) Effects of low ionic strength solutions on pH of acidic forested soils. *Soil Science Society of America Journal* 52:261–264

Riggan P, Lockwood R, Lopez E (1985) Deposition and processing of airborne nitrogen pollutants in Mediterranean-type ecosystems of southern California. *Environmental Science and Technology* 19:781–789

Rochette E, Drever J, Sanders F (1988) Chemical weathering in the West Glacier Lake drainage basin, Snowy Range, Wyoming: Implications for future acid deposition. *Contributions to Geology, University of Wyoming* 26:29–44

Rost-Siebert K (1983) Aluminium-Toxizitat und -Toleranz und Keimpflanzenvon Ficte (*Picea abies* Karst.) und Buche (*Fagus silvatica* L.) *Allgemeine Forstzeitschrift* 38:686–689

Schulze E-D (1989) Air pollution and forest decline in a spruce (*Picea abies*) forest. *Science* 244:776–783

Schulze E-D, Lange OL, Oren R (eds) (1989) *Forest Decline and Air Pollution*. Springer-Verlag, New York

Sharp R (1972) *Geology Field Guide to Southern California*. Kendall/Hunt, Dubuque, IA

Singleton G, Lavkulich L (1987) A soil chronosequence on beach sands, Vancouver Island, British Columbia. *Canadian Journal of Soil Science* 67:795–810

Sollins P, Grier C, McCorison F, Cromack K Jr., Fogel R, Fredriksen R (1980) The internal element cycles of an old-growth Douglas-fir ecosystem in western Oregon. *Ecological Monographs* 50:261–285

Tabatabai MA (1985) Physicochemical fate of sulfate in soils. *Journal of the American Physical and Chemical Association* 37:34–38

Tamm CO, Hällbacken L (1986) Changes in soil pH over a 50-year period under different forest canopies in SW Sweden. *Water, Air, and Soil Pollution* 31:337–341

Thornton FC, Schaedle M, Raynal DJ (1986a) Effects of aluminum on growth, development, and nutrient composition of honeylocust (*Gleditsia triacanthos* L.) seedlings. *Tree Physiology* 2:307–316

Thornton FC, Schaedle M, Raynal DJ (1986b) Effect of aluminum on the growth of sugar maple in solution culture. *Canadian Journal of Forest Research* 16:892–896

USDA Forest Service (no date) *Soil Survey of San Bernardino National Forest Area, California*

USGS (1970) *National Atlas of the United States.* U.S. Government Printing Office, Washington, DC

Van Miegroet H, Cole DW, Binkley D, Sollins P (1989) The effect of nitrogen accumulation and nitrification on soil chemical properties in alder forests. In: Olson R, Lefohn A (eds)*Effects of Air Pollution on Western Forests.* Transactions Series, No. 16, Air and Waste Management Association, Pittsburgh, pp 515–528

Vertucci F (1990) Methods of detecting and quantifying lake acidification. *Proceedings of the International Conference on Mountain Watersheds,* University of California Press, in press

Vitousek P, Gosz J, Grier C, Melillo J, Reiners W, Todd R (1979) Nitrate losses from disturbed ecosystems. *Science* 204:469–474

Walthall PM (1985) *Acidic Deposition and the Soil Environment of Loch Vale Watershed in Rocky Mountain National Park.* PhD dissertation, Colorado State University, Ft. Collins, 148p

Weetman GF (1988) Nutrition and fertilization of lodgepole pine. In: Schmidt W (compiler) *Proceedings—Future Forests of the Mountain West: A Stand Culture Symposium.* USDA General Technical Report INT-243, pp 231–239

Weintraub J (1986) *An Assessment of the Susceptibility of Two Alpine Watersheds to Surface Water Acidification: Sierra Nevada, California.* MS thesis, Department of Geology, Indiana University, 186p

Wolt J (1987) *Effects of Acidic Deposition on the Chemical Form and Bioavailability of Soil Aluminum and Manganese.* NCASI Technical Bulletin #518, National Council of the Paper Industry for Air and Stream Improvement, New York

Wyels W (1986) *The Buffering Capabilities of Sierra Nevadan Soils Exposed to Simulated Acid Precipitation.* MS thesis, University of California, Davis, 188p

Zöttl HW, Hüttl RF (1986) Nutrient supply and forest decline in Southwest-Germany. *Water, Air, and Soil Pollution* 31:449–462

5

Physiological Effects of Air Pollutants on Western Trees

A. Bytnerowicz and N. E. Grulke

Introduction

General physiological responses of forest trees to atmospheric pollution are well-documented (Smith 1974, Heath 1980, Guderian 1985, McLaughlin 1985, Wellburn 1988, Darrall 1989). Trees have a higher threshold of tolerance to air pollutants than do herbaceous species (Darrall 1989). Physiological responses of forest trees to air pollution in western North America have many similarities to the responses observed in eastern North America and in Europe. However, the combination of unique air pollution signatures compounded with high concentrations at some sites may result in unique physiological responses in the West. The objective of this chapter is to review recent knowledge on the effects of atmospheric pollutants on Western trees, with a focus on conifers.

Considerable variations in the response of trees to air pollutants are caused by differences in the pollutant dose, phenological stage, age of leaves exposed, seed source, nutritional status of plants, and/or the integrated effects of multiple stresses. Under field conditions, detecting physiological changes, and identifying the causes is difficult at best. Visible symptoms are most commonly used to detect pollutant damage, but changes in physiological processes may occur prior to visible, morphological damage or in the absence of visible damage. Phytotoxic effects may also be confused with other types of environmental stresses such as chlorosis caused by nutrient deficiencies and dieback associated with frost or drought. Fertilization effects caused by moderate levels of atmospheric deposition or elevated carbon dioxide may be confused with other ameliorating climatic (temperature, moisture) or biotic (local nutrient inputs, tree release) effects. Chronic exposure to pollutants may predispose foliage to other environmental stresses and pathogens (Miller 1983).

The literature is dominated by seedling and sapling responses to pollutant exposures with only a few extrapolations to mature trees. The model presented in Figure 5.1 provides a framework for this chapter, and describes pollutant deposition and transfer into the mesophyll, the suite of biochemical reactions that can take place and their effects on the cell, detectable leaf responses, shifts in within-plant resource allocation and their timing, and possible population responses. These effects will be discussed in detail in this chapter.

Phytotoxic Air Pollutants

General characteristics of atmospheric deposition in western North America are described in Chapter 3. The importance of ozone, nitrogen oxides, sulfur dioxide, and acidic precipitation as the main pollutants affecting Western forests has been shown. However, little is known about the effects of other air pollutants such as peroxyacetyl nitrate (PAN), volatile organic compounds (VOCs), fluorides (HF), particulate pollution, and their multiplicative effects. Atmospheric carbon dioxide is known to be increasing (Keeling et al. 1976), but because of its positive effects on plant carbon gain it is rarely considered to be a pollutant. The potential effects of elevated carbon dioxide will be discussed in Chapter 13.

On a regional scale, ozone is the only air pollutant that has proved to be clearly phytotoxic at ambient levels in some areas of western North America. A marked example of this is decreased growth, reduced foliar retention, and visible injury of ponderosa and Jeffrey pines (*Pinus ponderosa, Pinus jeffreyi*) on northwestern slopes of the San Bernardino Mountains in response to exposure to elevated ozone for more than 40 years (Miller et al. 1963, 1969). Elevated ozone concentrations have been correlated with reduced radial growth of Jeffrey pines (Peterson et al. 1988), and foliar injury of ponderosa and Jeffrey pines on the western slopes of the Sierra Nevada (Durisco and Stolte 1989). Elevated concentrations of ozone have also been recorded in some mountain locations near Seattle but with no accompanying visible injury (Basabe et al. 1989).

At present, ambient levels of sulfur dioxide and hydrogen fluoride do not cause visible damage or reduction of growth of forest trees in the West, although such effects were seen in the past in the vicinity of some of the industrial sources of pollution (Miller and McBride 1975). However, heavy metal deposition near smelters in central and eastern Mon-

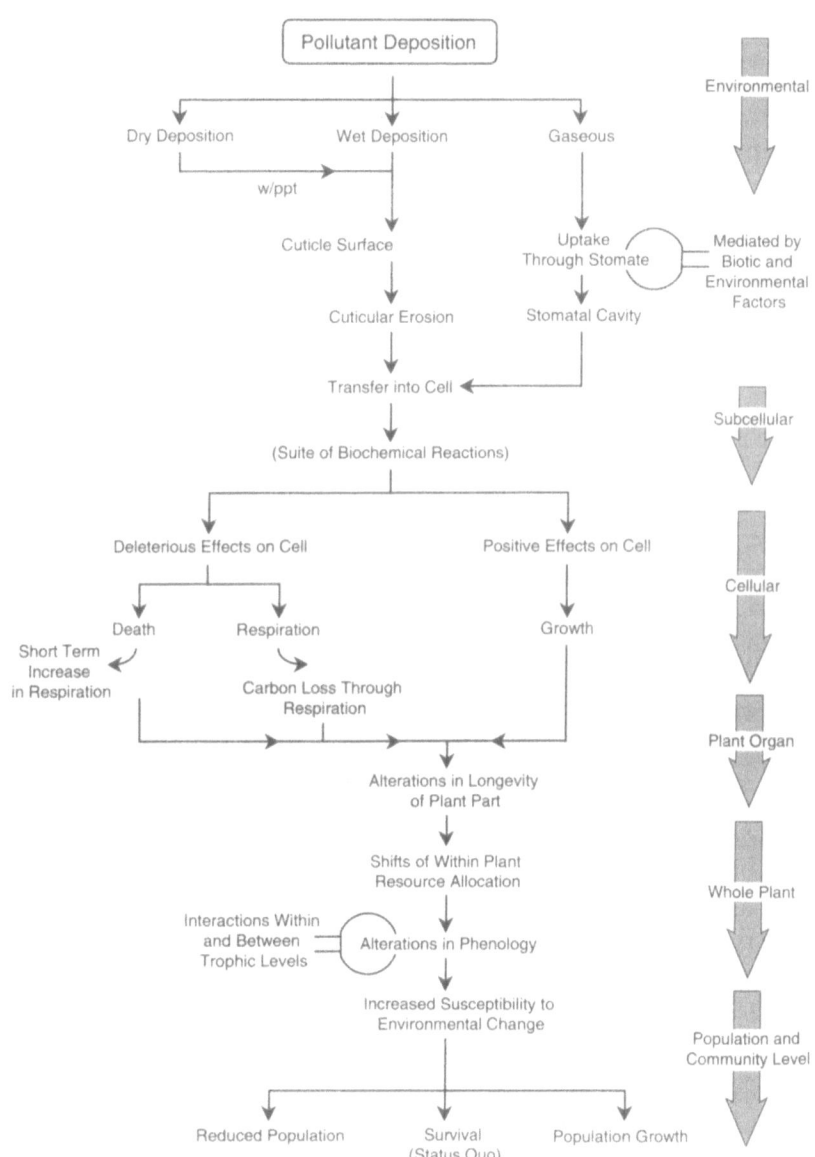

Figure 5.1 Pollutant transfer and effects at subcellular, cellular, plant organ, whole plant, and population levels. Note that at each level, effects may not be deleterious nor detectable. Compensatory responses at one level may prevent effects at higher levels.

tana, and in northern Idaho have caused local tree mortality (T. Weaver, pers. comm.). Despite the lack of general evidence of phytotoxicity of these air pollutants, effects on Western forest trees may occur alone, or in combination with other pollutants or environmental stresses.

Acidic precipitation in the form of cloud and fog deposition has perhaps the highest potential for phytotoxicity, but this has not yet been shown for ambient conditions. Several controlled experiments on forest trees determined toxic effects of fogs at acidities < pH 3 (NEG, unpubl. data; PR Miller et al., in prep.; Takemoto and Bytnerowicz 1992). Fog and cloud water of similar or even higher acidities were found in some of the mountain and coastal locations of California and Washington states (Brewer et al. 1983, Waldman et al. 1982, 1985).

In some mountain locations of southern California, gaseous nitric acid concentrations are very high, with 12 hour daytime averages reaching 10 ppb (Bytnerowicz et al. 1987, Solomon et al. 1989, AB, unpubl. data), which may have toxic effects on plants. Although no direct injury to plant foliage was found in short term experiments with high nitric acid concentrations (Hanson et al. 1989, Marshall and Cadle 1989), it is suspected that long term exposures to high ambient concentrations of nitric acid vapor may adversely affect foliage. Approximately 90% of nitric acid is deposited to plant surfaces (Hanson et al. 1989), and cuticular damage may be the primary effect of exposure. Interactions between nitric acid and ozone may be of special importance because of differences in the mechanisms by which injury occurs. Nitric acid is deposited primarily to cuticles and stomatal cavities and may potentially alter stomatal function. One possible effect of a change in stomatal control may be increased conductance leading to more ozone uptake and damage to mesophyll cells. Therefore, plants exposed to elevated concentrations of nitric acid may be predisposed to the phytotoxic effects of other air pollutants through changes in stomatal conductance or changes in the integrity of the cuticle leading to water loss from foliage.

Nitrogen oxides may also be responsible for forest decline at some mountain locations in the West. Although high concentrations of nitrogen oxides are not considered to be phytotoxic (NAS 1977, Amundson and MacLean 1982), experimental studies indicate that the products of nitrogen dioxide and especially nitric oxide metabolism may be extremely toxic to plants (Wellburn 1990). Modification of ozone phytotoxicity by nitrogen oxides should also be considered, because changes in their phytotoxicity with their interaction have been found (Reinert et al. 1975).

Two products of photochemical smog reactions, peroxyacetyl nitrate and peroxypropionyl nitrate (PPN), are more phytotoxic than ozone (Mudd 1982), even though they occur in much lower concentrations. Formaldehyde, which is present in photochemical smog and is produced in large quantities during methanol fuel combustion, is very reactive in biological systems, and also has a strong phytotoxic potential (de Konig and Jagier 1970). Volatile organic compounds (VOCs) and products of their reactions with ozone have received little attention as potential phytotoxins, although these compounds are considered to be phytotoxic (Masuch et al. 1986, Rennenberg and Polle 1989, Ennis et al. 1990). Their action is similar to that of some of the plant hormones (Garrec and Berteigne 1987) and can contribute to the formation of hydroperoxides (Becker et al. 1990). The effects of all of the compounds mentioned above on the physiology of Western trees, when known, will be discussed.

Stomatal Aperture and its Control

Although there are other pathways for pollutant transfer to the site of injury, the degree of stomatal opening largely determines the extent and duration of pollutant dose, and subsequent plant response (Wellburn 1988). Transfer of pollutants from the atmosphere to the cell follows the same pathways as carbon dioxide, but each pollutant has a different diffusivity constant for movement through air, solubility constant for movement across the apoplastic water, and hydrophobic or -phylic properties that affect the rate of transfer across cell walls and membranes (Figure 5.2).

Mechanisms by which atmospheric ozone induces stomatal closure have been described for herbaceous species and probably apply to vascular plants in general (Mudd and Kozlowski 1975, Heath 1975). In order for turgor to be maintained in the guard cells of the stomata, high intracellular concentrations of potassium ions must be maintained. Ozone alkalizes the cell membrane, increasing cell membrane permeability and loss of intracellular potassium ions, resulting in stomatal closure. In carefully controlled laboratory studies, this result is obtained in crop species that have been previously drought stressed (Unsworth et al. 1972). However, many of the experimental studies on the singular effects of elevated ozone on Western tree species show *increased* stomatal conductance. Extreme acidity also results in stomatal closure, but via a different mechanism. An excess of hydrogen ions causes cell walls to loosen (Cleland 1971), decreases turgidity of guard cells, and decreases stomatal conductance.

Sulfur dioxide has two effects on stomatal aperture depending on the humidity. Wellburn (1988) suggests that at high humidities, sulfur dioxide acts to increase stomatal aperture via increased turgidity of the guard cells. The opposite effect takes place when humidity is low (Mansfield and Majernik 1970). Experimentally enhanced sulfur dioxide

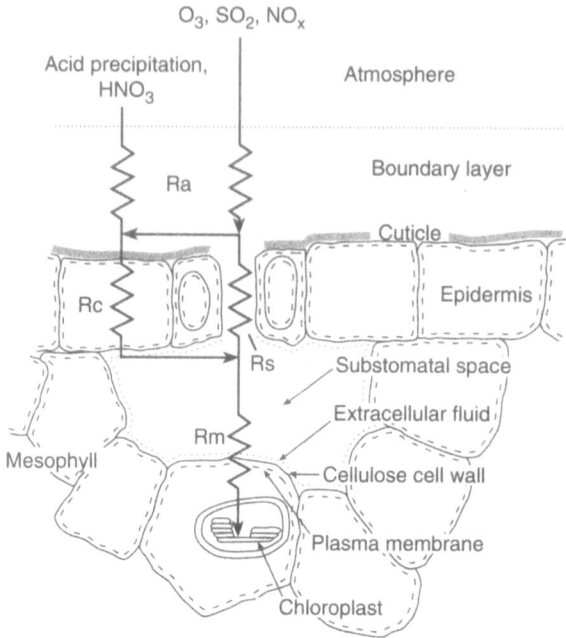

Figure 5.2 Possible pathways of entry of pollutants from atmosphere into the mesophyll (modified from Wellburn 1988). The boundary layer imposes a resistance (Ra), which depends on a number of factors including wind speed. Most movement into the leaf is then through stomates (Rs), and is largely a function of aperture. The resistance to pollutant movement across the cuticle and epidermis (Rc) is much greater than stomatal resistance. The mesophyll resistance (Rm) consists of a number of different components before the major sites of reaction are encountered.

levels (150 ppb) have been found to reduce stomatal conductance of water-stressed ponderosa pine, leading to restricted influx of sulfur dioxide into leaves. However, these effects were not apparent until the twelfth month of exposure (Houpis 1989). A northeastern conifer, jack pine (*Pinus banksiana*), exposed to elevated sulfur dioxide (0.1–1.0 ppm for 96 hours) had decreased stomatal conductance relative to controls (L'Hirondelle and Addison 1985).

There appears to be some genetic component to stomatal response to ozone exposure. Studies have examined individual tree sensitivity (Patterson and Rundel 1989) and family sensitivity (Beyers et al. 1991) to atmospheric ozone. "Sensitivity" was identified in both studies as greater chlorosis or necrosis in leaf tissue, and may be genetic or environmentally induced by nutrient deficiencies. In a field study of Jeffrey pine, stomatal conductance of sensitive trees was greater during late summer allowing a greater effective pollution dose. Possible impairment of stomatal control in sensitive trees was substantiated by controlled experiments of stomatal response to steady-state changes in vapor pressure deficit (Figure 5.3, Patterson and Rundel 1989). In an experimental study of 18 families of ponderosa pine, stomatal conductance was also greater in sensitive families (Beyers et al. 1991).

Despite the lack of a mechanistic explanation, experimental evidence generally shows enhanced stomatal aperture in response to short-term elevated ozone. In a four month experiment, seedlings of four common Western conifers, white fir (*Abies concolor*), Engelmann spruce (*Picea engelmannii*), Douglas-fir (*Pseudotsuga menziesii*), and ponderosa pine, were exposed to elevated concentrations of ozone (140 ppm-hour, 1.8x profile of Hogsett et al. 1989). Stomatal conductance of the current year foliage of exposed plants was greater during mid-season measurements than that of control plants receiving charcoal filtered air (NEG, unpubl. data). These results are consistent with those of Jeffrey pine (Patterson and Rundel 1989) for impairment of stomatal control under conditions of high vapor pressure deficits. In another study, stomatal conductance was also greater in ponderosa pine exposed to elevated ozone relative to charcoal-filtered air controls (Leininger and Fenn, in prep.).

An Eastern conifer, Fraser fir (*Abies fraseri*), (Tseng et al. 1988), and Norway spruce (*Picea abies*), (Keller and Hasler 1987, Freer-Smith and Dobson 1989) had enhanced stomatal conductance or transpirational losses in elevated ozone relative to controls. Two broadleaf trees, black willow (*Salix nigra*) and hazel alder *(Alnus serrulata)* (Greitner and Winner 1989), also showed enhanced stomatal conductance for trees exposed to elevated ozone concentrations.

There are exceptions to the generalized enhanced conductance in response to elevated ozone. In the Pacific Northwest with adequate soil moisture through the growing season, Douglas-fir showed a reduction in stomatal conductance with elevated ozone (charcoal filtered air plus 300 ppb ozone) (W. Schapp, pers. comm.). In an experimental study of well-watered ponderosa pine, saplings grown in ambient levels of ozone and particulates had reduced stomatal conductance relative to those exposed to ambient levels of ozone without particulates, and clean, charcoal-filtered air (Bytnerowicz and Takemoto 1989). Long-term ozone exposure of ponderosa pine in situ resulted in reduced stomatal conductance in a field study in the San Bernardino Mountains east of Los Angeles (Coyne and Bingham 1981). It is not clear whether long term exposure

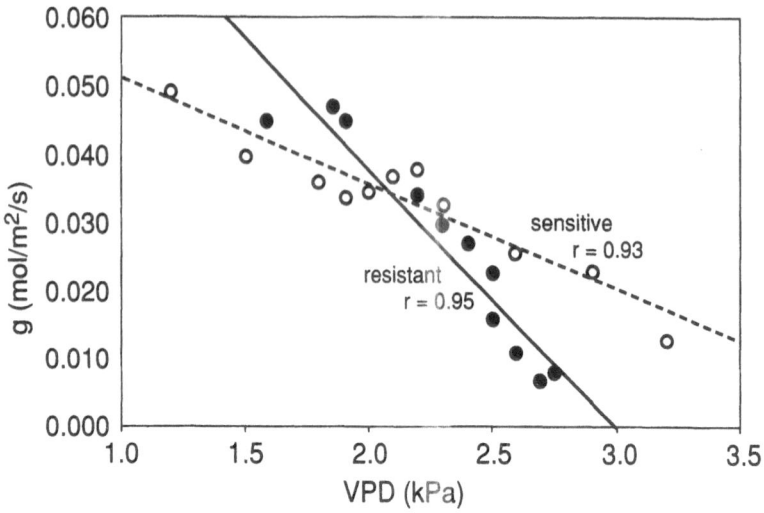

Figure 5.3 Stomatal responses (g) to changes in vapor pressure deficits (VPD). Sensitive individuals have a more sluggish response to changes in VPD than resistant trees, suggesting possible impairment of guard cell function. The r values presented are from least square regressions (from Patterson and Rundel 1989).

directly affected stomatal aperture or whether differential mortality through time selected for individuals with reduced stomatal conductance (lower pollutant dose). In another study, there was no apparent effect of elevated ozone on stomatal conductance of a Rocky Mountain species, subalpine fir (*Abies lasiocarpa*, NEG, unpubl. data).

Acidic precipitation in the range of pH 2.5 to 4.0 increased stomatal conductance of red spruce (*Picea rubens*, Eamus and Fowler 1990). With increased acidity (pH 2.0–2.5), conductance decreased in ponderosa pine (Takemoto and Bytnerowicz 1992), Jeffrey pine (Temple 1988), and Douglas-fir seedlings (NEG, unpubl. data). There is no information on the specific effects of other pollutants on stomatal aperture.

Mechanisms of Air Pollution Toxicity

Once pollutants enter the plant cell, a suite of biochemical reactions take place. The biochemical mechanisms through which air pollutants affect a plant depend on the chemical properties of the pollutant, secondary reactions of pollutants in plant tissues, and defense abilities of the plant.

An excellent review of biochemical changes caused by air pollution has been done by Wellburn (1988). Heath (1988) reviewed the chronic stress of the long-term effects of low ozone concentrations on plant systems. Another review paper describing ozone phytotoxicity was published by Krupa and Manning (1988). There have been many reviews and experimental papers dealing with mechanisms of sulfur dioxide phytotoxicity and related sulfur metabolism (e.g., Ziegler 1973, Rennenberg et al. 1990, Wellburn 1990). Despite these papers, our knowledge of the mechanisms of air pollutant phytotoxicity is incomplete and continues to develop. This is especially true for processes involving formation and reactions of free radicals.

Free Radicals

Ozone, peroxyacetyl nitrate, and sulfur dioxide increase the potential for the formation of highly phytotoxic oxygen radicals in plant cells. After entering the leaves through stomata, gases move into the intercellular spaces, and then have to pass through a thin layer of water surrounding individual cells. It is postulated that reactions of ozone and its alterations start taking place in this region of plant systems (Figure 5.4, Heath 1988). Ozone may form toxic hydrogen peroxide or free radicals through ozonolysis or peroxidation, which may reduce cell integrity (Heath 1988, Krupa and Manning 1988, Wellburn 1988). Ozone readily dissolves in water and, in an aqueous solution, it can break down to several types of products (Heath 1987). In an acidic solution, ozone is reasonably stable (Heath 1979). However, when solutions become more alkaline, reactions with the hydroxide ion lead to autocatalytic reactions that produce peroxyl radical ($HO_2^\bullet$) and superoxide (O_2^-). The superoxide anions can be formed from peroxyl-radicals by deprotonation. It is likely that the ozone molecule reacts with the superoxide anion to generate oxygen and an O_3^- radical, which in turn reacts with the proton. The resulting protonated radical, ozonide ($HO_3^\bullet$) decomposes rapidly and releases oxygen to form a hydroxyl radical ($OH^\bullet$), which is a very reactive chemical species. The cycle continues with another ozone reaction, forming the protonated peroxyl radical. This radical can restart the cycle, thus inducing an autocatalysis (Heath 1988). Ozone reactions in aqueous solutions of plant systems and formation of oxygen free radicals are presented in Figure 5.4.

Another theory explaining the participation of free radicals in ozone phytotoxicity is offered by Hewitt et al. (1990). These authors suggest that ethylene or other hydrocarbons (which are produced in plants from methionine in response to environmental stresses) react with ozone entering the plant system and produce free radicals. Organic hydroperoxides are another group of compounds that may play a role in ozone

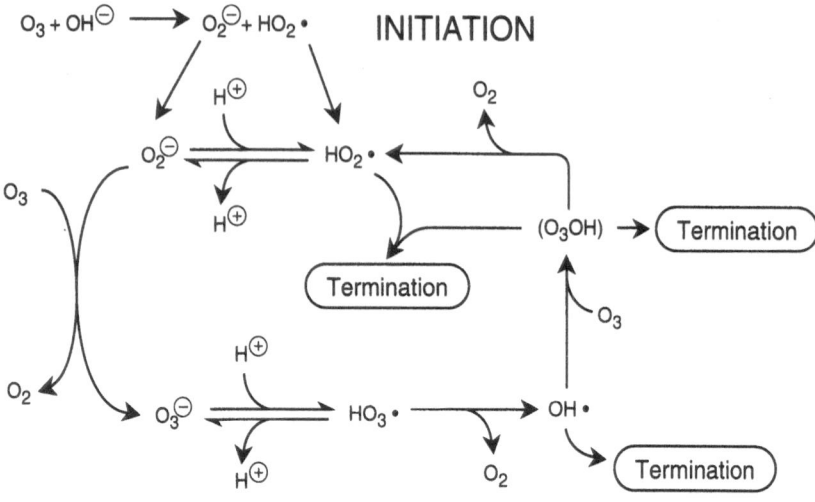

Figure 5.4 Reaction scheme of ozone interactions with an aqueous solvent (from Heath 1988). The top reaction is the initiation step of the total cycle, in which the hydroxyl ion catalyses the production of both peroxyl-radical and superoxide. Several termination steps are shown without details, but involve the production of hydrogen peroxide from the hydroxyl radical (HO·).

toxicity. These relatively stable compounds produced during a reaction of ozone with alkenes inhibit enzyme activity. They are found in the leaves of isoprene-emitting plants exposed to ozone, but not in control plants grown in clean air (Hewitt et al. 1990).

After sulfur dioxide entry into plant systems, bisulfite and sulfite are formed. These compounds may be readily oxidized in a series of reactions involving formation or consumption of free radicals such as $HSO_3^·$, $O_2^·$, $OH^·$, and $SO_2^{-·}$. However, formation of free radicals from sulfite should be considered as side reactions, because most of the sulfite is rapidly photo-oxidized to sulfate in the chloroplast and oxidized by the enzyme sulfite oxidase in mitochondria, which prevents a build-up of free radicals. Therefore, the contribution of free radicals in sulfur dioxide phytotoxicity is still debated (Wellburn 1988).

Transport of ozone and free radicals into the cell interior is poorly understood, although some of its effects on membranes have been described. As a hydrophylic molecule, ozone does not easily enter the hydrophobic regions of the membranes containing unsaturated lipids. Under chronic stress, lipid alteration by ozone does not occur readily

(Heath 1988). However, reactions of ozone with saturated lipids of plant membranes have been reported (Yoshida and Vemura 1984, Fangmeier et al. 1990). It is postulated that in the process of lipid peroxidation, the free radicals rather than ozone itself provide the initial attack (Roehm et al. 1971, after Wellburn 1990, Hewitt et al. 1990). Ozone also reacts with the membrane proteins, resulting in oxidation of cysteine and methionine residues (Heath 1987). Mudd (1982) indicated that attack by ozone occurs more readily on proteins than on lipids. These changes in plant membranes lead to damage of the cellular permeability barrier, increasing the possibility of transport of toxic substances into the cells. Inside the cells, ozone in the aqueous environment may undergo an array of reactions leading to formation of reactive radicals and also hydrogen peroxide (similarly as described above) that may damage most of the cell components.

Other Phytotoxic Chemical Species

Although the phytotoxicity of ozone, peroxyacetyl nitrate, and to some extent sulfur dioxide can be explained by reactions of free radicals within plants, most of the nitric oxide and nitrogen dioxide as well sulfur dioxide toxicity can be attributed to other products of their metabolism.

Sulfur dioxide molecules readily dissolve in water to form sulfite and bisulfite ions. The predominant site for sulfite photooxidation is the chloroplast. Photooxidation is believed to be the most important system for oxidation of sulfite (Sekiya et al. 1982). Plant metabolism of sulfur dioxide is complex, involving both photo-dependent and non-photodependent oxidation and reduction reactions and a potential for re-emission of sulfur in the form of reduced gases. In addition, sulfur may be incorporated into plant constituents such as amino acids and proteins (Ziegler 1973, Guderian 1977, Rennenberg 1984). Phytotoxicity of sulfur dioxide mainly results from accumulation of the sulfur dioxide intermediate metabolite, sulfite (Ziegler 1973, Miller and Xerikos 1979). Secondary sulfur metabolites such as sulfoxides (R-SO-R') and sulfones (R-SO$_2$-R') are also considered highly phytotoxic (Gietko 1976). The chloroplast is considered to be a primary site of many of the disturbances caused by sulfur dioxide or its products in aqueous solution. There is an array of potential detrimental effects of sulfur dioxide causing reduction of rates of various biosynthetic activities, particularly protein or starch synthesis and carbon dioxide fixation (Wellburn 1988). A summary of sulfur metabolism in plants indicating chemical species believed to cause phytotoxicity is presented in Figure 5.5.

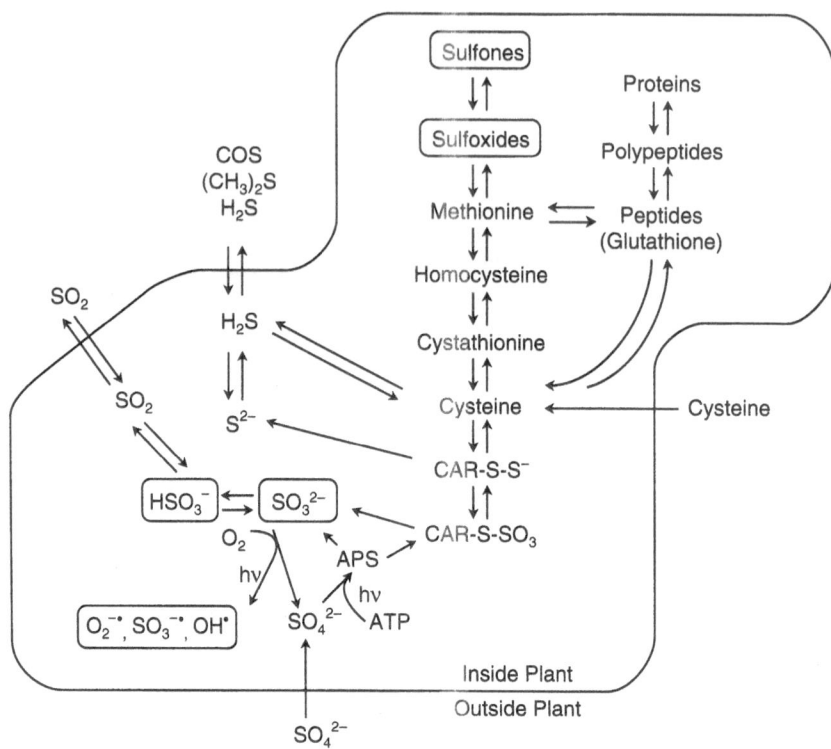

Figure 5.5 Diagram of steps in sulfur uptake metabolism and release. Circled compounds are phytotoxic. Adapted from Bytnerowicz and Molski (1978), Rennenberg (1982), and Tingey and Olszyk (1982).

Zeevaart (1976) suggested that nitrogen dioxide entering a leaf through stomata or the epidermis dissolves in the extracellular water of the sub-stomatal cavity or mesophyll and forms both nitrous and nitric acid which then dissociate to form nitrate, nitrite and free protons. Nitric oxide dissolves poorly in water and the chemical form of this gas in water is less certain. In the extracellular water of a plant, nitric oxide may form nitric acid, nitrite, nitrate and free protons, similarly to their formation from nitrogen dioxide but at a much slower rate. Solubility of nitric oxide in water is higher at low temperatures. The ratio of nitric oxide to nitrogen dioxide in ambient air also increases at low tempera-tures (Wellburn 1990). Most plants appear to tolerate an accumulation of high nitrate concentrations, but accumulation of nitrite may seriously impair plants. Therefore, most phytotoxic effects of nitric oxide and nitrogen dioxide are likely to arise from nitrite.

The effects of nitrous acid are unknown and the stresses of cellular acidification are not fully evaluated. However, it seems that extra nitrite coming into leaves by foliar uptake can disturb nitrogen metabolism, especially in plants where most of the nitrogen reduction occurs in roots. Nitrite also affects various membrane and metabolic functions. These include proton pumping and the sulfhydryl groups of regulatory proteins. Disturbance of these various processes in photosynthesis is most likely a reason for reduction of growth caused by atmospheric oxides of nitrogen (Wellburn 1990). The nitrogen ions are transported through the cell wall into the cytoplasm where they undergo a series of enzymatic reactions. Once pollutant-derived nitrogen has been reduced, most accumulates as amino acids (Durmishidze and Nutshubidze 1976, Yoneyama and Sasakawa 1979). Responses of plants to atmospheric nitrogen dioxide are very different according to whether the nitrogen supply is limiting or adequate (Wellburn 1990).

At concentrations of nitrogen dioxide considerably above normal ambient, cellular plasmolysis results because of lipid breakdown in membranes (which produces a water-soaked appearance). Unsaturated lipids in monolayers readily bind molecules like nitrogen dioxide. Two types of reactions take place within fatty acids; attachment of nitrogen dioxide to a double bond may cause a cis- to trans- isomerization or removal of hydrogen from methylene groups. Both processes may also initiate lipid peroxidation and cause changes in the surface properties of monolayers (Wellburn 1990). However, according to Mudd et al. (1984), current ambient levels of nitrogen dioxide would be too low to cause such effects. There are strong indications that atmospheric nitrogen dioxide inhibits lipid biosynthesis rather than causing damage to existing lipids in membranes. Fumigation of jack pine seedlings with 2 ppm nitrogen dioxide for two days caused inhibition of the biosynthesis of phospholipids and galactolipids (Malhotra and Kahn 1978).

Very little is known on the effects of gaseous nitric acid on plants. Elevated concentrations of nitrate and changes in activity of nitrate reductase (NaR) have been observed following exposures to nitric acid (Aber et al. 1989, Norby et al. 1989). It is known that concentrations of NaR are low in foliage of most of the woody plant species (Kramer and Kozlowski 1979). However, synthesis of this enzyme has been observed to be stimulated by an increase of nitrate in some plant species (Zielke and Filner 1971). Therefore, the extent of metabolic disturbance from uptake of nitric acid would most likely depend upon the degree to which NaR synthesis has been stimulated.

Morphological Changes

Visible Symptoms of Injury

For gaseous pollutants, the degree of pollutant injury depends to a large extent on the effective dose which is a function of concentration, the length of exposure, and stomatal aperture (Kozlowski and Constantinidou 1986). For wet and dry deposition on leaf surfaces, transfer rates into plant cells are more difficult to estimate, and therefore correlation with visible injury is more difficult to determine. Specific patterns of injury vary among pollutants, plant species, and site of injury. Visible symptoms of pollutant injury include mottling, chlorosis, necrotic lesions, and browning, which may induce early senescence and abscission of foliage (see Chapter 7).

Ozone injury of cells and tissues is essentially the same in woody and herbaceous plants, and the symptoms produced are similar. Injury occurs first in the most photosynthetically active tissues; in woody species, this may be either the palisade or mesophyll tissue. Symptoms consist of minute chlorotic flecks developing on the upper surface of sensitive leaves as chloroplasts become disrupted. Epidermal cells overlaying the injured area usually remain uninjured. As more cells become affected, the lesions increase in size and coalesce, and appear chlorotic, bleached or brownish. Severe lesions may extend through the leaf, with symptoms on both surfaces (Taylor 1984). For coniferous trees, chlorotic mottling appears on the older needles and progresses to the current season's needles. Another symptom of injury is a tan necrosis extending from the needle tip inwards. Premature abscission of needles progressing from older to younger is typical on ozone-damaged trees (Miller et al. 1963, Richards et al. 1968, Miller and Millecan 1971).

Pines tend to be among the most ozone-sensitive trees in Western forests; ponderosa, Jeffrey, and western white pine (*P. monticola*) are believed to be the most sensitive (Miller 1983). Sensitive species of pines exhibit a larger number of stomata per cross-sectional area of mesophyll cells. More stomata per unit mesophyll cell or cell area allow greater penetration of ozone. The number and thickness of epidermal and hypodermal cells is negatively correlated with ozone sensitivity. The most ozone-resistant pine species have the thinnest and lowest number of hypodermal cell layers (Evans and Miller 1972a).

Peroxyacetyl nitrate results in a glazed appearance of the undersurface of herbaceous plants by causing a collapse of protoplasts of the mesophyll tissue, creating air pockets under the cuticle. These symptoms have not been observed in conifer species. Needles of susceptible pine species develop chlorotic mottle on needle surfaces exposed to direct sun when exposed to very high (1 ppm) concentrations of peroxyacetyl nitrate. Peroxyacetyl nitrate typically affects rapidly expanding leaf tissue so that on pine, the mottle develops near the needle base. The symptoms of peroxyacetyl nitrate injury have not been observed on needles of pine growing in the field and injury has been produced only by fumigation with concentrations many times greater than has been measured in ambient air (Taylor 1984).

Classic sulfur dioxide injury on broadleaf species is distinct and characterized by areas of injured leaf tissues between the healthy tissue along the veins. If injury is slight, the damaged tissue is yellow but as injury becomes more severe, the interveinal areas turn brown as the tissue dies. In conifer needles the mesophyll cells are most susceptible to sulfur dioxide, but other cells are sometimes injured as well (Kozlowski and Constantinedou 1986).

Development of injury on plants exposed to nitric acid vapor has not been investigated. However, the cuticular layer and substomatal cavities of the leaf may be particularly susceptible to injury by nitric acid. Transport of nitric acid into the stomatal chamber and through cuticular channels has been demonstrated, but very little is known about the penetration of nitric acid into the leaf interior (Taylor et al. 1989, Marshall and Cadle 1989). Although it has not yet been shown, we can anticipate the effects of nitric acid on wax fibrils (Turunen and Huttunen 1990) and stomatal cavities (Taylor et al. 1989) to be similar to those caused by acidic precipitation.

For acidic precipitation (rain, fog, clouds), the contact time of acidic droplets or films on the leaf surface determines the degree of damage (Wellburn 1988). Because penetration of rain or leaf surface solutions through stomata is not likely, solutions are believed to move into the leaf interior through cuticular micropores (Evans 1982). Most of the damage occurs where moisture normally accumulates on leaves: along the side of the leaf veins or margins, at bases of trichomes, hydathodes, glandular hair, and on stigma tips (Evans 1982, Wellburn 1988). In most of the studies on the effects of acidic precipitation on plants, visible injury starts at pH < 3.0–3.6 (Shriner et al. 1990).

There is very little information on acidic precipitation injury symptoms on Western trees. In a greenhouse experiment, severe foliar necrosis of Jeffrey pine and giant sequoia (*Sequoiadendron giganteum*) seedlings were observed after one day of exposure to pH 2.0 acidic mist. The injury started as dull, grey-brown water-soaked lesions, and then developed into reddish-brown necrotic areas covering 10–50% of the needle surfaces of both species (Temple 1988). Turner et al. (1989) exposed seedlings of four Western conifers to episodic fog events of pH 2.1, 3.1 or 5.6

over a 60 day period. Foliar injury was observed only in western hemlock (*Tsuga heterophylla*; pH 2.1 and 3.1) and western redcedar (*Thuja
plicata*; pH 2.1); Douglas-fir and ponderosa pine showed no foliar injury.
In a field study, sixteen two-hour-long acidic fog exposures at pH < 3
induced brown tip burn injury on the current and previous year needles
of white fir. In the same study no significant increase of injury was
observed for ponderosa pine seedlings (Takemoto and Bytnerowicz
1992).

Effects on Cuticles

Air pollution-induced wax erosion resembles accelerated natural weathering of the needle surfaces, which differs from mechanical, fungal, and
insect damage. In general, an accelerated fusion of wax tubes in the
stomatal areas of the needle surfaces is the most common effect of air
pollution and acidic precipitation in conifers. The specificity of the
symptoms in the wax structures to different air pollutants is limited
because of natural variability (even within similar-aged trees of the same
provenance), and complexity of environmental factors affecting the wax
tubes (Turunen and Huttunen 1990).

Erosion of epicuticular waxes decreases the cuticular diffusion resistance
to water vapor (Fowler et al. 1980, Cape and Fowler 1981), causing
increased permeability of the cuticle and increased cuticular transport. In
northern conditions, air pollution can induce winter erosion of needle
surfaces, increasing the risk of drought stress during the late winter
when frozen soil prevents water uptake (Huttunen and Laine 1983).
Although no visible injury symptoms were seen on Scots pine (*Pinus
sylvestris*) needles exposed to simulated acid rain of pH 3 and 4, effects
on plant surfaces were found with scanning electron microscopy. These
symptoms included delayed development of epicuticular waxes, and
four forms of deformed stomatal complex: narrow, half-formed, occluded, and double-sized (Turunen and Huttunen 1991).

Stomatal occlusions as a result of progressive aggregation and melting of
wax tubes decrease gas exchange in spruce and fir needles during the
growing season (Rinallo et al. 1986, Sauter and Voss 1986). Ponderosa
and Coulter pines (*Pinus coulteri*) exposed to ambient levels of photochemical smog in the San Gabriel mountains in southern California
showed smoothing of young wax crystals: this effect was less apparent in
the clean-air treatment (Bytnerowicz et al. 1989).

In the mountains of southern California, high concentrations of trace
organic pollutants have been measured (Helmig and Array 1991). It has
been postulated that these compounds, as well as ozone and oxygen free

radicals, may be especially effective in producing changes in aromatic compounds and unsaturated hydrocarbons of the cuticular waxy layer (Mudd et al. 1984, Hewitt et al. 1990).

Effects on Cell Ultrastructure

Air pollutants may induce changes in cell ultrastructure long before visible symptoms appear (Soikkeli 1980), and the effects of low concentrations of air pollutants on evergreen plants over prolonged periods are perhaps most effectively determined from ultrastructure observations (Huttunen 1984). The development of ultrastructure injury is most pronounced in actively metabolizing tissues (Karenlampi and Soikkeli 1980). The most frequently observed changes in the features of chloroplasts are swelling, rounding or decrease in size; apparent doubling of the chloroplast envelopes; stretching of chloroplast envelopes; swelling of chloroplast thylakoids; reduction of grana lamellae; and granulation of the chloroplast stroma (Huttunen 1984).

Changes in the chloroplast stroma, such as the granulation observed in response to air pollution and other stresses, would be expected to reflect changes in the structure and catalytic activity of the enzymes of C3 metabolism (Parry and Whittingham 1984). Exposure to air pollutants can also disrupt the structure of the thylakoids and grana within the chloroplasts, and such disruptions are likely to have important consequences on the activity of Photosystem (PS) I and II (Sugahara, 1984).

In ponderosa pine, the most sensitive cells to ozone were mesophyll cells. The greatest injury occurred in cells adjacent to the endodermis. In elongating needles, the injury was most prevalent in internal mesophyll cell layers (Evans and Miller 1972b). No discernible effects of ozone were found in other cells within which chlorotic mottle developed. Chloroplasts and carbohydrate stain aggregated in the peripheral portions of the mesophyll cells. Aggregation of cytoplasmic nucleic acids and proteins took place in the damaged cells. Mesophyll cells were abnormally folded and twisted following chloroplast aggregation. They collapsed after their intercellular contents disappeared. Extensive wall deformations in the mesophyll usually resulted in a collapse of leaf tissue external to the endodermis (Evans and Miller 1972b, 1975). Karenlampi (1986) examined ponderosa pine needles collected from air pollution-sensitive and resistant individuals in Sequoia National Park and in the San Bernardino Mountains. In that study, only some of the cells in chlorotic tissue were disintegrating. Most of the affected cells retained their integrity, but had less starch, central vacuoles with coarsely granular contents or electron-dense deposits accumulated along the tonoplasts, and generally poorly defined cytoplasm and chloroplasts.

In the later stages of chlorosis, degenerated cells were observed which had a simplified thylakoid structure, abundant plastoglobuli, and droplets of lipid-like material. The same author found similar changes in the ultrastructure of Monterey pine (*Pinus radiata*), Italian stone pine (*Pinus pinea*) and Aleppo pine (*Pinus halepensis*) in the highly polluted areas of the Los Angeles Basin (Karenlampi 1987).

Carbon Balance and Allocation

In general, exposure to pollutants changes the net carbon balance of a plant through effects on the light reactions or enzymatic functions, increased respiration from reparative activities, or decreases in stomatal conductance. However, different responses to similar pollutant exposures have been described in the literature, even within species. For example, the effects of low pollutant levels may be stimulatory but pollutant levels over a certain threshold may become deleterious. Young tissue may be affected more than older tissue because of incomplete or thin cuticles, resulting in increased effective dose. In other cases, older tissue may be more susceptible because of overall lower net carbon balance and damaged cuticles. The phenological and ontological stage of the plant is often overlooked in experimental exposures. Obviously, if the plant is not physiologically active, it may be less responsive to pollutant exposure. Pollutants may alter phenology in the current year, and in subsequent years, even without continued exposure. The nutritional status of the foliage, leaf turnover rates, and other environmental stresses, whether induced or artifactual, can reverse experimental results when compared between studies. Lastly, the amount of tissue measured (single fascicle or a portion of branch vs. all current year foliage) and the reporting units of the measurement (extrapolation to whole seedling or "canopy") can significantly alter the conclusions of studies. Often variables are not known, measured, reported, or extrapolated in a fashion that allows comparisons between studies. Although visible symptoms may not be present, biochemical and physiological processes may have been affected.

Photosynthesis

Exposure to greater concentrations of ozone and sulfur dioxide are necessary before their effects are detected in coniferous trees relative to broadleaf species, and in trees relative to herbaceous species (Darrall 1989). In general, photosynthesis in Western trees shows a measurable response to ozone at concentrations >80 ppb, depending on the duration of the exposure and environmental conditions during fumigations.

Similar to pollutant effects on stomatal conductance, field and experimental studies at the single fascicle level have shown the full range of responses: deleterious, no effect, or enhanced net assimilation rate (NAR).

Associated with reduced stomatal conductance, gross photosynthesis was reduced with long-term exposure to elevated ozone in a field study of mature ponderosa pines in the San Bernardino Mountains (Coyne and Bingham 1981). In all retained needle age classes, losses in gross photosynthesis exceeded reduction in stomatal conductance, suggesting that injury to the mesophyll, photochemical mechanism, or carboxylation, was greater than injury to the stomata. This was supported by a study in which ponderosa pine seedlings were exposed to 200% ambient ozone concentrations for one season in the southern Sierra Nevada (seasonal 24 hour average concentrations of 0.09–0.10 ppm; peak hourly values <0.22 ppm). NAR of the current year needles was reduced with no decrease in stomatal conductance. In the second season of exposure, a significant reduction of stomatal conductance occurred and caused an even greater reduction of NAR than in the first season (Bytnerowicz et al. 1991).

Patterson and Rundel (1989) found during the 1987 growing season that Jeffrey pine trees resistant to ozone maintained lower stomatal conductance and consequently lower maximum NAR than sensitive trees in Sequoia National Park. However, the relationship was reversed during the following summer. Maximum NAR dropped in both groups of trees during the summer of 1988, but sensitive trees consistently maintained lower NAR (by 20–30%) than the resistant individuals (stomatal conductance values for the two groups of trees were similar). The initially higher NAR of the sensitive trees may reflect higher intrinsic stomatal conductances of sensitive individuals, or may result from increased carbon fixation as a result of greater carbon sinks in injured trees. Sensitive trees showed a 25% slower initial increase in NAR in response to increasing internal carbon dioxide, indicating a lower carboxylation efficiency. NAR at light saturation was 20% lower for the sensitive trees, suggesting possible injury to the electron transport or RuBP regeneration mechanisms of the photosynthetic apparatus. These effects of ozone on carboxylation efficiency and RuBP regeneration are consistent with that of an Eastern tree, loblolly pine (*Pinus taeda*) exposed to elevated ozone (Teskey et al. 1986, Sasek and Richardson 1989).

In a study of ponderosa pine seedlings exposed to elevated ozone, stomatal conductance was greater in exposed plants relative to controls. NAR, when measured on current year foliage at the same conductance values, was lower in exposed plants that in controls for current year foliage ("whole canopy"; NEG, unpubl. data). However, NAR (Beyers et al. 1991, Leininger and Fenn, in prep.) and photosynthetic efficiency

(Hom and Riechers 1991) of single fascicles of ponderosa pine seedlings were enhanced by elevated ozone in similar exposure studies. The discrepancy in response can be explained by significant reductions of older foliage, nitrogen retranslocation from senescing to retained needles, and subsequent enhanced NAR in retained needles in the "whole canopy" study. In another study, reductions in NAR in Douglas-fir were greater than accounted for by the accompanying reduction in conductance, and light compensation point was also greater in ozone-exposed saplings (Wieger, pers. comm.).

Giant sequoia is one of the only species to date where several tree age classes have been tested for sensitivity to elevated ozone (Figure 5.6). All comparisons were made between treatments on current year foliage. There was a significant change in foliage morphology from single needles on seedlings to appressed scales in more mature giant sequoia trees. In symptomatic current year seedlings, NAR was reduced significantly by 44% in elevated ozone (150% ambient) vs. charcoal-filtered controls after three months of exposure (monthly 24-hour average, 69 ppb for June through September, Sequoia National Park, CA; Grulke et al. 1989). In the same study, rooted cuttings of 12-year-old giant sequoia exposed to elevated ozone had a 20% reduction in NAR relative to controls, but differences between treatments were not significant. Branches of 120 year old giant sequoia were exposed in situ to several levels of ozone (20%, 100%, 200%, and 300%), but no significant differences in NAR between treatment levels were evident (NEG, unpubl. data). In another study, 2-year-old giant sequoia seedlings exposed to two months of elevated ozone showed no differences in growth (Temple 1988).

In a four month experimental exposure to elevated ozone (140 ppm-hour, 1.8x profile of Hogsett et al. 1989), NAR of four common Western conifers (white fir, Engelmann spruce, Douglas-fir, and ponderosa pine) was not affected relative to controls exposed to charcoal-filtered air (NEG, unpubl. data). NAR of pond pine (*Pinus serotina*, Barnes 1972) and red spruce (Kohut et al. 1990) was also not affected by elevated ozone. NAR was enhanced in two seed sources of eastern white pine (*Pinus strobus*, Barnes 1972) and in Norway spruce (Freer-Smith and Dobson 1989). Reductions in NAR were not necessarily accompanied by reductions in biomass for an Eastern conifer, Fraser fir (Tseng et al. 1988).

Houpis (1989) found reductions in NAR of ponderosa pine saplings at 150 ppb sulfur dioxide. For sulfur dioxide, the lowest concentrations reported to inhibit NAR in long-term fumigations is 60 ppb over 28 days for silver birch (*Betula pendula*, Freer-Smith 1985). In western Canada, the seasonal mean NAR of the current and previous year foliage of lodge-pole pine (*Pinus contorta*) x jack pine was significantly lower at the sites

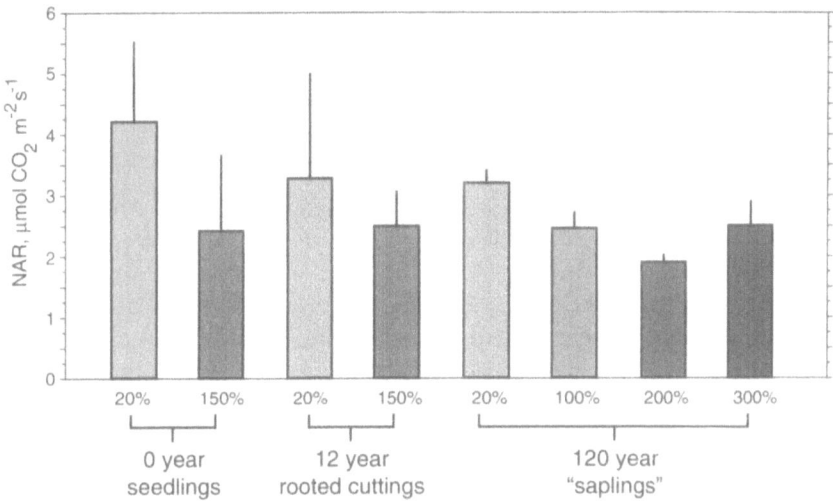

Figure 5.6 Comparison of the physiological response to ozone of current year leaf tissue attached to different aged stems of giant sequoia. Current year seedlings (n=8) and 12 year rooted cuttings (n=5) were both exposed in the same experiment in 1988; branches on the 120 year saplings (n=3) were exposed for two months in 1990. Treatments are % ambient ozone. Ozone concentrations during 1988 averaged 69 ppb for June through September (24 hour average), Sequoia National Park, CA.

frequently exposed to sulfur gases than at the sites infrequently exposed. Current year foliage at the sites infrequently exposed to sulfur gases achieved positive NAR 1–3 weeks earlier than foliage from the frequently exposed sites. Reduced NAR, and shorter periods of positive NAR combined with shorter needle retention and foliar chlorosis partly explained the previously reported lower productivity of trees at the sites with high frequency of sulfur gas exposures (Amundson et al. 1986).

In general, exposures of plants to sulfur dioxide cause reductions in photosynthesis. Sulphur dioxide directly affects photophosphorylation of ATP, especially at low phosphate levels, which is attributed to sulfate competing for phosphate sites during the ATP formation. This may in turn lead to a shortage of ATP, which is needed for carbon dioxide fixation, export of metabolites, and for the formation of proteins (Wellburn 1988). In comparison with the effects of ozone and sulfur dioxide alone, fumigations with a mixture of these gases have been found to have greater inhibitory effects on a variety of species. It is also apparent that different levels of nitrogen dioxide can bring about both increases and decreases of NAR within the same species (Wellburn 1988).

Acidic fog in the range of pH 2.5 to 4.0 tends to act as a fertilizer and increases NAR (red spruce, Eamus and Fowler 1990; loblolly pine, Hanson et al. 1988, Reich et al. 1987). However, effects on NAR were not necessarily reflected in biomass for red spruce (Kohut et al. 1990). At pH 2.4 and below, the deleterious effects of acidity overrode the fertilization effect, and reduced NAR in ponderosa pine (Takemoto and Bytnerowicz 1992), Jeffrey pine (Temple 1988), and red spruce (Kohut et al. 1990). In studies testing for the effects of the interaction of acidic fog and ozone, little (Reich et al. 1987) or no significant interactions were found in gas exchange measures (Skeffington and Roberts 1985, Taylor et al. 1986, Edwards et al. 1990). Winter exposure to acidic fog resulted in reduced conductance in white fir (similar winter and summer conductance values). This was reflected in reduced NAR for plants previously exposed to charcoal-filtered air, but not for plants previously exposed to elevated ozone (NEG, unpubl. data). Ponderosa pine and Douglas-fir had relatively inactive stomata during winter exposure to acidic fog, thus effectively avoiding this pollutant (NEG, unpubl. data).

Respiration

If pollutant concentration is high enough to cause cell death, it can be assumed that respiration is also increased (whether through short term increases as cells die, or through subsequent repair), but the amount of cellular damage will obviously determine whether or not increased respiration is detectable. Dark respiration appeared higher in all five Western conifers exposed to elevated ozone for one growing season, but no differences between the ozone exposed and control plants were significant (NEG, unpubl. data). Exposure of seedlings of giant sequoia to elevated ozone resulted in significant increases in dark respiration, and directly accounted for differences observed in diurnal carbon balance between elevated ozone and control plants (Grulke et al. 1989). However, dark respiration was not increased in mature foliage of the same species (NEG, unpubl. data). According to Wellburn (1990), it is unlikely that nitric oxide and nitrogen dioxide exposures at realistic concentrations have a direct and permanent effect on dark respiration. In a study by Coyne and Bingham (1981), although light respiration was inversely related to needle injury, the ratio of net photosynthesis to gross photosynthesis tended to decrease with ozone injury. The stressed trees apparently not only experienced reduced rates of carbon dioxide fixation, but also retained a smaller proportion of the assimilated carbon after respiration loss.

Allocation of Photosynthates

Any external influence that affects carbon acquisition, leaf turnover rates, nutrient content of retained foliage, and hormonal concentrations will likely alter within plant carbon allocation. In many experiments, exposure to elevated ozone and sulfur dioxide led to reduced root biomass, presumably because of foliage loss, large reparative costs to the foliage, reduced total photosynthate, and reduced allocation of photosynthate to below-ground tissues (McLaughlin and McConathy 1983). Reduction of root biomass may in turn lead to reduced ability to absorb soil nutrients, accelerating the stress imposed on a tree by ozone. Much of the literature describes pollutant influence on total biomass and above-ground plant parts (reviewed in Guderian 1977), and comparisons between studies are inconclusive without a clear analysis of relative shifts of within-plant resources (i.e., statistical analysis conducted on % biomass of roots vs. component changes or root to shoot ratios). The focus of the following section will be on relative changes in biomass allocation, because these shifts may lead to changes in ecosystem function via alterations in above- and below-ground allocation and ecosystem dynamics, and are thought to be a sensitive biological indicator of pollution stress (Jones and Mansfield 1982).

Two of the most extensive tests of biomass allocation in response to exposure to elevated ozone followed by acidic fog were conducted by PR Miller et al. (in prep.) for ponderosa pine, Douglas-fir, Engelmann spruce, subalpine fir, and white fir (ranked from sensitive to resistant), and Hogsett et al. 1989 for ponderosa pine, western hemlock, Douglas-fir, western redcedar, and lodgepole pine (ranked from sensitive to resistant). In both studies, inconsistent changes in the biomass of plant parts occurred in response to ozone exposure. In Miller's study, only two species showed shifts in biomass allocation (i.e., increase in one plant part at the "cost" of reduction in the biomass of another plant part). Ponderosa pine and Engelmann spruce both had reduced root to needle ratios in elevated ozone relative to charcoal-filtered air. Following ozone exposure, root-to-needle ratio of ponderosa pine declined in the pH 2.1 acidic fog treatment, but no biomass shifts were apparent in the acidic fog treatment for Engelmann spruce. In Hogsett et al.'s (1989) study, acidic fog reduced root to needle ratio in western redcedar and acidic fog alone and in combination with ozone reduced root-to-needle ratio in western hemlock. Elevated sulfur dioxide alone also reduced root-to-needle ratio in western hemlock. Most growth effects in Hogsett et al.'s study occurred during bud elongation in the spring following exposure. As part of this study, ponderosa pine exposed to elevated ozone were found to have reduced root starch reserves available just prior to and during bud break (Andersen et al. 1991).

The ratio between "recovered" root dry weight and needle dry weight in loblolly pine decreased by 20% over a range of a seasonal 24 hour mean ozone concentration of 22 to 92 ppb (ratio not calculated in paper; Shafer and Heagle 1989). Slash pine (*Pinus elliotii*) exposed to elevated ozone vs. charcoal-filtered control plants had reduced root weight relative to shoot weight, and the reduction in root growth rate was greater than the reduction in shoot growth rate (Hogsett et al. 1985). Root to leaf biomass ratio in paper birch (*Populus papyrifera*) was not altered when exposed to elevated ozone and acidic fog singly or in combination despite decreases in both leaf and root biomass in elevated ozone and increases in root weight in the pH 3.5 acidic fog (Keane and Manning 1988). This is the only study to report a positive, synergistic increase in leaf weight in a pH 3.5 acidic fog treatment in combination with elevated ozone exposure (elevated ozone, seven hours per day, five days per week, for 12 weeks). Ozone and sulfur dioxide alone, but not in combination, altered root-to-shoot ratios in loblolly pine (increase in some families, decrease in others; Winner et al. 1987).

In their study of the interaction between ozone and nitrogen dioxide, (Kress and Skelly 1982) found that nitrogen dioxide alone decreased root and total biomass of sweetgum (*Liquidambar styraciflua*), and ozone and nitrogen dioxide in combination resulted in less than additive reduction (possibly negatively synergistic) in root and total biomass. It is likely that in response to reduced needle function and biomass, root biomass was re-allocated to maintain damaged needles since no new needle tissue was produced. However, loblolly pine exposed to elevated ozone levels had reduced, labelled photosynthate allocation to fine roots (Adams et al. 1990).

Photosystems

Changes in the photosystems of plants caused by air pollution can be identified from deterioration of photosynthetic pigments and reduction in the efficiency of the photochemical reactions. In several studies, decreases in chlorophyll a and b, the ratio of chlorophyll a to b, and carotenoid concentration have been associated with pollutant exposure. Few studies have specifically examined changes in chlorophyll fluorescence and apparent quantum efficiency as a result of pollutant exposure.

Pigment Concentration

There are two theories that address changes in chlorophyll content after exposure to sulfur dioxide. One theory suggests that the destruction of chlorophyll is catalyzed by chlorophyll-specific enzymes that convert chlorophyll a and b to chlorophyllide by removing the phytol group (Malhotra 1977). Another theory suggests that chlorophyll is converted to pheophitin by replacing magnesium with hydrogen on the chlorophyll molecule (Hallgren 1978). Preferential destruction of chlorophyll a results from lower activation energy required for removal of the magnesium atom and its greater hydrophylic characteristics. There are no studies that indicate how sulfur dioxide may mediate these processes.

The effects of sulfur dioxide exposures (150 ppb) on ponderosa pine foliage pigments were not observed until after eight months of exposure. Chlorophyll a and carotenoids were primarily affected, while chlorophyll b concentrations were only slightly affected. The reductions in pigment concentrations were expected to result in reduced photosynthesis and increased susceptibility to photoinhibition (Houpis 1989).

Little is known about the mechanisms of pigment destruction by ozone. However, decline of chlorophyll concentration has been proposed as an indicator of ozone phytotoxicity (Knudson et al. 1977). In a short-term experiment with elevated ozone concentrations, significant reductions in chlorophyll content of Rocky Mountain provenances of ponderosa pine seedlings were found (Aitken et al. 1984). Foliar tissue from mature grafted scions of ponderosa pine showed decreased concentrations of chlorophyll a following 10 months of exposure to 2x ambient ozone (Neuman et al. 1991). After two seasons of exposure, ponderosa pine seedlings exposed to 200% ambient ozone concentrations in the southern Sierra Nevada had significantly reduced concentrations of chlorophyll a and b, chlorophyll a/b, and carotenoids; the effects of ozone increased as the season progressed (Figure 5.7, Bytnerowicz et al. 1991).

Contrary to the effects of ozone and sulfur dioxide, plants exposed for short periods to nitric oxide and nitrogen dioxide are often greener than plants grown in clean air. After longer periods of exposures, the stimulatory effect of nitrogen dioxide disappears and inhibition of pigment biosynthesis takes place (Zeevaart 1976, Sandhu and Gupta 1989). The effects of nitrogen dioxide on pigments in herbaceous species depend on pollutant concentration (Sabaratnam et al. 1988). Nitrate ions reduce the pH gradient across the thylakoid membranes of oats (Robinson and Wellburn 1983), possibly mediated by a free radical mechanism because there are many similarities between the effects of ozone alone and combined effects of sulfur dioxide and nitrogen dioxide (Reinert et al. 1975).

Pigment concentration in foliage of white fir seedlings was not affected by acidic fog (pH 4, 3, and 2) exposures. However, concentrations of chlorophyll and carotenoids were increased in ponderosa pine foliage in the pH 2.0 treatment (Takemoto and Bytnerowicz 1992). One season

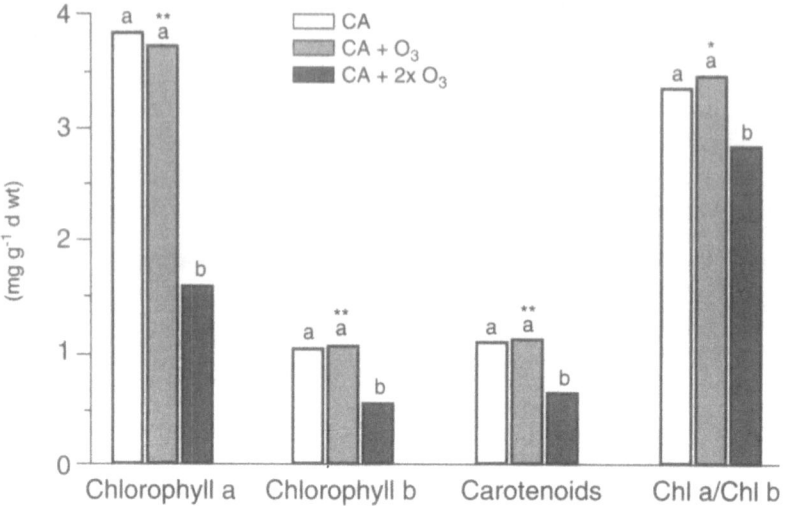

Figure 5.7 Effect of ozone on pigment concentrations of previous year foliage of ponderosa pine seedlings in the second season of exposure. Different letters over the bars indicate significant differences either at p≤0.05(*), or p≤0.01(**). CA= charcoal-filtered air; CA + O₃ = charcoal-filtered air with an addition of ozone at ambient concentrations; CA + 2x O₃ = charcoal-filtered air with an addition of ozone at double ambient concentrations.

exposure to ambient levels of photochemical smog in the San Gabriel Mountains did not affect chlorophyll concentrations in foliage of ponderosa, Jeffrey, Coulter, and Scots pines (Bytnerowicz et al. 1989).

Chlorophyll Fluorescence

Chlorophyll fluorescence is a sensitive indicator of photosynthetic energy conversion (Papageorgiu 1975). The specific site of ozone damage is believed to be on the PS II donor site (water-splitting enzyme system) prior to any decrease in energy transfer efficiency within the pigment system. With increasing exposure of bean plants (*Phaseolus vulgaris* L.) to ozone, the electron transport from PS II to PS I also became inhibited (Schreiber et al. 1978). Changes in chlorophyll fluorescence took place during a long-term experiment in which ponderosa pine seedlings were exposed to ambient and 200% ambient ozone concentrations in the southern Sierra Nevada. In the second year of exposure, significant reductions of the ratio of variable fluorescence to maximal fluorescence (Fv/Fm) were determined (Figure 5.8, Bytnerowicz et al. 1991), indicating changes in the efficiency of PS II and transport of energy between PS II and PS I (Oquist and Wass 1988). No changes in Fv/Fm ratio were apparent in a short-term ozone exposure of mature giant sequoia foliage (NEG, unpubl. data), but perhaps refraction of light from the scale-like needles confounded the results.

Schmidt et al. (1990) also used chlorophyll fluorescence to study the toxic effects of sulfur dioxide and nitrogen dioxide on photosynthetic systems of plants. The mechanism of sulfur dioxide effects on fluorescence is believed to be the weakening of the PS II donor site because of its acidifying action, and inhibition of the Calvin cycle activation. Nitrogen dioxide when dissolved in water produces nitric and nitrous acid, which may cause acidification of the systems in a similar way to sulfur dioxide. However, in light, nitric and nitrous acid are rapidly reduced to ammonia and the hydroxyl ion is produced as a co-product. In such conditions, the hydroxyl ion prevents net acidification of the system and changes in the photosystems.

Apparent Quantum Efficiency

Apparent quantum efficiency is limited by the light-harvesting system and the efficiency of the photochemical reactions. It is essentially an integration of pigment concentration, fluorescence, and electron transfer. Apparent quantum efficiency was significantly reduced for ponderosa pine after four months of exposure to elevated ozone (140 ppm-hour, 1.8x profile of Hogsett et al. 1989), but was not affected in white fir or Douglas-fir seedlings (NEG, unpubl. data). Reduced apparent quantum efficiency resulted in loblolly pine after 200% ambient ozone exposure for five months (Sasek and Richardson 1989).

Effects on Plant Water Relations

Few studies have examined pollutant effects on plant water status. Ozone and sulfur dioxide do not appear to affect the water relations of plant tissues in short-term experimental studies. In a branch fumigation study of mature foliage of giant sequoia, there was no apparent effect of 300% ambient ozone relative to ambient ozone levels on needle osmotic + matric potential or turgor for current and previous year's foliage (NEG, unpubl. data). Based upon xylem pressure potentials and leaf diffusive resistances, water relations of lodgepole pine x jack pine and other conifers in the vicinity of the Whitecourt Gas Plant in Alberta, Canada, were not measurably affected by the sulfur gas emissions. Therefore, the observed effects of sulfur gas emissions on photosynthesis of these species (Amundson et al. 1986) were not caused by drought stress, and

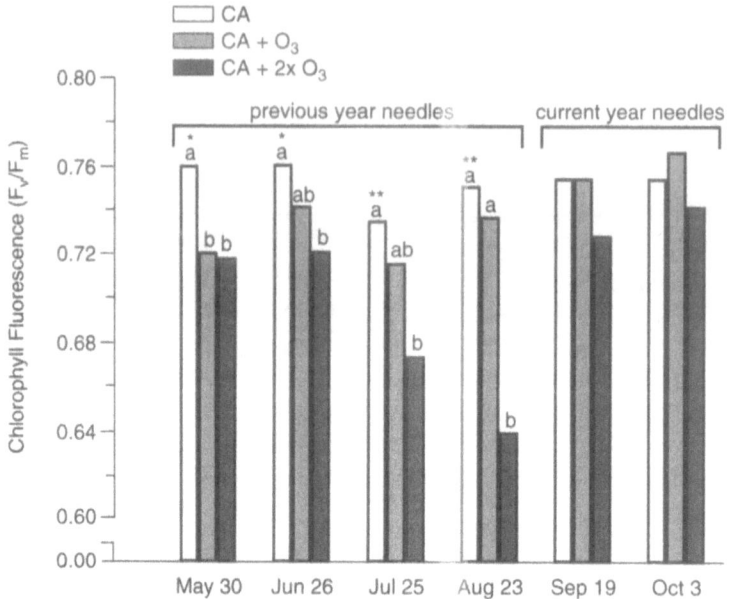

Figure 5.8 Effect of ozone on chlorophyll fluorescence of ponderosa pine seedling foliage in the second season of exposure. Different letters over the bars indicate significant differences either at p≤0.05(*), or p≤0.01(**). CA= charcoal-filtered air; CA + O₃ = charcoal-filtered air with an addition of ozone at ambient concentrations; CA + 2x O₃ = charcoal-filtered air with an addition of ozone at double ambient concentrations.

cannot be explained by changes of stomata or tree water status (Mayo et al. 1986), as was found by Coyne and Bingham (1981) and Bytnerowicz et al. (1991) for ponderosa pine affected by ozone. As mentioned above, short-term experimental exposure to ozone generally results in enhanced stomatal conductance, and if exposure is prolonged, could eventually result in increased drought stress.

Increasing water stress moderates the detrimental effects of atmospheric pollutants because of reduced stomatal conductance, and consequently reduced pollutant dose. In a study by Houpis (1989), decreased effects of sulfur dioxide on ponderosa pine were due to decreasing stomatal conductance in the water stressed plants, leading to the restricted influx of sulfur dioxide into the leaves.

Elemental Status

Air pollution may affect the nutritional status of forest trees. Gaseous pollutants and pollutants dissolved in rain and fog may be absorbed by foliage or by roots from the soil. Elements contained in particulate pollutants may be absorbed into foliage only after being dissolved in water (dew or fog), and mechanisms of this uptake are not known. A good example of this is foliage uptake of copper and zinc applied at three levels to lodgepole pine in a greenhouse experiment (Jacobs and Weaver, in prep.). A significant amount of metal strongly adhered to cuticles which could be removed only with chelation. Damage was detected at metal concentrations of ≥ 0.001 M. These results contrast with the general belief that most of the particulate pollution accumulated on foliar surfaces will be taken up through the roots, after being resuspended or washed off the leaves.

Limited information is available on the effects of photochemical pollutants on nutrient uptake. Since ozone exposures reduce the allocation of carbon to roots, a typical reaction of plants is a decrease of root biomass (often described as reduction of the root to shoot ratio). Therefore, uptake of nutrients by plants exposed to ozone may be affected. Tingey et al. (1986) found that ozone exposures decreased concentrations of calcium, magnesium, iron, and manganese in foliage, and increased the concentrations of potassium, phosphorous, and molybdenum in pods of bean plants. No significant changes in nutrient content of stems and roots were determined. The decreased concentrations of nutrients in foliage appeared to be the result of reduced transport into the leaves, and reallocation from older to younger leaf tissue rather than reduced uptake or leaching. Results from the ozone and acidic precipitation experiments in which foliar nutrient and net throughfall nutrient content were studied provide further supporting evidence for such changes (loblolly pine, Edwards et al. 1991).

Nitrogen concentrations were reduced, but no changes in concentrations were found for phosphorous, calcium, and potassium in the foliage of ponderosa pine seedlings after one season of exposure at ambient levels of photochemical smog in the San Gabriel Mountains (Bytnerowicz et al. 1990). No clear effects of ozone exposures on elemental composition of Jeffrey pine and giant sequoia seedlings were determined (Westman and Temple 1989), and Anderson and Houpis (1991) saw no significant effect of a ten month ozone exposure (2x ambient) on the foliar nutrient content of mature branches of ponderosa pine.

Exposure of plants to nitrogen dioxide, nitric oxide, and nitric acid vapor may increase the concentration of nitrate in plant tissue. However, concentrations of total nitrogen within plant shoots generally decline following exposures to nitrogen dioxide. The reason for this decline is not clear, but translocation of additional nitrogen from shoots to roots appears to offer a partial explanation (Wellburn 1990). Changes in uptake of nitrogen may also cause alterations in the uptake of other nutrients.

Nitrate reductase (NaR), an enzyme that catalyzes reduction of nitrate to nitrite, is substrate-induced. Current evidence indicates that the activity of NaR in higher plants is regulated by new enzyme synthesis and breakdown (Wellburn 1990). Wingsle et al. (1987) has shown that the NaR activity of Scots pine seedlings increased significantly because of short term fumigation with nitrogen dioxide, while no changes occurred in the activity of control seedlings with nitrate supplied through the soil. Norby et al. (1989) observed an increase of NaR activity in red spruce seedlings exposed to low concentrations of either nitrogen dioxide or nitric acid vapor for one day. Nitric oxide produces both nitrite and nitrate ions in aqueous fluids, but the ratio of nitrate to nitrite is very low. Thus the plants exposed to high nitric oxide levels may have elevated nitrite concentrations if additional nitrite reductase (NiR) is not introduced in the chloroplast fast enough (Wellburn 1990).

Exposures to sulfur dioxide increase the sulfur content of both herba-ceous and woody plants (Ziegler 1973, Godzik 1976, Guderian 1977). The needle sulfur content of Scots pine from the highly polluted areas of Poland was 3–4 fold higher than that of needles collected in clean control areas (Bytnerowicz et al. 1980, 1981/1982, Dmuchowski et al. 1981/1982). Accumulation of total sulfur, sulfate-sulfur, and organic-sulfur in tree foliage caused by sulfur air pollution exposures has served as a basis for bioindication of sulfur dioxide pollution (Molski et al. 1983, Huttunen et al. 1985, Legge et al. 1988). The median sulfate-sulfur/organic-sulfur ratio for the current three years of growth of lodgepole x jack pine foliage varied from 0.29 at a reference location to 0.88 at a location with high sulfur deposition. This elevated ratio of sulfate-sulfur/organic-sulfur was accompanied by reduction of net photosynthesis of the current year foliage and lowered soil pH (Legge et al. 1988).

Acidic precipitation may cause both leaching and absorption of plant mineral components with subsequent changes in plant nutritional status (Scherbatskoy and Klein 1983, Jacobson et al. 1989). Exposures of Jeffrey pine and giant sequoia seedlings to acidic mist (pH 3.4–2.0) resulted in an increase of foliar nitrogen and sulfur concentrations, and increases or decreases in concentrations of various metallic cations (Westman and Temple 1989). Acidic precipitation in the range of pH 5.6 to 3.0 caused a significant increase in the needle content of nitrogen, phosphorus, potassium, calcium, manganese, cadmium, and chromium in eastern white pine (Reich et al. 1988), probably as a result of increased reallocation of elements from older to younger tissue. Garten and Hanson (1990) found that the leaves of red maple (*Acer rubrum*) and white oak (*Quercus alba*) exposed to acidic rain absorbed more ^{15}N-ammonium than ^{15}N-nitrate. Greater retention of ammonium is consistent with field observations of ammonium preferentially retained in forest canopies. In a field experiment in which ponderosa pine and white fir seedlings were exposed to acidic fogs of pH 2, 3, and 4, no significant changes in foliar concentrations of nitrogen, phosphorus, potassium, calcium and magnesium were found (Takemoto and Bytnerowicz 1992). Exposure of Douglas-fir seedlings to pH 3.1 fog (24 events over an 84 day period for a total of 96 hours) did not affect foliar nutrient contents (Turner et al. 1989, Turner and Tingey 1990).

Changes in the nutritional status of forest trees, especially following increases in nitrogen deposition, may cause serious ecological consequences in some of the forested areas in the western United States. Increased leaf turnover and higher foliar nutrient content in high air pollution sites of the San Bernardino Mountains indicated that litter decomposes more quickly, accelerating nutrient cycling in these forests (Fenn and Dunn 1989). Changes like these affect carbon and nutrient cycling, and may cause alterations of plant phenology, resistances to various environmental stresses, and eventually changes of species composition in various forest ecosystems (Chapter 6).

Defense Mechanisms

Various mechanisms have evolved to protect plants from the toxic effects of air pollutants. Plants can avoid, tolerate, or compensate for damage. In the case of some levels of wet or dry deposition, plants may even benefit from the nutrient additions.

Avoidance

Except for nitrogen oxides, stomatal mechanisms largely determine the exposure of internal tissues to pollutants. Resistance to nitrogen oxide movement into the leaf may be equal between the cuticle and stomata (Wellburn 1988). Stomatal mechanisms have been discussed extensively earlier in this chapter. Other reactions to pollutants that may constitute avoidance are an early onset of dormancy or senescence of affected plant parts. If pollutants are at high enough levels to result in reduced positive carbon balance, less growth will occur, and there will be a smaller sink to which nutrients are drawn. This process will inevitably lead to further reductions in positive carbon balance and will lead to needle or whole branch death. Younger tissue that may be able to procure enough "defense" via antioxidants will be retained.

Tolerance

Detoxification of photooxidants is especially important in Western forests because of the high pollutant levels and long summer season exposures. The strategy evolved in detoxification of photooxidants involves enzymatic processes and chemical reactions with anti-oxidants. Hydrogen peroxide, organic peroxides, and superoxide radicals can be detoxified enzymatically, whereas ozone or hydroxyl radicals can only be removed in plant cells by chemical reactions with antioxidants (Rabinowitz and Fridowich 1983).

Mehlhorn et al. (1986) determined that in the needles of European white fir (*Abies alba*) and Norway spruce exposed to low concentrations of ozone and sulfur dioxide, glutathione concentrations increased most rapidly, and were followed by increases in other anti-oxidants, vitamin E, and C. Tolerant clones of eastern white pine had greater glutathione reductase concentrations than did sensitive clones following ozone exposures (Chevonne et al. 1991). Induction of ascorbate peroxidase and glutathione reductase activities in pea plants (*Pisum sativum*) resulted from exposure to a mixture of sulfur dioxide, ozone, and nitrogen dioxide (Mehlhorn et al. 1987). In the same experiment, superoxide dismutase activity increased in ozone sensitive clones, but not in tolerant ones.

Mehlhorn and Wellburn (1987) and Mehlhorn et al. (1991) described another possible mechanism of plant tolerance to photooxidants related to the formation of free radicals from stress ethylene and ozone. They found that short term fumigation of pea, bean, and tobacco seedlings (*Nicotiana tabacum*) to 150 ppb ozone caused immense visible leaf injury if the plants were pre-grown in clean air. However, if the seedlings were grown in air containing elevated ozone concentrations, the subsequent exposures did not cause any foliar damage. The authors relate this phenomenon to the fact that the preconditioned seedlings produced only small amounts of ethylene (an adjustment to the stress), whereas those injured by the burst of ozone produced large quantities of "stress" ethylene which subsequently reacted with ozone and produced toxic free radicals.

Tolerance of plants to air pollutants also depends on foliage age. Young poplar leaves (*Populus robusta*) were more resistant to 2 ppm of sulfur dioxide than older leaves. Young leaves contained approximately five times more superoxide dismutase (SOD) than older ones (Tanaka and Sugahara 1980). The subcellular location of detoxification compounds and enzymes are presented in Figure 5.9 (Rennenberg and Polle 1989).

Increase of free proline concentrations in plants experiencing environmental stress seems to be another mechanism of tolerance. Such increases have been seen in plants under the influence of sulfur dioxide, other toxic gases and drought stress. After treating poplar leaves with proline solutions, sulfur dioxide phytotoxicity was reduced: degradation of chlorophyll a and b was smaller, and visible injury of leaves was not as apparent (Karolewski 1985).

For sulfur dioxide, conversion to relatively non-toxic compounds (sulfate, cysteine, glutathione) seems to be a main mechanism of plant protection. Tolerant clones of eastern white pine had appreciably more sulfur in plant tissue than susceptible clones (Roberts 1976), probably because of an enhanced ability to convert sulfur dioxide into non-toxic sulfur. Storage of excess sulfur as total sulfate is the best known example of this mechanism (Ziegler 1973, Legge et al. 1988). The reduction of

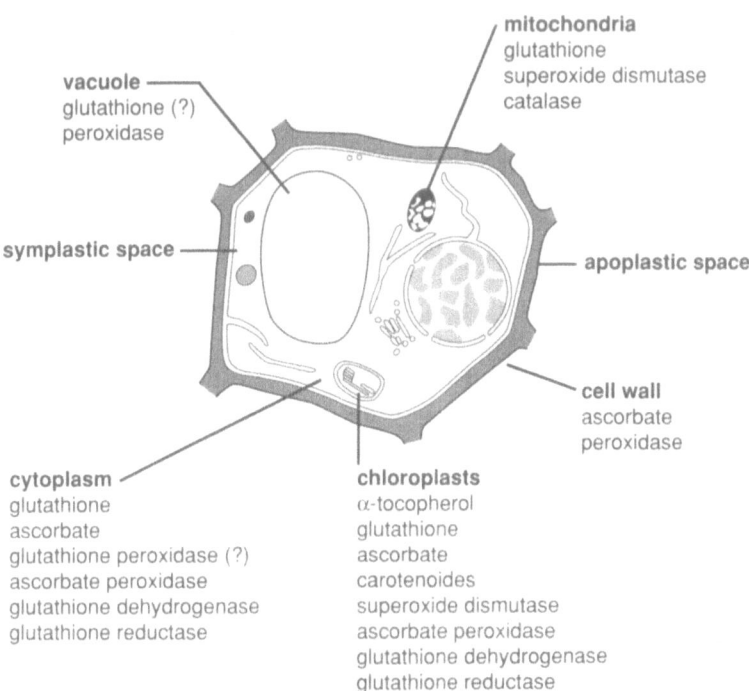

Figure 5.9 Detoxification of photooxidants in plants (adapted from Rennenberg and Polle 1989).

sulfite in chloroplasts produces sulfide, especially hydrogen sulfide and sulfydryl groups. The release of hydrogen sulfide in light by leaves of various tree species appears to be a route by which plants may release into the atmosphere the excess of sulfur accumulated from exposures to sulfur dioxide (Spaleny 1977, Wilson et al. 1978, Hallgren and Fredriksson 1982, Filner et al. 1984, Weigel et al. 1989). Other reduced sulfur compounds (dimethylsulfide and carbonylsulfide) are also emitted by plants exposed to sulfur dioxide. Sulfur dioxide itself may be re-emitted from plants (Nikolaevskii et al. 1976, Rennenberg et al. 1990). Emission of sulfur dioxide from Norway spruce foliage occurred mainly at night while the hydrogen sulfide gas was emitted continuously, with high rates during the day and low rates at night. The total amount of sulfur dioxide emitted from the needles was correlated with the availability of sulfate in the soil (Rennenberg et al. 1990). Metabolism of sulfur in plants, including detoxification of toxic compounds is presented in Figure 5.5.

Many studies on the effects of nitrogen dioxide indicate that plant detoxification of nitrogen dioxide occurs via nitrate reduction followed by synthesis of amino acids and proteins. However, it is possible that other natural metabolic processes could detoxify the products of atmospheric nitric oxide and nitrogen dioxide. One obvious pathway is polyamine production (Wellburn 1990). Other means of detoxification, such as the release of nitrogen-containing gases may also be important. Natural emissions of nitrogen dioxide, nitric oxide, and ammonia from plant tissues and canopies have been reported (Hill 1971, Farquhar et al. 1979).

Compensation

In order for a response to be compensatory, there must be both evidence for damage, whether in the form of reduced biomass or altered physiological function of a particular tissue type, and reallocation of resources from one tissue to another in response to that damage, and not simply reduction in biomass of the affected part. Total plant, root, and shoot biomass, and root-to-shoot ratios generally decrease in response to elevated pollutant levels. The best example of a compensatory response is decreased root biomass as an indirect response caused by reallocation of carbon from roots to the foliage (Reich et al. 1987, Temple 1988). Another example of compensatory response is within branch translocation of leaf nitrogen from older, more heavily damaged tissue to younger tissue in ponderosa pine, resulting in enhanced photosynthetic capacity (Hom and Riechers 1991). Compensatory responses may not be able to occur in nutrient-stressed plants (NEG, unpubl. data). Increased production of detoxifying compounds may also be considered a compensatory response to elevated pollutant levels.

Response at the Population Level

Where air pollution concentrations are high, a strong selection through time is imposed on individuals to avoid, tolerate, or compensate for pollutant effects. There is considerable variability in individual responses to air pollution stress. Differences in population sensitivity to ozone were found in different seed sources of eastern white pine (Barnes 1972), and in different clones of Norway spruce (Fuhrer et al. 1990). In a study of the effects of the interaction of ozone, acidic fog, and drought stress on 18 half-sib and full-sib families of ponderosa pine, there were strong family differences in response (Temple, pers. comm.). Several studies have specifically addressed ecotypic variation in response to pollution (Taylor 1978, Ernst et al. 1985). Stomatal closure has been associated with long-term exposure to ozone (Coyne and Bingham 1981), and this may reflect long-term selection for individuals with lower conductance (lower effective dose) that allowed them to survive high pollution in the San Bernardino Mountains over the last 40 years. In addition to drought stress, nutritional deficiencies, and differences in phenology, genetic differences may affect individual sensitivity to pollutants.

Conclusions

There is relatively little knowledge on the physiological effects of air pollution on Western trees. Effects of ozone are best understood, and with new sets of experimental data, better understanding of these effects, especially at the seedling level, is becoming available. More research is needed on mechanisms within the context of carefully controlled experiments.

Very little is known with regard to the physiological effects of other air pollutants that are important in this part of the country. The effects of nitrogenous air pollutants have been neglected for many years; however, results of recent studies indicate the importance of some of the nitrogenous pollutants, such as nitric acid, nitrogen dioxide, ammonia, peroxyacetyl nitrate, or particulates. In some of the forested areas of California, in particular the San Bernardino and San Gabriel Mountains, and the western side of the Sierra Nevada, high levels of nitrogen deposition have been detected. A wide range of effects from direct toxicity of peroxyacetyl nitrate, nitrogen dioxide, or nitric acid to overloading of ecosystems with nitrogen may be expected.

Only a small portion of the Western tree species have been included in air pollution studies. Ponderosa and Jeffrey pines, Douglas-fir and white fir, giant sequoia, and some species of oak have been studied most frequently, but even for these species only very limited information exists on their physiological responses to air pollutants.

More studies aimed at linking seedling responses with those of mature trees are essential for a better understanding of potential long-term changes in natural stands. Such sets of data are also required for models dealing with changes in forests caused by air pollution and other stresses, including potential climatic changes.

More studies aimed at establishing a link between the genetic origin of trees and their sensitivity to air pollution are needed. Results of such studies may help in selecting the genotypes best suited for reforestation in the regions under severe air pollution stress.

There is still a lack of simple and reliable physiological or biochemical indicators of photochemical air pollution effects on trees. Early awareness of the symptoms of air pollution phytotoxicity could serve as a warning and a stimulus for regulatory action to protect forests in the regions endangered by air pollution.

References

Aber JD, Nadelhoffer KJ, Steudler P, Melillo JM (1989) Nitrogen saturation in a northern forest ecosystem. *BioScience* 39:378–386

Adams MB, Edwards NT, Taylor GE Jr, Skaggs BL (1990) Whole-plant ^{14}C-photosynthate allocation in *Pinus taeda*: Seasonal patterns at ambient and elevated ozone levels. *Canadian Journal of Forest Research* 20:152–158

Aitken WM, Jacobi WR, Staley JM (1984) Ozone effects on seedlings of Rocky Mountain ponderosa pine. *Plant Disease* 68:398–400

Amundson RG, Maclean DC (1982) Influence of oxides of nitrogen on crop growth and yield: An overview. In: Schneider T, Grant L (eds) *Air Pollution by Nitrogen Oxides.* Elsevier, pp 501–510

Amundson RG, Walker RB, Legge AH (1986) Sulphur gas emissions in the boreal forest: The West Whitecourt case study. VII: Pine tree physiology. *Water, Air, and Soil Pollution* 29:129–147

Andersen CP, Hogsett WE, Wessling R, Plocher M (1991) Ozone decreases spring root growth and root carbohydrate content in ponderosa pine the year following exposure. *Canadian Journal of Forest Research* 21:1288–1291

Anderson PD, Houpis JLJ (1991) Foliar nutrient status of *Pinus ponderosa* exposed to ozone and acid rain. Supplement to *Plant Physiology* 96:173

Barnes RL (1972) Effects of chronic exposure to ozone on photosynthesis and respiration of pines. *Environmental Pollution* 3:133–138

Basabe FA, Edmonds RL, Chang WL, Larson TV (1989) Fog and cloud chemistry in western Washington. In: Olson RK, Lefohn AS (eds) *Effects of Air Pollution on Western Forests*. Transactions Series, No. 16, Air & Waste Management Association, Pittsburgh, pp 33–49

Becker KH, Brockman KJ, Bechara J (1990) Production of hydrogen peroxide in forest air by reaction of ozone with terpenes. *Nature* 346:256–258

Beyers JL, Riechers GH, Temple PJ (1991) Photosynthetic capacity of ponderosa pine seedlings exposed to different levels of atmospheric ozone and drought stress. *Ecological Society of America Bulletin* 72(2) (Supplement):69

Brewer RL, Gordon LS, Shepard LS, Ellis EC (1983) Chemistry of mist and fog from the Los Angeles urban area. *Atmosphere and Environment* 17:2267–2270

Bytnerowicz A, Molski B (1978) Metabolism siarki w roslinach. *Wiadomosci Botaniczne* 22:17–29

Bytnerowicz A, Takemoto BK (1989) Effects of photochemical smog on the growth and physiology of ponderosa pine seedlings grown under nitrogen and magnesium deficiencies. In: Olson RK, Lefohn AS (eds) *Effects of Air Pollution on Western Forests*. Transactions Series, No. 16, Air & Waste Management Association, Pittsburgh, pp 455–467

Bytnerowicz A, Dmuchowski W, Molski B (1980) The air pollution accumulation capabilities of some tree species in the vicinity of the chemical plant in Torun. *Rocznik Dendrologiczny* 33:15–28

Bytnerowicz A, Dmuchowski W, Molski B (1981/1982) Effects of needle harvest time, age of needles and age of Scots pine (*Pinus silvestris* L.) trees on the accumulation of total sulphur. *Rocznik Dendrologiczny* 34:51–68

Bytnerowicz A, Miller PR, Olszyk DM, Dawson PJ, Fox CA (1987) Gaseous and particulate air pollution in the San Gabriel Mountains of southern California. *Atmospheric Environment* 21:1805–1814

Bytnerowicz A, Olszyk DM, Huttunen S, Takemoto BK (1989) Effects of photochemical smog on growth, injury, and gas exchange of pine seedlings. *Canadian Journal of Botany* 67:2175–2181

Bytnerowicz A, Poth M, Takemoto BK (1990) Effects of photochemical smog and mineral nutrition on ponderosa pine seedlings. *Environmental Pollution* 67:233–248

Bytnerowicz A, Dawson PJ, Morrison CL (1991) Physiological and growth responses of ponderosa pine seedlings to ambient and elevated ozone concentrations in southern Sierra Nevada. Proceedings of the 84th Annual Meeting, Air and Waste Management Association, 16–21 July 1991, Vancouver, British Columbia, AMWA, Pittsburgh, PA

Cape JN, Fowler D (1981) Changes in epicuticular wax of *Pinus sylvestris* exposed to polluted air. *Silva Fennica* 15:457–458

Chevonne BI, Anderson JV, Hess JL (1991) Seasonal changes in components of the antioxidant system in eastern white pine and ozone effects on antioxidant metabolism. Proceedings of the 84th Annual Meeting, Air and Waste Management Association, 16–21 July 1991, Vancouver, British Columbia, AMWA, Pittsburgh, PA

Cleland R (1971) Cell wall extension. *Annual Review of Plant Physiology* 22:192–222

Coyne PI, Bingham GE (1981) Comparative ozone dose response of gas exchange in a ponderosa pine stand exposed to long-term fumigations. *Journal of the Air Pollution Control Association* 31:38–41

Darrall NM (1989) The effect of air pollutants on physiological processes in plants. *Plant, Cell, and Environment* 12:1–30

Dmuchowski W, Bytnerowicz A, Molski B (1981/1982) The influence of boreal sites on the accumulation of total sulphur in Scots pine (*Pinus silvestris* L.) needles. *Rocznik Dendrologiczny* 34:69–77

Durisco DM, Stolte KW (1989) Photochemical oxidant injury to ponderosa (*Pinus ponderosa* Laws.) and Jeffrey pine (*Pinus jeffreyii* Grev. and Balf.) in the national parks of the Sierra Nevada of California. In: Olson RK, Lefohn AS (eds) *Effects of Air Pollution on Western Forests*. Transactions Series, No. 16, Air & Waste Management Association, Pittsburgh, pp 261–278

Durmishidze SV, Nutshubidze NN (1976) Absorption and conversion of nitrogen dioxide by higher plants. *Doklardie Biochemie* 227:104–107

Eamus D, Fowler D (1990) Photosynthetic and stomatal conductance responses to acid mist of red spruce seedlings. *Plant, Cell, and Environment* 13:349–357

Edwards GS, Edwards NT, Kelly JM, Mays PA (1991) Ozone, acidic precipitation, and soil Mg effects on growth and nutrition of loblolly pine seedlings. *Environmental and Experimental Botany* 31(1):67–78

Edwards NT, Taylor GE Jr, Adams MB, Simmons GL, Kelly JM (1990) Ozone, acidic rain, and soil magnesium effects on growth and foliar pigments of *Pinus taeda* L. *Tree Physiology* 6:95–104

Ennis CA, Lazrus AL, Zimmerman PR (1990) Flux determinations and physiological response in the exposure of red spruce to gaseous hydrogen peroxide, ozone, and sulphur dioxide. *Tellus* 42B:183–199

Ernst WHO, Tonnejick AEC, Pasman FJM (1985) Ecotypic response of *Silene cucubalus* to air pollutants (SO_2, O_3). *Journal of Plant Physiology* 188:439–450

Evans LS (1982) Biological effects of acidity in precipitation on vegetation: A review. *Environmental Experimental Botany* 22:135–169

Evans LS, Miller PR (1972a) Comparative needle anatomy and relative ozone sensitivity of four pine species. *Canadian Journal of Botany* 50:1067–1071

Evans LS, Miller PR (1972b) Ozone damage to ponderosa pine: A histological and histochemical appraisal. *American Journal of Botany* 59:297–304

Evans LS, Miller PR (1975) Histological comparison of single and additive O_3 and SO_2 injuries to elongating ponderosa pine needles. *American Journal of Botany* 62:416–421

Fangmeier A, Kress LW, Lepper P, Heck WW (1990) Ozone effects on the fatty acid composition of loblolly pine needles (*Pinus taeda*). *New Phytologist* 115:639–647

Farquhar GD, Wetselaar R, Firth PU (1979) Ammonia volatilization from senescing leaves of maize. *Science* 203:1257–1258

Fenn ME, Dunn PH (1989) Litter decomposition across an air-pollution gradient in the San Bernardino Mountains. *Soil Science Society of America Journal* 53:1560–1567

Filner P, Rennenberg H, Sekiya J, Bressan RA, Wilson LG, LeCureux L, Shimei T (1984) Biosynthesis and emission of hydrogen sulfide by higher plants. In: Koziol MJ, Whatley FR (eds) *Gaseous Air Pollutants and Plant Metabolism*. Butterworths, London, pp 291–312

Fowler D, Cape JN, Nicholson IA, Kinnaird JW, Paterson IS (1980) The influence of a polluted atmosphere on cuticle degradation in Scots pine (*Pinus silvestris*). In: Drablos D, Tollan A (eds) *Proceedings: "Ecological Impacts of Acidic Precipitation."* SNSF Project, NHL, Sandefjord, Norway, March 11–14, pp 14

Freer-Smith PH (1985) The influence of SO_2 and NO_2 on the growth development and gas exchange of *Betula pendula* Roth. *New Phytologist* 99:417–430

Freer-Smith PH, Dobson MC (1989) Ozone flux to *Picea sitchensis* (Bong) Carr and *Picea abies* (L.) Karst during short episodes and the effects of these on transpiration and photosynthesis. *Environmental Pollution* 59:161–176

Fuhrer G, Dunkl M, Knoppik D, Selinger H, Blank LW, Payer HD, Lange OL (1990) Effects of low-level long-term ozone fumigation and acid mist on photosynthesis and stomata of clonal Norway spruce (*Picea abies* (L.) Karst.) *Environmental Pollution* 64:279–293

Garrec JP, Berteigne M (1987) Effects of organic micropollutants on vegetation. In: *Direct Effects of Dry and Wet Deposition on Forest Ecosystems—in Particular Canopy Interactions.* CEC-Air Pollution Research Report 4, pp 236–244

Garten CT Jr, Hanson PJ (1990) Foliar retention of ^{15}N-nitrate and ^{15}N-ammonium by red maple (*Acer rubrum*) and white oak (*Quercus alba*) leaves from simulated rain. *Environmental Experimental Botany* 30:333–342

Gietko NV (1976) Osobiennosti nakopleniia siernistyh i azotistih soiedinienii v listiah niekotoryh vidov topolia v usloviiah zadymlieniia atmosfiernogo vozduha dvuokisiu siery. Rastieniia i Promyshliennaia Srieda, Naukova Dumka, Kiev, 63–64

Godzik S (1976) External needle waxes of *Pinus silvestris* and their modification by air pollution. In: *Abstracts of Papers, 3rd International Congress of Plant Pathology, Munich.* 16–23 August, 1978, Berlin, pp 349

Greitner CS, Winner WE (1989) Effects of O_3 on alder photosynthesis and symbiosis with *Frankia*. *New Phytologist* 111:647–656

Grulke NE, Miller PR, Wilborn RD, Hahn S (1989) Photosynthetic response of giant sequoia seedlings and rooted branchlets of mature foliage to ozone fumigation. In: Olson RK, Lefohn AS (eds) *Effects of Air Pollution on Western Forests.* Transactions Series, No. 16, Air & Waste Management Association, Pittsburgh, pp 429–442

Guderian R (1977) Air pollution, phytotoxicity of acidic gases and its significance in air pollution control. *Ecological Studies 22,* Springer-Verlag, Berlin

Guderian R (ed) (1985) Air pollution by photochemical oxidants. *Ecological Studies 52,* Springer-Verlag, Berlin, 296p

Hallgren J-E (1978) Physiological and biochemical effects of sulfur dioxide on plants. In: Nriagu (ed) *Sulfur in the Environment, Part II: Ecological Impacts.* John Wiley and Sons, New York, pp 163–209

Hallgren J-E, Fredriksson S-A (1982) Emission of hydrogen sulfide from sulfur dioxide-fumigated pine trees. *Plant Physiology* 70:456–459

Hanson PJ, McLaughlin SB, Edwards NT (1988) Net CO_2 exchange of *Pinus taeda* shoots exposed to variable ozone levels and rain chemistries in field and laboratory settings. *Physiologia Plantarum* 74:635–642

Hanson PJ, Rott K, Taylor GE Jr, Gunderson CA, Lindberg SE, Ross-Todd BM (1989) NO_2 deposition to elements representative of a forest landscape. *Atmospheric Environment* 23:1783–1794

Heath RL (1975) Ozone. In: Mudd JB, Kozlowski TT (eds) *Responses of Plants to Air Pollution.* Academic Press, New York, pp 23–55

Heath RL (1979) Breakdown of ozone and formation of hydrogen peroxide in aqueous solutions of amine buffers exposed to ozone. *Toxicological Letters* 4:449–453

Heath RL (1980) Initial events in injury to plants by air pollutants. *Annual Review of Plant Physiology* 31:395–431

Heath RL (1987) The biochemistry of ozone attack on the plasma membrane of plant cells. *Recent Advances in Phytochemistry* 21:29–54

Heath RL (1988) Biochemical mechanism of pollutant stress In: Heck WW, Taylor OC, Tingey DT (eds) *Assessment of Crop Loss form Air Pollutants.* Elsevier Applied Sciences

Helmig D, Array J (1992) Organic chemicals in the air at Whitaker's Forest/Sierra Nevada Mountains, California. *Science of the Total Environment* (in press)

Hewitt CN, Lucas P, Wellburn AR, Fall R (1990) Chemistry of ozone damage to plants. *Chemistry & Industry* 15:478–481

Hill AC (1971) Vegetation, a sink for atmospheric pollutants. *Journal of Air Pollution Control Association* 21:341–346

Hogsett WE, Plocher M, Wildman V, Tingey DT, Bennett JP (1985) Growth reponse of two varieties of slash pine to chronic ozone exposures. *Canadian Journal of Botany* 63:2369–2376

Hogsett WE, Tingey DT, Hendricks C, Rossi D (1989) Sensitivity of western conifers to SO_2 and seasonal interactions of acid fog and ozone. In: Olson RK, Lefohn AS (eds) *Effects of Air Pollution on Western Forests*. Transactions Series, No. 16, Air & Waste Management Association, Pittsburgh, pp 469–492

Hom JL, Riechers GH (1991) Carbon assimilation, allocation, and nutrient use efficiency of ponderosa pine as a compensatory response to ozone fumigation and water stress. *Ecological Society of America Bulletin* 72(2) (Supplement):144

Houpis JLJ (1989) *Seasonal Effects of Sulfur Dioxide on the Physiology and Morphology of* Pinus ponderosa *Seedlings*. PhD dissertation, University of California, Berkeley, 256p

Huttunen S (1984) Interactions of disease and other stress factors with atmospheric pollution. In: Treshow M (ed) *Air Pollution and Plant Life*. John Wiley & Sons, Ltd, New York, pp 321–356

Huttunen S, Laine K (1983) Effects of air-borne pollutants on the surface wax structure of *Pinus sylvestris* needles. *Annals Bot. Fennici* 20:79–86

Huttunen S, Laine K, Torvela H (1985) Seasonal sulphur contents of pine needles as indices of air pollution. *Annals Bot. Fennici* 22:343–359

Jacobson JS, Lassoie JP, Osmeloski J, Yamada K (1989) Changes in foliar elements in red spruce seedlings after exposure to sulfuric and nitric acid mist. *Water, Air and Soil Pollution* 48:141–159

Jones T, Mansfield TA (1982) Studies on dry matter partitioning and distribution of ^{14}C-labelled assimilates of *Phleum pratense* exposed to SO_2 pollution. *Environmental Pollution Series A* 28:199–207

Karenlampi L (1986) Relationship between macroscopic symptoms of injury and cell structural changes in needles of ponderosa pine exposed to air pollution in California. *Annals Bot. Fennici* 23:255–264

Karenlampi L (1987) Visible symptoms and mesophyll cell structural responses to air pollution in two lowland pines (*Pinus radiata* and *P. halepensis*) in southern California. *Savonia* 9:1–12

Karenlampi L, Soikkeli S (1980) Morphological and fine structural effects of different pollutants on plants: Development and problems of research. In: *Effects of Airborne Pollution on Vegetation Symposium*. United Nations, Warsaw, Poland, pp 92–99

Karolewski P (1985) The role of free proline in the sensitivity of poplar (*Populus* "Robusta") plants to the action of SO_2. European Journal of Forest Pathology 15:199–206

Keane DT, Manning WJ (1988) Effects of ozone and simulated acid rain on birch seedlings growth and formation of ectomycorrihizae. *Environmental Pollution* 53:55–65

Keeling CD, Bacastow RB, Bainbridge AE, Ekdahl CA Jr, Guenther PR, Waterman LS, Chin JFS (1976) Atmospheric carbon dioxide variations at Mauna Loa Observatory, Hawaii. *Tellus* 28:779–788

Keller T, Hasler R (1987) Some effects of long-term ozone fumigation on Norway spruce, I: Gas-exchange and stomatal response. *Trees* 1:129–133

Knudson LL, Tibbits TW, Edwards GE (1977) Measurement of ozone injury by determination of leaf chlorophyll concentrations. *Plant Physiology* 60:606–608

Kohut RJ, Laurence JA, Amundson RG, Raba RM, Melkonian JJ (1990) Effects of ozone and acidic precipitation on the growth and photosynthesis of red spruce after two years of exposure. *Water, Air, and Soil Pollution* 51:227–286

de Konig H, Jagier Z (1970) Effects of aldehydes on photosynthesis and respiration of *Euglena gracilis*. *Archives of Environmental Health* 20:720–722

Kozlowski TT, Constantinedou HA (1986) Responses of woody plants to environmental pollution. *Forestry Abstracts* 47:1–51

Kramer PJ, Kozlowski TT (1979) *Physiology of Woody Plants*. Academic Press, Inc, Orlando, 811p

Kress LW, Skelly JM (1982) Response of eastern forest tree species to chronic doses of ozone and nitrogen dioxide. *Plant Diesease* 66:1149–1152

Krupa SV, Manning WJ (1988) Atmospheric ozone: Formation and effects on vegetation. *Environmental Pollution* 50:101–137

Legge AH, Bogner JC, Krupa SV (1988) Foliar sulphur species in pine: A new indicator of a forest ecosystem under air pollution stress. *Environmental Pollution* 55:15–27

L'Hirondelle SJ, Addison PA (1985) Effects of SO_2 on leaf conductance, xylem tension, fructose and sulphur levels of Jack pine seedlings. *Environmental Pollution* 39:373–386

Malhotra SS (1977) Effects of aqueous sulphur dioxide on chlorophyll destruction in *Pinus contorta*. *New Phytologist* 78:101–109

Malhotra SS, Khan AA (1978) Effects of sulfur dioxide fumigation on lipid biosynthesis in pine needles. *Phytochemistry* 17:241–244

Mansfield TA, Majernik O (1970) Can stomata play a part in protecting plants against air pollutants? *Environmental Pollution* 1:149–154

Marshall JD, Cadle SH (1989) Evidence for trans-cuticular uptake of HNO_3 vapor by foliage of eastern white pine (*Pinus strobus* L.). *Environmental Pollution* 60:15–28

Masuch G, Kettrup A, Mallant RKAM, Slanina J (1986) Effects of H_2O_2 containing acidic fog on young trees. *International Journal of Environmental and Analytical Chemistry* 27:183–213

Mayo JM, Hartgerink AP, Legge AH (1986) Sulphur gas emissions in the boreal forest: The West Whitecourt case study, VI: Woody plant water stress. *Water, Air, and Soil Pollution* 29:113–127

McLaughlin SB (1985) Effects of air pollution on forests: A critical review. *Journal of Air Pollution Control Association* 35:512–534

McLaughlin SB, McConathy RK (1983) Effects of SO_2 and O_2 on allocation of ^{14}C-labeled photosynthate in *Phaseolus vulgaris*. *Plant Physiology* 73:630–635

Mehlhorn H, Wellburn AR (1987) Stress ethylene formation determines plant sensitivity to ozone. *Nature* 327:417–418

Mehlhorn H, Seufert G, Schmidt A, Kunert KJ (1986) Effects of SO_2 and O_3 on production of antioxidants in conifers. *Plant Physiology* 82:336–338

Mehlhorn H, Cottam DA, Lucas PW, Wellburn AR (1987) Induction of ascorbate peroxidase and glutathione reductase activities by interactions of mixtures of air pollutants. *Free Radical Research Communications* 3:1–5

Mehlhorn H, O'Shea JM, Wellburn AR (1991) Atmospheric ozone interacts with stress ethylene formation by plants to cause visible plant injury. *Journal of Experimental Botany* 42:17–24

Miller JE, Xerikos PB (1979) Residence time of sulfite in SO_2 "sensitive" and "tolerant" soybean cultivars. *Environmental Pollution* 18:259–264

Miller PR (1983) Ozone effects in the San Bernardino National Forest. In: *Air Pollution and the Productivity of the Forest, Symposium Proceedings*. Pennsylvania State University Press, State College, PA, pp 161–197

Miller PR, McBride JR (1975) Effects of air pollutants on forests. In: Mudd JB, Kozlowski TT (eds) *Responses of Plants to Air Pollution.* Academic Press, New York, pp 196–235

Miller PR, Millecan AA (1971) Extent of oxidant air pollution damage to some pines and other conifers in California. *Plant Disease Report* 55:555–559

Miller PR, Parmeter JR Jr, Taylor OC, Cardiff EA (1963) Ozone injury to the foliage of *Pinus ponderosa. Phytopathology* 53:1072–1076

Miller PR, Parmeter JR Jr, Flick BH, Martinez CW (1969) Ozone dosage response of ponderosa pine seedlings. *Journal of Air Pollution Control Association* 19:435–438

Molski B, Bytnerowicz A, Dmuchowski W (1983) Mapping air pollution of forests and agricultural areas of Poland by sulfur accumulation in pine *(Pinus sylvestris* L.) needles. *Aquilo Seria Botanica* 19:326–331

Mudd JB (1982) Effects of oxidants on metabolic function. In: Unsworth MH, Ormrod DP (eds) *Effects of Gaseous Air Pollution in Agriculture and Horticulture.* Butterworth, London, pp 189–203

Mudd JB, Kozlowski TT (1975) *Responses of Plants to Air Pollution.* Academic Press, NY

Mudd JB, Banerjee SK, Dooley MM, Knight KL (1984) Pollutants and plant cells: Effects on membranes. In: MJ Koziol, FR Whatley (eds) Gaseous Pollutants and Plant Metabolism. Butterworth, London, pp 105–116

National Academy of Sciences (1977) *Nitrogen Oxides.* Committee on Medical and Biological Effects of Environmental Pollutants Washington, DC, 333p

Neuman LE, Houpis LJL, Anderson PD (1991) Trends in *Pinus ponderosa* foliar pigment concentration due to chronic exposure of ozone and acid rain. Supplement to *Plant Physiology* 96:172

Nikolaevskii VC, Kaziekina LP, Vidiakina OA (1976) Traslokacija siery rastieniiami pri pogloschcheni siernistogo gasa listiami. Rastieniia i Promyschliennaia Srieda, Naukova Dumka, Kiev, 112–114

Norby RJ, Weeresurija Y, Hanson PJ (1989) Induction of nitrate reductase activity in red spruce needles by NO_2 and HNO_3 vapor. *Canadian Journal of Forest Research* 19:889–896

Oquist G, Wass R (1988) A portable microprocessor instrument for measuring chlorophyll fluorescence kinetics in stress physiology. *Physiologia Plantarum* 73:211–217

Papageorgiu G (1975) Chlorophyll fluorescence: An intrinsic probe of photosynthesis. In: Govindjee R (ed) *Bioenergetics of Photosynthesis*. Academic Press, New York, pp 319–371

Parry MAJ, Whittingham CP (1984) Effects of gaseous air pollutants on stromal reactions. In: Koziol MJ, Whatley FR (eds) *Gaseous Air Pollutants and Plant Metabolism*. Buttersworth, London, pp 161–168

Patterson MT, Rundel PW (1989) Seasonal physiological responses of ozone stressed Jeffrey pine in Sequoia National Park, California. In: Olson RK, Lefohn AS (eds) *Effects of Air Pollution on Western Forests*. Transactions Series, No. 16, Air & Waste Management Association, Pittsburgh, pp 419–428

Peterson DL, Arbaugh MJ, Robinson LJ (1988) The effects of ozone stress on tree growth and vigor in the Sierra Nevada of California, USA. In: Bucher JB, Bucher-Wallin I (eds) *Air Pollution and Forest Decline*. IUFRO P2.05, Interlaken, Switzerland, pp 289–294

Rabinowitz HD, Fridowich J (1983) Superoxide radicals, superoxide dismutases and oxygen toxicity in plants. *Photochemical Photobiology* 37:679–690

Reich PB, Schoettle AW, Stroo HF, Troiano J, Amundson RG (1987) Effects of ozone and acid rain on white pine (*Pinus strobus*) seedlings grown in five soils, I: Net photosynthesis and growth. *Canadian Journal of Botany* 65:977–987

Reich PB, Schoettle AW, Stroo HF, Troiano J, Amundson RG (1988) Effects of ozone and acid rain on white pine (*Pinus strobus*) seedlings grown in five soils, III: Nutrient relations. *Canadian Journal of Botany* 66:1517–1531

Reinert RA, Heagle AS, Heck WW (1975) Plant responses to pollutant combinations. In: Mudd JB, Kozlowski TT (eds) *Responses of Plants to Air Pollution*. Academic Press, New York, pp 159–187

Rennenberg H (1982) Glutathione metabolism and possible biological roles in higher plants. *Biochemistry* 21:2771–2781

Rennenberg H (1984) The fate of excess sulfur in higher plants. *Annual Review of Plant Physiology* 35:121–153

Rennenberg H, Polle A (1989) Effects of photooxidants on plants. In: Georgii H-W (ed) *Mechanisms and Effects of Pollutant-Transfer into Forests*. Kluwer Academic Publishers, pp 251–258

Rennenberg H, Huber B, Schroder P, Stahl K, Haunold W, Georgii H-W, Slovik S, Pfanz H (1990) Emission of volatile sulfur compounds from spruce trees. *Plant Physiology* 92:560–564

Richards BL Sr, Taylor OC, Edmunds GF Jr (1968) Ozone needle mottle of pine in southern California. *Journal of Air Pollution Control Association* 18-73-77

Rinallo C, Raddi P, diLondardo V (1986) Effects of simulated acid deposition on the surface structure of Norway spruce and silver fir needles. *European Journal of Forest Pathology* 16:440–446

Roberts BR (1976) The response of field-grown white pine seedlings to different sulphur dioxide environments. *Environmental Pollution* 11:175–180

Robinson DC, Wellburn AR (1983) Light-induced changes in the quenching of 9-amino-acridine fluorescence by photosynthetic membranes due to atmospheric pollutants and their products. *Environmental Pollution* 32:109–120

Roehm JN, Hadley JG, Menzel DB (1971) Oxidation of unsaturated fatty acids by ozone and nitrogen dioxide. *Archives of Environmental Health* 23:142–148

Sabaratnam S, Gupta G, Mulchi C (1988) Effects of nitrogen dioxide on leaf chlorophyll and nitrogen content of soybean. *Environmental Pollution* 51:113–120

Sandhu R, Gupta G (1989) Effects of nitrogen dioxide on growth and yield of black turtle bean (*Phaseolus vulgaris* L.) cv Domino. *Environmental Pollution* 59:337–344

Sasek TW, Richardson CJ (1989) Effects of chronic doses of ozone on loblolly pine: Photosynthetic characteristics in the third growing season. *Forest Science* 35(3):745–755

Sauter JJ, Voss U (1986) SEM-observations on the structural degradation of epistomatal waxes in *Picea abies* L. Karst. —and its possible role in the "Fichtensterben." *European Journal of Forest Pathology* 16:408–423

Scherbatskoy T, Klein RM (1983) Response of spruce and birch foliage to leaching by acidic mists. *Journal of Environmental Quality* 12:189–195

Schmidt W, Neubauer C, Kolbowski J, Schreiber U, Urbach W (1990) Comparison of effects of air pollutants (SO_2, O_3, NO_2) on intact leaves by measurements of chlorophyll fluorescence and P700 absorbance changes. *Photosynthetic Research* 25:241–248

Schreiber U, Vidaver W, Runeckles VC, Rosen P (1978) Chlorophyll
    fluorescence assay for ozone injury in intact plants. *Plant
    Physiology* 61:80–84

Sekiya J, Wilson LG, Filner P (1982) Resistance to injury by sulfur
    dioxide. *Plant Physiology* 70:437–441

Shafer SR, Heagle AS (1989) Growth responses of field-grown loblolly
    pine to chronic doses of ozone during multiple growing seasons.
    Canadian Journal of Forest Research 19:821–831

Shriner DS, Heck WW, McLaughlin SB, Johnson DW, Irving PM, Joslin
    JD, Peterson CW (1990) Response of vegetation to atmospheric
    deposition and air pollution. NAPAP SOS/T Report 18. In:
    Acidic Deposition: State of Science and Technology, Volume III,
    National Acid Precipitation Assessment Program, 722 Jackson
    Place NW, Washington, DC 20503

Skeffington RA, Roberts TM (1985) The effects of ozone and acid mist on
    Scots pine saplings. *Oecologia* 65:201–206

Smith WH (1974) Forest and air quality. *Journal of Foresty* 83:82–92

Soikkeli S (1980) Ultrastructure of the mesophyll in Scots pine and
    Norway spruce: Seasonal variation and molarity of the fixative
    buffer. *Protoplasma* 103:241–252

Solomon PA, Fall T, Salmon L, Lin P, Vasquez F, Cass GR (1989)
    *Acquisition of Acid Vapor and Aerosol Concentrations Data for Use in
    Dry Deposition Studies in the South Coast Air Basin.* Report to the
    California Air Resources Board, Contract No. A4-144-32

Spaleny J (1977) Sulphate transformation to hydrogen sulphide in spruce
    seedlings. *Plant Soil* 48:557–563

Sugahara K (1984) Effects of air pollutants on light reactions in
    chloroplasts. In: Koziol MF, Whatley FR (eds) *Gaseous Air
    Pollutants and Plant Metabolism.* Buttersworth, London, pp 169–180

Takemoto BK, Bytnerowicz A (1992) Effects of acidic fog on growth,
    physiology and biochemistry of ponderosa pine (*Pinus ponderosa*)
    and white fir (*Abies concolor*) seedlings. *Environmental Pollution*
    (in press)

Tanaka K, Sugahara K (1980) Role of superoxide dismutase in the
    defense against SO_2 toxicity and induction of superoxide
    dismutase with SO_2 fumigations. In: *Studies on the Effects of Air
    Pollutants on Plants and Mechanisms of Phytotoxicity.* Research
    Report National Institute Environmental Studies No. 11,
    pp 155–179

Taylor GE Jr (1978) Genetic analysis of ecotypic differentiation within an annual plant species, *Geranium carolinianum* L., in response to sulfur dioxide. *Botanical Gazette* 139:326–368

Taylor GE Jr, Norby RJ, McLaughlin SB, Johnson AH, Turner RS (1986) Carbon dioxide assimilation and growth of red spruce (*Picea rubens* Sarg.) seedlings in response to ozone, precipitation chemistry, and soil type. *Oecologia* 70:163–171

Taylor GE Jr, Hanson PJ, Baldocchi DD (1989) Pollutant deposition to individual leaves and plant canopies: Sites of regulation and relationship to injury. In: Heck WW, Taylor OC, Tingey DT (eds) *Assessment of Crop Loss from Air Pollutants.* Elsevier Applied Sciences, London

Taylor OC (1984) Organismal responses of higher plants to atmospheric pollutants: Photochemical and other. In: Treshow M (ed) *Air Pollution and Plant Life.* John Wiley & Sons Ltd, pp 215–238

Temple PJ (1988) Injury and growth of Jeffrey pine and giant sequoia in response to ozone and acidic mist. *Environmental Experimental Botany* 28:323–333

Teskey RO, Bongarten BC, Cregg BM, Dougherty PM, Hennessey TC (1986) Stomatal and non-stomatal limitations to net photosynthesis in *Pinus taeda* L. under different environmental conditions. *Tree Physiology* 3:41–61

Tingey DT, Olszyk D (1982) Intraspecific variability in metabolic responses to SO_2. In: Winner WE, Mooney HA, Goldstein R (eds) *Sulfur Dioxide and Vegetation: Physiology, Ecology, and Policy Issues.* Stanford University Press, Stanford, CA

Tingey DT, Rodecap KD, Lee EH, Moser TJ, Hogsett WE (1986) Ozone alters the concentrations of nutrients in bean tissue. *Angew. Botanik* 60:481–493

Tolley LC, Strain BR (1985) Effects of CO_2 enrichment and water stress on gas exchagne of *Liquidambar styraciflua* and *Pinus taeda* seedlings grown under different irradiance levels. *Oecologia* 65:166–172

Tseng EC, Seiler JR, Chevone BI (1988) Effects of ozone and water stress on greenhouse-grown Fraser fir seedling growth and physiology. *Environmental Experimental Botany* 28(1):37–41

Turner DP, Tingey DT (1990) Foliar leaching and root uptake of Ca, Mg and K in relation to acid fog effects on Douglas-fir. *Water, Air, and Soil Pollution* 49:205–214

Turner DP, Tingey DT, Hogsett WE (1989) Acid fog effects on conifer seedlings. In: Bucher JB, Bucher-Wallin I (eds) *Air Pollution and Forest Decline, Proceedings of the 14th International Meeting for Specialists in Air Pollution Effects on Forest Ecosystems.* IUFRO P2.05, Interlaken, Switzerland, October 2–8, 1988, Birmensdorf, pp 125–129

Turunen M, Huttunen S (1990) A review of the response of epicuticular wax of conifer needles to air pollution. *Journal of Environmental Quality* 19:35–45

Turunen M, Huttunen S (1991) Effects of simulated acid rain on the epicuticular wax of Scots pine needles under northerly conditions. *Canadian Journal of Botany* 69:412–419

Unsworth MH, Bisce PV, Pinckney HR (1972) Stomatal responses to SO_2. *Nature* 239:458

Waldman JM, Munger JW, Jacob DJ, Flagan RC, Morgan JJ, Hoffman MR (1982) Chemical composition of acid fog. *Science* 218:677–680

Waldman JM, Munger JW, Jacob DJ, Hoffman MR (1985) Chemical characterization of stratus cloudwater and its role as a vector for pollutant deposition in a Los Angeles pine forest. *Tellus* 37B:91–108

Wallick K (1990) Basil chlorosis: a physiological disorder in CO_2-enriched atmospheres. Plant Disease 74:171–173

Weigel HJ, Halbwachs G, Jager HJ (1989) The effects of air pollutants on forest trees from a plant physiological view. *Journal of Plant Disease Protection* 96:203–217

Wellburn AR (1988) *Air Pollution and Acid Rain: The Biological Impact.* Longman Scientific & Technical, Burnt Mill, England

Wellburn AR (1990) Tansley Review No. 24. Why are atmospheric oxides of nitrogen usually phytotoxic and not alternative fertilizers? *New Phytology* 115:395–429

Westman WE, Temple PJ (1989) Acid mist and ozone effects on the leaf chemistry of two western conifer species. *Environmental Pollution* 57:9–26

Wilson LG, Bressan RA, Fielner P (1978) Light-dependent emission of hydrogen sulfide from plants. *Plant Physiology* 61:184–189

Wingsle G, Nasholm T, Lundmark T, Ericsson A (1987) Induction of nitrate reductase in needles of Scots pine by NO_x and NO_3^-. *Physiologia Plantarum* 70:399–403

Winner WE, Cotter IS, Powers HR, Skelly JM (1987) Screening loblolly pine seedling responses to SO_2 and O_3: Analysis of families differing in resistance to fusiform rust disease. *Environmental Pollution* 47:205–220

Yoneyama T, Sasakawa H (1979) Transformation of atmospheric NO_2 absorbed by spinach leaves. *Plant and Cell Physiology* 20:263–266

Yoshida S, Vemura M (1984) Protein and lipid compositions of isolated plasma membranes from orchard grass (*Dactylis glomerata* L.) and changes during cold acclimation. *Plant Physiology* 75:31–37

Zeevaart AJ (1976) Some effects of fumigating plants for short periods with NO_2. *Environmental Pollution* 11:97–108

Zielke HR, Filner P (1971) Synthesis and turnover of nitrate reductase induced by nitrate in cultured tobacco cells. *Journal of Biological Chemistry* 246:1772–1779

Ziegler (1973) The effects of air-polluting gases on plant metabolism. *Environmental Quality Safety* 2:182–208

6

Pollution Impacts at the Stand and Ecosystem Levels

D. Binkley, T. D. Droessler, and J. Miller

Introduction

The responses of forests to pollutants depend on the sensitivity of individual trees and species to the pollutants, and on complex interactions between trees, the environment, and other organisms. Impaired growth in one individual or species may represent an opportunity for increased growth for other individuals or species. An ecosystem perspective is necessary for a full assessment of pollutant impacts on forests. However, knowledge is strongest about the relation between pollution and individual trees, weaker at the level of stands of trees, and weakest at the ecosystem level (Smith 1990). The effects of pollutants at an ecosystem level must be evaluated in terms of deviations from expected patterns of change within forests over time. Defining "expected patterns of change" is a major challenge, as pervasive change is basic to the nature of forests (Oliver and Larson 1990), and classic theories regarding ecosystem succession toward predictable future states are seriously in doubt (Botkin 1989). In this chapter, we discuss important aspects of forests at the level of stands and ecosystems, as a foundation for the regional case studies presented in later chapters.

Current perspectives on the nature of forests differ greatly. For example, Oliver and Larson (1990) advocate a "mechanistic" view of forests in contrast to a "systems" view; they prefer the term "stand" to "ecosystem" (which does not appear in their index). In contrast, Smith (1990) states that the goal of his book is expressly to describe the interactions of pollutants with forests at the ecosystem level. This chapter takes a middle ground, focusing on stands when discussing patterns of forest change over time, and on ecosystems when discussing complex interactions between species, pollution, and environmental factors. Given the current state of knowledge, we focus on describing the important pieces of the puzzle, with only a few examples available to illustrate many points. Basic patterns of tree growth and stand development in the absence of pollution impacts are described, and then contrasted with pollution-affected forests. We conclude with a synthesis of current thinking about stand development and ecosystem responses to varying degrees of pollution stress.

Growth

The production of biomass in forests depends on the interception of light by leaves, and on the conversion of intercepted light energy into carbohydrates. The efficiency of this conversion depends largely on the availability of water since leaves under drought stress intercept light, but fail to produce carbohydrates. Some of the carbohydrates produced from photosynthesis are respired as carbon dioxide as part of the cost of synthesizing and maintaining plant tissues; the remaining carbohydrates form new leaves, stems, branches, and roots. The productivities of forests differ because of differences in stand leaf area and resultant light interception and in the efficiency of producing carbohydrates from intercepted light. These differences depend in large part on climate, soils, species composition, and history of disturbances.

Measurements of tree growth include changes in tree diameter (or radius), height, volume, and biomass. Volume and biomass growth continue at high rates through most of the lifespan of trees that maintain dominant positions in the canopy. The growth of suppressed trees usually declines as the canopies of dominant trees expand, often leading to death of the suppressed trees. The increase in height, radius, and volume of a Douglas-fir (*Pseudotsuga menziesii*) tree on a dry site in Oregon illustrates a common pattern (Figure 6.1a; data provided by Joe Means, USDA Forest Service). Trees increase in height relatively rapidly; the Douglas-fir at age 150 years had attained half of the total height reached by age 400 years. The midpoint in radius did not occur until about age 200 years, which is later than expected and may indicate that the tree experienced strong competition from other trees during the first century. The accumulation of stem volume was very slow for almost 300 years, with about half the volume accumulating during the last century. In terms of annual growth, the maximum height growth occurred at about age 50 (Figure 6.1b), while the maximum growth rate for radius (age 200) and volume (age 300) occurred much later. Foresters typically

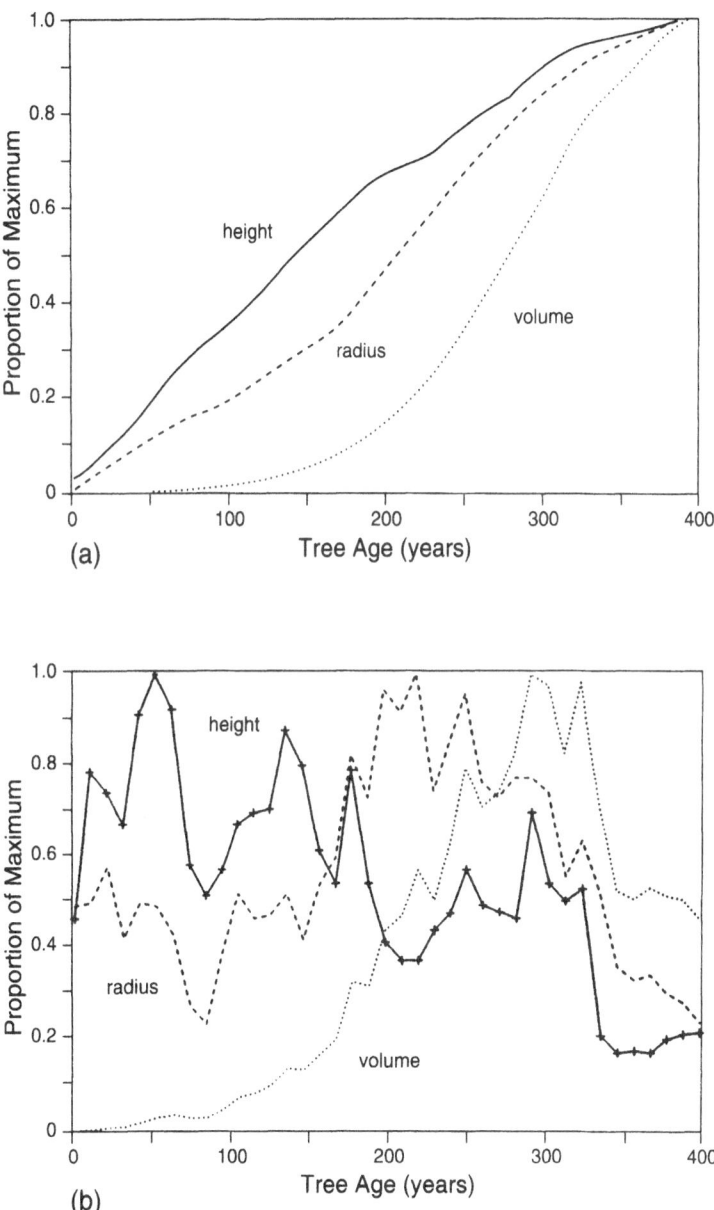

Figure 6.1 Accumulation (a) and growth rate (b) of height, radius, and volume for a single Douglas-fir tree on a dry site in Oregon (from data supplied by J. Means, U.S.D.A. Forest Service).

evaluate growth in terms of when the maximum mean annual increment (volume divided by age) occurs. Even after 400 years of growth, the mean annual increment of this Douglas-fir was increasing.

Patterns in stand growth differ substantially from those of individual dominant trees. Annual growth of dominant trees often continues to increase long after the growth rate of the stand has reached its maximum and begun to decline. The current volume increment for the Douglas-fir tree in Figure 6.1 can be contrasted with the expected current volume increment for the whole stand (Figure 6.2; derived from normal stand tables of McArdle et al. (1961) for a site index of 40 m at 100 years). The peak volume growth for such stands occurs near age 50, contrasted with the single tree's peak near age 300. This example illustrates the complexities faced in determining trends in growth for single trees or for stands of trees, and in determining any role that pollution may play in driving these trends.

The rates and patterns of stand growth also depend on the species present, further complicating the task of differentiating pollutant impacts from normal trends. For example, a stand of nitrogen-fixing red alder (*Alnus rubra*) on a very fertile site on the coast of Oregon showed very

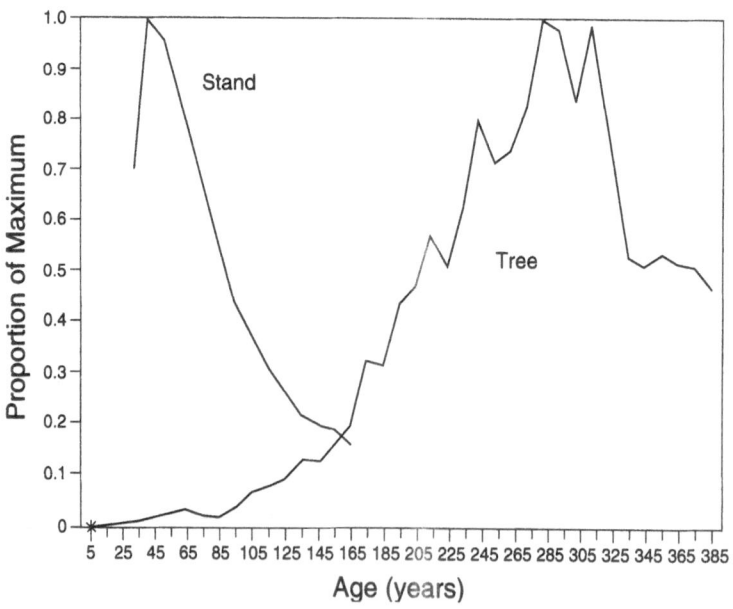

Figure 6.2 Relative rate of volume growth for the Douglas-fir tree from Figure 6.1 compared with expected rate for an entire stand with a site index of 40 m at 100 yr (from yield tables of McArdle et al. 1961).

rapid biomass accumulation and growth soon after stand initiation; stand growth rates began to decline at about age 30 (Figure 6.3, from Binkley and Greene 1983). Despite the reduction in stand growth rate after age 30, the growth rate of dominant trees continued to increase with no peak in sight by age 55 (data not shown). A conifer stand on the same site grew more slowly initially, but maintained high growth rates for a longer period, resulting in substantially more biomass accumulation by age 50. The pattern for a stand containing both conifers and alder showed annual production that matched that of the conifer stand, but much lower biomass accumulation. The difference in the accumulation of biomass between the stands with similar growth rates reflects the greater mortality of trees in the mixed stand. These shifts in dominance in multi-species stands, coupled with the importance of species in affecting rates of production and mortality, also underscore the challenge of deducing pollution impacts on forests.

Stand Development

Stand development involves changes in forest structure (size, number of trees, species) over time. Trends depend on conditions at the time of stand origin, but typically follow general patterns that are mediated by disturbance, management, site quality, stand density and other factors. Most stands progress through recognizable stages as they age and develop, and a 4-stage model of stand development can be applied to most of the coniferous forests in the West (e.g., Oliver 1981, Peet and Christensen 1980, 1987, Long and Smith 1984, Peet 1988). Each stage is associated with characteristic patterns of tree establishment, growth, and death. A stand moves from the initiation stage immediately following a major disturbance, through thinning and transition stages, to reach the old-growth stage (Peet and Christensen 1987).

These normal stages of stand development are relevant to the evaluation of pollutant impacts for several reasons. The impacts of pollutants likely differ among the stages because of changes in tree vigor and species composition between stages, and because key life stages (such as seed production and seedling establishment) are more critical in some stages of stand development than in others. The assessment of pollution impacts is also greatly complicated by the dynamic nature of stands progressing through the stages of development; the identification of pollutant impacts must be made on top of a very dynamic set of normal changes occurring within forests.

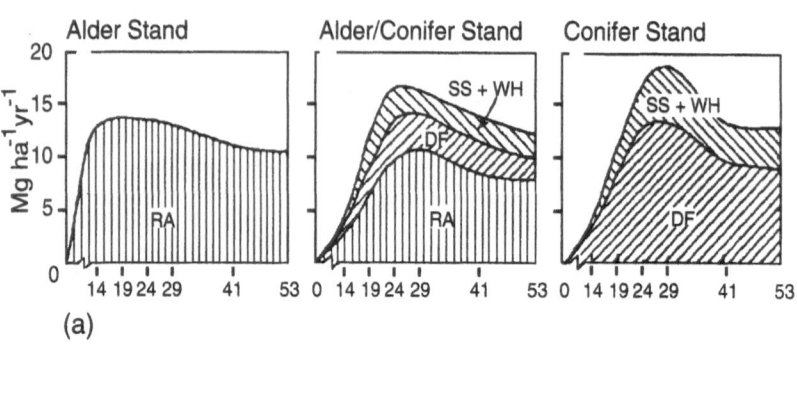

(a)

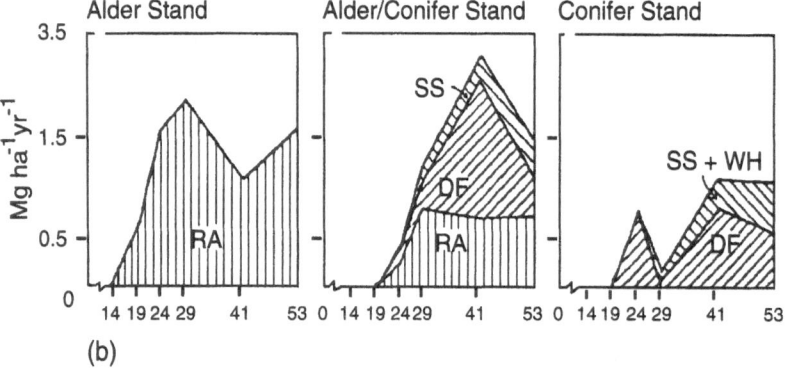

(b)

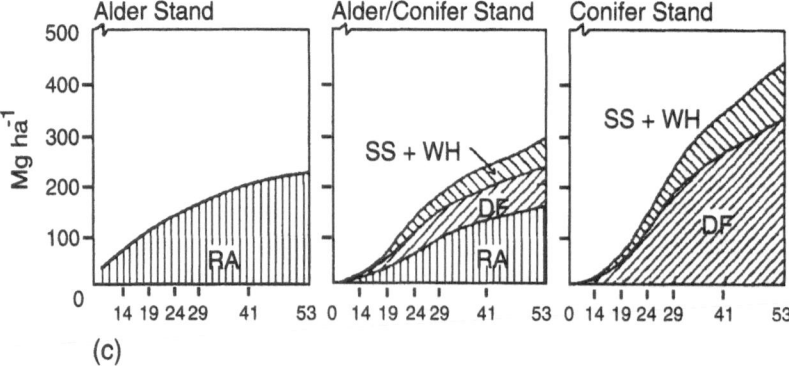

(c)

Figure 6.3 Annual rate of biomass accumulation (top), mortality (middle), and total biomass accumulation (bottom) for adjacent forests of red alder, conifers, and mixed alder and conifers. RA=red alder, DF=Douglas-fir, SS=Sitka spruce, WH=western hemlock (from Binkley and Greene 1983).

Stand Initiation Stage

Following a major disturbance that removes most or all of the canopy of the previous forest, resource availability to establishing trees increases greatly. The rate of recolonization of the site by trees depends on the presence of surviving root systems, buried seeds, or input of seeds from surrounding forests. The relative importance of these sources of new trees depends on species, site conditions, and the nature of the disturbance. Competition between trees is slight (Long and Smith 1984), but competition with non-tree understory plants is often important. Productivity tends to increase rapidly as re-establishing vegetation quickly occupies the site.

The primary productivity of the ecosystem is divided among tree and non-tree vegetation. High rates of productivity found on disturbed fertile sites may occur largely in the non-tree component of the vegetation. Tree leaf area reaches a maximum at the end of the stand initiation stage, and is a key determinant of progress toward the thinning stage (Long and Smith 1984).

The development of non-tree understory plants is probably greatest in the early period of the initiation stage, declining as the leaf area of trees increases. Understory species may increase again through the later transition and old-growth stages.

In the West, the initiation stage typically lasts about ten years when the spatial extent of the disturbance is small enough to allow reseeding from surrounding forests. For larger disturbances, the stand initiation period often lasts more than fifty years (Peet 1988). On the most severe sites, especially near the elevational limit of trees, stand initiation may be limited to episodes of favorable weather conditions for tree establishment. These favorable weather conditions may occur only sporadically, with intervals as long as 100 years (Means 1982).

Thinning Stage

As stand leaf area reaches a maximum, domination of site resources by the most vigorous trees leads to mortality of less vigorous trees. The same pattern pertains to the decline of non-tree vegetation as the tree canopy and root system increase their utilization of site resources. Where site conditions and the nature of disturbance leads to a prolonged stand initiation stage, the thinning stage may be bypassed. Examples include natural stands with high mortality during site initiation and natural stands or plantations where the initial spacing of seedlings is wide.

Stand development during the thinning stage is characterized by declining growth rates for many trees as competition among trees becomes more important, and by death of suppressed trees. In a stand that passes relatively rapidly from the initiation stage to the thinning stage, the sudden decline in annual ring widths could be misinterpreted as the onset of stress caused by pollution. In this case, the stress results from competition among trees and the decrease in growth of individual trees does not represent a change in the rate of growth of the stand. In fact, growth rates of stands tend to reach a maximum during the thinning stage.

The death rate of trees typically continues at a constant per capita rate in unmanaged stands during this stage (Peet and Christensen 1980, 1987). The probability that a tree will survive (or die) during a given period is independent of the number of trees in the stand (the total number of tree deaths/ha would be greater in dense stands). In Douglas-fir forests, the annual rate of mortality ranges from about 5 to 8 trees per thousand, compared with 3 to 5 per thousand for ponderosa pine (*Pinus ponderosa*) stands (Franklin et al. 1987).

With the full utilization of site resources, establishment of new trees is uncommon during the thinning stage, unless cutting of trees substantially increases the resources available to establishing seedlings. As a stand progresses through the thinning stage, deaths of single trees create islands of resource availability that are too large to be readily occupied by surviving trees; this allows regeneration of species that tolerate low availability of resources (called shade-tolerant species, or simply tolerant species), and moves the stand into the transition stage. Unmanaged stands in the West commonly spend 50 to 100 years in the thinning stage.

Transition Stage

The transition stage is characterized by increasing unevenness in the stand, including uneven distributions of leaf area, age classes of trees, and site resources. Information on the dynamics of stand leaf area is sparse, but a decline relative to the thinning stage has been found in some cases (Aplet et al. 1989, Ryan 1990). The death of individual large trees takes on increasing importance during this stage. Stands dominated by single species may shift to multi-species, multi-storied structures. In some extreme cases, such as the old Douglas-fir forests of the Pacific Northwest, the death of a single dominant tree may represent an annual rate of mortality per hectare that matches or exceeds the rate of biomass increment on all other trees. As the accumulation of biomass in living trees declines, the old-growth stage is reached. The approach to old-growth conditions may take 200 years to over 500 years (Oliver 1981). In many parts of the West, the time between major disturbances is too short to allow most stands to move beyond the transition stage (Peet 1988).

Old-growth Stage

The definition of old-growth is widely debated, but certain characteristics are commonly associated with old-growth stands. The oldest age-class dominates the site, but as large trees die and fall, openings are created that allow the understory to flourish and grow into the canopy. The old-growth forest becomes a mosaic of large, old trees interspersed with the initiation, thinning and transition stages, which may persist in varying proportions until a large-scale disturbance occurs. The nature of old-growth mosaics is one of great variability, and classic ideas about steady state forest conditions at landscape scales may not be valid (Botkin 1989).

The accumulation of biomass in the old-growth stage is expected to be zero, or to oscillate between periods of increase and decrease, but few long-term studies are available to document this expectation. Grier and Logan (1977) estimated that the biomass increment of trees in a 450-year-old Douglas-fir forest in Oregon was about -4,000 kg/ha annually because of substantial mortality of large trees. Franklin and DeBell (1988) characterized the dynamics of an old-growth forest of Douglas-fir and western hemlock (*Tsuga heterophylla*) in Washington for a period of 36 years, and concluded that the distribution of size classes and species remained relatively constant, despite the death of 22% of the trees. The major causes of tree death were wind (accounting for 46% of the deaths) and suppression (40% of the deaths).

Stand Development and Ecosystem Processes in the West

Empirical data are not available to synthesize a picture of stand development through the four stages for most forest types in the West. Peet (1981, 1988) summarized patterns of stand development in the Front Range of Colorado for a favorable site (Figure 6.4a) and an unfavorable site (Figure 6.4b). On the favorable site, the initiation period is characterized by high rates of tree establishment, increasing biomass and production, and high species diversity. With the onset of the thinning stage, production plateaus as biomass continues to accumulate and species diversity and rates of tree establishment decline sharply. In the transition stage, productivity and biomass decline largely due to the mortality of large trees. Species diversity and tree establishment increase. The creation of larger gaps during the old growth stage allows a resurgence of

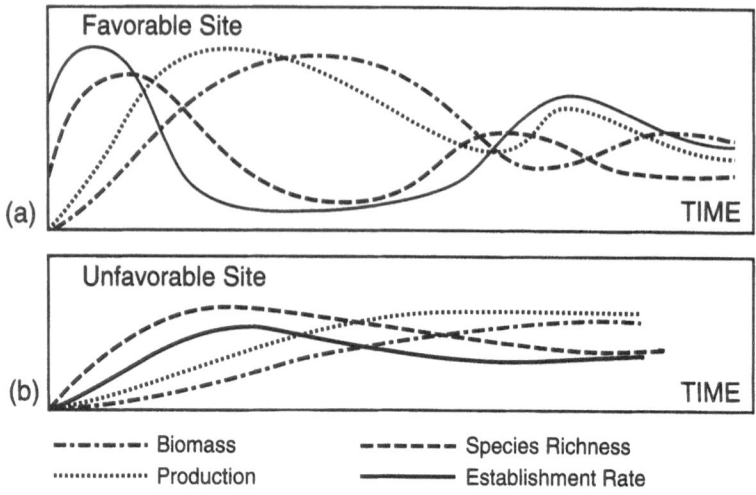

Figure 6.4 Generalized development patterns for Colorado Front Range forests; (a) is typical of lodgepole pine and Douglas-fir forests, and (b) is more representative of high-elevation spruce-fir forests (from Peet 1988).

productivity and biomass accumulation. On a less favorable site, the maximum rates of all four features of the stand are lower and the distinctions between stages become muted.

Changes in species composition are expected to be dramatic during stand development in many of the major forest types of the West, often including replacement of early dominant species such as aspen (*Populus tremuloides*) in parts of the Rocky Mountains or red alder in parts of the Northwest. In some cases, such as ponderosa pine forests in the Southern Rockies, a single tree species dominates all stages of stand development. Environmental changes that would accompany these stand development sequences have not been well characterized, but some generalizations are possible based on a variety of studies from around the West.

Several studies in the West support a primary role of leaf area and light interception in determining tree productivity and changes in understory light environments with stand development (Brix 1981, 1983, Waring 1983, Binkley and Reid 1984, Oren et al. 1987, Smith and Long 1989). For example, Gholz (1982) documented a strong relationship between aboveground net primary productivity and stand leaf area in a transect in Oregon of forests ranging from rain forests on the coast to pinyon/juniper stands east of the Cascade Mountains. He concluded that the effects of the diverse environments on forest productivity were mediated largely through effects on canopy leaf area. The leaf area of forests in the West ranges from near 0 m² of leaves (on a projected, or shadow-cast basis) per m² of ground area for newly regenerating forests, up to a maximum of about 10 to 13 m²/m² for stands of shade-tolerant species such as Sitka spruce (*Picea sitchensis*) (Franklin 1988). The rate of leaf area development following major disturbances is highly variable, ranging from 6 to 50 years to reach maximum levels in many Western forests (Zavitkovski and Newton 1971, Long and Smith 1984). Understory leaf area declines as the overstory canopy expands (Turner et al. 1978).

Changes through stand development also strongly influence temperatures within ecosystems. Canopies moderate fluctuations in temperatures at the soil surface, and reduce both daily highs and lows in comparison to disturbed sites. In a classic study of a Douglas-fir forest at the Lubrecht Experimental Forest in Montana, Hungerford (1980) measured the maximum and minimum temperatures on August 29, 1978 in stands that had received a variety of silvicultural treatments. The intact forest showed a maximum temperature of 35°C at the soil surface, and a daily minimum of 2°C, while the clearcut site had a high of 56°C (possibly lethal to seedlings), and a low of -5°C. The period between the last freezing temperature of spring and first freeze of autumn in the uncut forest was 112 days, compared with only 20 days in the clearcut site (Hungerford 1980).

Water use by forests also changes with stand development; studies of changes in forest hydrology following harvest document these changes. In Arizona, harvesting all ponderosa pine trees in a watershed increased streamflow by about 125 mm (equivalent to a layer of water 125 mm deep across the entire surface of the watershed; Ffolliott and Thorud 1974). Factors contributing to the increase in runoff after harvesting include reduced interception loss from the canopy (evaporation before precipitation reaches the ground), and reduced transpiration by trees. Both factors lead to wetter soils, and presumably less moisture stress for remaining or newly established trees. At the Coram Experimental Forest in Montana, Newman and Schmidt (1980) followed the seasonal changes in soil moisture in plots with intact larch/Douglas-fir forests and in plots that had received various levels of harvesting. The intact forest depleted soil moisture to below 15% (water weight/dry soil weight) during the summer, compared with a minimum value of 25% for the young forest regenerating in the clearcut. Increased soil moisture and water run-off may persist through decades of stand development (Troendle and Kaufmann 1987). Adams et al. (1991) compared soil moisture levels in a clearcut/burn and control area in the Oregon Cascade Range. The clearcut/burn area averaged higher moisture levels than the control area for the first few years. Within four years, however, the surplus moisture in the clearcut/burn area had declined to less than the control moisture level and remained less over the next decade of monitoring. The deficits were attributed to a rapid increase in plant cover after the burn.

The rate of nutrient availability in forests depends primarily on the chemical quality of organic matter present in the forest floor and mineral soil, and on moisture and temperature conditions that affect the activity of decomposer microbes (Binkley 1986). All of these factors probably change with stand development, but few studies have characterized how changing microenvironmental conditions alter nutrient availability through stand development in the West. General expectations are that nutrient availability should increase after forest removal due to wetter and warmer soil conditions (Edmonds et al. 1989), and so far all studies in the West have shown higher nitrogen availability in younger forests (Binkley 1984, Matson and Boone 1984, Hart and Firestone 1989, Arthur 1990, Frazer et al. 1990, Page-Dumroese 1991).

Pollution Impacts at the Stand and Ecosystem Levels

The impacts of pollution on the growth of individual trees can be dramatic, ranging from moderate reductions in annual ring widths to complete absence of growth rings in the lower stem for some years (Chapter 7). Altered growth rates can lead to changes in patterns of tree death, species regeneration, and species composition. The effect of air pollution on the growth of stands is more complicated than the effects on single trees, because the reductions in growth of sensitive individuals may be partially offset by increased growth of less sensitive trees (Chapter 7).

Differential Impacts on Species, Trees, and Stands

Pollutant impacts across stand development stages depend in part on the relative sensitivities of species to the pollutants. General sensitivity rankings have been developed for some groups of species. For example, in the San Bernardino Mountains (Chapter 12), species sensitivity to ozone decreases in the order of: ponderosa pine, Jeffrey pine (*Pinus jeffreyi*), white fir (*Abies concolor*), Coulter pine (*Pinus coulteri*), incense cedar (*Libocedrus decurrens*), big-cone Douglas-fir (*Pseudotsuga macrocarpa*), and sugar pine (*Pinus lambertiana*).

In a study of ponderosa pine in the San Bernardino Mountains, annual radial growth declined from 5 mm yr^{-1} in healthy trees to 3 mm yr^{-1} in pollution impacted trees (McBride et al. 1975). However, the stands examined by McBride et al. (1975) differed substantially in structure, so differences in radial growth may not correspond to differences in growth at a stand level. The height reduction caused by ozone was about 2 m in otherwise comparable 30-year old trees. Height and diameter growth reductions combined to reduce merchantable wood volume from 0.55 m³/tree to 0.14 m³/tree (McBride et al. 1975). However, widths of annual rings from 1950 to 1975 increased in approximately one-third of the ponderosa pines measured in the San Bernardino Mountains at two locations that received high ozone exposures (Miller 1983). These growth increases may have resulted from decreased competition after the death of neighboring trees. Ponderosa pine trees were highly variable in sensitivity to ozone, and less resistant genotypes may have been eliminated from the forests over the past 30 years. This pollutant-induced thinning may have enhanced the vigor of the remaining, more resistant, individual ponderosa pine (Miller 1983) and trees of other species.

The effects of pollutants on stand structure, especially upon relationships between overstory and understory plants, may be complex. Early expectations were that large, dominant trees might receive the greatest exposure to pollutants and show the greatest impacts; understory plants were expected to be less exposed and less affected (Woodwell 1970, Treshow and Anderson 1988). However, species sensitivity is often more important than pollutant dose, and generalizations about responses of overstory and understory plants may not be warranted. For example, Anderson (1966, cited in Treshow and Anderson 1988) examined the impacts of fluoride pollution around a phosphate plant in Idaho. Oregon grape (*Berberis repens*), an understory herb, showed injury at lower exposure levels than the overstory Douglas-fir.

Pollutant Interactions with Pests

Air pollutants can play an important role in mediating or increasing pest and pathogen attacks in the West. James et al. (1980) showed that colonization of stumps and roots by the root rot fungus, *Fomes annosus*, was 30 to 50% faster in pine trees showing chronic ozone injury than in asymptomatic trees; controlled ozone fumigation of 6-year-old ponderosa and Jeffrey pine seedlings also increased fungal infection. In an area with high ozone levels, bark beetles (*Dendroctonus brevicomis*) were important agents in over 80% of the pine trees that died (Miller 1983). Bark beetles and defoliators can also be important pests because of other factors such as normal aging or changes in stand composition; pest presence does not necessarily indicate air pollution damage.

Pollutant Impacts and the Four Stages of Stand Development

Air pollution can affect stand initiation by influencing cone production and the likelihood of seed regeneration, and through direct effects on seedlings. Seed production in conifers varies with genotypes and level of stress. Stress may lead to short-term increases in seed production in some species whereas prolonged stress may reduce seed crops. Hedgecock (1912, cited in Smith 1990) found that conifers bore few or no cones near the Washoe smelter at Anaconda, Montana. Other studies have documented reduced cone crops near point sources of sulfur dioxide, and regions of high ozone pollution (Smith 1990). In the San Bernardino Mountains, dominant ponderosa pine and Jeffrey pine trees showing severe ozone damage produced significantly fewer cones than uninjured dominant trees (Luck 1980). However, acorn production by California black oak (*Quercus kelloggii*) was not reduced by ozone exposure (Miller et al. 1980), suggesting the possibility of better regeneration of oak than pine in areas of high ozone. Seed germination is not likely to be directly affected by either acid deposition or ozone (Smith 1990). Following germination, growth and development of seedlings may be affected by both gaseous and aqueous pollutants (Chapter 5).

Indirect effects of pollution may also be important in seedling establishment and growth, which are strongly influenced by the micro-environment at the soil surface. For example, reductions in canopy leaf area in forests of southern California following ozone damage allowed more solar irradiance to reach the forest floor, raising seed bed temperatures and seedling transpiration demands (Miller 1983). However, McBride et al. (1985) found that the level of ozone exposure had no effect on the establishment of seedlings of ponderosa pine or white fir during a 10-year period in permanent plots along an ozone gradient in the San Bernardino Mountains.

During the thinning stage, pollution impacts may be particularly important. Tree-to-tree competition is probably the most important determinant of individual tree growth and mortality. Pollution-induced stress on trees of sensitive genotypes and species is likely to impair their ability to compete with less sensitive trees. No clear examples of pollution impacts at the thinning stage are available, but susceptibility of stressed trees to bark beetle attack was demonstrated experimentally for lodgepole pine in Oregon. Waring and Pitman (1983) used pheromones to attract large numbers of beetles to trees that had received treatments of fertilizer (to reduce stress) or carbohydrate (to tie up soil nutrients and increase stress). They found that the death of trees depended on the number of beetle attacks and on the vigor of the tree. Tree vigor was indexed as the grams of wood produced per m^2 of tree foliage. High vigor trees survived relatively high numbers of attacks, whereas low-vigor trees were killed even by low numbers of attacks. A similar pattern may develop if stands in a thinning stage were exposed to pollution-induced stress.

During the thinning stage, trees impacted by pollutants during the preceding stand initiation stage would likely be at a competitive disadvantage. A drought or other stress would make them especially vulnerable. These trees would likely die during the thinning stage, but the growth of associated surviving trees might increase, keeping stand growth high.

During the transition stage, pollutant impacts might accelerate stand dynamics by increasing the mortality rate of overstory trees, which in turn would favor the establishment of new individuals. In the West, pollution-induced changes in species composition in the San Bernardino Mountains (see Chapter 12) fit this pattern, where impaired growth and greater mortality of ponderosa and Jeffrey pines may benefit regeneration of incense cedar. In this case, pollution stress increased susceptibility of the pines to bark beetles and root pathogens which were the direct causes of death. High mortality of pines may lead to dominance of the stand by species more tolerant of ozone, with potentially little effect on the total productivity of the stand.

Little empirical information is available on pollution effects on old-growth forests. In theory, pollution effects might accelerate normal processes of tree death and replacement in a mosaic of stands in the old growth stage. Pollution stresses could affect the survival and growth of both old, established trees and of newly regenerating seedlings. Pollutant exposure may reduce seed sources; however, accelerated death of dominant trees could reduce the environmental limitations to seedling establishment as the canopy opens.

Pollution Impacts on Environmental Factors

The role of pollution in altering regimes of light and temperature in forest ecosystems depends strongly on changes in forest canopies. Reductions in canopy leaf area should reduce light interception and allow greater soil temperatures, and greater fluctuations in diurnal and seasonal temperatures. The impact of pollution on environmental factors would likely differ across stand development stages; however, insufficient information is available to support a discussion of these differences.

As with temperature, the effect of pollutants on forest hydrology would be expected to relate strongly to changes in the forest canopy. Miller et al. (1980) noted that reductions in the leaf area of ponderosa pine trees resulted in greater penetration of rainfall to the soil surface, lowering interception losses from the canopy but probably increasing evaporation from the surface of the forest floor. They also pointed out that ozone-impacted needles showed reduced stomatal conductance per unit of leaf area, reducing transpiration by impacted trees. Whether transpiration by the entire stand would be reduced depends of course on the ability of neighboring trees to increase their rates of transpiration. If overall stand transpiration were lower, increased soil water availability may allow less affected plants to transpire more, increasing their exposure (via open stomata) to ozone (Miller 1983).

The direct effects of air pollution on nutrient availability have received little attention in the West, and for the most part only speculations are possible. A wide range of effects can be postulated. For example, deposition of nitric acid to nitrogen limited soils might increase nitrogen availability (Chapter 4). Nitrogen deposition is poorly characterized in the West (Chapter 3), though we do know that deposition of nitrate and ammonium in rain and snow is generally low. No dramatic, or perhaps even measurable, effects on nutrient cycling are currently expected from acid deposition, except perhaps in southern California where nitrogen deposition rates are high (Fenn and Dunn 1989).

Some indirect effects of oxidant pollution on nutrient cycling may also occur, including greater needlefall rates (at least temporarily) and altered concentrations of nutrients in litter (Chapters 4 and 12). The importance of such changes to the rate of nutrient availability and other processes remains unknown. Long-term changes in ecosystem structure following air pollution exposure can also influence nutrient cycling. Changes in species composition could lead to species-driven changes in nutrient cycling patterns; but again no evidence is available to support speculations. Perhaps more critical would be increased fire risk that might result from greater accumulation of fine fuels in pollution stressed stands. Greater fuel loads coupled with greater drying due to increased light penetration could alter both the frequency and intensity of fires. Changes in fire characteristics would have large impacts on nutrient losses, cycling and availability (Binkley and Christensen 1992).

Synthesis

A classification scheme was developed by Smith (1974, with modifications in Smith 1981, 1990, and Bormann 1985, 1990) to describe classes of forest response to increasing pollution stress. In Class I, pollutant levels are generally low, and ecosystems serve as both sources (hydrocarbons) and sinks for low levels of pollutants. Forests remain relatively unaffected, or may even benefit from slight fertilization effects.

Early in Class II pollutant levels begin to interfere with some aspects of the life cycle of sensitive species or individuals. Plants are either adversely affected or their growth is enhanced because of adverse effects on other plants. Pollutants inhibit photosynthesis and reduce available carbohydrate reserves needed for reproduction or defensive mechanisms. The normal dynamics of forests are large enough relative to such pollution-induced changes that detection of early Class II effects is often difficult.

With increasing pollution, Class II responses would include declines in populations of sensitive species. Some individuals of sensitive species may survive, and decreases in pollution levels would allow sensitive species to recover. Shifts in species dominance may maintain total ecosystem function (such as net primary production) at levels similar to unpolluted conditions, although structural changes (such as species composition or canopy distribution) would occur.

Bormann (1985, 1990) stressed that forests in Class II begin to lose redundancy, or the capacity to respond flexibly to new stresses. For example, pollution stress may reduce the genetic variability within populations by removing sensitive individuals, and this reduction in genetic diversity may impair the population's ability to cope with other stresses such as drought or insect outbreaks. Similarly, the removal of sensitive species from a diverse forest may constrain the ability of the forest to respond to changes in climate. Losses in redundancy may be difficult (or impossible) to detect, but may be fundamentally important to the ecosystem response to both natural and pollution-related stresses.

In Class III, greater pollution stress results in major disruptions of the forest. Productivity declines, biotic regulation of nutrient cycles weakens, and some species disappear. Later recovery of the forest (if pollution levels declined) could be impaired by a lack of propagules for some species.

The symptoms characterizing the stages of forest ecosystem decline depend in part on the stage of stand development. A matrix can be constructed to show the interactions between stage of stand development and stage of ecosystem stress (Table 6.1). The movement of a forest ecosystem through time may be vertically through Table 6.1 if pollution levels are constant, horizontal to the right or diagonally to the lower right if pollution levels are increasing, or potentially to the left or lower left if pollution levels are declining. Trends for ecosystems within Class I are summarized from the discussion earlier, and those in Class II represent slight modifications that include primarily shifts in species dominance and perhaps composition. Pollution effects may be severe enough in Class III to shorten or preclude some of the stand development stages.

Table 6.1 Interaction of stages of stand development and ecosystem stress. See text for descriptions of stand development stages and forest stress classes.

Stand Stage	Forest Stress Class I	Forest Stress Class II	Forest Stress Class III
Stand Initiation	Regeneration by seed, sprouting; competition between trees and non-tree vegetation; rapid expansion of canopies and root systems	Subtle shifts in species success in regeneration; shifts in competitive success alter growth rates, with some species improving at expense of others	Substantial changes in species success, with dominance of pollution resistant species; reduced growth of most species; slower expansion of canopies and roots
Thinning	Maximum leaf area attained; mortality increasing from competition between trees; maximum productivity	Maximum leaf area attained, but proportion from resistant species greater; competition induced mortality with greater rates in sensitive species but similar overall rate	Leaf area reduced; mortality driven by competition, pollution stress and pathogens; in severe case, thinning stage may be bypassed
Transition	Increased variation in canopies allowing reinitiation of seedling establishment; high productivity	Very similar to Class I , with possible differences in species composition and dominance	May occur sooner as pollution stresses disaggregate canopy; low productivity
Old-growth	Mosaic of age classes, diverse stand structure, low or negative net productivity	Very similar to Class I, with possible differences in species composition and dominance	Uneven structure with partial or complete removal of dominant trees; stage may not occur if stress is severe enough, with site moving to stand initiation stage

Conclusions

Two major challenges confront research and management of forests in polluted environments:

—Detecting shifts of ecosystems from Class I to Class II; and

—Predicting (and modifying?) further changes in ecosystems that fall within Class II.

As a forest shifts from Class I to Class II, substantial pollution-induced changes may occur that are difficult to separate from the normal changes associated with stand development, including patterns for single trees, for species within stands, and for entire forests. Bormann (1985, 1990) raises a series of sobering questions about the impacts of a 10% reduction in productivity resulting from pollution stress, including issues of species diversity, soil erosion, and nutrient retention. In most cases, a 10% reduction in net productivity may go undetected, and we cannot predict whether consequences for ecosystem structure and function would be small or large. The second issue of predicting further changes where pollution impacts are detected is also largely beyond the current state of science.

References

Adams PW, Flint AL, Fredriksen RL (1991) Long-term patterns in soil moisture and revegetation after a clearcut of a Douglas-fir forest in Oregon. *Forest Ecology and Management* 41:249–263

Aplet GH, Smith FW, Laven R (1989) Stemwood biomass and production during spruce-fir stand development. *Journal of Ecology* 77:70–77

Anderson FK (1966) *Air Pollution Damage to Vegetation in Georgetown Canyon, Idaho.* MSc thesis, University of Utah, Salt Lake City

Arthur M (1990) *Biogeochemistry of the Loch Vale Watershed, Rocky Mountain National Park.* PhD dissertation, Cornell University, Ithaca

Binkley D (1984) Does forest removal increase rates of decomposition and nutrient release? *Forest Ecology and Management* 8:229–233

Binkley D (1986) *Forest Nutrition Management.* Wiley, New York

Binkley D, Christensen N (1992) Canopy fire effects on nutrient cycling and productivity. In: Laven R, Omi P (eds) *Pattern and Process in Crown Fire Ecosystems.* Princeton University Press, in press

Binkley D, Greene S (1983) Production in mixtures of conifers and red alder: The importance of site fertility and stand age. In: Ballard R, Gessel S (eds) *IUFRO Symposium on Forest Site and Continuous Productivity.* USDA Forest Service General Technical Report PNW-163, Portland, Oregon, pp 112–117

Binkley D, Reid P (1984) Long-term responses of stem growth and leaf area to thinning and fertilization in a Douglas-fir plantation. *Canadian Journal of Forest Research* 14:656–660

Bormann FH (1985) Air pollution and forests: An ecosystem perspective. *BioScience* 35(7):434–441

Bormann FH (1990) Air pollution and temperate forests: Creeping degradation. In: Woodwell G (ed) *The Earth in Transition: Patterns and Processes of Biotic Impoverishment.* Cambridge University Press, Cambridge, pp 25–44

Botkin D (1989) *Discordant Harmonies: A New Ecology for the Twenty-First Century.* Oxford University Press, Oxford

Brix H (1981) Effects of nitrogen fertilizer source and application rates on foliar nitrogen concentration, photosynthesis and growth of Douglas-fir. *Canadian Journal of Forest Research* 11:775–780

Brix H (1983) Effects of thinning and nitrogen fertilization on growth of Douglas-fir: Relative contribution of foliage quantity and efficiency. *Canadian Journal of Forest Research* 13:167–175

Edmonds R, Binkley D, Feller MC, Sollins P, Abee A, Myrold DD (1989) Nutrient cycling: Effects on productivity of Northwest forests. In: Perry D et al. (eds) *Maintaining the Long-Term Productivity of Pacific Northwest Forest Ecosystems.* Timber Press, Portland, pp 17–35

Ffolliott PF, Thorud DB (1974) *Vegetation Management for Increased Water Control in Arizona.* Arizona Agricultural Experiment Station Technical Bulletin 215

Fenn M, Dunn P (1989) Litter decomposition across an air pollution gradient in the San Bernardino Mountains. *Soil Science Society of America Journal* 53:1560–1567

Franklin JF (1988) Pacific Northwest forests. In: Barbour MG, Billings WD (eds) *North American Terrestrial Vegetation.* Cambridge University Press, Cambridge, pp 102–130

Franklin JF, DeBell DS (1988) Thirty-six years of tree population change in an old-growth *Pseudotsuga-Tsuga* forest. *Canadian Journal of Forest Research* 18:633–639

Franklin JF, Shugart HH, Harmon ME (1987) Tree death as an ecological process. *BioScience* 37:550–556

Frazer D, McColl J, Powers R (1990) Mineralization of soil nitrogen in a managed, mixed-conifer forest in northern California. *Soil Science Society of America Journal* 54:1145–1152

Gholz HL (1982) Environmental limits on aboveground net primary production, leaf area, and biomass in vegetation zones of the Pacific Northwest. *Ecology* 63:469–481

Grier CC, Logan RS (1977) Old-growth *Pseudotsuga menziesii* communities of a western Oregon watershed: Biomass distribution and production budgets. *Ecological Monographs* 47:373–400

Hart SC, Firestone M (1989) Evaluation of three in situ soil nitrogen availability assays. *Canadian Journal of Forest Research* 19:185–191

Hedgecock GG (1912) Winter-killing and smelter-injury in the forests of Montana. *Torrya* 12:25–30

Hungerford RD (1980) Microenvironmental response to harvesting and residue management. In: *Environmental Consequences of Timber Harvesting in Rocky Mountain Coniferous Forests.* USDA Forest Service General Technical Report INT-90, Ogden, pp 37–74

James RL, Cobb FW, Miller PR, Parmeter JR Jr (1980) Effects of oxidant air pollution on susceptibility of pine roots to *Fomes annosus*. *Phytopathology* 70:560–563

Long JN, Smith FW (1984) Relation between size and density in developing stands: A description and possible mechanisms. *Forest Ecology and Management* 7:191–206

Luck RF (1980) Impact of air pollution on ponderosa and Jeffrey pine cone production. In: Miller PR (ed) *Proceedings Effects of Air Pollutants on Mediterranean and Temperate Forest Ecosystems.* USDA Forest Service General Technical Report PSW-43, pp 240

Matson P, Boone R (1984) Natural disturbance and nitrogen mineralization: Wave-form dieback of mountain hemlock in the Oregon Cascades. *Ecology* 65:1511–1516

McArdle RE, Meyer WH, Bruce D (1961) *The Yield of Douglas-fir in the Pacific Northwest.* USDA Technical Bulletin #201, Washington, DC

McBride JR, Semion VP, Miller PR (1975) Impact of air pollution on the growth of ponderosa pine. *California Agriculture* 29:8–9

McBride JR, Laven R, Miller PR (1985) Effects of oxidant air pollutants on forest succession in the mixed conifer type of southern California. In: *Proceedings Air Pollutant Effects on Forest Ecosystems*. Acid Rain Foundation, St. Paul, Minnesota, pp 157–167

Means JE (1982) Developmental history of dry coniferous forests in the Western Oregon Cascades. In: Means JE (ed) *Forest Succession and Stand Development Research in the Northwest*. Forest Research Laboratory, Oregon State University, Corvallis, pp 142–158

Miller PR (1983) Ozone effects in the San Bernardino National Forest. In: *Proceedings: Air Pollution and the Productivity of the Forest*. Izaak Walton League and Penn State University, pp 161–197

Miller PR, Longbotham GJ, Van Doren RE, Thomas MA (1980) Effect of chronic oxidant air pollution exposure on California black oak in the San Bernardino Mountains. In: Plumb T (ed) *Proceedings: Ecology, Management, and Utilization of California Oaks*. USDA Forest Service General Technical Report PSW 44, Riverside, pp 220–229

Newman HC, Schmidt WC (1980) Silviculture and residue treatments affect water use by a larch/fir forest. In: *Environmental Consequences of Timber Harvesting in Rocky Mountain Coniferous Forests*. USDA Forest Service General Technical Report INT-90, Ogden, pp 75–115

Oliver CD (1981) Forest development in North America following major disturbances. *Forest Ecology and Management* 3:153–168

Oliver CD, Larson BC (1990) *Forest Stand Dynamics*. McGraw-Hill, New York.

Oren R, Waring RH, Stafford SG, Barrett JW (1987) Twenty-four years of ponderosa pine growth in relation to canopy leaf area and understory competition. *Forest Science* 33:538–547

Page-Dumroese D (1991) Organic horizons of western-montane forest soils. In: Harvey A, Neuenschwander L (eds) *Management and Productivity of Western-Montane Forest Soils*. USDA Forest Service General Technical Report, Ogden, in press

Peet RK (1981) Forest vegetation of the Colorado Front Range: Composition and dynamics. *Vegetatio* 45:3–75

Peet RK (1988) Forests of the Rocky Mountains. In: Barbour MG, Billings WD (eds) *North American Terrestrial Vegetation*. Cambridge University Press, Cambridge, pp 63–102

Peet RK, Christensen NL (1980) Succession: A population process. *Vegetatio* 43:131–140

Peet RK, Christensen NL (1987) Competition and tree death. *BioScience* 37:586–595

Ryan MG (1990) Growth and maintenance respiration in stems of *Pinus contorta* and *Picea engelmannii. Canadian Journal of Forest Research* 20:48–57

Smith FW, Long JN (1989) The influence of canopy architecture on stemwood production and growth efficiency of *Pinus contorta* var. *latifolia. Journal of Applied Ecology* 26:681–691

Smith WH (1974) Air pollution—effects on the structure and function of the temperate forest ecosystem. *Environmental Pollution* 6:111–129

Smith WH (1981) *Air Pollution and Forests: Interactions between Air Contaminants and Forest Ecosystems.* Springer-Verlag, New York

Smith WH (1990) *Air Pollution and Forests: Interactions Between Air Contaminants and Forest Ecosystems.* Springer-Verlag, New York

Treshow M, Anderson FK (1988) *Plant Stress from Air Pollution.* Wiley, New York

Troendle C, Kaufmann M (1987) Influence of forests on the hydrology of the subalpine forest. In: *Management of Subalpine Forests: Building on 50 Years of Research.* USDA Forest Service General Technical Report RM-149, Ft. Collins, Colorado, pp 68–78

Turner J, Long JN, Backiel A (1978) Under-storey nutrient content in an age sequence of Douglas-fir stands. *Annals of Botany* 42:1045–1055

Waring RH (1983) Estimating forest growth and efficiency in relation to canopy leaf area. *Advances in Ecological Research* 13:327–354

Waring RH, Pitman GB (1983) Physiological stress in lodgepole pine as a precursor for mountain pine beetle attack. *Zeitschrift für angewandte Entomologië* 96:266–270

Woodwell GM (1970) Effects of pollution on the structure and physiology of ecosystems. *Science* 168:429–433

Zavitkovski, J, Newton M (1971) Litterfall and litter accumulation in red alder stands in western Oregon. *Plant and Soil* 34:257–268

7

Methods of Assessing Responses of Trees, Stands and Ecosystems to Air Pollution

K. W. Stolte, D. M. Duriscoe, E. R. Cook, S. P. Cline

Introduction

The responses of forests to air pollution in the western United States range from cellular injury on individuals to alteration of forest communities. These responses have been observed in field studies and controlled exposure studies. A variety of methods have been developed by the USDA Forest Service, the USDI National Park Service, the US Environmental Protection Agency, academic institutions, and private consultants to determine the nature, incidence, and severity of pollution effects on Western forests.

The choice of indicators to be measured depends on the objectives of the study and the pollutants of interest (Miller 1989). Some pollutants leave elemental signatures, such as increased concentrations or presence of novel chemicals. Fluoride (Carlson and Dewey 1972) and sulfur dioxide (Nriagu 1978, Smith 1990) leave elemental signatures as well as producing distinct visible injury symptoms. Acid rain and toxic metals (e.g., zinc, vanadium, arsenic) can leave elemental signatures and can produce visible symptoms on foliage at high concentrations, with subsequent effects on vigor, growth, and productivity (Lepp 1981a, b, Wellburn 1989, MacKenzie and El-Ashry 1989). Naturally occurring stable isotopes of elements are sometimes greatly enriched or depleted in anthropogenic air masses compared to natural background levels; differences in isotope ratios may serve as markers of anthropogenic deposition, uptake by biological systems, and alteration of nutrient cycles (Krouse 1974, Jackson and Gough 1989).

Ozone and some other gaseous oxidants (e.g., hydrogen peroxide) leave no discernible residue or elemental signature, and the only signature to be detected is the biological response of the plant. The responses of conifers in the western United States to oxidants include:

- Cellular injury (Evans and Miller 1972a, b, Carlson and Dewey 1972) and change in metabolites (Tingey et al. 1976)

- Altered physiological processes (e.g., photosynthesis, respiration, water and nutrient cycles) (Coyne and Bingham 1982, Patterson and Rundel 1989, Grulke et al. 1989)

- Macroscopic (visible) injury (Miller et al. 1963, Richards et al. 1968)

- Altered phenological cycles (e.g., increased abscission rates, altered needle initiation) (Miller and Van Doren 1982, Duriscoe and Stolte 1989, Ewell et al. 1989)

- Altered growth (increased or decreased) (Peterson et al. 1987, Miller et al. 1989; see chapters 11 and 12)

- Community perturbations (e.g., change in species diversity, importance, and cover) (Miller et al. 1982)

This chapter focuses on the three main types of assessments of pollution effects used in the case studies chronicled in Chapters 8 through 12: measures of (1) crown condition of individual trees; (2) impacts on populations and communities; and (3) temporal patterns in radial growth. The concepts behind the development of each approach are introduced with references to previous work, leading to a discussion of the state of the science. The importance of quality assurance techniques to the success of any assessment of air pollution effects is also discussed.

Crown Condition Assessments

Crown injury symptoms are the most common kind of visible pollutant injury recorded in laboratory and field air pollution studies in the West. Crown injury symptoms are measured by assessments of the stem and foliage portions of the trees. Gaseous pollutants pass through the stomata of conifer foliage and cause direct damage to photosynthetic cells, sometimes producing a diagnostic, visible foliar injury pattern. Impairment of essential biological processes in the needles may eventually lead to other crown responses such as premature needle abscission, reduced crown vigor, increased susceptibility to pathogens, and tree death (Miller 1977). Acidic deposition (wet and dry) and toxic metals are more likely to initially impact root systems and associated nutrient cycling processes.

Unless concentrations of acidic deposition or metals are very high, crown degeneration occurs secondarily from the alteration of nutrient and water cycles (Wellburn 1989, Smith 1990).

Crown response variables are macroscopic morphological alterations of the crown and stem features by air pollutants that are quantifiable within an acceptable degree of precision. Crown response variables that have been measured in Western pollution studies include foliar discoloration (visible injury), foliage longevity, needle length, foliage production, crown density, live crown ratio, cone production, stem growth, and toxic element accumulation. These variables can be quantified by visual observation, imaging and subsequent image processing, elemental analyses, and common forest mensuration techniques. Determining whether changes in these morphological features are caused by air pollution may require ancillary pathological and environmental evaluations (Wellburn 1989, Treshow and Anderson 1989, Smith 1990).

Description and Diagnosis of Gaseous Pollutant Injury

A description of the effects of gaseous pollutants on the foliage of conifers is essential to understanding the methods that have evolved in the western United States to identify and quantify pollution injury. Gaseous air pollutants found in the western United States initially injure foliage by entering intercellular spaces when stomata are open as part of the normal processes of exchanging atmospheric gases (Chapter 5). Cellular injury occurs as pollutants degrade cell membranes (Evans and Miller 1972a, b) and toxic substances accumulate. Chlorophyll degradation, cell desiccation, and cell death occur on the needle surface. Controlled fumigation studies (Miller et al. 1963, Treshow 1970, Gordon 1974) have established that ozone, sulfur dioxide, and hydrogen fluoride cause injury to cells in needles (histological injury) that result in visible foliar discolorations that are distinct and readily discernable. Foliar injury commonly includes chlorosis (yellowing) or necrosis (browning) of needle tissue.

Visible foliar injury is sometimes correlated with other pollution effects on tree crown condition and with alterations of population and community-level forest dynamics (Carlson and Dewey 1972, Miller 1977, Taylor 1980). Visible injury indicates physiological impairment that may result in measurable reductions in net photosynthesis, increases in respiration, and alteration of nutrient and water cycles within the foliage (see Chapter 5, also Patterson and Rundel 1989). However, visible injury is most effective in identifying the occurrence of phytotoxic levels of a pollutant; higher order effects (such as altered photosynthesis, growth, reproduction, or fitness) may or may not co-occur with visible injury. The presence of visible injury symptoms indicates only that ambient levels of pollutants are sufficiently elevated to injure cells within needles. Conversely, a lack of visible foliar injury does not prove there is no pollution impact (Pye 1988; see Grulke et al. 1989 for an example with giant sequoia [*Sequoiadendron giganteum* Bucch.]).

Premature needle abscission correlates with ozone exposure and degree of foliar injury in some Western conifer species (Miller et al. 1983, Duriscoe and Stolte 1989). Chronic pollution stress has also been correlated in some species with the size, numbers, and structure of needles produced (Coyne and Bingham 1982, Ewell et al. 1989). Responses among individual trees and species are highly variable and the thresholds where visible injury is severe enough to initiate higher order effects under field conditions are poorly known (Smith 1990).

Ozone

Controlled exposure studies and field studies of ozone impacts on Western conifer species have demonstrated that a distinct visible discoloration known as chlorotic mottle typically occurs on needle surfaces (Miller et al. 1963, Richards et al. 1968, Pronos et al. 1978, Duriscoe and Stolte 1989). Chlorotic mottle begins as a degradation of mesophyll cells below the epidermis (Evans and Miller 1972a, b, Rice et al. 1983) that includes amorphous staining of cellular contents, plasmolysis of cell walls, and cell necrosis. As the cell membrane degrades, cellular contents are lost and chlorophyll degrades. The degradation of chlorophyll is manifested on the needle surface as irregular, amorphous, chlorotic blotches with diffuse borders (Figure 7.1). This symptom is visually distinct from other nonpollutant foliar discolorations.

Chlorotic mottle or mottling first appears near the tips of needles (Miller et al. 1963, Richards et al. 1968) and progresses basipetally (toward the base) until 50% or more of the needle surface is affected, at which point necrosis of the needle tips is initiated and progresses basipetally. Trees with tip necrosis or necrotic band symptoms are more severely injured than trees with only chlorotic mottling (Miller et al. 1963). As both mottle and tip necrosis intensify, needle abscission occurs. In contrast, a healthy branch with four annual whorls of needles (whorls 1–4) is initially free of chlorotic mottle. The progression of chlorotic mottle injury, without the occurrence of tip necrosis, on the foliage of ponderosa pine is illustrated graphically in Figure 7.2. As ambient levels of ozone reach phytotoxic levels, a small amount of chlorotic mottle appears on the oldest (fourth) whorl, but abscission rates of needles have not been altered (Figure 7.2a). Increased exposure to ozone results in 30% or more of the surface area of whorl 4 covered with chlorotic mottle and a small amount of mottle on whorl 3 (Figure 7.2b). Continued exposure to ozone results in the devel-

Figure 7.1 Foliar discoloration symptoms commonly observed on symptomatic ponderosa and Jeffrey pines in the western United States: two left needles show chlorotic mottle symptoms (amorphous chlorotic blotches) from ozone injury that are observed on all needle surfaces; right needle illustrates winter fleck symptoms (smaller oval-shaped tan lesions with distinct margins) that are most frequently found on the upper surfaces of needles.

opment of mottle on needles of whorl 2 (one year old needles) and abscission of whorl 4 needles (Figure 7.2c). Still longer exposure to ozone will produce mottle on whorl 1 needles (current year's growth) and lead to the abscission of whorl 3 needles (Figure 7.2d). In the most severe cases of injury, only the current year's needles remain and have over 50% mottle (Figure 7.2e).

Chlorotic mottle has been observed on many Western conifer species and has been shown to have relatively consistent symptoms for species with long cylindrical leaves, short flat leaves, and small scalelike leaves (Richards et al. 1968, Miller et at. 1983). Chlorotic mottle appears to be the main symptom on secondary needles of ponderosa pine; tip necrosis is a more common symptom on cotyledons and primary needles. Needle length and the total number of fascicles produced may also be reduced (Miller et al. 1963, Ewell et al. 1989) when trees reach the injury stages depicted in Figures 7.2d–e (Miller et al. 1963, Miller 1977, Duriscoe and Stolte 1989).

Chlorotic mottle from ozone injury can be differentiated from foliar discoloration from pests such as fungi, scale insects, and chewing and sucking insects (Hill et al. 1970, Miller 1977, Pronos et al. 1978) based on the appearance of the symptom and the pattern of development of the symptoms. *Elytroderma deformans* (elytroderma needle cast fungus) causes a reddish-brown foliar discoloration and premature abscission of ponderosa pine needles but can be differentiated from ozone-induced chlorotic mottle by the brownish color of the needles, the presence of fruiting bodies on the needles, and the presence of witches brooms (tightly packed profusions of short branches in the crown). Similarly, *Lophodermella cerina* (pine needle cast) causes reddish-brown discoloration of ponderosa pine needles with a distinct band near the middle, and premature needle abscission which results in thin crowns (Walters 1978).

Abiotic foliar injuries such as winter fleck (Miller and Evans 1974), salt, drought, lightning and winter injury (Walters 1978) can be visually distinguished from ozone injury by the appearance of the symptom, the pattern of development on the needles, and effects on the tree crown. Drought stress on conifers causes an overall discoloration of all the needles, with the tips sometimes turning brown, and death of the crown starting at the top and working down (Walters 1978). Ozone injures the older whorls of needles first and typically causes crown death, starting at the bottom and progressing upward (Miller 1977).

Sulfur dioxide

Exposures of Western conifer species to sulfur dioxide in the field and under controlled conditions have typically shown discoloration of foliage (tip necrosis) on needle surfaces (Barrett and Benedict 1970, Treshow and Anderson 1989). Tip necrosis, usually without any dark band separating green (uninjured tissue) from brown (necrotic tissue), initially begins within the mesophyll parenchyma cells as chloroplasts become granulated (amorphous staining), followed by cell plasmolysis, collapse and necrosis. This initial pattern of injury is histologically similar to ozone injury. Mesophyll injury due to sulfur dioxide, in contrast to ozone injury, is followed by degradation of the vascular tissues. Albuminous cells of the phloem elements, and transfusion parenchyma, hypertrophy and eventually collapse (Carlson 1974). Endodermal cells eventually collapse as necrosis becomes more severe.

The tips of the needles are affected first because sulfur oxides that cannot be assimilated by the needle into amino acids are transported to the needle tips as toxic degradation products which cause cell death. Tip necrosis first appears at the tips of older needles and progresses basipetally until much of the needle surface is affected, and abscission of the needle begins. Branch mortality begins in the lower crown and progresses to the apex. The inner portions of the crown die before outer portions of the crown (Carlson 1974). Tip necrosis symptoms have been observed on many Western conifer species (Davis and Wilhour 1976) with relatively consistent expression for species with different needle shapes. Although tip necrosis appears to be the predominant symptom on primary and secondary needles of Western conifers, it cannot by itself be considered a diagnostic symptom to distinguish sulfur dioxide foliar injury from other visible foliar symptoms. Tip necrosis of conifer needles can be caused by a variety of other biotic and abiotic agents including drought, nutrient stress, and insects (Barrett and Benedict 1970).

Because sulfur dioxide injury results from the bioaccumulation of sulfur products in the needle tips, analyses of sections of the necrotic needle tips should reveal higher levels of these sulfur compounds than other sections of the same needle or unaffected needles in the same whorl. Severely damaged Douglas-fir needles (severe foliar discoloration) close to a pulp and paper mill in Montana had sulfur concentrations of 29% compared to needles with moderate foliar discolorations (22% sulfur), slight foliar discoloration (18% sulfur), and no visible foliar discoloration (3% sulfur) (Carlson 1974).

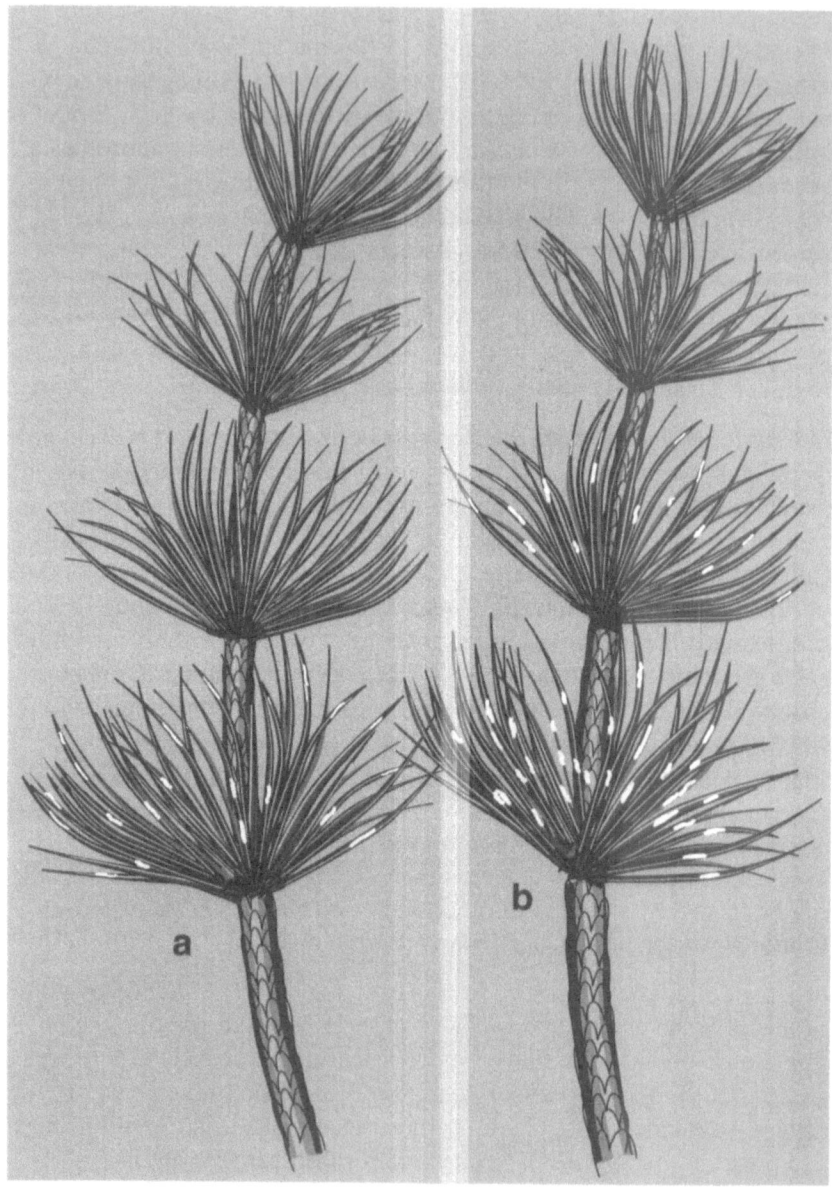

Figure 7.2 Progression of ozone-induced chlorotic mottle injury on ponderosa or Jeffrey pines. (a) Injury is first manifested on oldest whorl (4th whorl) of needles as chlorotic (yellow) blotches. (b) Chlorotic mottle injury intensifies on whorl 4 and begins to affect whorl 3.

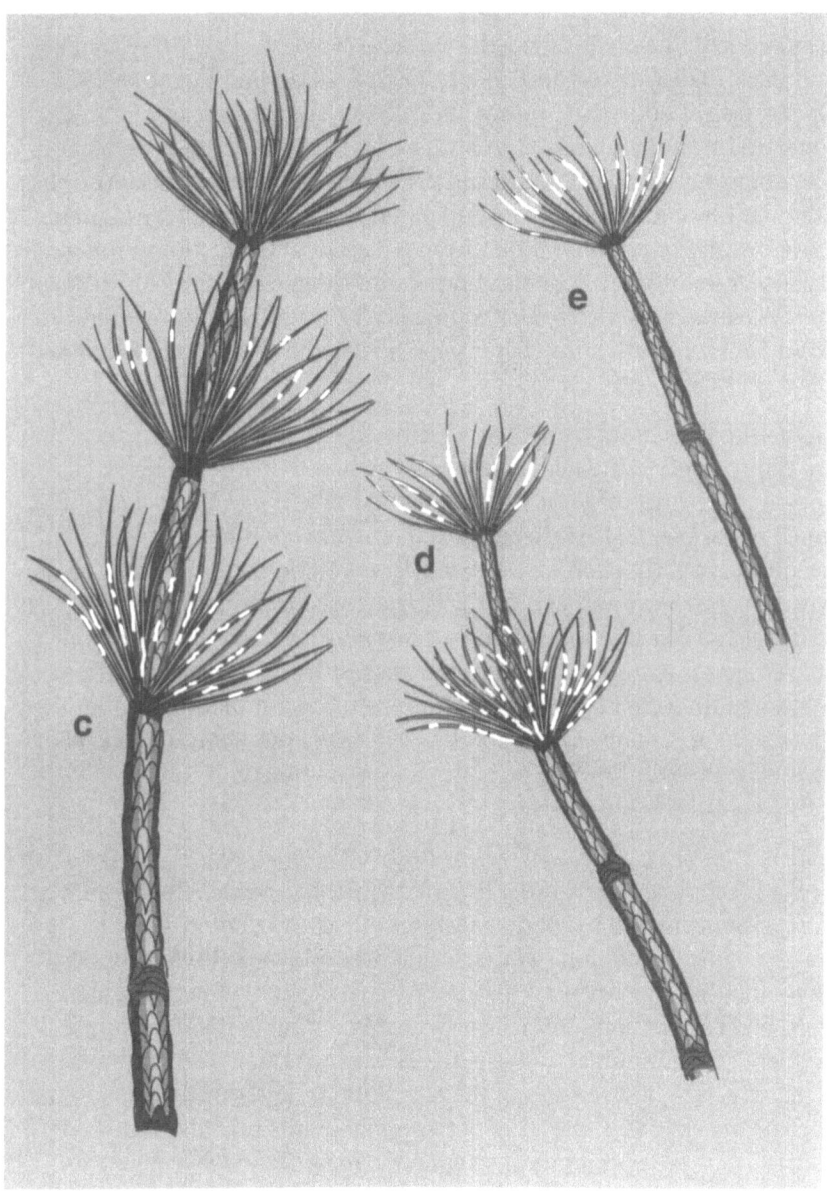

Figure 7.2 (continued) *(c) Whorl 4 has abscised, injury on whorl 3 intensifies, and whorl 2 becomes injured. (d) Whorl 3 abscises as injury on whorl 2 intensifies and whorl 1 (last initiated whorl) becomes injured. (e) Severely injured branch; intense chlorotic mottle on whorl 1; all older whorls have abscised.*

Hydrogen fluoride

Field and controlled exposure studies of hydrogen fluoride effects on Western conifer species have demonstrated that visible discoloration appears as tip necrosis on needle surfaces (Solberg et al. 1957, Treshow and Pack 1970, Carlson and Dewey 1972, Carlson and Hammer 1975, Davis and Wilhour 1976, Bunce 1979, 1984). Tip necrosis is commonly separated from green healthy tissue by a distinct reddish-brown band. The injury observed on needle surfaces initially begins with chloroplast granulation of mesophyll palisade parenchyma cells near the stomata. Hypertrophy of phloem, transfusion parenchyma, and albuminous cells crushes transfusion tracheids and phloem elements. Enlarged nuclei are always associated with hypertrophied cells. Resin canals are often occluded by hypertrophied epithelial nuclei and tissue (Carlson and Dewey 1972).

Needle tips are first affected because hydrogen fluorides are toxins that are transported to the needle tips and cause cell death. Tip necrosis first appears at the tips of younger needles and progresses basipetally until much of the needle surface is affected, at which point abscission of the needle begins. Tip necrosis has been observed on many Western conifer species (Treshow and Pack 1970). Tip necrosis of conifer needles similar to hydrogen fluoride injury can be caused by a variety of other biotic and abiotic agents including drought, nutrient stress, and insects. Because hydrogen fluoride injury results from the bioaccumulation of fluoride in the needle tips, analyses of necrotic needle sections should reveal elevated levels of fluoride.

Other pollutants

In the West, other air pollutants such as nitrogen oxides (NO_x), ammonia/ammonium (NH_x), peroxyacetylnitrate (PAN), chlorides (Cl_2, HCl, NaCl), acidic deposition (wet, dry, and aerosol), and ethylene (C_2H_4) have not been documented to cause extensive visible foliar discolorations on Western conifers. This is probably owing to the relatively low ambient concentrations of these pollutants in Western forests, the lack of research near point sources of these pollutants, and other intrinsic factors such as the apparent tolerance of perennial plants to PAN. When concentrations of these pollutants are high enough to cause foliar injury to coniferous or deciduous species, symptoms include (Jacobson and Hill 1970):

- Nitrogen oxides: interveinal necrotic lesions (broadleaf plants) similar to sulfur dioxide injury; tip necrosis of conifers.

- Ammonia/ammonium: interveinal and marginal chlorosis/necrosis similar to sulfur dioxide injury; tip necrosis of conifers.

- Peroxyacetylnitrate: bronzing of the lower leaf surfaces of herbs.

- Chlorides: marginal, interveinal, and tip chlorosis/necrosis (chlorine); random necrotic lesions (hydrogen chloride); tip necrosis (sodium chloride).

- Acidic deposition: random necrotic lesions.

- Ethylene: epinasty of leaves/stems; injury to flower sepals.

Ancillary Factors

The success of any assessment method depends on more than just technically correct implementation of the method. The effect of where and when the method is applied must be considered in evaluating results, and effects of confounding factors need to be considered. This section discusses some of the important ancillary considerations in conducting crown condition assessments.

Timing of injury evaluations

The timing of field injury evaluations is important in quantifying pollution injury in natural stands. Although not well documented, annual cycles of the severity of foliar discoloration and foliage retention occur, influenced by seasonal peaks of ozone exposure, phenological cycles of the plant, and environmental factors. Field evaluations of the crown conditions of Jeffrey and ponderosa pines in the western United States are often conducted in the early fall when symptom severity is high (following high summer ozone concentrations) and needle retention is relatively low (due to summer drought) (Miller and Millecan 1971, Pronos et al. 1978, Vogler 1982, Duriscoe and Stolte 1989).

Selection of species

The selection of pollutant sensitive species, if the information is available, is an important method in any field study of air pollutants and conifers. The sensitivity of a species to air pollution is commonly determined by chamber fumigations or results from field studies (sites across a pollution gradient) so that target species are selected before the beginning of field injury determinations (see Chapter 5). Varieties or populations of a species can have different sensitivities to pollutants. For example, ponderosa pine (var. *scopulorum*) in the Front Range of the Rocky Mountains in Colorado is relatively tolerant to ozone (Spotts 1969, James and Staley 1980, Miller et al. 1983, Aitken et al. 1984), while the same variety growing in the mountains of southern Arizona appears to be more sensitive (Duriscoe 1987a; Chapter 10). Species that have been documented as the most sensitive to ozone include ponderosa pine in California (*Pinus ponderosa* var. *ponderosa*), ponderosa pine from central-southern Arizona (vars. *scopulorum* and *arizonica*), Jeffrey pine (*Pinus jeffreyi*), western white pine (*Pinus monticola*), and white fir (*Abies concolor*) (Richards et al. 1968, Miller et al. 1983). Sulfur dioxide sensitive species include: western larch (*Larix occidentalis*), ponderosa pine (vars. *ponderosa* and *scopulorum*), Douglas-fir (*Pseudotsuga menziesii*), and subalpine fir (*Abies lasiocarpa*) (Barrett and Benedict 1970, Davis and Wilhour 1976). Fluoride sensitive species include Douglas-fir, lodgepole pine (*Pinus contorta*), western larch, ponderosa pine (var. *scopulorum*), western white pine and blue spruce (*Picea pungens*) (Treshow and Pack 1970, Gordon 1974, Carlson and Hammer 1975, Davis and Wilhour 1976, Bunce 1979, 1984).

Foliage factors affecting sensitivity

Evans and Miller (1972a, b) indicated that differences in the sensitivity of conifers to ozone in California could be explained in part by the internal morphology of needles. This difference in tree sensitivity due to leaf morphology may explain why some populations of conifer species known to be sensitive to ozone appear to be more tolerant when grown in areas of extreme winter temperatures (e.g., *Pinus ponderosa* var. *scopulorum*). Changes in leaf morphology, such as increased sclerophylly as a result of low winter temperatures, may account for the relative higher tolerance of the Colorado population of *Pinus ponderosa* var. *scopulorum* to ozone (Aitken et al. 1984; Chapter 9) compared to the southern Arizona population (Duriscoe 1987a; Chapter 10). Additional work is needed to determine if the relationship between sclerophylly and increased foliar tolerance to air pollutants is due to reduced rates of gas exchange or increased tolerance within the foliage.

The phenological development of foliage alters its sensitivity to air pollutants (Coyne and Bingham 1982, Miller et al. 1982, Patterson and Rundel 1989). Newly initiated leaves are typically more tolerant of air pollutants because stomata have not become fully functional and inter-cellular spaces have not yet developed. Older foliage is also relatively resistant because of reduced gas exchange (Patterson 1991) and increased suberization of cell walls (Evans and Miller 1972a, b). However, chronic air pollution stress on conifers is cumulative, and older age classes of needles have the greatest amount of injury (Miller et al. 1963, Pronos et al. 1978, Duriscoe and Stolte 1989).

Growth stage

Growth stage and foliage age within any growth stage also affect relative pollutant sensitivity (Miller et al. 1963, Coyne and Bingham 1982). Recently germinated seedlings are sometimes the most sensitive growth stage (Stolte 1982). The epigeous (cotyledons above the ground surface) conifers appear to be an especially sensitive plant group (Smith 1990). In species where the older age classes are sensitive to pollutants, seedlings are only slightly more sensitive than the older age classes (Miller 1977, Stolte 1982).

Other response factors

Ancillary biological, physical, and environmental measurements at a site are needed to evaluate the response of trees to pollution. Controlled fumigation exposures as well as field studies have provided insight into the role of environmental factors such as radiation, temperature, relative humidity, and soil moisture in plant responses to pollution (Heck et al. 1965, Taylor 1974, Huttunen 1984, Miller et al. 1989). These environmental factors affect pollutant response primarily by influencing stomatal functioning (Chapter 5). Correlations between other site factors and pollution injuries indicate that microsite differences in topography near point sources can influence the pollutant exposure of groups of trees (Carlson 1974, Hutchinson and Whitby 1976). River drainages, ridge tops, and saddles are topographic features that tend to enhance pollutant flux through conifer crowns, while ridges may block the flow of polluted air to other sites (see Chapter 3). Characteristics of the site, such as slope, aspect, elevation, percent bare rock, and species composition can be used to estimate relative temperature and moisture differences between sites. Site information on latitude and climatic information on winter minimums, as related to the morphology of conifer foliage, can aid in interpretations of pollutant stress on the forest.

Quality Assurance of the Measurement Process

The measurement process involves selection of the method(s), including sample collection and analysis, use of the method(s), and verification and validation of the resulting data. Measurement errors can be relatively large for subjective or complicated methods (Gumpertz et al. 1982, Cline et al. 1989). Measurement errors often are not quantified or reported, especially for many biological variables assessed during air pollution effects studies. Data collected with a faulty measurement process may lead to erroneous inferences about forest condition.

Faulty or poor quality data may result from inherent inaccuracies of methods (e.g., low repeatability), misapplication of the methods (e.g., deviation from standard protocols), and mismanagement of data (e.g., failure to check for transcription errors). Application of quality assurance procedures to the measurement process can reduce the collection of faulty data. Quality assurance is a system of activities designed to assure that products, services, or data meet stated standards of quality within stated levels of tolerance or uncertainty (Taylor 1985). Quality assurance includes quality control and quality assessment. Quality control represents all of the activities carried out to control the quality of the produced item as specified by the user or client. Important elements of quality control include selection and training of a competent staff, use of reliable instruments and raw materials, methods testing, development and use of standardized procedures, calibration, and control charting (Cline and Burkman 1989). Quality assessment consists of activities to assure that the quality control system is working properly. Key elements of quality assessment include internal checks such as systematic analysis of reference materials, and occasional checks by external sources, including equipment challenges and analysis of blind samples.

Data quality can be evaluated by the accuracy of identification of the property measured and the accuracy of quantification of that property (Taylor 1985). For example, the quality of data on visible foliar symptoms depends upon accurate identification of the type of injury (e.g., ozone injury versus winter desiccation) and the accurate quantification of the level of injury (e.g., 5% versus 20% of needle surface area). The accuracy of quantification (i.e., precision and bias) is an attempt to estimate the true value of a measurement property and always involves some level of uncertainty. Measurements are unacceptable unless the resulting data can be assigned a level of uncertainty within a stated probability. For example, Cline et al. (1989) reported that visual estimates of forest canopy condition at sites in a particular state or province (based on a system of 12 canopy condition classes) failed to meet a project precision goal of ±1 class deviation for 90% of the measurements on a plot when made by field crews from other states or provinces. In contrast, measurements made by crews from the state or province containing the study site met the precision goal. Thus, by comparing the precision of measurements made by two different types of crews, it was possible to determine which type of crew was more reliable for detecting real changes in forest canopy condition.

The objectives of individual research projects on pollutant effects influence the selection of methods used, including issues of time (e.g., sampling frequency) and space (e.g., number and types of sample sites, treatments, and observations). Ideally, the response variables that are the most indicative of air pollution stress are selected for measurement.

Appropriate techniques are characterized by the ability to assess key characteristics with 1) high selectivity (e.g., have unique or unambiguous responses to specific pollutants), 2) high sensitivity (consistently responds or measures a response at low concentrations), 3) high precision (measurements are reproducible), and 4) low bias (operates without artifacts such as contamination, loss, and calibration error) (Tingey et al. 1979, Taylor 1987). All of these characteristics are judged relative to the data requirements of the study. For example, a technique may not be the most precise of those available, but it can still be adequate relative to the precision needed to address the research question.

Written, standardized procedures are particularly important for difficult methods (e.g., foliar injury assessment), when multiple crews are collecting data, and when long-term data collection is envisioned (e.g., forest health monitoring). Proper execution of a measurement procedure involves operating the method in a "state of statistical control," which means that the measurement system is stable and data are reproducible within statistically defined limits. Statistical control is needed to evaluate the precision of a measurement process and is a prerequisite for estimating bias. Attainment and maintenance of statistical control is best verified with control charts, which ideally are real-time graphical plots of sequential test results in relation to statistically derived warning and control limits (usually ±2 and ±3 standard deviations, respectively) (Taylor 1987).

High quality data from a measurement system can be compromised by errors introduced during data handling and processing. Data handling errors can be particularly serious because they usually are not random, e.g., glitches in blocks of data, incorrect assignment of plot or treatment codes. Consequently, a systematic approach to data handling and database management is required, and data verification and validation procedures must be documented and checked as part of the quality assurance process.

Thorough documentation has both short- and long-term benefits to the project. In the short-term, documentation of research activities, events, and decisions that seem obvious or even trivial may turn out to be important to the proper interpretation of the data. In addition, the long-term integrity and usefulness of a research data set is increased greatly by thorough documentation, including journal publication of the methods separately or as part of the research results.

Methods for Evaluating Crown Condition

Visual estimates

The incidence and severity of foliar injury to Western conifers from air pollutants has been evaluated at the needle, whorl, branch, or tree level by visual observation of the foliage. The presence of injury on needles of different ages has been used as an indicator of the severity of injury at the branch and tree level (Pronos et al. 1978, Williams and Williams 1986); injury may take several years to appear if the dose (foliar uptake of ozone) is slight and may occur on current year's growth if the dose is high. The severity of injury has also been measured by visual estimate of the percentage of the total leaf area exhibiting air pollution injury. However, visible expression of pollutant injury may overlap with other abiotic (e.g., winter fleck) or biotic (e.g., scale) injuries that complicate visual estimates of percent cover. A greater degree of judgement on the part of the observer is required for this type of measurement as opposed to mere presence or absence, and the possibility for observer bias must be strictly controlled using quality assurance/control procedures, such as thorough training and an ongoing remeasurement program.

Measurements of the total leaf area affected by visible injury symptoms are more useful than simple presence or absence data because they correlate more closely with effects on photosynthesis, foliar water relations, and growth (Horsfall and Cowling 1978). Net photosynthesis of Jeffrey pines in the Sierra Nevada of California has been shown to decrease as the percentage of leaf area with visible ozone injury increases (Figure 7.3).

Visual estimation of the relative proportions of areas of injured and uninjured foliage has been addressed by Horsfall and Barratt (1945). They demonstrated that the human eye was most perceptive at differentiating disease injury on foliage when the amount of injury discoloration was very small or very large compared with the surrounding colors. The Horsfall-Barratt system (Horsfall and Barratt 1945) of estimating plant disease divides the total range of injury percentages into twelve classes (Table 7.1) and is based on the logarithmic response of visual acuity (Horsfall and Cowling 1978).

Preliminary data on the relationship between percent chlorotic mottle and reductions in net photosynthesis (Figure 7.3) suggest that twelve classes of injury may be more precise than needed to represent our current understanding of the biological significance of chlorotic mottle. Jeffrey pine trees in the Sierra Nevada with foliar ozone injury had an approximately linear decrease in net photosynthetic rates when percent chlorotic mottle increased from 0 to approximately 30% of the leaf area. Chlorotic mottle injury above 30% of the surface area resulted in only minor further decreases in net photosynthetic rates.

Based on this information, a simplification of the Horsfall-Barratt system has been proposed for rating the severity of chlorotic mottle injury to Jeffrey and ponderosa pines in the West. The Western Pine Method (WPM; Stolte and Miller 1991) is the result of a multiagency collaboration (US Forest Service, National Park Service, Environmental Protection Agency, and the California Air Resources Board) consisting of workshops and field tests. Severity of injury is scored in six injury classes (Table 7.1) using an arc-sine square root transformation of percentage data (Muir and McCune 1987). Retention of six injury classes gives WPM the flexibility to respond to improvements in our understanding of the relationships between visible injury and physiological responses of Western pines.

The precision and accuracy of visual estimates of foliar discoloration were investigated by Gumpertz et al. (1982) on four broadleaf crops. They found that the precision of three people who estimated total percent leaf area injured was high when compared with other sources of variation, particularly the variation in injury among the plants. Observations were most accurate when they could be compared with a quantifiable standard (obtained by overlaying a transparent grid on the leaves and counting the squares that fell over injured areas) and adjusted for observer bias. They pointed out, however, that this is necessary only if the desired precision has biological meaning. In many cases, the precision level for most visual estimates is greater than the corresponding biological response, such as the relationship between percent chlorotic mottle and reductions in net photosynthesis discussed above (Figure 7.3). They also concluded that estimating the percent injury of the three most injured leaves was a good predictor of the percent area injured of all leaves if the three most injured leaves were not completely injured (>95%). The proportion of leaves injured was not a good predictor of the percent area injured of all leaves.

In a field study of ozone-injured ponderosa pine, Muir and Armentano (1987) compared the results of foliar ozone injury evaluations made with pruned branches examined "in-hand" and remote assessments made with optical instruments (two types of binoculars and a spotting scope). They concluded that remote assessment methods did not yield sufficient precision and accuracy to warrant their use in describing severity of injury on an individual tree, and were only marginally appropriate in assessing stand or population mean values of injury severity. Poor correlations ($r^2 < 0.31$) between optical instruments and in-hand assessments (which were treated as "true" values) of foliar injury were found for most variables. Two experienced observers performed the in-hand evaluation, and their results were found to be comparable; significant differences were found for estimates of needle length ($p=0.029$), but not for percent chlorosis ($p=0.446$), percent necrosis ($p=0.352$), percent other injury ($p=0.907$), and percent needle retention ($p=0.231$). This study suggests that visual estimates of injury severity have the highest quality when made with foliage in-hand and with known standards available for visual comparison, rather than with binoculars or whole-crown observations from a distance.

Spectral analyses

Spectral evaluation of air pollution impacts on vegetation has made rapid advancements in recent years as ground and airborne sensors have increased in spatial and spectral resolution. Spectral evaluation is based

Table 7.1 *The Horsfall-Barratt system (Horsfall and Barratt 1945) for classifying visible injury on foliage is based on the logarithmic response of visual acuity and includes twelve injury classes. The Western Pine Method (WPM; Stolte and Miller 1991) has only six injury classes based on the relationship between the percentage of upper surface area of needles covered with chlorotic mottle and net photosynthesis in Jeffrey pine. See text for details.*

Horsfall-Barratt Injury Class	Percent Injury	WPM Injury Class	Percent Chlorotic Mottle
1	0	1	0
2	1–3	2	1–6
3	4–6	3	7–25
4	7–12	4	26–50
5	13–25	5	51–75
6	26–50	6	76–100
7	51–75		
8	76–88		
9	89–94		
10	95–97		
11	98–99		
12	100		

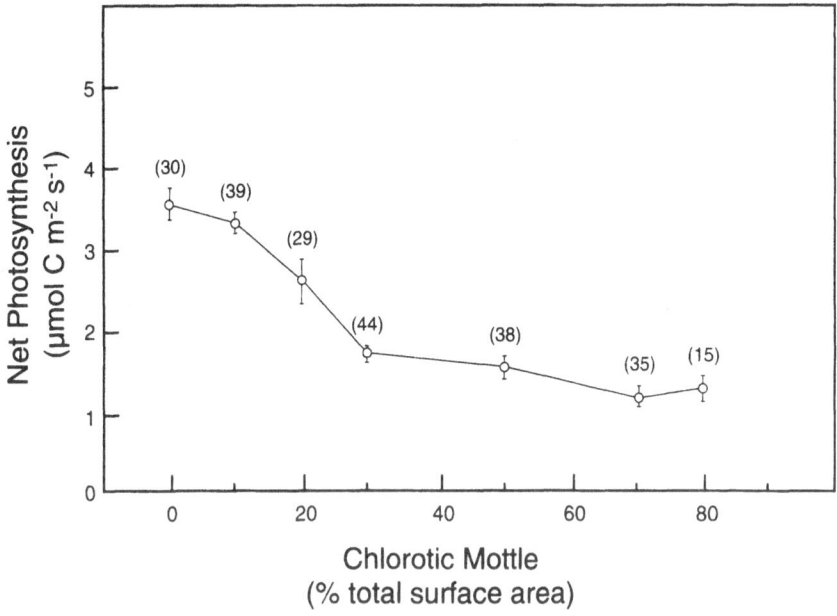

Figure 7.3 Relationship between photosynthetic rate of Jeffrey pine needles and percentage of needle surface area with ozone-induced chlorotic mottle. Numbers in parentheses are the number of measurement replicates. From Stolte and Miller (1991).

upon analysis of reflected electromagnetic radiation; these spectra vary by material based upon their chemical and structural composition. Use of airborne and satellite sensors to detect the effects of air pollution on forests is based upon predictable changes in the reflectance characteristics as foliage becomes chlorotic and senescent (Ustin et al. 1989). The ability to detect changes in chlorophyll concentration, water content, elemental content, and structural features of conifer needles before visible injury appears may allow detection of early stages of injury. Analyses of spectral reflectance in discrete bands or combinations of bands can give information on shifts in chlorophyll a and b ratios, the moisture content of the tissues, and other chemical compounds. Research is underway to develop techniques to improve the spatial resolution of remote sensing to evaluate individual tree canopies, differentiate pollutant-caused chlorosis from chlorosis due to natural senescence or other injuries, increase spectral sensitivity, and improve remote sensing of mixed conifer stands under diverse topographical conditions. Spectral analysis of crown images would help to eliminate the subjectivity of human eye assessments.

Ustin et al. (1989) and Rock et al. (1986, 1988, 1989) have described methods for detection of air pollution injury to forests using spectral sensing techniques. They found that forest decline symptoms produce changes in the spectral properties of tree canopies, such as red and blue spectrum shifts, that may be detectable with high resolution spectrometry. More recently, camera and video capture of reflected electromagnetic radiation has been digitized for computer analysis. Camera and videographic techniques may improve estimates of foliage injury under field conditions and provide quality assurance and control audits of visual estimates. Additional research is needed to delineate the precision, accuracy, and cost of these methods. Remote sensing techniques are most valid when verified by extensive ground truthing using hands-on verification of tree crown conditions.

Remote sensing using visible or infrared spectral measurements has been used in a number of pollutant studies in the West to delineate pollutant injury over large areas. Carlson and Dewey (1972) used aerial photography (Ektachrome Aero film) to detect crown injury to conifers from fluoride where foliar fluoride levels exceeded 60 mg/kg (dry weight) and mortality of conifers where foliar fluoride levels exceeded 300 mg/kg. Larsh et al. (1970) used aerial photographs to establish the extent of ozone injury along a pollution gradient in the San Bernardino and San Gabriel Mountains of southern California.

Sampling of conifer crowns

When sampling branches from individual trees for symptom evaluation, it is important to minimize or account for the expected between or within-tree variance of symptom expression. The source of this variability includes between-tree genetic differences in sensitivity to pollution (Scholz et al. 1989), within-tree differences in shading of foliage, microhabitat differences in the availability of moisture and nutrients, and differences in micro-meteorology which may influence canopy exposure to pollutants (see Chapter 3).

Within-tree variance in foliar expression of ozone chlorotic mottle on 20 ponderosa pine trees was examined by Muir and Armentano (1987) in an area of the southern Sierra Nevada with moderate ozone injury symptoms (chlorotic mottle on the second or older whorls of needles; Pronos et al. 1978). They analyzed data on chlorotic mottle, necrosis, fascicle retention, needle length, and other injuries (biotic and abiotic) from hands-on evaluation of five pruned branchlets (smallest terminal leader or side branch unit containing a full complement of needle whorls relative to other branches in the same tree). They found that data from five branchlets was not sufficient to separate tree means (p<0.05) at a preset level of precision (usually 10%) selected for the sample area for any variable except necrosis because the within-tree variances were too great. Using an allowable error of 10% and a 95% confidence coefficent in a standard sample size formula (e.g., Snedecor and Cochran 1967), they estimated that the required sample sizes, including all whorls on each branchlet, for each measurement for a given level of precision (in parentheses) would be:

- chlorotic mottle (±10%), 8 branchlets

- necrosis (±10%), 3 branchlets

- needle retention (±10%), 42 branchlets

- needle length (±3 cm), 11 branchlets

- "other" injuries (±10%), 7 branchlets

These relatively large sample sizes (number of branchlets) may cause concern to field scientists interested in describing the crown condition of individual trees in natural stands. Pruning eight or more branchlets from a mature conifer can be difficult, and long-term monitoring of individual trees may lead to severe pruning of the lower crown foliage. The use of injury rating systems with fewer injury classes (e.g., Western Pine Method (six classes) instead of the Horsfall- Barratt system (twelve classes) (Table 7.1)), and the use of branch and tree level indices, should require fewer branchlets to define tree means, because the individual observations of branch symptoms will fall into fewer injury classes and result in less variance around the mean value of each branch. Additionally, indices emphasize common signals among symptoms and thus reduce the "noise" associated with the individual symptoms (Muir and McCune 1987). In most cases, researchers are interested in describing air pollution effects on stands of trees or on populations, and between-tree variances will be more important than the within-tree variances. Estimates of the between tree variances can be improved by sampling more trees per stand.

Whole-crown assessments in areas with slight to moderate injury, with or without optical aid, are a poor alternative to branch sampling (Innes 1988, Miller et al. 1989) because the correct diagnosis of the cause of foliage discoloration can only be accomplished with foliage in-hand. Whole crown assessments may be appropriate for smaller trees where the crown is readily visible without optical aid, or in areas of severe air pollution impacts where there is little doubt that an air pollutant is the cause of crown discoloration and pollution injury is severe.

Selection of branches from different portions of the tree crowns can be done systematically, randomly, or by stratification. Stolte and Bennett (1984) recommended sampling branchlets from the lower crown at evenly spaced azimuths (for five branches these would be about 72° apart). Implementing this systematic approach was not always practical in the field (lower crowns of trees are particularly prone to variability in the spacing of branches) but provided a useful suggestion for addressing branch variation in the lower crown. In practice, most sampling is done randomly from a population of terminal, secondary, or tertiary branches distributed among the more accessible larger branches in the lower crown. The sampling of branches up to 15 m above the ground may be accomplished with some precision with a pole pruner. At greater heights, tree climbing, or less accurate means (rope saw, shotgun) may be necessary. The effects of pruning on the vigor of dominant and codominant conifers that exceed 0.1 m diameter at 1.4 m height should be negligible. Negative effects on carbon budgets are generally difficult to detect until 30% or more of the crown is removed (Muir and Armentano 1987). However, pruning of smaller, subdominant trees can more readily have negative effects, minimizing the usefulness of this canopy class for long-term monitoring.

Foliage longevity

In Western pine species, maximum foliage longevity ranges from two years for Monterey pine (*Pinus radiata*) to more than 40 years for bristlecone pine (*Pinus aristata*), and is related to the elevation at which the tree grows (Ewers and Schmidt 1982). In unpolluted environments in California and in the absence of other stresses affecting needle retention, ponderosa pines retain foliage for three to five years (a tree average of three to five annual whorls retained at any time) and Jeffrey pines six to eight years (Munz and Keck 1973).

Exposure to ozone can reduce the duration of needle retention in ponderosa and Jeffrey pines. In a study by Miller and Van Doren (1982) in the San Bernardino Mountains of southern California, the number of needles or needle fascicles within reach of the evaluator were counted on the branches of 13 trees at each of five sites along a pollution gradient (west to east) over a three year period. There was a positive correlation between abscission rate and seasonal ozone exposure (Figure 7.4). Abscission was most dramatic when visible needle injury had reached 80–90 percent of the needle surface area. In a study by Duriscoe and Stolte (1989), ponderosa and Jeffrey pines with moderate chlorotic mottle retained three or fewer annual whorls of needles (Figure 7.5). Older whorls of needles abscised as younger whorls of needles became injured.

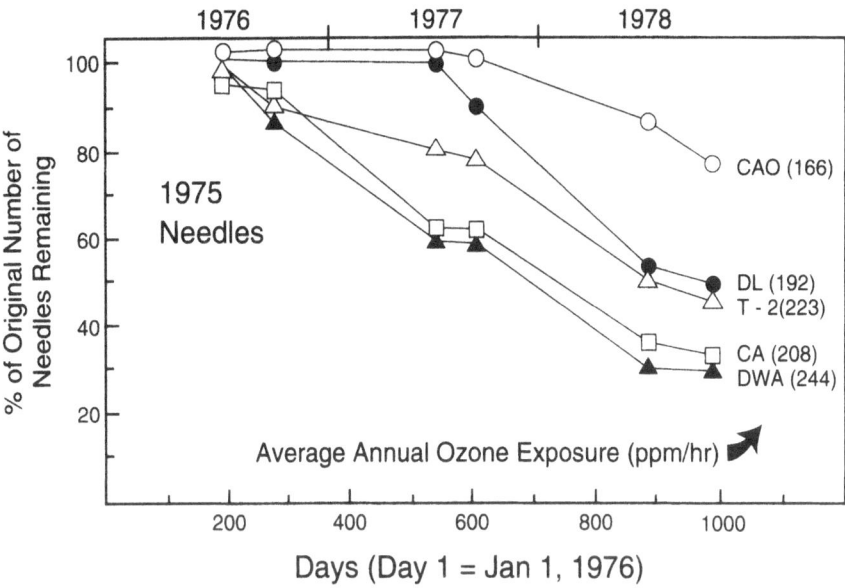

Figure 7.4 Percent abscission of 1975 needle age class in ponderosa and Jeffrey pine trees from 1976–1978 at five sites along a pollution gradient in the San Bernardino Mountains, California. Annual cumulative ozone exposures (ppm-hour) at each site are in parentheses; letters to the left of parentheses are plot identification codes. Adapted from Miller and Van Doren (1982).

Reduction in needle longevity is an indicator of air pollution stress for conifers when other factors leading to premature abscission of needles are eliminated or taken into account. Foliage longevity (Figure 7.6) may be determined by counting nodes (separation of whorls of needle fascicles produced each year), starting from the branch tip (most recent whorl) and proceeding to the last remaining whorl of needles (oldest whorl). Annual whorls are counted on branchlets (usually a more recent side branchlet) or on the main stem of seedlings or saplings. On most pines, the annual "nodes" are relatively discrete and this type of measurement can be very objective and free of observer bias. A more precise quantification of the total foliage remaining on the branch may be made by estimating the percentage of the needle complement remaining within each whorl, in classes following the Horsfall-Barratt system, at systematic intervals (incremental percentages) (Stolte and Bennett 1984) or at coarser intervals (thirds or halves of the full whorl) (Stolte and Miller 1991). Additionally, length of the foliated portion of the branchlet has been used to describe needle longevity (e.g., Stolte and Miller 1991). This approach is especially suited to spruce and fir species.

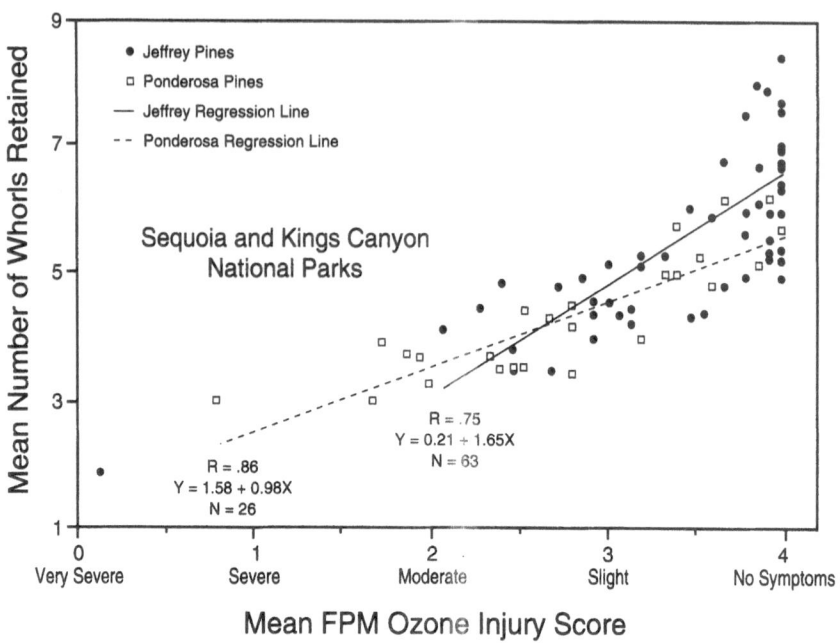

Figure 7.5 Relationship between severity of chlorotic mottle injury (mean FPM scores—low scores equal high injury) and retention of needle whorls for Jeffrey and ponderosa pines in Sequoia and Kings Canyon National Parks. Adapted from Duriscoe and Stolte (1989).

Crown vigor

The quantity of live foliage, expressed as live crown ratio or the percentage of live crown, can provide information on the degree of pollutant stress to an individual tree. Pine species typically self-prune the lowest branches as the trees grow, producing a crown form that may be free of live branches in the lower third (or more) of the stem at maturity (particularly in more closed stands). Ozone stress to ponderosa and Jeffrey pines in the San Bernardino Mountains has a similar effect, decreasing the vertical length of the live crown (Parmeter and Miller 1968). Ozone-induced lower crown branch mortality is preceded by a decline in vigor of the lower crown that may be evidenced as a reduction in needle length (Parmeter et al. 1962, Miller et al. 1963) and the production of fewer numbers of needle fascicles (Miller et al. 1963, Ewell et al. 1989). The total dry weight of the current year foliage may be reduced by exposure to ozone (Miller 1977). Specific leaf weight of ponderosa pine needles can also be reduced, which may lower the photosynthetic capacity of needles (Coyne and Bingham 1981, 1982, Ewell et al. 1989). The appearance of shorter ponderosa pine needles in the lower crown,

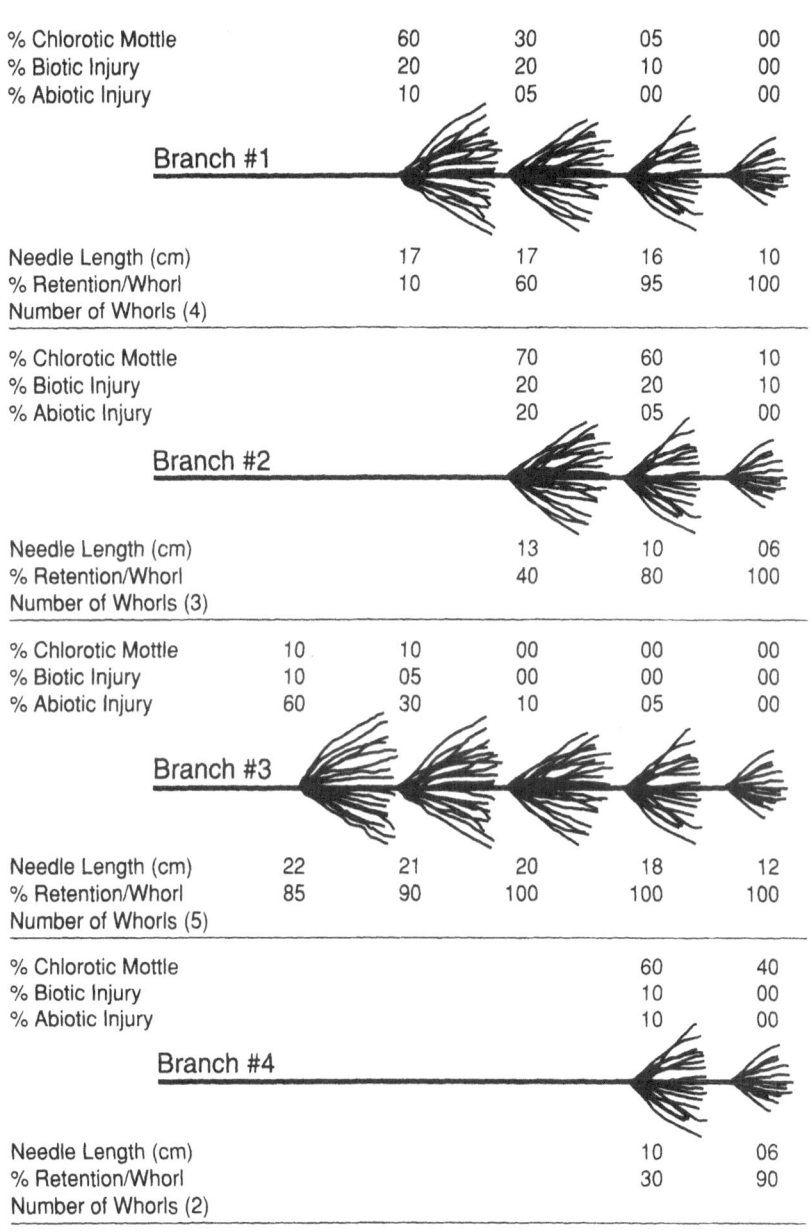

% Chlorotic Mottle		60	30	05	00
% Biotic Injury		20	20	10	00
% Abiotic Injury		10	05	00	00

Branch #1

Needle Length (cm)		17	17	16	10
% Retention/Whorl		10	60	95	100
Number of Whorls (4)					

% Chlorotic Mottle			70	60	10
% Biotic Injury			20	20	10
% Abiotic Injury			20	05	00

Branch #2

Needle Length (cm)			13	10	06
% Retention/Whorl			40	80	100
Number of Whorls (3)					

% Chlorotic Mottle	10	10	00	00	00
% Biotic Injury	10	05	00	00	00
% Abiotic Injury	60	30	10	05	00

Branch #3

Needle Length (cm)	22	21	20	18	12
% Retention/Whorl	85	90	100	100	100
Number of Whorls (5)					

% Chlorotic Mottle				60	40
% Biotic Injury				10	00
% Abiotic Injury				10	00

Branch #4

Needle Length (cm)				10	06
% Retention/Whorl				30	90
Number of Whorls (2)					

Figure 7.6 Variables measured to determine air pollution impacts (chlorotic mottle, biotic injury, abiotic injury, needle length, percent fascicle retention on each year's whorls, and the number of whorls per branch) on ponderosa and Jeffrey pine branches. These variables can be evaluated separately or combined with whole crown evaluations (e.g., live crown ratio) into additive indexes to indicate the degree of pollution stress at the tree level. Examples illustrate ozone injury from slight (Branch #3) to moderately severe (Branch #4). Adapted from Stolte and Bennett (1984).

relative to upper crown needle length, results from chronic exposure to ozone (Miller 1977). Axelrod et al. (1980) estimated the leaf area index for whole trees and found that severely injured trees had less than half the leaf area available for photosynthesis than slightly injured trees. These effects, combined with a decrease in foliage longevity of the remaining live branches, may lead to a drastic reduction in crown fullness or density, readily observed even at a distance (Figure 7.7).

Mortality of lower branches of pines from air pollution can be distinguished from branch mortality caused by limb rust (*Peridermium filamentosum*) because limb rust often starts at midcrown and progresses upward and downward (Walters 1978). Fluoride injury on Douglas-fir in Montana can cause high branch mortality at the apex of the trees, in contrast to the lower crown defoliation caused by sulfur dioxide (Carlson and Dewey 1972, Carlson 1974, Carlson and Hammer 1975) and ozone (Miller 1977, Duriscoe and Stolte 1989).

Indicators of crown vigor include needle length, number of fascicles per whorl, and foliar biomass, and may be measured from a sample of pruned branches, or from evaluations of reachable branches on saplings and open-grown mature trees. Another approach is whole-tree measurement of crown vigor using live crown ratio (or percent live crown) and crown density (Figure 7.8). Live crown ratio is measured as the portion of the stem that has live branches relative to the portion of the stem without branches (Figure 7.8b). This may be visually estimated by an observer or measured using a clinometer or similar device. Reduced live crown ratios may indicate abnormal lower branch mortality resulting from air pollution stress, particularly if the live crown shows evidence of ozone stress as measured by the number of needle age classes retained, the severity of foliar discoloration on remaining whorls, and the shortness of the needles. Trees with dominant and co-dominant crown positions are most easily evaluated for live crown ratio. Sub-dominant trees already have reduced live crowns because of self-pruning due to shading.

Crown density, the fullness of the existing live crown, has also been visually estimated based on comparison of sample trees with a reference tree from the same stand, a reference photograph, or a reference drawing (Duriscoe 1987b, Innes 1988). The density of the whole crown can be evaluated or the upper and lower crowns can be evaluated individually (Figure 7.8c). Evaluation of both the upper and lower crowns may provide information on causal agents of crown thinning because certain agents, e.g., ozone, are likely to degrade the lower crown more rapidly than the upper crown. Crown assessments that combine observations on live crown ratio and crown density provide a more complete picture of crown vigor (Figure 7.8d).

Figure 7.7   The effects of chronic ozone injury on crown vigor (center tree) reflected in needle retention, needle length, crown density and percent live crown on a symptomatic (ozone sensitive) ponderosa pine in the western United States. Asymptomatic (ozone tolerant) genotype of ponderosa pine (tree on right) with relatively good crown vigor.

(a) Yellow Pine Crowns

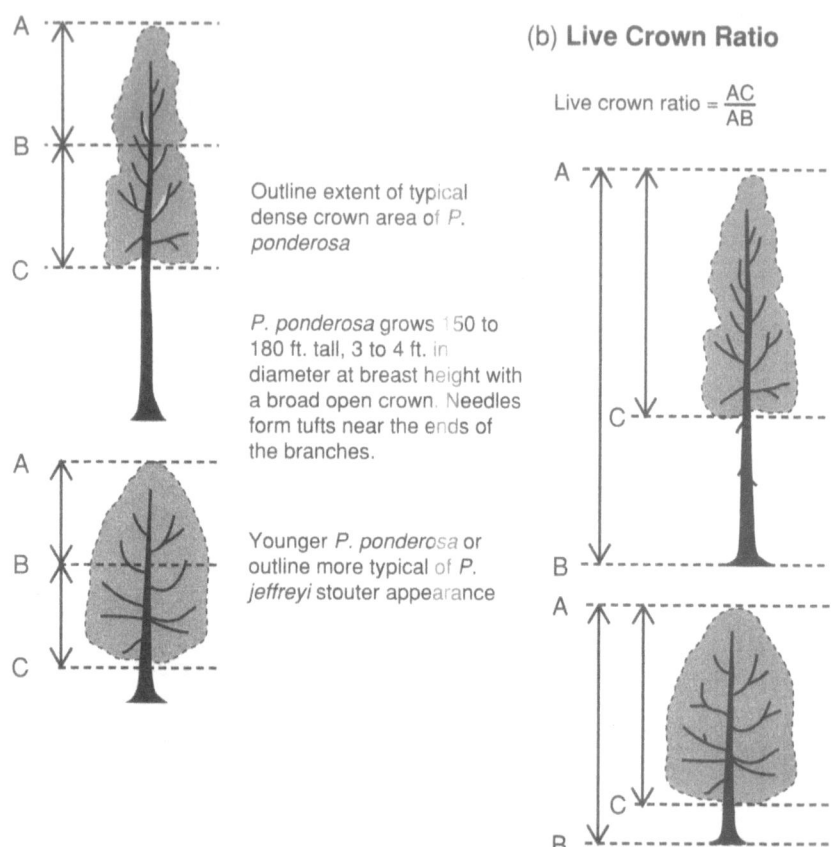

Outline extent of typical dense crown area of *P. ponderosa*

P. ponderosa grows 150 to 180 ft. tall, 3 to 4 ft. in diameter at breast height with a broad open crown. Needles form tufts near the ends of the branches.

Younger *P. ponderosa* or outline more typical of *P. jeffreyi* stouter appearance

(b) Live Crown Ratio

$$\text{Live crown ratio} = \frac{AC}{AB}$$

Figure 7.8 Rating whole crown vigor on ponderosa (PP) or Jeffrey (JP) pine trees in the western United States using crown density (CD) or live crown ratio (LCR) methods. (a) Habit of mature and younger PP or JP and the visual or measured delineation of the crown into upper (AB) and lower (BC) halves. (b) Mature and younger PP or JP and the delineation of live crown ratio (AC/AB). (c) Methods of rating upper and lower crown density on trees with different branching habits. (d) Crown density and live crown ratio measurements on PP or JP. Adapted from Duriscoe (1987b).

(c) **Crown Density Rating**

Estimate fullness and percentage of live branches in each half and assign a numerical rating on a scale of zero to ten.

(1) and (2) have different growth patterns, but the same relative density of foliage.

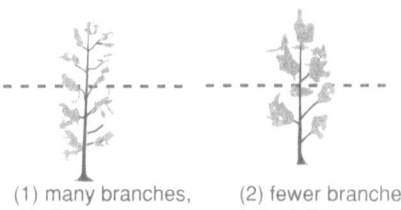

(1) many branches, thin foliage

(2) fewer branches, fuller foliage

Numerical ratings for (3) and (4) depend on the species; that is, the greatest fulness, density, and shape typically found in that species.

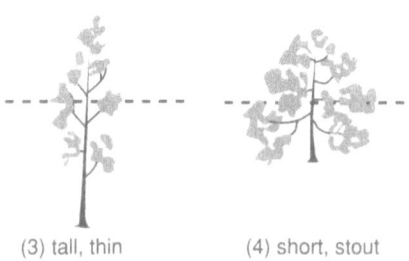

(3) tall, thin

(4) short, stout

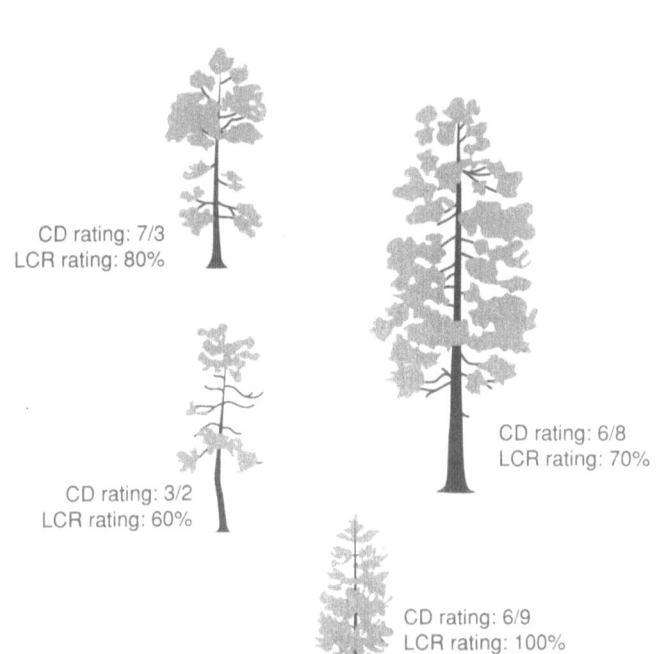

CD rating: 7/3
LCR rating: 80%

CD rating: 3/2
LCR rating: 60%

CD rating: 6/8
LCR rating: 70%

CD rating: 6/9
LCR rating: 100%

Stem growth

Studies on ponderosa pine in southern California (Miller 1977, Miller et al. 1989) and in the Sierra Nevada (Vogler 1982) have demonstrated a relationship between incremental growth of stems and tree sensitivity to ozone. The radial growth of stems (including the bark) can be measured using a forester's stem diameter tape, while vertical growth of saplings can be measured between branch nodes on the main stem. Miller et al. (1989) found that the relative diameter growth of the boles of symptomatic ponderosa pines and Jeffrey pines in the San Bernardino Mountains was less than asymptomatic trees of the same size class. Radial growth assessments, without radial ring width determinations, can only be made by repeated measurements on the same trees (e.g., trees in plots). Trees to be remeasured for diameters can be precisely relocated by attaching a numbered tree tag with a nail. Vogler (1982) found that the incremental height growth of symptomatic ponderosa pine saplings growing in the Sierra Nevada was reduced relative to asymptomatic saplings in the same stand. The apical leaders of severely injured ponderosa pines in the San Bernardino Mountains have been observed growing horizontally while under no obvious light stress conditions (Miller 1973). Height growth can be evaluated by repeated annual measurements using a height pole or clinometer. Incremental growth of stem leaders of saplings can be measured using a height pole or clinometer, by repeated measurements of total height, or by measurements of annual height increments between major branch whorls.

Tree-level injury indices

Observations of crown response variables made at different levels of biological organization may be compiled in an index to describe the severity of pollution injury to a particular tree (Muir and McCune 1987). Because trees are a fundamental biological unit of forest ecosystems, most injury response variables are appropriately summarized at this level and then summarized at the plot or stand level to give information on the incidence and severity of injury. Indices can be used to quantify pollution stress, estimate the potential for growth effects, and reduce statistical "noise" associated with individual symptoms (e.g., reduce the number of sample branches required for measuring foliar variables as discussed earlier in this chapter). Indices can weight symptoms according to their importance in diagnosing air pollution response and in estimating potential growth and physiological effects. Ideal indices produce high correlations between crown response variables and ambient pollutant exposures. Atmospheric monitoring of ambient concentrations of pollutants is generally sparse in Western forests (Chapter 3), so assessments of crown injury may help delineate relative pollution concentration gradients.

Controlled exposure studies and field studies of ozone injury on ponderosa and Jeffrey pines have led to the development of several ozone injury indices. An additive index known as the Oxidant Injury Score (OIS) was developed to monitor crown condition in permanent plots in the San Bernardino Mountains of southern California in the early 1970's (Miller et al. 1989). The index combines data from binocular evaluations of needle discoloration, longevity, and relative length in the upper and lower crowns with visual evaluations of lower branch mortality. The OIS index is additive (data from all variables are added together) and foliage longevity receives the greatest weight. Lower scores indicate more severe injury; a tree with a score of 0 is dead while trees with scores of 36 and greater have no symptoms of ozone injury. Plot-level averages of OIS values have correlated well with measurements of ozone concentrations in mountainous terrain in southern California (Miller 1977).

In the Sierra Nevada, US Forest Service pathologists initiated the monitoring of trends in ozone injury to ponderosa and Jeffrey pines by establishing an extensive network of plots in 1974 (Pronos et al. 1978, Vogler 1982, Allison 1984, Williams and Williams 1986). They used a basic tree-level index, the Forest Pest Management (FPM) index, that determines the number of annual whorls free of chlorotic mottle. The FPM index score is low for severely injured trees, similar to the OIS index. Trees are assumed to have no significant ozone injury if they possess four or more annual whorls with no chlorotic mottle on the youngest four whorls. Partial whorls retained (less than 100% fascicle retention per whorl) are counted as whole whorls. Five different scores are possible: 4, 3, 2, 1, and 0 (4 = 4 or more annual whorls free of ozone injury; 0 = no annual whorls free of chlorotic mottle).

The National Park Service (NPS) devised a method for measuring the crown condition of pines in surveys and permanent plots (Stolte and Bennett 1984). The method calls for evaluating variables, measured at the whorl, branch, and tree level, that are affected directly or indirectly by ozone. The whorl and branch level variables are measured from five pruned branches from the lower crown of trees of all size classes and include:

- percent foliar surface area with chlorotic mottle
- percent surface area with necrosis
- percent surface area with other injury (biotic and abiotic)
- modal needle length per whorl
- percent needle fascicles remaining per whorl
- number of whorls per branch

The tree level variables are:

- percent live crown ratio
- crown density (upper and lower crown)
- cone crop (1–10 scale)
- type and severity of bole injury (1–10 scale)
- bole diameter at breast height

Some of the NPS variables are then combined in a nonadditive index to give a tree level score for the amount of ozone injury. Tree level scores are averaged (n=15) to give plot level scores.

Using the NPS methods for data collection, Duriscoe (1991) devised an additive index known as the Eridanus Injury Index (EII). The EII has been used in quantifying ozone effects on pines in the West by the US Forest Service, National Park Service, and the California Air Resources Board. The EII possesses all the desirable characteristics for injury indices established by Muir and McCune (1987), and has been modified and described as part of the Western Pine Method (Stolte and Miller 1991). It is a composite score consisting of four variables representing four primary effects of ozone on pines:

- Whorl retention: increased senescence of needles and reduced whorl retention that reduces the amount of carbon fixation;
- Chlorotic mottle: the appearance of chlorotic mottle symptoms on remaining needles that further reduces photosynthesis;
- Percent live crown: reduced live crown ratio as lower branches die first in declining trees; and
- Needle length: reduced length of emergent needles as carbon reserves become limiting.

The relative contribution of each variable to the composite score is weighted based on its relative importance to carbon fixation and growth. Whorl retention receives the greatest weighting, with chlorotic mottle, percent live crown, and needle length receiving proportionately less weight. The index has a range of 0 to 100 (0 = a tree with no chlorotic mottle; 100 = a severely injured tree retaining only the current year's foliage, more than 75% of the foliar surface area with chlorotic mottle, an average modal needle length of less than 1 cm, and a percent live crown of 10% or less). A tree must have a measurable amount of chlorotic mottle to receive a non-zero score, regardless of the condition of the crown.

Population and Community Responses

Crown condition assessments of individual trees in many stands are needed to detect the effects of air pollutants on populations of trees; evaluations of forest condition over large areas are commonly called surveys. A survey frequently consists of a one-time sample of a forest population with the intent of describing the incidence and severity of air pollution effects. Surveys are particularly useful in describing the spatial distribution of injury within the total areal extent of the population. With no intention to resample the same trees, it is not necessary to tag and map trees. Surveys designed to estimate spatial extent can not be used to deduce temporal trends unless very large numbers of trees are evaluated at each site.

Surveys may be developed into a system of plots to be monitored for trends over time (Figure 7.9) (Miller 1977, Duriscoe and Stolte 1989, Miller et al. 1989; see Chapter 12). Plots are relocatable areas containing systematically or randomly selected individual trees within a defined area (fixed or variable size). Trees are marked for relocation. Plots in forests have traditionally been located and relocated from site descriptions, photographs, and topographical maps. Land navigation systems such as the LORAN (land-based radio signals) or Global Positioning Systems (GPS) (satellite-emitted radio signals) have had only limited use in locating plots, but this use should increase in the future as more satellites are put into orbit and the cost of instruments decreases.

Assessing Responses of Populations

The initial steps in selecting methods for assessing air pollution effects on forest stands are to develop the hypotheses to be tested, determine the growth stage(s) of each species to be evaluated, ascertain the nature and concentration of the ambient pollutant(s) of concern, and delineate the response variables to be measured. In many Western studies, the focus of the research has been on the mature growth stages of pollutant-sensitive species at sites along steep (point source) or gradual (regional) pollutant gradients. The response variables most often evaluated were:

- presence, severity, and incidence of visible foliar injury
- reduced retention of needles
- accumulation of toxic substances in foliage
- stem growth
- resistance of the trees to natural biotic and abiotic stresses
- tree mortality

(a)

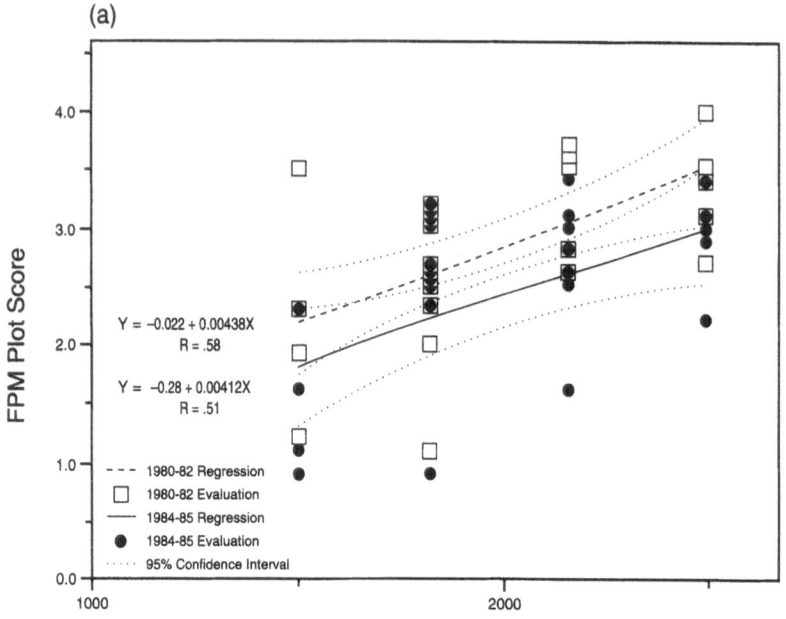

(b)

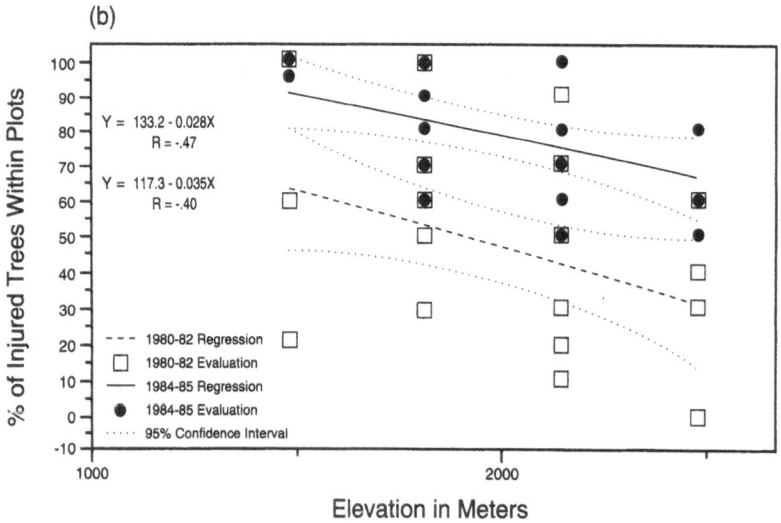

Elevation in Meters

Figure 7.9 Trends in (a) the severity of ozone-induced foliar injury as measured by Forest Pest Management (FPM) scores (low scores equal high injury), and (b) the incidence of foliar injury on ponderosa and Jeffrey pines in plots at different elevations in Sequoia and Kings Canyon National Parks in 1980–82 and 1984–85. Substantial increases in the incidence of injury were observed between the two periods. From Duriscoe and Stolte (1989).

Other injury responses such as altered needle growth, increased branch mortality, and reduced live crown ratio or density have also been measured (Pronos et al. 1978, Miller 1989, Duriscoe and Stolte 1989, Ewell et al. 1989, Miller et al. 1989). The hypotheses focused on the nature, severity, and incidence of injury; spatial and temporal trends in injury; relationships between pollutant injury and natural forest stresses; and the ecological ramifications of acute or chronic pollution injury (Miller 1977, Taylor 1974, 1980, Wellburn 1989, Smith 1990).

Area sampling and survey designs

The sampling of trees within forests is a critical element in the accurate quantification of the condition of the population. Procedures for sampling spatial phenomena are described in reference texts on the subject (e.g., Haggett et al. 1977, Taylor 1977). Six basic methods are available: systematic, simple random, stratified random, stratified systematic unaligned, multistage, and multifactor. Forest plots or survey points have traditionally used the systematic design in forest health surveys (Innes 1988). While systematic sampling with an appropriately small grid size may yield a very accurate estimate of the true population values, it may not be the most efficient method. If prior knowledge of the population is available, a stratified sample will usually yield comparable or better accuracy for less effort (Taylor 1977).

The distribution of tree species within the forest area is a fundamental spatial factor known to most forest managers. This information may be derived from vegetation type maps or a Geographic Information System (GIS) and used as the first level of stratification, defining the target population. Some type of randomization of point locations and tree selection within plots should be employed when locating survey or plot sites, as random samples have the advantage of being independent of one another, a desirable characteristic for many statistical procedures. Simple random sampling of the target population, however, allows the possibility of large contiguous areas being left out of the sample and thus limiting spatial resolution of the data. As a compromise, a spatially stratified random sample outlines the target population or sites where the species of interest occurs (Figure 7.10). After defining the target population and species to be evaluated, a systematic grid divides the area to be sampled into a given number of equal areas, and a random point is selected within each grid square that contains a specified areal coverage of the target species (Figure 7.10a). This method allows for the quantification of the incidence, severity, and spatial extent of the measured crown response variable(s) (Figure 7.10b) (Duriscoe 1990).

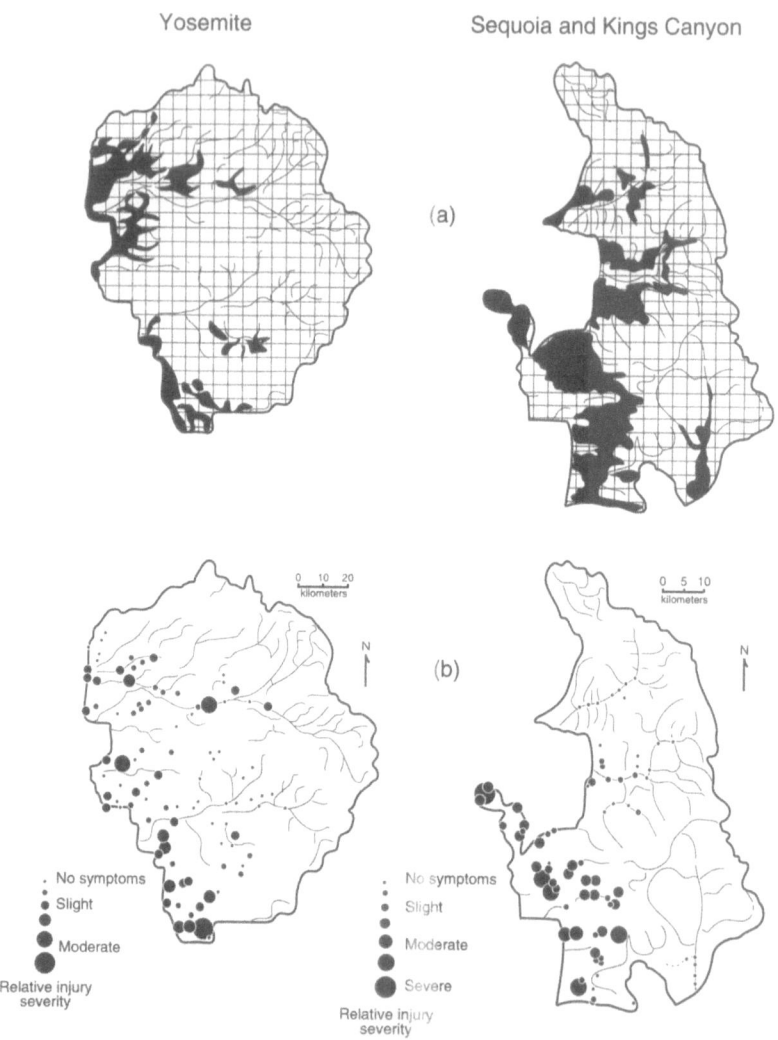

Figure 7.10 *Stratified random sampling design for evaluating air pollution injury on ponderosa and Jeffrey pine trees in Sequoia and Kings Canyon National Parks and Yosemite National Park. (a) Distribution of mixed conifer type containing ponderosa and Jeffrey pine overlaid with a 3.2 km density grid (grids not drawn to scale). Trees were scored for ozone injury at randomly selected sample points within the grid units containing the mixed conifer type. (b) Distribution of the severity of ozone injury on ponderosa and Jeffrey pines in both parks (note that injury severity scales are different for each park). From Duriscoe and Stolte (1989).*

Further levels of stratification can be based on knowledge of the spatial distribution of air pollutant(s) concentrations and durations. In many air pollution-forest response studies the objectives are to examine trends in effects along a pollution gradient (measured or inferred) (Miller 1977, Miller et al. 1989, Chapters 8–12 in this volume), or to compare impacted areas (based on the assessment of crown response variables) with areas that are relatively free of pollution impacts (Peterson et al. 1987, 1989; Chapter 11). In complex mountain areas, topography strongly influences the mixing of air and transport patterns of pollutants into the forests (Chapter 3). Portions of drainage basins may respond differently, so stratification into relatively similar groupings may be helpful. Evaluating pollution injury by drainage basin may be an appropriate method to obtain a more accurate estimate of the response of the entire population. Duriscoe (1990) found large differences in the severity and incidence of injury to the crowns of ponderosa and Jeffrey pines in different river basins in Sequoia and Kings Canyon National Parks in California.

The type of survey depends on the nature and source of the pollutant (regional or point-source). In surveys designed to evaluate the effects of a regional pollutant such as ozone on a major forest area (Miller et al. 1989, Duriscoe 1990), the stratified random design is very effective at determining incidence, severity, and spatial extent of injury (Arbaugh et al. 1991). Near point sources, stratified random sampling (using smaller grids) is sometimes used, but more typically radial transects (Figure 7.11) (Carlson and Dewey 1972, Severson et al. 1990) or less frequently zonal methods (Figure 7.12) (Gordon 1974, Carlson 1974) are used to map the severity, incidence, and extent of injury around point sources of fluorides, sulfur oxides, or toxic metals (Carlson and Hammer 1975, Bunce 1979, 1984).

The influence of environmental factors on the response of trees to pollutants may be reduced by stratifying the sample to include only certain micro-site types, such as south-facing, open-grown stands. If this information is available on a population-wide level, this method may reduce the statistical static resulting from the inherent variability of the population. Of course, conclusions based upon a stratified sample apply only to the portion of the population sampled, just as population samples apply only to the whole and not to selected subsets. For example, Duriscoe and Stitt (1990) used a stratified-random sample design to evaluate the accumulation of sulfur and toxic metals in relation to the subalpine fir population growing within the boundaries of North Cascades National Park Complex. The areal extent of subalpine fir was identified using digitized information on a GIS. Further stratifications of the subalpine fir population using the GIS included only open canopy stands, south-to-west facing aspects, and slopes less than 80%. The stratified stands were then located in seven smaller areas based on

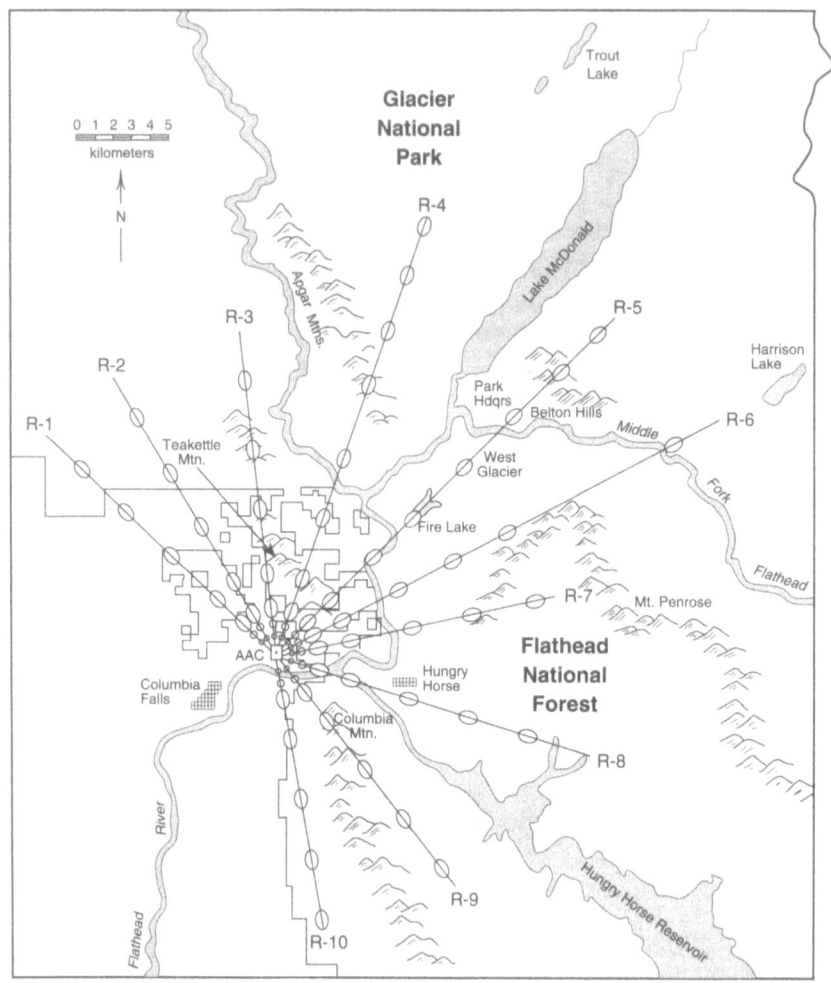

Figure 7.11 Radial transect sampling design to detect the impacts of fluoride emissions from the Anaconda Aluminum Company (AAC) on the surrounding mixed conifer forests in Montana. Fluoride concentrations in trees and other biological targets at the sample points (circles) on transect lines were used to construct pollution isopleths. Adapted from Carlson and Dewey (1972).

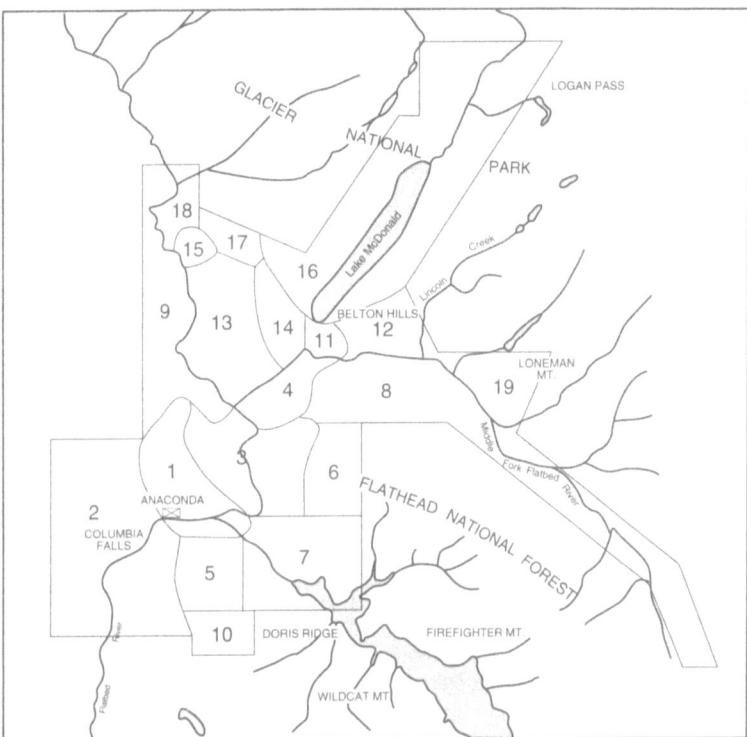

Figure 7.12   Zonal sampling design to evaluate the impacts of fluoride emissions from the Anaconda Aluminum Company (   ) on the surrounding mixed conifer forests. Same area as in Figure 7.11. Adapted from Gordon (1974).

topography that delineated the seven primary water drainage basins in the area. Air basins were assumed to be the same as the water basins in this area of mountainous topography. Three sample points were then selected at random from the population of stratified stands in each of the seven air basins. This method gave every stratified stand within each drainage basin an equal chance of being selected, maintained some degree of spatial resolution, and allowed for a means of differentiating air pollution deposition and plant uptake between drainage basins.

Sample size

The number of samples required to estimate the incidence and severity of air pollution effects on the crowns of a population of trees depends on the following: the allowable error of the estimate (the confidence interval), the probability that the true population value will lie within the bounds of the confidence interval (the confidence coefficient), and the actual proportion of injured trees (p—for incidence) or natural variability (σ^2—for severity) in the population. The first two factors are set by the researcher, and may be based on research objectives and available resources, while the latter factors may be determined by estimation from previously collected data or a theoretical consideration of the "worst case."

Incidence is the frequency of occurrence of a variable such as the number of injured trees in a stand. Incidence is an important statistic in areas of low pollution exposure or in wilderness areas where the existence of any trees showing visible pollution injury, even if the injury is very low in severity, may be interpreted as a significant deterioration of resource values. Inferences made about proportions are based on binomial sampling as described in various statistical texts (e.g., Snedecor and Cochran 1967, Box et al. 1978). Assumptions which must be met to use binomial sampling theory are that observations should be random, independent, and from a population of infinite size (about 400+ individual trees in practice). If an objective of a study is to identify the incidence of injury in stands of less than 400 trees, total enumeration should be considered rather than sampling.

The peculiarities of the binomial distribution are such that the greatest numbers of samples are required when the true proportion (p) of injured trees in the population is 0.5 for a given allowable error in the estimate (half-width of the confidence interval = d) and the confidence coefficient. Table 7.2 illustrates the required sample sizes for estimating selected values of p using a 95% confidence coefficient and one of three different values for d. It is apparent that if the true value of p is completely unknown, at least 100 observations should be taken to ensure an estimate with a precision of ±0.10 or 10%.

One index of the intensity of injury to populations is calculated by multiplying the mean severity of injury in the injured portion of the population by the incidence of injury (Muir and McCune 1987). Various indicators of injury severity may be used, such as foliage retention or percent leaf area covered with visible damage, but data forming highly skewed distributions should be avoided in order to apply parametric statistics in the analyses. A synthetic index such as the EII will usually satisfy this condition. The required sample size for estimating the mean severity of injury to the injured portion of the population may then be calculated using the following equation (e.g., Snedecor and Cochran 1967):

Table 7.2 *The effect of differences in the true proportion (p) of injured trees in a population on the sample size required to estimate p with a given allowable error (d, equal to the half-width of the confidence interval) and a 95% confidence coefficient. Derived from Box et al. (1978)*

Allowable error (d)	Approximate number of samples required for p=						
	0.05	0.10	0.20	0.50	0.80	0.90	0.95
0.05	130	200	350	400	350	200	130
0.10	45	60	80	100	80	60	45
0.15	30	35	40	45	40	35	30

$$n=(Z^2 * \sigma^2)/d^2 \qquad \text{(Equation 7.1)}$$

where

σ^2 = the population variance

$Z = 1.96$ (critical value of standard normal Z and confidence coefficient of 95%)

d = the allowable error in the sample estimate (half-width of the confidence interval)

The use of Table 7.2 and Equation 7.1 may be illustrated with the following "worst case" example which is based upon studies in the Sierra Nevada (Duriscoe 1991). Prior knowledge of the worst case suggests that 50% of the trees will be injured and that the injured trees will have a mean injury index of 50 (0 to 100 scale) with a population standard deviation (σ) of 15. The sample size required to estimate the incidence of injury in the population (confidence interval ±0.05 or 5%, confidence coefficient 95%) would be 400 trees (Table 7.2). For severity, if we require a confidence interval of ±5 index units (5% of the total range) and a 95% confidence coefficient for an estimated population value, 35 *injured* trees must be sampled (Equation 7.1). In this example, when analogous half-widths of the confidence interval (5 units out of 100) are used in each calculation, far more samples are required to accurately estimate the incidence of injury in a population than to estimate the mean severity of injury to the injured portion of the population.

It is often necessary to identify differences in the incidence or severity of injury between populations or within a population over time. This usually involves statistical testing, wherein two or more mean injury levels are compared, and requires that sample sizes be determined using equations other than Equation 7.1 (e.g., chapter 6 in Snedecor and Cochran 1967). Alternatively, the sample size needed to differentiate among sample means by a set difference can be determined through the results of sampling studies. For example, in a study of ponderosa and Jeffrey pine, Duriscoe (1991) determined the number of sample trees required per plot to differentiate among high, moderate, and low levels of ozone injury using plot-level averages of EII injury index scores and a confidence coefficient of 95%. Approximately 30 trees per plot should be sampled to detect a statistical difference of 10–15 units among the mean EII scores (Figure 7.13). Only small increases in precision were gained by sampling more than 30 trees in each plot.

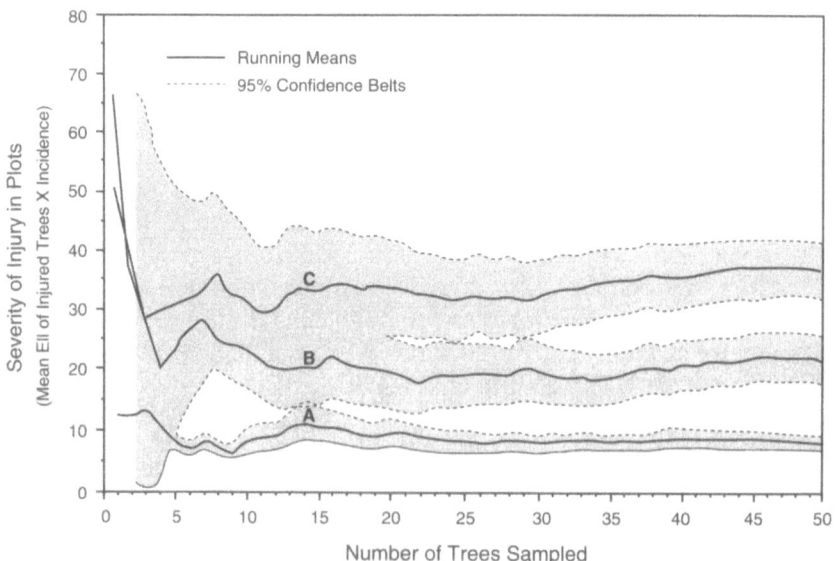

Figure 7.13 Running means of injury severity index (solid lines) and 95% confidence intervals (shaded areas) for single plots with (a) slight, (b) moderate, and (c) severe pollution injury. Sample sizes of at least 30 trees per plot are recommended to differentiate injury classes between plots or changes in the mean injury level within a plot over time. Adapted from Duriscoe (1991).

Summarizing results

Incidence of foliar injury (Pronos et al. 1978, Williams and Williams 1986) or an injury index combining severity of injury to individuals with incidence of injury can both be used as measures of injury to a population. However, important information can be lost when data are summarized as a single index value. In many cases it is desirable to report the distribution of individuals among injury classes as well as the synthetic indices.

For example, Table 7.3 presents hypothetical data on the incidence and severity of injury in 8 different populations of ponderosa or Jeffrey pine. Injury scores are tallied in one of six injury classes numbered 0 to 5 (low to high pollution injury), and then compared using frequency histograms. Increasing population injury is indicated by a shift in the peaks of the histogram from left to right. The histograms illustrate differences in the condition of populations that have very similar injury index scores.

Assessing Effects on Communities

Determination of mortality due to chronic air pollution can be accomplished by monitoring individual trees in plots over time. Chronic exposures to high levels of air pollutants can ultimately result in degradation of plant communities. As increasing numbers of sensitive genotypes of pollutant-sensitive species are killed, shifts in the composition and structure of plant communities may occur. Severe injury to individuals (Figure 7.7) does not always translate into detectable effects at the community level (Chapter 6), since impairment of some individuals may be compensated for by increased functioning of other individuals (e.g., release of neighbors following gap formation). In most cases, however, as pollutant deposition continues, more individuals of sensitive species are affected and more tolerant species begin to show evidence of pollutant stress. In extreme cases, gene pools are altered, mineral cycling is affected, insect, fire, and other stress cycles are altered, and successional patterns change (Miller 1977, Smith 1990).

The most serious effect of air pollution on conifer communities has been the degradation of forests around point sources. Intensive sampling using arithmetic or geometric transects, grids, or zonal sampling designs around point sources can elucidate the pronounced injury and mortality gradients that often occur around these pollutant sources. Stands of conifers close to the source (often within a 5–10 km radius) may have high mortality rates, leading to devegetation of the site. More commonly, severely impacted conifer communities near point sources of pollutants experience destruction of the more pollutant-sensitive species but remain

Table 7.3 *Method of describing injury severity of trees in plots or populations using hypothetical data, based on the incidence of injury in six injury classes. Incidence of injury in all injury classes is multiplied by the mean injury index to produce a weighted injury index (data for calculation of mean injury index not shown). Injury class 0 = no injury, 1 = very slight, 2 = slight, 3 = moderate, 4 = severe, 5 = very severe.*

Class 0	Class 1	Class 2	Class 3	Class 4	Class 5	Frequency distribution of injury	Mean injury index (injured trees only)	Incidence of injury (%)	Weighted injury index (population)
100	0	0	0	0	0		—	0	—
90	10	0	0	0	0		12	10	1.2
60	30	10	0	0	0		18	40	7.2
30	20	30	20	0	0		30	70	21.0
20	10	20	40	10	0		50	80	40.0
0	10	20	50	20	0		65	100	65.0
10	10	20	30	20	10		65	90	58.5
0	10	10	30	40	10		75	100	75.0

The column headers over Class 0 through Class 5 are bracketed as "Trees per injury class (100 trees per sample)".

vegetated by the more pollutant-tolerant species (Carlson and Dewey 1972, Carlson 1974, Gordon 1974, Carlson and Hammer 1975, Bunce 1979, 1984).

Van Hook (1974) studied the effects of fluoride emissions from a phosphate plant in Montana by sampling along 12 radial transects centered on the plant. This approach is particularly appropriate if little is known about the prevailing wind patterns in the area. Van Hook found decreasing levels of fluoride in the foliage of conifers, in the soils, and in forage grasses with increasing distance from the source. Bunce (1979) utilized a grid design to locate 64 plots within the fume zone of an aluminum smelter in British Columbia to determine the effects of fluoride emissions on the radial ring growth of hemlock trees. Stem diameters and heights were also measured and foliage fluoride concentrations were determined. He estimated that 2,700 m^3 of wood were lost annually from 1954–1976 due to the operation of the smelter. Although the smelter caused a decrease in wood production, community analyses indicated that species regeneration within the impacted zone of the forest was unaffected.

Taylor et al. (1986) used a radial transect sampling design to evaluate the effects of chronic ozone (regional source—San Joaquin Valley of central California) and sulfur oxides (point source—an oil extraction facility in Bakersfield, California) on the crown condition of ponderosa and Jeffrey pines, the concentration of sulfur in the associated soil horizons, and the concentration of sulfur in associated lichen species in the southern Sierra Nevada. They found that concentrations of extractable sulfate in the 0–5 cm horizon decreased with elevation and distance from the source. A similar trend was observed for total sulfur in the foliage of pine needles and lichens. Crown condition assessments indicated that trees in plots nearest the valley had more visible foliar discoloration and the lowest retention of needle fascicles.

Determination of air pollutants as the causal agent of conifer community degradation can be a difficult process because subtle, chronic impacts of pollutants must be differentiated from the natural biotic and abiotic stresses on the trees and the natural successional processes that take place within the communities (Miller 1977, Treshow and Anderson 1989, Wellburn 1989, Smith 1990). Pollution effects on conifer communities from point sources are typically easier to identify and quantify because of the strong pollutant gradient surrounding point sources (National Research Council of Canada 1939, Whitby and Hutchinson 1974, Hutchinson and Whitby 1976). The most effective way of delineating chronic effects of pollutants is the examination of many variables using a variety of methods designed to differentiate pollutant effects from natural processes at different levels of biological organization. For example, evaluation of foliage for visible foliar discoloration (branch level), the crown for patterns of thinning, dieback, and branch mortality (tree level), and the foliar elemental content of different species (community level) would help to quantify fluoride stress on a conifer community near a point source. The establishment of long-term ecological monitoring plots is essential to elucidate the effects of chronic pollution on community structure.

The best documented case of regional air pollution causing serious impacts on Western conifers comes from the work of Paul Miller, US Forest Service, and associates in southern California (Chapter 12). This research included analyses of visual symptoms, physiology (photosynthesis and respiration), foliage retention and initiation, stem growth, susceptibility to pine beetles (*Dendroctonus brevicomis*) and root rot (*Fomes annosus*), nutrient cycling, cone production, and rates of mortality for ponderosa and Jeffrey pines to determine the effects of chronic oxidant (primarily ozone) exposure on a mixed-conifer forest. By evaluating the incidence and severity of ozone injury and mortality rates, they determined that impacts were greatest at plots closest to the Los Angeles air basin. Community analyses indicated that loss of pollutant-sensitive ponderosa and Jeffrey pines may be causing a shift in the plant community from an open canopy pine forest to a more closed-canopy oak-chaparral community.

Dendrochronology

This section provides an introduction to tree-ring analysis and a theoretical framework for the tree-ring studies of pollution impacts on forests that are presented later in this book (Chapters 8–11). To date, "no readily detectable, pollutant-specific single marker for identifying the effects of air pollution on forests or trees has been identified" (National Research Council 1989, p. 2). Yet, there are many non-specific indicators of stress that can detect anomalous growth patterns in trees, which could ultimately be related to air pollution. One such indicator, available in all temperate forests, is the annual tree-ring increment.

Tree rings are the only widely available source of long-term data on forest growth and productivity that may predate the present era of elevated atmospheric pollution. Year-to-year changes in ring width integrate past environmental and biological influences on tree growth, allowing detection of anomalous changes in tree rings that may be consistent with those expected from pollution stress.

Annual tree-ring increments can be analyzed in several ways for evidence of air pollution effects. These include using annual ring widths (e.g., Peterson et al. 1987, Johnson et al. 1988), ring areas (e.g., Hornbeck and Smith 1985, Phipps and Whiton 1988), and ring volumes (e.g., LeBlanc et al. 1987a, b), along with ring densities (e.g., Conkey 1988) and chemistry (e.g., Baes and McLaughlin 1984). The easiest measure of annual tree growth, and the one that will be emphasized here, is the radial ring-width series from the breast-height (1.4m) region of a tree.

Although tree-ring analysis based on breast-height ring-width series has obvious potential for the study of regional forest declines, its application in such studies has been difficult and controversial. One difficulty lies in the fact that tree rings are an integration of many environmental and biological influences on tree growth, including stand dynamics, tree maturation, and climate (Chapter 6). This means that any pollution signal may be small and embedded in a high level of natural environmental noise. The signal from air pollution may be weak relative to the noise of normal variations, giving low ratios of signal to noise. This problem is especially acute in the study of regional forest declines where air pollution levels are low compared to regions near pollution point-sources. Consequently, tree-ring studies of regional forest declines may need to examine hundreds or even thousands of ring-width series (e.g., Hornbeck and Smith 1985, Schweingruber 1986, McLaughlin et al. 1987) in order to reduce the noise level and increase the degrees-of-freedom for statistical tests.

Another difficulty is the lack of a generally applicable "normal" or expected growth model (c.f. Hyink and Zedaker 1987) for natural forests that can be used to identify abnormal growth patterns in trees. Such models are rare for all but the simplest cases, e.g., open-canopy forests and even-aged, single-species plantations. For mature closed-canopy forests, which are often a mixture of tree species and age classes, useful normal growth models rarely exist. In such environments, the evolution of tree rings through time is highly stochastic (Cook 1987b) and difficult to predict using deterministic, normal growth models.

Finally, what should an air pollution signal look like in tree rings? Anomalous reductions in ring-widths are not uniquely related to pollution effects on regional forest growth (Cook and Innes 1989). Current understanding of the physiological effects of various pollutants on tree growth is still poor, especially for relatively low doses (chronic exposures) of air pollution (see Chapter 5). There is also a lack of understanding of the interactions between pollution and climate, which may act synergistically or in opposition depending on the way in which climate is affecting tree growth at the time. Thus, without a good pollution response model for tree growth at the species level, there is little hope that tree-ring analysis can prove that pollution is directly responsible for a regional forest decline. Rather, tree-ring analysis is presently best suited for eliminating natural explanations of decline, such as climate, stand dynamics and maturation, and for discovering anomalous properties in the data that are consistent with the intervention of some new stress such as air pollution.

Proper cross-dating of tree-ring data is a first step in analysis. As used here, cross-dating is a quality assurance/control procedure that applies to any application of radial increment data, wherein each tree ring is correctly dated to its calendar year of formation. It is also used to correct the dating for anomalies such as locally absent rings and false rings (sensu Fritts 1976) that would cause a simple ring count to be incorrect. Thus, cross-dating is synonymous with the correct dating and calendar-year alignment of tree-ring series. The mechanics and methods of cross-dating, along with much more comprehensive reviews of dendrochronological methods, are described in Fritts (1976), Hughes et al. (1982), and Cook and Kairiukstis (1990). Cross-dating is necessary to obtain accurate mean-value functions of tree rings computed from individual tree-ring series that are properly matched in time. Correlating misdated tree-ring data with climate data would produce biased results and potentially incorrect interpretations because of the dating errors in the tree-ring data. Cross-dating eliminates these sources of bias. This issue applies equally to both mean-value functions and individual radial tree-ring series. Both kinds of tree-ring data have been used for analyses of tree growth, as presented in Chapters 8–11. Therefore, cross-dating should be regarded as the single most important quality assurance/control step in any dendrochronological study because the validity of all subsequent results and interpretations depend on correct cross-dating.

Another component of dendrochronology is site selection. Ideally, sites selected for tree-ring analyses should be based on an a priori set of criteria dictated by the scientific problem being investigated. For example, if climate is to be reconstructed from tree rings, then minimally-disturbed, climatically-stressed sites should be selected (Fritts 1976, Hughes et al. 1982). In the context of air pollution impacts on forests, site selection criteria might be determined by other factors such as the presence of air pollution pockets or gradients, the distribution of certain tree species sensitive to a particular air pollutant, and the spatial distribution of tree species or forests experiencing anomalous declines. Each of these criteria has been used in one way or another in the tree-ring applications described in other chapters in this book. A drawback to selecting sites based on criteria related directly to air pollution is that any result obtained from such analyses might be a biased representation of the overall health and vigor of the forest in which the sites are located. However, if the basic scientific problem is the documentation of a probable air pollution impact on the growth of a selected tree species, then such restricted sampling is valid. For the more general problem of regional impacts of air pollution on forests, site selection using random sampling methods is statistically more powerful.

The selection of individual trees for sampling at a tree-ring site again depends on the nature of the study and the scientific question being addressed. For climate studies, trees showing obvious injury from insects, disease, or fire would not normally be sampled because such injuries could mimic or obscure the desired climatic signal in the tree rings. The same argument applies to the study of air pollution impacts on tree growth. However, certain air pollutants like ozone visibly affect the foliage of sensitive tree species and genotypes in known ways, primarily the accelerated abscission of needle whorls and the production of distinct visible foliar discoloration symptoms on some or all of the remaining needles. It is therefore possible to purposely sample both symptomatic (air pollutant sensitive) and asymptomatic (air pollutant tolerant) trees as part of an air pollution study directed at looking at growth effects within a species. The premise is that symptomatic trees will show a stronger air pollution signal in their ring widths if the air pollution stress is severe enough to overcome the inherent resiliency of the tree to loss of fixed carbon. This stratified sampling approach has been used by Peterson et al. (1987) in their study of Jeffrey pine and by Peterson and Arbaugh (Chapter 11) for ponderosa pine (var. *ponderosa*) in the Sierra Nevada. Alternatively, Graybill, Peterson and Arbaugh (Chapter 9) were not able to use this approach in their studies in the Colorado Front Range because of lower ozone concentrations (Chapter 3), the relatively higher tolerance of Rocky Mountain ponderosa pine (var. *scopulorum*) to ozone, and the subsequent lack of foliar injury symptoms that facilitate the identification of pollutant-sensitive and pollutant-tolerant individuals. Similarly, it is possible to stratify the sampled trees by crown dieback class in the hope that trees with more crown canopy dieback might be more severely impacted by air pollutants. Unfortunately, crown canopy dieback can be caused by both natural and anthropogenic factors, with the former being much more common and likely than the latter. Therefore, it is dangerous to assume and difficult to prove an air pollution effect on trees experiencing some form of generic canopy dieback.

A Model for Analyzing Tree Rings

A tree-ring series can be considered as a linear aggregate of several unobserved subseries or signals. This aggregate can be expressed as

$$R = A + C + dD1 + dD2 + E \qquad\qquad \text{(Equation 7.2)}$$

where:

R is the observed ring-width series n years in length;

A is the age and size related trend in ring width;

C is the yearly variation in ring width due to climate;

D1 is the change in growth due to endogenous disturbance;

D2 is the change in growth due to exogenous disturbance; and

E is the unexplained year-to-year variability not related to the other signals.

This model indicates the components that contribute to the observed tree ring without implying linear additivity. The d associated with D1 and D2 is a binary indicator of the presence (d=1) or absence (d=0) of either class of disturbance in the ring widths. Thus, A, C, and E are assumed to exist in R for all n years, although their relative contributions to the total variance of R may vary with time. In contrast, D1 and D2 are assumed to be present only if either disturbance has occurred at some time t0 in the life of the tree. And, in most cases, we should also expect the effect of either disturbance to eventually disappear in the ring widths. When both of these properties are fulfilled, a disturbance effect may be formally called a pulse to differentiate it from the continuous A, C, and E. There may, of course, be more than one disturbance of either kind in a tree-ring series.

In theory, an endogenous disturbance can be distinguished from an exogenous disturbance by comparing ring-width patterns of trees in a stand. Endogenous disturbances result from stand dynamics in which individual or small patches of canopy trees are removed by processes that do not affect the stand as a whole (Chapter 6). The development of canopy gaps creates patterns of suppression and release in the ring widths of surviving trees adjacent to the gaps. Trees away from these gaps show no response. Therefore, assuming that endogenous disturbances are random events in both space and time, D1 should be strictly contemporaneous only in the ring widths of trees adjacent to the same gap, but not elsewhere in the stand. In contrast, exogenous disturbances are assumed to affect large areas of forest simultaneously. The pertinent feature of D2, then, is synchrony in time throughout the stand, which theoretically differentiates it from the temporal and spatial randomness of D1.

The impact of airborne chemicals on forest growth and productivity can be considered an exogenous disturbance if, for a given pollutant, the dose-response relationship is basically uniform across a sampled stand of trees. In describing the linear aggregate model in the context of forest-pollution studies, Cook (1987a) added a pollution term, dP, to Equation 7.2 to explicitly differentiate this kind of large scale anthropogenic disturbance from similar scale natural forest disturbances, such as those caused by insects, disease, and unusual meteorological events.

The growth trend A in Equation 7.2 is usually assumed to be a non-stationary process that reflects, in part, the geometrical constraint of adding a volume of wood each year to a stem of increasing radius and height. When this constraint is the principal source of the trend, A can often be modeled as a deterministic, negative exponential function of time (Fritts et al. 1969) having the form

$$A = a * e^{-bt+k} \hspace{4cm} \text{(Equation 7.3)}$$

where a, b, and k are coefficients estimated from the data and t is time. This model is often appropriate for estimating the ring-width trends of trees growing in open-canopy forests where between-tree competition is minimal and natural disturbances are rare. It also assumes that the observed ring-width series begins after the juvenile period of increasing radial growth found in many conifers because Equation 7.3 is a mono-tonic function that strictly decreases to the limit k. For those ring-width series with an early juvenile growth period, a more general exponential function can be used (e.g., Warren 1980). Eriksson (1989) has criticized the asymptotic behavior of Equation 7.3 because it implies that ring widths do not diminish to zero in old trees as they senesce and approach death. However, for old-age conifers growing on semi-arid sites, Fritts et al. (1969) found that a standard negative exponential model with a zero asymptote (k=0) commonly underestimated the observed ring widths in the mature growth period.

From the form of Equation 7.3, it is clear that a negative trend in ring width is an expected property when competition and disturbances do not appreciably affect tree growth. Thus, considerable care must be taken to properly discriminate between decreasing ring widths due to increas-ing tree age and size and reductions in growth that may be caused by air pollution. Indeed, the identification of D2 in R (Equation 7.2) may be very difficult if D2 is only expressed as an added reduction in ring width and does not depart strongly from that estimated by A. However, when deterministic growth trend models such as Equation 7.3 are appropriate for the tree-ring data being analyzed, a theoretical basis for their use can be claimed, which may make any identified pollution effects more tenable.

In closed-canopy forests, the effects of between-tree competition and disturbances can cause an observed trend in ring widths to depart significantly from that predicted by theoretical growth models such as Equation 7.3 (Cook 1987a, b). When this is the case, the expectation of decreasing ring width with increasing age is often inappropriate. This condition complicates the search for D2 because now the effects of both A and D1 have to be taken into account before the existence of D2 can be tested. The simultaneous estimation of A and D1 using flexible curve fitting and smoothing methods, such as orthogonal polynomials (Fritts 1976) and smoothing splines (Cook and Peters 1981), cannot differentiate between D1 and D2. Consequently, there is some danger that the pollution signal of interest will be inadvertently removed as part of the joint estimation of A and D1. However, there are ways of reducing the impact of D1 without directly estimating it by exploiting the spatial randomness of endogenous disturbances as they affect trees in a stand (Cook 1985).

Component C in Equation 7.2 represents the integrated effect of climatically-related environmental variables on tree growth, except those that are only associated with stand disturbances. Typical variables composing C are precipitation, temperature, and heat sums as they affect available moisture supply, evapotranspiration demand, phenology, and rates of reactions. These variables are assumed to affect all trees in a stand similarly. Thus, C is a signal in common to the sampled trees and is usually the principal source of marker years used for cross-dating tree rings. Some climatic variables that have been used to model C are monthly temperature and precipitation (Fritts 1976), and drought indices (Cook and Jacoby 1977). These variables are usually regarded as stationary stochastic processes although they may be persistent in an autoregressive sense (Gilman et al. 1963) and can possess trend (e.g., Jones et al. 1986, Bradley et al. 1987).

When possible, the influence of climate on ring width should be modeled as part of any study of pollution effects on forests (Cook 1987a). An unusual reduction in ring width that appears consistent with hypothesized pollution effects could, in fact, be caused by adverse climatic extremes or climatic change. If this were the case, then a pollution hypothesis could be falsely accepted. Alternately, it is possible that better-than-average climatic conditions might compensate for an air pollution effect and largely mask a growth reduction due to airborne chemicals. If this were the case, then a pollution hypothesis could be falsely rejected.

Puckett (1982) proposed that a change in the climatic response of trees may be an indicator of physiological dysfunction caused by air pollution; this change could occur even in the absence of an abnormal ring-width reduction. Therefore, it offers the possibility of a different kind of tree-ring indicator of pollution stress in trees. The physiological basis for expecting a change in climatic response from exposure to air pollutants has not yet been established, so time-dependence in any statistical properties of tree rings is not necessarily strong evidence for a pollution effect on tree growth. The statistical techniques for seeking such changes in tree rings are available (e.g., Visser and Molenaar 1986, 1988, Cook et al. 1987, Van Deusen 1987, 1989, 1990) and can be readily applied in ways that may yield some evidence of air pollution effects.

As specified in Equation 7.2, D1 or D2 are present in the ring widths only after some arrival time t0 and usually decay to zero (effectively) given sufficient time. This intervention model for disturbance signals in tree rings is appropriate if the physical or biological mechanisms of distur-bance can be modeled as impulses or spikes that shock a tree's physiol-ogy at time t0 and than disappear. For example, the formation of a gap by treefall in a closed-canopy forest occurs instantaneously, a ground fire may pass through a stand in a matter of minutes to hours, and insect defoliation may only occur over one growing season. Because these disturbances are confined to one growing season (i.e. the sampling interval of tree rings), they can be modeled as spikes having the general form [... 0 0 1 0 0 ...] with the sign of the spike being dependent on the effect of the disturbance on growth.

The general spike intervention model for disturbances proposed above is not appropriate if the intervention causes a long-term or permanent change of state in the forest environment. The hypothesized effect of airborne chemicals on forest ecosystems fits into this latter kind of intervention model because anthropogenic emissions of ozone precur-sors, ammonium, nitrogen oxides, and sulfur dioxide have been continu-ously added to the environment throughout the past century (Husar 1989) and especially since 1940 in certain parts of the United States. Consequently, if D2 is present in tree rings and is caused by air pollu-tion, its form is not likely to be that of a pulse. Rather, it may be more like a step having the general form [...0 0 1 1 1...] or even a ramp. Of course, a tree's response to a step intervention is unlikely to look like a pure step because of physiological persistence, delays in response, and the possible masking effects of other variables such as climate and endogenous disturbances. However, if persistence, delays, and masking are taken into account when modeling interventions, it is possible to recover the form of the original intervention to a reasonable approxima-tion. Box and Tiao (1975) provide details on the characterization of interventions in time series.

E represents the unexplained variance in the ring-width series after the contributions of A, C, D1, and D2 have been taken into account. Some possible sources of E are microsite variations within the stand, gradients in soil chemistry and hydrology, genetic variability within the sampled population, and measurement error. E is assumed to be serially uncorrelated within and spatially uncorrelated between trees in the stand, although the latter assumption may be difficult to satisfy without explicitly modeling the spatial autocorrelation of the errors. In fact, the difficulty in satisfying this ideal error structure using traditional dendro-chronological methods has lead to new and innovative techniques of simultaneous estimation that allow for much more general error struc-tures (Eriksson 1989, Van Deusen 1989).

Methods of Detecting Change in Tree-Ring Series

The conceptual model just described emphasized the notion that a pollution effect on tree growth should be viewed as an exogenous disturbance or an intervention of a new kind of stress into the forest ecosystem. The jargon of intervention analysis (sensu Box and Tiao 1975) has also been used as a convenient way of introducing this topic. How-ever, there are different ways of searching for the presence of an inter-vention in tree rings that may be due to pollution stress, and we do not necessarily advocate any particular method of analysis for this purpose. Indeed, the variety of statistical methods used in this book to analyze tree-ring data for possible pollution effects illustrates the rich variety of approaches one may take in such studies.

Consider now a suite of m cross-dated and measured ring-width series from a stand of trees that is hypothesized to be impacted by airborne chemicals. On the premise that this impact should be revealed by some anomalous property in the tree rings, what statistical methods might be applied to test this intervention?

A frequently applied approach, which is based on the experience of dendroclimatologists in modeling climatic signals in tree rings, has been to detrend and standardize (sensu Fritts 1976) each ring-width series and then average the standardized tree-ring indices into a mean-value function. This is accomplished by independently fitting a growth curve to each ring-width series as an estimate of A, and perhaps D1. The tree-ring indices are then calculated as

$$I = R \, / \, G \qquad\qquad\qquad \text{(Equation 7.4)}$$

where:

I is the series of dimensionless tree-ring indices having a defined mean of 1.0 and homoscedastic variance,

R is the measured ring-width series, and

G is a growth curve equal to A+dD1.

Of course, if d=0 or G is constrained to a simple deterministic family of growth curves (e.g., Fritts et al. 1969), G reduces to A. Otherwise, more flexible growth curves (e.g., Cook and Peters 1981) might be used that jointly estimate A+D1. Note that Equation 7.4 is still compatible with the linear aggregate model proposed earlier because the logarithmic transformation of R would linearize the operation of computing tree-ring indices.

Assume for now that d=0 for D1. Then, the tree-ring indices computed by Equation 7.4 are expressed as:

$$I = C + dD2 + E \qquad\qquad \text{(Equation 7.5)}$$

Assuming that C and D2 are common to all trees and the error variance E is serially and spatially uncorrelated within and between trees, the mean-value function of tree-ring indices will reduce E as a function of $1/m$ and produce better estimates of C and D2 than exist in any one series of indices. This should enhance the probability of finding a pollution signal, should it exist, providing that the estimation of G has not inadvertently removed it from the tree rings. Now if d=1 for D1 and G=A, then

$$I = C + D1 + dD2 + E \qquad\qquad \text{(Equation 7.6)}$$

This potentially complicates the search for D2 because, as described earlier, there may not be any way of differentiating D1 from D2 in individual tree-ring series. But given that the arrival time of D1 is assumed uncorrelated between trees not adjacent to the same gap, the variance accounted for by D1 can also be reduced in the mean-value function by averaging. The mean-value function can then be tested for anomalous residual negative growth departures and time-dependent climatic effects that may be indicative of pollution stress.

Mean-value functions of tree-ring indices have been used in various ways to study the impact of air pollution on tree growth. In studies where tree-ring sites were situated along well defined pollution gradients produced by intense point-sources such as copper smelters, changes in some attributes of mean tree growth have been correlated with the known gradients (e.g., Nash et al. 1975, Thompson 1981, McClenahen and Dochinger 1985, Fox et al. 1986). These changes included anomalous reductions in growth and a change in the ability of climate to predict tree-ring indices as the sites approached the pollution point-sources. Kincaid (1987) was even able to quantify growth loss as a function of changing sulfur dioxide emissions.

In other studies where air pollution is regional and the exposures of forests lower, the analyses have been more difficult and the interpretations more tenuous. Puckett (1982) used response function analysis (sensu Fritts 1976) to search for changes in climatic response, which could indicate a response to acid rain, by splitting the data into two equal time periods and comparing the fitted climate models. Although he found apparent evidence for a change in response between the two periods that he interpreted as evidence for anomalous behavior, the results were questionable because of uncertainties about the small sample stability of response functions.

As a means of reducing this uncertainty, Cook (1987a) devised a multiple linear regression method that uses forecasts or extrapolations of tree rings from a fitted climate model to test for changes in the predictive ability of tree rings by climate. A critical provision of this method was the use of a validation period for testing the stability of the climatic response model (i.e. the correctness of the selected climate variables) before it was used to extrapolate into the time period when the trees were suspected of being affected by some new stress like air pollution. This method was used to study the decline of red spruce (*Picea rubens* Sarg.) in the Appalachian Mountains of the eastern United States. In analyzing 35 mean index chronologies of red spruce, the validated climatic response models were not able to predict the standardized tree-ring indices after 1960, either in mean level (McLaughlin et al. 1987) or in changes from year-to-year (Johnson et al. 1988). These results suggested that a change had occurred in the environment of red spruce that was coincidental with the beginning of its documented decline.

About the same time, Visser and Molenaar (1986, 1988) and Van Deusen (1987, 1989, 1990) independently applied the Kalman filter (Harvey 1984) to the problem of estimating time-dependent or dynamic regression models in tree-ring analysis. Presently, this is the most powerful method available for that purpose because its mathematical theory is well developed and it does not require a priori information about the timing of pollution effects, as Cook's (1987a) method does. When Visser and Molenaar (1986) used the Kalman filter to reanalyze the same data used by Puckett (1982), they found that there was little evidence for time-dependence of the kind previously claimed. This result probably reflects the poor stability of the response function method for small sample sizes by Puckett (1982). Since then, Cook and Johnson (1989) have shown with the Kalman filter that there is an elevational dependence in the response of red spruce to summer and winter temperatures and that the change in climatic response in red spruce after 1960 may not have been as abrupt as previously thought. In addition, their results suggest that red spruce has been responding to changing climate and new climatic variables since the 1920s. These new insights would have been difficult to obtain without the Kalman filter, which is rapidly becoming the standard method of analysis for testing the presence of time-dependence in tree rings.

The use of mean-value functions is appealing on statistical grounds. The estimation of C and the search for D2 is made presumably easier by reducing the noise variance in the data due to E and D1 by ~1/m, and by reducing the dimensions of the problem from m to only one series. However, the use of averaged tree-ring data can be criticized for losing too much individual tree information. Using a tree-ring mean-value function relies on an implied assumption that the tree-ring series being averaged belong to the same population in terms of climatic response and genetic sensitivity to air pollution. To the extent that cross-dating implies a common environmental signal between trees that is often related to climate, it would seem that averaging is appropriate when cross-dating is strong. However, there may be tree-ring sites of interest where the trees cross-date weakly, suggesting considerable heterogeneity in climatic response between trees. In this case, it might be useful to model individual tree-ring series for climate and pollution effects. The use of individual radial increment series in this kind of study is comparatively new and could add substantial new information to our understanding of how individual trees and forests respond to environmental stress. However, it must also be recognized that the noise level in individual radial increment series is likely to be very high due to the unattenuated contributions of D1 and E, thus making the problem of differentiating D2 from D1 more formidable and generally less certain.

Genetic variation in ring widths could derive from either normal variations in growth within a gene pool, or from greater sensitivity of some genotypes to pollutants. An independent means of separating the trees into sensitivity classes (such as foliar symptoms; Peterson et al. 1987 and Chapter 11) is needed to distinguish between these two possibilities.

The sequential analysis of tree rings, where each stage of analysis is independent of all other stages (e.g., first detrend, then average, and finally regress), has been criticized by Van Deusen (1989) and Eriksson (1989) for not properly handling the error structure and aggregation of tree-ring series implied by the linear aggregate model. For example, Van Deusen (1989) noted that prewhitening tree rings before doing the regressions on climate, as done by Cook et al. (1987) and in Chapter 8 of this volume, is not appropriate because autocorrelation in the tree rings should be handled as part of the error structure of the model, not removed prior to regression. It is not clear that this theoretically valid criticism has any serious consequences in practice. For example, the climate model for red spruce appears to be robust with respect to the tree-ring and climate data being analyzed, the method of analysis, and the analyst (e.g., Cook et al. 1987, Federer et al. 1989, Van Deusen 1990).

Van Deusen (1989) and Eriksson (1989) also noted that detrending, climate modeling, and the search for time-dependent effects should all be done simultaneously across the full suite of tree-ring series from a stand. Van Deusen (1989) describes an exponential aggregate model for this purpose, and Eriksson (1989) proposes a random coefficient model specifically designed for analyzing longitudinal data (of which tree rings are a form). Each method requires some rather simple assumptions about the form of the growth trend model, and the full random coefficient model requires the estimation of a very large number of parameters. Neither method has been used enough on tree-ring data to tell if any significant benefits will arise from these more complicated approaches.

An alternative method for detecting changes in tree-ring series, either in individual records or in mean-value functions, is intervention detection (Downing and McLaughlin 1987). This method basically uses the time series outlier detection method of Chang (1982) coupled with autoregressive-moving average (ARMA) time series modeling (Box and Jenkins 1976) and intervention analysis (Box and Tiao 1975) to exhaustively search for unusual pulses or step-like changes in time series residuals. To date, the only application of intervention detection to the study of forest decline is that of McLaughlin et al. (1987) and Downing and McLaughlin (1987), who analyzed 2,433 red spruce tree-ring series for statistical evidence of anomalous ring-width declines. They found that the frequency of identified negative-step interventions peaked in the 1955-59 period, which generally agreed with previous estimates of when red spruce began to decline in the northern Appalachian Mountains (e.g., Johnson and McLaughlin 1986).

Conclusions

The methods of analysis described here provide an introduction to some of the approaches and methods used in the tree-ring studies described in this book (Table 7.4). It is clear that no consensus exists on how to statistically analyze tree rings, which is both a blessing and a curse. The paucity of useful biological theory for guiding tree-ring analyses through the thicket of available statistical methodologies means that there will always be justifiable concerns about the meaning of statistical associations and the validity of causal inferences drawn from them. This is especially true in the often controversial study of air pollution and its effect on tree growth. However, the fact that different data analysts and statisticians may analyze the same data in quite different ways, means that there is room for innovation, improvements in methods, and independent validation of results. Thus, tree-ring analyses can be very robust and provide new insights into how trees interact with their environment.

Table 7.4 *General tree-ring analysis approaches and methods used in Chapters 8–11 of this book. A ● means that the tree-ring analyses in the chapter explicitly examined the tree-ring data for certain classes of variance described in this section. See the specific chapters for details of the statistical methods used.*

	Chapter 8	Chapter 9 A	Chapter 9 B	Chapter 10	Chapter 11
Growth-trend estimation and analysis	●	●		●	
Single tree-ring index chronology	●		●		●
Mean tree-ring index chronology analysis	●	●	●	●	
Climatic response modeling	●	●	●	●	
Endogenous disturbance analysis	●				
Exogenous disturbance analysis: Climate model extrapolation tests	●	●		●	
Exogenous disturbance analysis: Time-dependent models using the Kalman filter			●		●

References

Aitken WM, Jacobi WR, Staley JM (1984) Ozone effects on seedlings of Rocky Mountain ponderosa pine. *Plant Disease* 68:398–401

Allison JR (1984) *An Evaluation of Ozone Injury to Pines on the Tahoe National Forest.* Forest Pest Management, Pacific Southwest Region, Report No. 84–30, San Francisco

Arbaugh M, Bednar L, Moore J, Cline S (1991) Statistical recommendations for index construction, plot design, and data gathering methods to monitor long-term effects of ozone on conifer forests of the Sierra Nevada. In: Stolte KW, Miller PR (eds) *Management Recommendations and Specifications for Plot Design and Sampling Methods to Monitor Long-term Effects of Ozone on Western Coniferous Forests.* Proceedings of a Pine Plot Workshop. March 14–15, 1989, Riverside, California. US Forest Service, Pacific Southwest Forest and Range Experiment Station, Riverside, CA, 88p

Axelrod MC, Coyne PI, Bingham GE, Kircher JR, Miller PR, Hung RC (1980) Canopy analysis of pollutant injured ponderosa pine in the San Bernardino National Forest. In: *Proceedings of the Symposium on Effects of Air Pollutants on Mediterranean and Temperate Forest Ecosystems.* USDA Forest Service General Technical Report PSW-43

Baes CF, McLaughlin SB (1984) Trace elements in tree rings: Evidence of recent and historical air pollution. *Science* 224:494–497

Barrett TW, Benedict HM (1970) Sulfur Dioxide. In: Jacobson JS, Hill AC (eds) *Recognition of Air Pollution Injury to Vegetation: A Pictorial Atlas.* Air Pollution Control Association, Pittsburgh

Box GEP, Tiao GC (1975) Intervention analysis with applications to environmental and economic problems. *Journal of the American Statistical Association* 70:70–79

Box GEP, Jenkins GM (1976) *Time Series Analysis: Forecasting and Control.* Holden-Day, San Francisco

Box GEP, Hunter WG, Hunter JS (1978) *Statistics for Experimenters.* John Wiley & Sons, New York, 653p

Bradley RS, Diaz HF, Eischeid JK, Jones PD, Kelly PM, Goodess CM (1987) Precipitation fluctuations over Northern Hemisphere land areas since the mid-19th century. *Science* 237:171–175

Bunce HWF (1979) Fluoride emissions and forest growth. *Journal of the Air Pollution Control Association* 29:642–643

Bunce HWF (1984) Fluoride emissions and forest survival, growth and regeneration. *Environmental Pollution (Series A)* 35:169–188

Carlson CE (1974) *Sulfur Damage to Douglas-fir Near a Pulp and Paper Mill in Western Montana.* USDA Forest Service, Division State and Private Forestry Publication Number 74-13

Carlson CE, Dewey JE (1972) *Environmental Pollution by Fluorides in Flathead National Forest and Glacier National Park.* US Department of Agriculture Forest Service, Northern Region Headquarters, Forest Insect and Disease Branch, Missoula

Carlson CE, Hammer WP (1975) Impact of fluorides and insects on radial growth of lodgepole pine near an aluminum smelter in northwestern Montana. Abstract from: *Proceedings of the Montana Academy of Sciences* 35:9

Chang I (1982) *Outliers in Time Series.* PhD dissertation. University of Wisconsin, Madison

Cline SP, Burkman WG (1989) The role of quality assurance in ecological research programs. In: Bucher JB, Bucher-Wallin I (eds) *Air Pollution and Forest Decline: Proceedings of the 14th International Meeting for Specialists in Air Pollution Effects on Forest Ecosystems.* IUFRO P2.05, Interlaken, Switzerland, 2–8 October, 1988, Birmensdorf, 1989, pp 361–365

Cline SP, Burkman WG, Geron CD (1989) Use of quality control procedures to assess errors in measuring forest canopy condition. In: Olson RK, Lefohn AS (eds) *Effects of Air Pollution on Western Forests.* Transactions Series, No. 16, Air and Waste Management Association, Pittsburgh, pp 379–387

Conkey LE (1988) Decline in old-growth red spruce in western Maine: An analysis of wood density and climate. *Canadian Journal of Forestry Research* 18:1063–1068

Cook ER (1985) *A Time Series Analysis Approach to Tree-Ring Standardization.* PhD dissertation. University of Arizona, Tucson

Cook ER (1987a) The use and limitations of dendrochronology in studying effects of air pollution on forests. In: Hutchinson TC, Meema KM (eds) *Effects of Atmospheric Pollutants on Forests, Wetlands, and Agricultural Ecosystems.* Springer-Verlag, Berlin, pp 277–290

Cook ER (1987b) The decomposition of tree-ring series for environmental studies. *Tree-Ring Bulletin* 47:37–59

Cook ER, Johnson AH, Blasing TJ (1987) Forest decline: Modeling the effect of climate in tree rings. *Tree Physiology* 3:27–40

Cook ER, Innes JL (1989) Tree-ring analysis as an aid to evaluating the effects of air pollution on tree growth. In: Woodwell GM (chairman) *Biologic Markers of Air-Pollution Stress and Damage in Forests.* National Academy Press, Washington, DC, pp 157–168

Cook ER, Jacoby GC (1977) Tree-ring-drought relationships in the Hudson Valley, New York. *Science* 198:399–401

Cook ER, Kairiukstis LA (eds) (1990) *Methods of Dendrochronology: Applications in the Environmental Sciences.* Kluwer Academic Publishers, Dordrecht

Cook ER, Peters K (1981) The smoothing spline: A new approach to standardizing forest interior tree-ring width series for dendroclimatic studies. *Tree-Ring Bulletin* 41:45–53

Cook ER, Johnson AH (1989) Climate change and forest decline: A review of the red spruce case. *Water, Air and Soil Pollution* 48:127–140

Coyne PI, Bingham GE (1981) Comparative ozone dose response of gas exchange in a ponderosa pine stand exposed to long-term fumigations. *Journal of the Air Pollution Control Association* 31:38–41

Coyne PI, Bingham GE (1982) Variation in photosynthesis and stomatal conductance in an ozone-stressed ponderosa pine stand: Light response. *Forest Science* 28:27

Davis DD, Wilhour RG (1976) *Susceptibility of Woody Plants to Sulfur Dioxide and Photochemical Oxidants.* EPA Ecological Research Series, EPA-600/3-76-102

Downing D, McLaughlin SB (1987) Intervention detection—a systematic technique for examining shifts in radial growth rates of forest trees. In: Jacoby GC, Hornbeck JW (eds) *Proceedings of the International Symposium on Ecological Aspects of Tree-Ring Analysis.* United States Department of Energy, Washington, DC, pp 543–554

Duriscoe DM (1987a) *Evaluation of Ozone Injury to Selected Tree Species in the Rincon Mountains of Arizona, 1985 Survey Results.* A report by Holcomb Research Institute, Butler University. USDI National Park Service, Air Quality Division, Denver, CO, 138p

Duriscoe DM (1987b) *Evaluation of Ozone Injury to Selected Tree Species in Sequoia and Kings Canyon National Parks, 1985 Survey Results.* A report by Holcomb Research Institute, Butler University. USDI National Park Service, Air Quality Division, Denver, CO, 228p

Duriscoe DM (1990) *Evaluation of Ozone Injury to Selected Tree Species in Sequoia and Kings Canyon National Parks, 1986 Survey Results.* Holcomb Research Institute, Butler University, Indianapolis. USDI National Park Service, Air Qulaity Division, Denver, CO

Duriscoe DM (1991) *Methods for Sampling of Pinus ponderosa and Pinus jeffreyi for the Evaluation of Oxidant-induced Foliar Injury,* Final Report Contract #CX-0001-4-0058, Work Assignment #14, submitted to USDI National Park Service, Air Quality Division, Denver, CO

Duriscoe DM, Stitt SCF (1990) *Sampling Design and Sample Acquisition for Elemental Baseline Study in North Cascades National Park Complex.* Final Report Contract #CX-0001-4-0058, Work Assignment #24, submitted to USDI National Park Service, Air Quality Division, Denver, CO

Duriscoe DM, Stolte KW (1989) Photochemical oxidant injury to ponderosa pine (*Pinus ponderosa* Laws.) and Jeffrey pine (*Pinus jeffreyi* Grev. and Balf.) in the national parks of the Sierra Nevada of California. In: Olson RK, Lefohn AS (eds) *Effects of Air Pollution on Western Forests.* Transactions Series, No. 16, Air and Waste Management Association, Pittsburgh, pp 261–278

Eriksson M (1989) *Integrating Forest Growth and Dendrochronological Methodologies.* PhD dissertation, University of Minnesota, St. Paul

Evans LS, Miller PR (1972a) Ozone damage to ponderosa pine: A histological and histochemical appraisal. *American Journal of Botany* 59:297–304

Evans LS, Miller PR (1972b) Comparative needle anatomy and relative ozone sensitivity of four pine species. *Canadian Journal of Botany* 50:1067–1071

Ewell DM, Mazzu LC, Duriscoe DM (1989) Specific leaf weight and other characteristics of ponderosa pine as related to visible ozone injury. In: Olson RK, Lefohn AS (eds) *Effects of Air Pollution on Western Forests.* Transactions Series, No. 16, Air and Waste Management Association, Pittsburgh, pp 411–418

Ewers FW, Schmidt R (1982) Longevity of needle fascicles of *Pinus longaeva* and other north American pines. *Oecologia* 51:107–115

Federer CA, Triton LM, Hornbeck JW, Smith RB (1989) Physiologically based dendroclimate models for effects of weather on red spruce basal-area growth. *Agricultural and Forest Meteorology* 46:159–172

Fox CA, Kincaid WB, Nash TH, Young DL, Fritts HC (1986) Tree-ring variation in western larch (*Larix occidentalis*) exposed to sulfur dioxide emissions. *Canadian Journal of Forest Research* 16:283–292

Fritts HC (1976) *Tree Rings and Climate*. Academic Press, London

Fritts HC, Mosimann JE, Bottorff CP (1969) A revised computer program for standardizing tree-ring series. *Tree-Ring Bulletin* 29:15–20

Gilman DL, Fuglister FJ, Mitchell JM (1963) On the power spectrum of "red noise." *Journal of the Atmospheric Sciences* 20:182–184

Gordon CC (1974) *Environmental Effects of Fluoride: Glacier National Park and Vicinity, 1974*. Final Report to the US Environmental Protection Agency—Region VIII. EPA-908/1-74-001

Grulke NE, Miller PR, Wilborn RD, Hahn S (1989) Photosynthetic response of giant sequoia seedlings and rooted branchlets of mature foliage to ozone fumigation. In: Olson RK, Lefohn AS (eds) *Effects of Air Pollution on Western Forests*. Transactions Series, No. 16, Air and Waste Management Association, Pittsburgh, pp 429–441

Gumpertz ML, Tingey DT, Hogsett WE (1982) Precision and accuracy of visual foliar injury assessments. *Journal of Environmental Quality* 11:549–553

Haggett P, Cliff AD, Frey A (1977) *Locational Methods*. Edward Arnold, London

Harvey AC (1984) A unified view of statistical forecasting procedures. *Journal of Forecasting* 3:245–275

Heck WW, Dunning JA, Hindawi CJ (1965) Interactions of environmental factors on the sensitivity of plants to air pollution. *Journal of the Air Pollution Control Association* 15: 511–515

Hill AC, Heggestad HE, Linzon SN (1970) Ozone. In: Jacobson JS, Hill AC (eds) *Recognition of Air Pollution Injury to Vegetation: A Pictorial Atlas*. Air Pollution Control Association, Pittsburgh, pp B1–B22

Hornbeck JW, Smith RB (1985) Documentation of red spruce growth decline. *Canadian Journal of Forest Research* 15:1199–1201

Horsfall JG, Barratt RW (1945) An improved grading system for measuring plant disease. *Phytopathology* 35:655

Horsfall JG, Cowling EB (1978) Pathometry: The measurement of plant disease. In: *Plant Disease: An Advanced Treatise. Vol. 2: How Disease Develops in Populations.* Academic Press, New York, pp 119–136

Hughes MK, Kelly PM, Pilcher JR, LaMarche JR (eds) (1982) *Climate from Tree Rings.* Cambridge University Press, Cambridge

Husar RB (1989) Air pollutant distribution and trends. In: Woodwell GM (chairman), *Biologic Markers of Air-Pollution Stress and Damage in Forests.* National Academy Press, Washington, DC, pp 29–46

Hutchinson TC, Whitby LM (1976) The effects of acid rainfall and heavy metal particulates on a boreal forest ecosystem near the Sudbury smelting region of Canada. In: Dochinger LS, Seliga TA (eds) *Proceedings of the First International Symposium on Acidic Precipitation and the Forest Ecosystem.* USDA Forest Service General Technical Report No. NE-23, Upper Darby, PA, pp 745–765

Huttunen S (1984) Interactions of disease and other stress factors with atmospheric pollution. In: Treshow M (ed) *Air Pollution and Plant Life.* Wiley, Chichester, pp 321–356

Hyink DM, Zedacker SM (1987) Stand dynamics and the evaluation of forest decline. *Tree Physiology* 3:17–26

Innes JL (1988) Forest health surveys: A critique. *Environmental Pollution* 54:1–15

Jackson LL, Gough LP (1989) The use of stable sulfur isotope ratios in air pollution studies: An ecosystem approach in south Florida. In: Rundel PW, Ehleringer JR, Nagey KA (eds) *Stable Isotopes in Ecological Research,* Ecological Studies Volume 68. Springer-Verlag, New York, pp 471–490

Jacobson JS, Hill AC (1970) *Recognition of Air Pollution Injury to Vegetation: A Pictorial Atlas.* Air Pollution Control Association, Pittsburgh

James RL, Staley JM (1980) *Photochemical Air Pollution Damage Survey of Ponderosa Pine Within and Adjacent to Denver, Colorado: A Preliminary Report.* USDA Forest Service Forest Insect and Disease Management, Biological Evaluation R2-80-6, Lakewood, CO

Johnson AH, Cook ER, Siccama TG (1988) Climate and red spruce growth and decline in the northern Appalachians. *Proceedings of the National Academy of Sciences* 85:5369–5373

Johnson AH, McLaughlin SB (1986) The nature and timing of the deterioration of red spruce in the northern Appalachian Mountains. In: *National Research Council, Acid Deposition Long-Term Trends*. National Academy Press, Washington, DC, pp 200–230

Jones PD, Raper SCB, Bradley RS, Diaz HF, Kelly PM, Wigley TML (1986) Northern hemisphere surface air temperature variations, 1851–1984. *Journal of Climate and Applied Meteorology* 25:161–179

Kincaid WB (1987) Dendrochronological analysis of ambient sulfur dioxide effects in western larch. In: Jacoby GC, Hornbeck JW (eds) *Proceedings of the International Symposium on Ecological Aspects of Tree-Ring Analysis*. United States Department of Energy, Washington, DC, pp 410–416

Krouse HR (1974) Sulphur isotope abundance elucidates uptake of atmospheric sulphur emissions by vegetation. *Nature* 265:45–46

Larsh RN, Miller PR, Wert SL (1970) Aerial photography to detect and evalute air pollution damaged ponderosa pine. *Journal of the Air Pollution Control Association* 20:289–292

LeBlanc DC, Raynal DJ, White EH (1987a) Acidic deposition and tree growth: I. The use of stem analysis to study historical growth patterns. *Journal of Environmental Quality* 16:325–333

LeBlanc DC, Raynal DJ, White EH (1987b) Acidic deposition and tree growth: II. Assessing the role of climate in recent growth declines. *Journal of Environmental Quality* 16:334–340

Lepp NW (ed) (1981a) *Effect of Heavy Metal Pollution on Plants. Volume 1. Effects of Trace Metals on Plant Function*. Pollution Monitoring Series, Applied Science Publishers, London

Lepp NW (ed) (1981b) *Effect of Heavy Metal Pollution on Plants. Volume 2. Metals in the Environment*. Pollution Monitoring Series, Applied Science Publishers, London

MacKenzie JJ, El-Ashry MT (eds) (1989) *Air Pollution's Toll on Forests and Crops*. Yale University Press, New Haven

McClenahen JR, Dochinger LS (1985) Tree ring response of white oak to climate and air pollution near the Ohio River Valley. *Journal of Environmental Quality* 14:274–280

McLaughlin SB, Downing DJ, Blasing TJ, Cook ER, Adams HS (1987) An analysis of climate and competition as contributors to decline of red spruce in high elevation Appalachian forests of the eastern United States. *Oecologia* 72:487–501

Miller PR (1973) Oxidant-induced community change in a mixed conifer forest. In: Naegele JA (ed) *Air Pollution Damage to Vegetation.* American Chemical Society, Washington, DC, pp 101–117

Miller PR (ed) (1977) *Photochemical Oxidant Air Pollutant Effects on a Mixed Conifer Forest Ecosystem.* Annual Progress Report, 1975–1976, EPA-600/3-77-104, US Environmental Protection Agency

Miller PR (1989) Biomarkers for defining air pollution effects in Western coniferous forests. In: Woodwell GM (chairman) *Biologic Markers of Air Pollution Stress and Damage in Forests.* National Academy Press, Washington, DC, pp 111–118

Miller PR, Evans LS (1974) Histopathology of oxidant injury and winter fleck injury on needles of western pines. *Phytopathology* 64:801–806

Miller PR, Millecan AA (1971) Extent of oxidant air pollution damage to some pines and other conifers in California. *Plant Disease Reporter* 55:555–559

Miller PR, Van Doren RE (1982) Ponderosa and Jeffrey pine foliage retention indicates ozone dose response. In: Conrad CE, Oechel WC (eds) *Proceedings of the Symposium on Dynamics and Management of Mediterranean-type Ecosystems.* USDA Forest Service General Technical Report PSW-58, Berkeley, CA

Miller PR, Parmeter JR Jr, Taylor OC, Cardiff EA (1963) Ozone injury to the foliage of *Pinus ponderosa. Phytopathology* 53:1072–1076

Miller PR, Taylor OC, Wilhour RG (1982) *Oxidant Air Pollution Effects on a Western Coniferous Forest Ecosystem.* Environmental Protection Agency, Environmental Research Brief EPA-600/D-82-276

Miller PR, Longbotham GJ, Longbotham CR (1983) Sensitivity of selected western conifers to ozone. *Plant Disease* 67:1113–1115

Miller PR, McBride JR, Schilling SL, Gomez AP (1989) Trend of ozone damage to conifer forests between 1974 and 1988 in the San Bernardino Mountains of southern California. In: Olson RK, Lefohn AS (eds) *Effects of Air Pollution on Western Forests.* Transactions Series, No. 16, Air and Waste Management Association, Pittsburgh, pp 309–324

Muir PS, Armentano TV (1987) *Evaluating Oxidant-induced Injury to Foliage of Pinus ponderosa (west) and Pinus strobus (east): A Comparison of Methods.* Holcomb Research Institute, Butler University, Interim Report to the USDI National Park Service, Air Quality Division, Denver, CO

Muir PS, McCune B (1987) Index construction for foliar symptoms of air pollution injury. *Plant Disease* 71(6): 558–565

Munz PA, Keck DD (1973) *A California Flora.* University of California Press

Nash TH, Fritts HC, Stokes MA (1975) A technique for examining non-climatic variation in widths of tree rings with special reference to air pollution. *Tree-Ring Bulletin* 35:15–24

National Research Council (1989) *Biologic Markers of Air-Pollution Stress and Damage in Forests.* National Academy Press, Washington, DC

National Research Council of Canada (1939) *Effect of Sulphur Dioxide on Vegetation.* The Associate Committee on Trail Smelter Smoke, Ottawa, Canada, N.R.C. No. 815

Nriagu JO (ed) (1978) *Sulfur in the Environment. Part II: Ecological Impacts.* Wiley, New York

Parmeter JR, Miller PR (1968) Studies relating to the cause of decline and death of ponderosa pine in southern California. *Plant Disease Reporter* 52:707–711

Parmeter JR, Bega RV, Neff T (1962) A chlorotic decline of ponderosa pine in southern California. *Plant Disease Reporter* 46: 269–273

Patterson MT, Rundel PW (1989) Seasonal physiological responses of ozone stressed jeffrey pine in Sequoia National Park, California. In: Olson RK, Lefohn AS (eds) *Effects of Air Pollution on Western Forests.* Transactions Series, No. 16, Air and Waste Management Association, Pittsburgh, pp 419–428

Peterson DL, Arbaugh MJ, Wakefield VA, Miller PR (1987) Evidence of growth reduction in ozone-injured Jeffrey pine (*Pinus jeffreyi* Grev. and Balf.) in Sequoia and Kings Canyon National Parks. *Journal of the Air Pollution Control Association* 37:906–912

Peterson DL, Arbaugh MJ, Robinson LJ (1989) Ozone injury and growth trends of ponderosa pine in the Sierra Nevada. In: Olson RK, Lefohn AS (eds) *Effects of Air Pollution on Western Forests,* Transactions Series, No. 16, Air and Waste Management Association, Pittsburgh, pp 293–307

Phipps RL, Whiton JC (1988) Decline in long-term growth trends of white oak. *Canadian Journal of Forest Research* 18:24–32

Pronos J, Vogler DR, Smith RS (1978) *An Evaluation of Ozone Injury to Pines in the Southern Sierra Nevada.* USDA Forest Service, Pacific Southwest Region, Forest Pest Management Report No. 78-1, San Francisco

Puckett LJ (1982) Acid rain, air pollution, and tree growth in southeastern New York. *Journal of Environmental Quality* 11:376–381

Pye JM (1988) Impact of ozone on the growth and yield of trees: A review. *Journal of Environmental Quality* 17:347–360

Rice PM, Boldi RA, Carlson CE, Tourangeau PC, Gordon CC (1983) Sensitivity of *Pinus ponderosa* foliage to airborne phytotoxins: Use in biomonitoring. *Canadian Journal of Forest Research* 13:1083–1091

Richards BL, Taylor OC, Edmunds GF (1968) Ozone needle mottle of pine in southern California. *Journal of the Air Pollution Control Association* 18:73–77

Rock BN, Vogelmann JE, Williams DL, Vogelmann AF, Hoshizaki T (1986) Remote detection of forest damage. *BioScience* 36:439–445

Rock BN, Hoshizaki T, Miller JR (1988) Comparison of in situ and airborne spectral measurements of the blue shift associated with forest decline. *Remote Sensing of Environment* 24:109–127

Rock BN, Vogelmann JE, Defeo NF (1989) The use of remote sensing for the study of air pollution effects in forests. In: Woodwell GM (chairman) *Biologic Markers of Air-Pollution Stress and Damage in Forests*. National Academy Press, Washington, DC, pp 183–194

Scholz FH, Gregorius R, Rudin D (1989) Genetic effects of air pollutants in forest tree populations. *Proceedings of the Joint Meeting of the IUFRO Working Parties in Grobhansdorf*. August 3–7, 1987

Schweingruber FH (1986) Abrupt growth changes in conifers. *IAWA Bulletin* 7:277–283

Severson RC, Crock JG, Gough LP (1990) *An Assessment of the Geochemical Variability for Plants and Soils and an Evaluation of Industrial Emissions near the Kenai National Wildlife Refuge, Alaska*. US Geological Survey Open File Report 90-306, Denver, CO

Smith WH (1990) *Air Pollution and Forests*. Springer-Verlag, New York

Snedecor GW, Cochran WG (1967) *Statistical Methods, 7th edition*. Iowa State University Press, 50p

Solberg RA, Adams DF, Ferchau HA (1957) Some effects of hydrogen fluoride on the internal structure of *Pinus ponderosa* needles. *Proceedings of Natural Air Pollution Symposium* 3:164–176

Spotts RA (1969) *Environmental Factors of Pine Tip Burn*. Masters thesis. Colorado State University, Fort Collins, CO

Stolte KW (1982) *Effects of Ozone on Chaparral Species in the South Coast Air Basin.* Masters thesis. University of California at Riverside, Riverside, CA

Stolte KW, Bennett JP (1984) *Standardized Procedures for Establishing and Evaluating Pollution Injury on Pines in Permanent Plots.* USDI National Park Service, Air Quality Division, Denver, Colorado

Stolte KW, Miller PR (eds) (1991) *Management Recommendations and Specifications for Plot Design and Sampling Methods to Monitor Long-term Effects of Ozone on Western Coniferous Forests.* Proceedings of a Pine Plot Workshop, March 14–15, 1989, Riverside, California. US Forest Service, Pacific Southwest Forest and Range Experiment Station, Riverside, CA, 88p

Taylor JK (1985) What is quality assurance? In: Taylor JK, Stanley TW (eds) *Quality Assurance for Environmental Measurements.* ASTM STP No. 867, pp 5–11

Taylor JK (1987) *Quality Assurance of Chemical Measurements.* Lewis Publishers, Inc, Chelsea, MI

Taylor PJ (1977) *Quantitative Methods in Geography.* Houghton Mifflin, Boston

Taylor OC (1974) Air pollution effects influenced by plant-environment interaction. In: WM Duggar (ed) *Air Pollution Effects on Plant Growth,* A. Chem. Coc. Symp Ser. 3, Washington, DC, pp 1–7

Taylor OC (1980) *Photochemical Oxidant Air Pollution Effects on a Mixed Conifer Forest Ecosystem.* Final report. USEPA, Ecological Research Series, EPA-600/3-80-002

Taylor OC, Miller PR, Page AL, Lund LJ (1986) *Effects of Ozone and Sulfur Dioxide Mixtures on Forest Vegetation of the Southern Sierra Nevada.* Final Report to California Air Resources Board. ARB Contract No. AO-135-33, Sacramento, CA

Thompson MA (1981) Tree rings and air pollution: A case study of *Pinus monophylla* growing in east-central Nevada. *Environmental Pollution, Series A* 26:251–266

Tingey DT, Wilhour RG, Standley C (1976) The effect of chronic ozone exposures on the metabolic content of ponderosa pine seedlings. *Forest Science* 22:234–241

Tingey DT, Wilhour RG, Taylor OC (1979) The measurement of plant response. In: Heck WW, Krupa SV, Linzon SN (eds) *Handbook of Methodology for the Assessment of Air Pollution Effects on Vegetation.* Air Pollution Control Association, Pittsburgh

Treshow M (1970) Ozone damage to plants. *Environmental Pollution* (1):155–161

Treshow M, Anderson FK (1989) *Plant Stress from Air Pollution.* Wiley, New York

Treshow M, Pack MR (1970) Fluoride. In: Jacobson JS, Hill AC (eds) *Recognition of Air Pollution Injury to Vegetation: A Pictorial Atlas.* Air Pollution Control Association, Pittsburgh

Ustin SL, Curtiss B, Martens SN, Vanderbilt VC (1989) Early detection of air pollution injury to coniferous forests using remote sensing. In: Olson RK, Lefohn AS (eds) *Effects of Air Pollution on Western Forests.* Transactions Series, No. 16, Air and Waste Management Association, Pittsburgh, pp 351–378

Van Deusen PC (1987) Testing for stand dynamics effects in red spruce growth trends. *Canadian Journal of Forest Research* 17:1487–1495

Van Deusen PC (1989) A model-based approach to tree ring analysis. *Biometrics* 45:763–779

Van Deusen PC (1990) Evaluating time-dependent tree ring and climate relationships. *Journal of Environmental Quality* 19:481–488

Van Hook C (1974) Fluoride distribution in the Silverbow, Montana, area. *Fluoride* 7:181–199

Visser H, Molenaar J (1986) *Time-dependent Responses of Trees to Weather Variations: An Application of the Kalman Filter.* Report 50385-MOA 86-3041, N.V. KEMA, Arnhem, the Netherlands

Visser H, Molenaar J (1988) Kalman filter analysis in dendroclimatology. *Biometrics* 44:929–940

Vogler DR (1982) *Ozone Injury and Height Growth of Planted Ponderosa Pines on the Sequoia National Forest.* USDA Forest Service, Pacific Southwest Region, Forest Pest Management Report No. 82-18, San Francisco

Walters JW (1978) *A Guide to Forest Diseases of Southwestern Conifers.* USDA Forest Service, Southwestern Region, Forest Insect and Disease Management, R3 78-9, Albuquerque, NM

Warren WG (1980) On removing the growth trend from dendrochronological data. *Tree-Ring Bulletin* 40:35–44

Wellburn A (1989) *Air Pollution and Acid Rain: The Biological Impact.* Longman Group UK Limited

Whitby LM, Hutchinson TC (1974) Heavy-metal pollution in the Sudbury mining and smelting region of Canada. II: Soil toxicity tests. *Environmental Conservation* 1:191–200

Williams WT, Williams JA (1986) Effects of oxidant air pollution on needle health and annual-ring width in a ponderosa pine forest. *Environmental Conservation* 13:229–233

Section II

Regional Studies
of Forest Growth
and Condition

8

Old-Growth Douglas-Fir
in Western Washington

L. B. Brubaker, S. Vega-Gonzalez, E. D. Ford,
C. A. Ribic, C. J. Earle, G. Segura

Dense conifer forests dominated by Douglas-fir (*Pseudotsuga menziesii* [Mirb.] Franco) cover lowlands and mid-elevations of the western Pacific Northwest. Old-growth stands in these forests are renowned for their massive trees and complex vertical structure (Waring and Franklin 1979, Franklin 1988). The Pacific Northwest is a leading timber producing region within the western conifer forest type (Ulrich 1984), and the economic importance of these forests is projected to continue in the future (Kulp 1986).

The Seattle-Tacoma-Everett metropolitan complex in the central Puget Lowland is the largest population and industrial center in the Pacific Northwest. Prevailing winds transport atmospheric emissions from this area to forest lands on the western slopes of the Cascade Mountains. Recent monitoring in Cascade forests east and southeast of the greater Seattle area has recorded ozone concentrations (> 80 ppb) that are known to cause visible damage in laboratory exposures (Böhm 1989).

Remnants of extensive, old-growth Douglas-fir forests remain at scattered locations in the western Cascade and Olympic Mountains, providing a rare opportunity to describe the characteristics of tree-growth variations during long periods known to be free of anthropogenic influences such as combustion of fossil fuels. This information can be used to determine whether recent growth trends are unusual and potentially related to air pollution.

This chapter describes temporal and spatial patterns in ring-width variations of Douglas-fir in stands of 400 to 600 years in age. Recent growth variations are evaluated in light of longer-term growth patterns, recent climate patterns, and regional pollution sources.

Old-growth Douglas-fir Forests

Regional Setting

Douglas-fir is the most abundant and widely distributed tree species in western regions of the Pacific Northwest, occupying extensive areas of the western slopes of the Cascade Range, the Olympic Mountains, and the Coastal Ranges of Oregon and British Columbia (see Chapter 1). The climate of this region is characterized by wet, mild winters and cool, relatively dry summers (see Chapter 2). Precipitation ranges from 800–3000 mm with 75–80% occurring between October and March. Mean annual temperatures are 8–10°C. Minimum winter temperatures rarely fall below -4°C and the frost free season is 5–10 months long. Soils of the Olympic Peninsula, the North Cascades and northern Puget Trough were formed primarily in late-Wisconsin glacial drift and Holocene tephras. Soils in the Central Cascades and southern Puget Trough have originated in parent materials of glacial drift, tephra, basalt, andesitic, and mixed lithologies, and span a range of ages from a few thousand to more than a million years (Franklin and Dyrness 1973, Ugolini 1990 personal communication). Soils in the glaciated terrain are moderately deep, acidic, and well-drained; those south of the Wisconsin glacial moraine are deep, clay-rich, and leached except in valley bottoms (see Chapter 4).

Composition and Successional Patterns

Douglas-fir, an early seral species in this region, typically occurs in association with more-shade-tolerant species such as western hemlock (*Tsuga heterophylla* [Raf.] Sarg.) and western redcedar (*Thuja plicata* Donn) at low elevations and western hemlock and Pacific silver fir (*Abies amabilis* [Dougl.] Forbes) near its upper elevational limits (Franklin and Dyrness 1973). Douglas-fir is a long-lived species, persisting as a dominant component of many stands for up to 1000 years (Hemstrom and Franklin 1982). Because stand-replacing disturbances normally occur at intervals shorter than Douglas-fir's life span (Hemstrom and Franklin 1982, Agee and Flewelling 1983, Agee 1991), Douglas-fir is seldom completely replaced by shade-tolerant species.

Douglas-fir is the most abundant colonist on most sites after severe disturbances, although western hemlock is also a common colonist on moist and middle-elevation sites. Recruitment of new individuals ceases at crown closure, which generally occurs within 40 years after disturbance (Franklin and Hemstrom 1981). However, age structures in some old-growth forests suggest that re-establishment may continue beyond

250 years after catastrophic disturbances (Franklin and Hemstrom 1981). Virtually all stands in the Puget Sound region achieve crown closure, and enter prolonged periods (100 years or more) of intense competition and crown differentiation. Tree mortality increases due to competition and stress-induced susceptibility to other factors such as insects and diseases.

Tree mortality during stand development increases the structural diversity of forest stands (Oliver 1981). Depending on the size and number of trees involved in mortality events, canopy openings (gaps) of different dimensions and shapes are formed (Spies and Franklin 1989). Small gaps stimulate the growth of neighboring individuals, whereas large gaps also allow the recruitment of new stems. New individuals entering stands are typically shade tolerant species such as western hemlock, western redcedar and Pacific silver-fir.

The density of Douglas-fir declines substantially after 200–300 years, but its basal area stays relatively constant because surviving individuals continue to increase in diameter (Agee 1981, Huff 1984, Spies and Franklin 1991). The slower, but continued death of canopy dominants creates gaps that allow the recruitment of shade tolerant species into the stand, increasing the range of tree sizes in the stand (Spies and Franklin 1991). This stage of stand development is termed old-growth (see Chapter 6). In the lowlands of the Puget Sound region, old-growth forests are dominated by Douglas-fir, western hemlock, and western red cedar. At the upper elevation limit of Douglas-fir, Pacific silver fir is common in the understory of old-growth forests.

Disturbance Patterns

Fire and wind are the most important natural disturbances in the Puget Sound region (Franklin 1988). Large windthrows are most common in coastal areas and decrease in importance in interior forests. Fire increases in frequency from coastal to interior areas, and from north to south (Agee 1991).

High intensity windstorms cause catastrophic stand disturbances primarily in Sitka spruce (*Picea sitchensis* [Bong.] Carr.) forests along the Pacific coast. Stand-destroying windstorms are less common in Douglas-fir forests that dominate more interior portions of the Puget Sound region. Here windthrow events typically kill single trees or small groups of trees (Spies and Franklin 1989). The importance of wind as a gap-forming agent in such forests has recently been recognized by studies showing that windthrow can be responsible for up to 40% of the individual tree mortality observed in old-growth Douglas-fir stands (Franklin and DeBell 1988).

Fire is by far the most important stand-level disturbance in forests of the Douglas-fir region (Hemstrom and Franklin 1982). Fires are infrequent, large, and typically kill most trees in the stand (Hemstrom and Franklin 1982, Agee 1991). This contrasts with the higher fire frequencies in ponderosa pine (*Pinus ponderosa* Dougl. ex Loud.) forests of the east slope of the Cascades and in mixed conifer forests of the Rocky Mountains, Sierra Nevada, and southern California (Agee 1981, Franklin 1988, see Chapters 9, 11, and 12). Although catastrophic fires can destroy forest across extensive areas in the Douglas-fir region, the occurrence of a fire in a given stand is strongly influenced by the stand's location in the landscape. Valley bottoms, river drainages, alluvial terraces and north-facing slopes are generally least susceptible to fire, and the oldest stands of Douglas-fir generally occur on these sites (Hemstrom and Franklin 1982).

Research Design

No signs of extensive growth declines have been reported for forests in western Washington. We concentrated our efforts in surveying sites around the industrialized Puget Sound Basin and pollution free Olympic Peninsula using dendrochronological methods.

Sequences of ring widths from old-growth Douglas-fir trees provide an excellent opportunity to characterize tree-growth variations over long periods of time. The primary objective of our work was to use these long records to investigate natural variations in tree growth at different temporal and spatial scales in the Puget Sound Basin. By "natural" we mean variations in tree growth free of anthropogenic influences. Anomalous patterns of recent growth can then be identified.

Because of the great stand ages, tree-ring samples provide growth records well past the stages of juvenile growth and intense self-thinning that strongly influence individual tree growth and complicate analyses of exogenous influences such as pollution (see Chapter 7). Over the period of study, short-term variations (annual) in ring width are most likely the result of annual climatic variations, whereas intermediate-term variations (one to several decades) are due to disturbance, competition and slowly varying climate. In order to depict these components of growth, the linear aggregate model described in Chapter 7 was used as a conceptual framework.

Linear Aggregate Model

According to the linear aggregate model, the radial growth of trees in forest stands can be viewed as a composite of diverse signals (see Chapter 7, Equation 7.2). Ring width, or a mathematical transformation of ring width, can be modeled as the sum of terms representing factors expected to influence tree growth. Such factors include age, endogenous and exogenous disturbances, climate, and anthropogenic effects.

In order to evaluate potential air pollution effects, at least some understanding of the way other factors affect variability in growth is required. If some of these factors can be controlled or measured accurately, the variability due to the remaining factors can be investigated. For example, the temporal characteristics of growth before the onset of pollution can be used as a basis to monitor changes thought to be pollution-induced. Spatial patterns of growth can similarly be examined to infer causal factors. Disturbances, for example, are usually site-specific events, whereas climatic variations should affect growth over large regions. Like climate, pollution damage from urban centers will show large spatial patterns, but we expect these to be geographically related to source areas and wind patterns, especially in complex terrain characteristic of western Washington (see Chapter 3). In the Puget Sound region, weather station records represent the only measured environmental data that can be compared to ring-width series in an investigation of causal factors.

The aggregate model requires a consideration of forest dynamics, climate, and pollution sources to represent tree-growth variations. Characteristics of the Puget Sound region that were particularly important in the development of our research design and choice of model structure are described in the following paragraphs.

Forest dynamics

The closed-canopy, old-growth status of Douglas-fir stands has important implications for our analyses. First, age-related trends, which are difficult to define in young trees, can be readily modeled in our long core records (generally 450 years). Second, the most common exogenous disturbances in these stands, windstorms and fires, are rare events that kill most trees; thus intermediate-term growth variations probably reflect less-catastrophic processes such as slowly changing climate, gap replacement, or pollution. Gap-forming events will affect individual trees or small groups of trees. Responses to climate and pollution, on the other hand, should be expressed by most trees in a stand given the same genetic sensitivity and exposure to pollutants.

Climate

Several investigations have documented the importance of climatic variations on tree growth in forest zones of western Washington (Brubaker 1980, Graumlich and Brubaker 1986, Graumlich 1987, Graumlich et al. 1989, Brubaker et al. 1990). In general, these studies have shown that temperature variation exerts a more important influence on tree growth than does precipitation, although snow depth is the predominant controlling factor near treeline. Because both intermediate-term and short-term climatic variations influence ring-width patterns, it was necessary for us to examine climate-growth relationships at differing temporal frequencies.

Pollution

The major sources of air pollutants in the Puget Sound region are the Seattle-Tacoma-Everett metropolitan complex and point emission sources in Bellingham, Anacortes, and Centralia (Böhm 1989, Edmonds and Basabe 1989). Prevailing airflows across the region typically advect pollutants eastwards toward the Cascade Mountains. Forests in the Central Cascades to the southeast of the Seattle-Tacoma corridor probably experience higher pollution levels in summer than other forests in the region. Northwesterly winds predominate during summer, bringing air from polluted lowlands to the Central Cascades. Available air quality data support this idea, showing higher concentrations of ozone, and solutes in rain, snow, and cloud in areas east and southeast of metropolitan Seattle than in areas to the northeast or west (see Chapter 3, Basabe et al. 1989, Edmonds and Basabe 1989). We consider 1960 as the beginning of potential pollution effects on tree growth because the region experienced its most rapid population growth after that time.

Summary of Approach

In terms of the components of the conceptual model of radial growth, we assume that the major natural influences on tree growth over the period and area of study are: 1) gap replacement events, which affect intermediate-term variations of one to a few trees, 2) short and intermediate-term climatic variations, which should have stand-level as well as regional effects, and, potentially, 3) pollution, which should predominantly affect forests in recent decades southeast of the greater Seattle area. The age-related trends should be easily modeled due to the old-growth status of the stands. Natural exogenous disturbances are not important to our study because such disturbances typically kill most or all trees in a stand, and thus seldom leave records in growth-ring sequences. Our analyses address individual tree, stand, and regional patterns of variations, because variations at different spatial scales can suggest the effect of different causal factors. We describe short and intermediate-term variations, as well as potential causal relationships of these variations to climate.

Data Base

Ring-Width Data

Increment cores were collected from nine sites (Figure 8.1) representing a wide range of pollution levels, temperature, and moisture conditions. Three sites were located in the Olympic Peninsula (Big Quilcene River (BQR), Olympic Road Trail 3116 (ORT), Staircase Loop Trail (SLT)), and are assumed to represent low air pollution concentrations (see Chapter 3); three sites were located along western slopes of the North Cascade Mountains (Deer Creek Pass (DCP), San Juan Hill (SJH), and Annette Lake Trail (ALT)); and three sites were on the western slopes of the Central Cascades (Tahoma Creek Trail (TCT), Carbon River (CAR), and Silver Creek (SIL)). The six Cascades sites represent a range of pollution levels with the most severe conditions in the Mount Rainier region.

All sites are old-growth stands dominated by Douglas-fir older than 450 years with western hemlock or Pacific silver fir. Due to extensive logging in the region, site selection in each area was constrained primarily by the availability of old-growth stands. At each sampling location, two increment cores were collected from each of 20–30 dominant Douglas-fir trees at least 80 m from the stand edge which was typically a clearcut boundary. The sampling sites were generally 5 hectares in area and showed no evidence of major disturbance, such as extensive windthrow or fire. Similarly, none of the sampled trees showed evidence of physical damage to crowns or stems. Site and stand characteristics are presented in Table 8.1.

In the laboratory, all cores were cross-dated and measured according to quality control and quality assurance guidelines (Chapter 7). Only accurately dated ring-width data were entered into the analysis of temporal and spatial patterns of tree growth variations.

Climatic Data

Climatic stations were selected according to two criteria: 1) proximity to core collection sites and 2) length and quality of record (Table 8.1). Climatic data for each station were retrieved from the computerized data

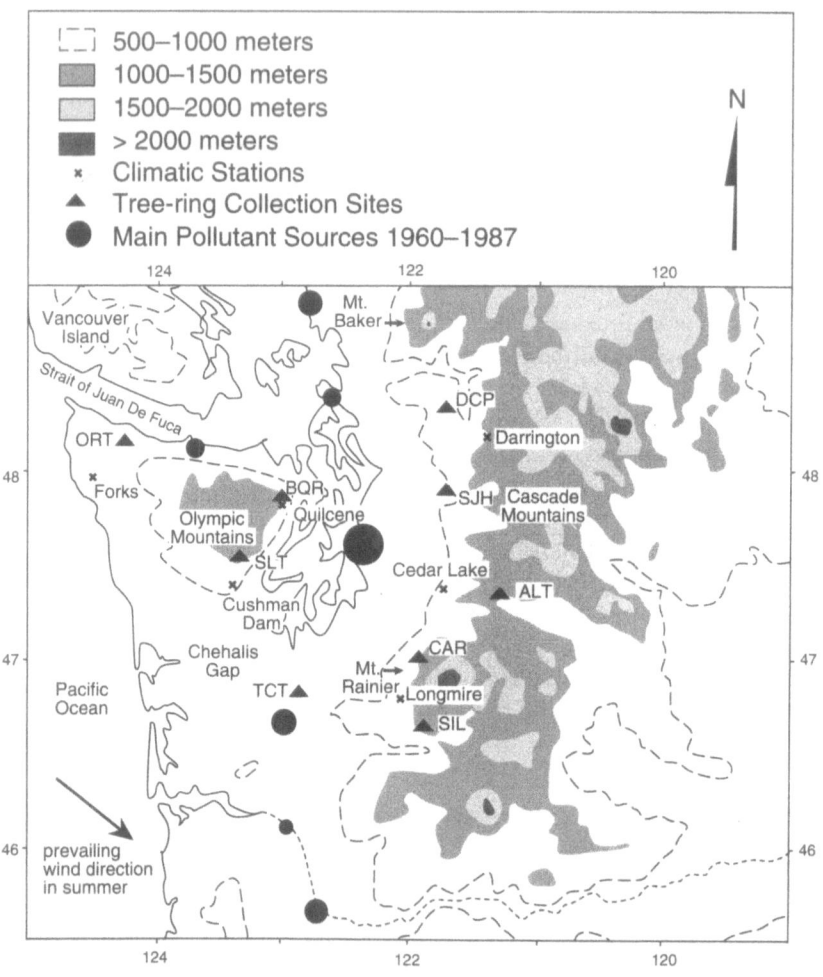

Figure 8.1 Map of western Washington with major geographic features, urban pollution sources, tree-ring collection sites (BQR: Big Quilcene River, ORT: Olympic Road Trail 3116, SLT: Staircase Loop Trail, DCP: Deer Creek Pass, SJH: San Juan Hill, ALT: Annette Lake Trail, TCT: Tahoma Creek Trail, CAR: Carbon River, SIL: Silver Creek), and climatic stations.

Table 8.1 Site locations and associated weather stations used in analyses.

Name	Tree-ring sites Lat	Long	Elev(m)	Weather Stations Name	Lat	Long	Elev(m)	Climatic Analysis Periods
ALT	47:22	121:20	745–850	Cedar Lake	47:25	121:44	475	1931–1987
DCP	48:20	121:43	945–1160	Darrington	48:15	121:36	168	1931–1987
SJH	47:53	121:43	925–1010	Darrington				1931–1987
BQR	47:50	123:02	850–885	Quilcene	47:49	122:55	37	1948–1987
ORT	48:10	124:17	225–310	Forks	47:57	124:22	107	1931–1987
SLT	47:31	123:20	245–380	Cushman	47:22	123:10	7	1931–1987
TCT	46:48	122:53	960–980	Longmire	46:45	121:49	841	1931–1987
SIL	46:38	121:50	850–950	Longmire				1931–1987
CAR	47:00	121:53	601–604	Longmire				1931–1987

base published by the National Climatic Data Center. Daily precipitation and maximum temperature records were retrieved and monthly or seasonal averages computed. Missing values were filled by using the overall monthly mean for the missing datum.

Temporal and Spatial Patterns of Growth Variations

Methods

Temporal variations

The general procedure for analyzing each ring-width sequence consisted of the sequential extraction of the different components as outlined by the linear aggregate model (Figure 8.2). In the context of this study, an alternate, simpler stochastic model for R_{it}, the logarithm of the observed ring width at the time t for the i-th tree is:

$$R_{it} = A_{it} + S_t + T_{it} + E_{it}$$

where A_{it} is an age-related growth component, possibly different for different trees; S_t is a site-specific component of intermediate-term variation that encompasses components C and D2 from Equation 7.2 in Chapter 7 (the exogenous disturbances term); T_{it} is a tree-specific inter-mediate-term trend component (the endogenous disturbances term); and E_{it} is a stochastic term representing short-term tree growth responses to short-term climatic variations, unexplained growth variations unique to individual trees, and the error term.

We used this model to extract the growth components (Figures 8.2 and 8.3). First, we hypothesized an exponential model for age, thus allowing the modeling and extraction of A_{it} from each tree sequence. A common smooth component S_t was then fitted to a series formed by averaging the tree sequences adjusted for age (Figure 8.3b). This series represents the average yearly radial growth of Douglas-fir at each site and, in this context, is called the stand average. After the A_{it} and S_t components were removed from each tree series, a smooth curve representing unique intermediate-term tree variations, T_{it}, was fitted to each individual series. We use the term "smooth" to refer to a series containing no oscillations with frequencies smaller than 60 years. The residuals from the fitting, E_{it}, were then analyzed for autocorrelation (Figure 8.3c). The final series adjusted for autocorrelation are termed the individual prewhitened sequences (Figure 8.3d). The stand averages (representing intermediate and short-term growth variations in common for most trees at a site) and the individual prewhitened sequences (representing short-term growth variations of individual trees) were used to analyze the effects of climate on tree growth where "growth" represents ring width data adjusted for sources of variation defined by the linear model. The details of these analyses can be found in Brubaker et al. (1989).

Spatial analyses

We considered three levels of spatial variation in ring width data: between cores within a tree, between trees within a stand, and between stands across the region. The sampling unit in our study is the tree, and each core sequence represents a replicate measure of the process of interest, namely radial tree growth. Building on the assumption of a multiplicative model, we took the geometric averages of contemporaneous core measurements to represent the tree-sequences. The spatial variability between cores was explored in detail for two of the sites: ORT in the Olympic Mountains and DCP in the North Cascades. For these sites, the temporal analyses described above were carried out for individual cores, with similar results for cores from the same tree.

We investigated individual tree growth variations on the premise that gap-forming events have individualistic effects within stands. An individualistic approach was also warranted because responses to natural and anthropogenic factors may vary among trees within stands owing to genetic and microsite characteristics. Purely statistical considerations suggest that investigations of individual tree variations should be made before stand-level summaries are justified. The importance of this approach is borne out by the data from individual trees, which show considerable differences at intermediate temporal frequencies (Figure 8.4).

After removing autocorrelation from the individual tree-sequences, we computed the cross-correlation coefficients for all pairs of sequences to investigate the spatial structure within stands. Positive correlations among these individual tree series are indirect evidence for the influence of short-term climatic variation on tree growth.

Product-moment correlation coefficients were computed among the stand averages and these were plotted against intersite distance. This analysis reveals spatial patterns of correlation related to distance across the region. Strong statistical correlations among sites would suggest strong climatic controls over growth.

Pre-1960 vs post-1960 comparisons

For each individual tree sequence adjusted for age related effects, the mean for the period from 1934–1960 (low levels of air pollution) was compared with the mean for the period 1961–1987 (higher levels of air

Model:
$$R_{it} = A_{it} + S_t + T_{it} + E_{it}$$

1. Take logarithms of raw ring - width sequences
$$R_{it} = \log(r_{it})$$

2. Estimate and remove age - related term
$$\hat{A}_{it} = k_i + h_i\, t$$

3. Estimate and remove trend common to all trees
$$\hat{S}_t = \text{smooth}\left[\frac{1}{n}\sum_{i=1}^{n}(R_{it} - \hat{A}_{it})\right]$$

4. Estimate and remove tree - specific trends
$$\hat{T}_{it} = \text{smooth}\left[R_{it} - \hat{A}_{it} - \hat{S}_t\right]$$

5. Fit AR process to residuals
$$\hat{E}_{it} = R_{it} - \hat{A}_{it} - \hat{S}_t - \hat{T}_{it}$$

6. Obtain prewhitened residuals to be used in climate modeling
$$\hat{\varepsilon}_{it} = \hat{E}_{it} - a_{i1}\hat{E}_{i(t-1)} - \ldots - a_{ir}\hat{E}_{i(t-r)}$$

The subscript *it* represents the observation time *t* for the *i*-th tree.

R_{it}: logarithm of observed ring width (r_{it})

A_{it}: growth trend related to tree age, size and stage of stand development. The estimate is a straight line fitted to the logarithms, R_{it} with coefficients k_i (y intercept) and h_i (slope), different for each tree.

S_t: "smooth" (see text for definition) trend, representing growth variation common to all trees in the stand (*Signal*). It includes climatic effects, exogeneous disturbances, and possibly, pollution effects.

T_{it}: another smooth trend specific to each tree in the stand. This represents low frequency variation particular to individual trees, i.e. due to processes like competitive interactions among trees, and tree response to gap-forming events.

E_{it}: unexplained year-to-year variability not related to the other signals, particular to individual trees. Stochastic component including measurement error.

e_{it}: Random error. White noise.

The coefficients a_{ir}, r=1, 2, ... r are those corresponding to the fitted autoregressive process of order r, AR(r).

Figure 8.2 Sequence of steps used in ring-width detrending procedure.

pollution) by taking the simple difference of post-1960 mean growth minus the pre-1960 mean. We considered differences greater or less than .05 standard units (growth adjusted for age effects) to be sufficiently different from zero to represent a "change" in growth. The .05 value roughly corresponds to one standard error for the post-1960 minus pre-1960 differences.

Results

General
Figure 8.5 shows comparative boxplots for the variation of ring widths about the corresponding median value for each site. Table 8.2 presents a summary of site statistics. In general, the variability and absolute size of ring width in our samples increased from the North Cascades group (DCP, SJH, ALT) to the Olympic group (ORT, BQR, SLT), to the Central Cascades group (CAR, TCT, SIL). CAR showed the most variability in ring width, but also had the smallest sample size.

Tree-specific intermediate-term and short-term variations
Intermediate-term trends constituted an important component of the variance in individual tree growth at all sites, suggesting that competition and gap-forming processes are important within the stands (Figure 8.4). Strong dependencies exist between each year's growth and growth of the preceding year, i.e., autocorrelation accounts for a large proportion of ring-width variability (Table 8.2). The Olympic and Central Cascades sites are represented by first order autoregressive models, while the North Cascades sites are represented by second and third order models, indicating longer temporal dependencies. The occurrence of autoregressive components is common in tree ring data , and is usually assumed to indicate that biological processes cause growth in one year to influence growth in subsequent years (Fritts 1976, Meko 1981, Hughes et al. 1982, Monserud 1986). For example, a year that is beneficial for growth increases the size or capacity of the foliage and the root system, with benefits continuing in the following year.

Stand-level intermediate term variation
Figure 8.6 presents the stand averages of age-detrended tree-ring sequences. Each site shows a generally similar pattern of increasing growth in the 20th-century that persists through at least the mid-20th century, a decline and subsequent rise after ca. 1950, and then a continued ascent (e.g., ALT, SIL) or a period of relative stability. For several sites (ALT and Central Cascades sites), the growth increases during the last 100 years

Olympic Road Trail: TREE 081

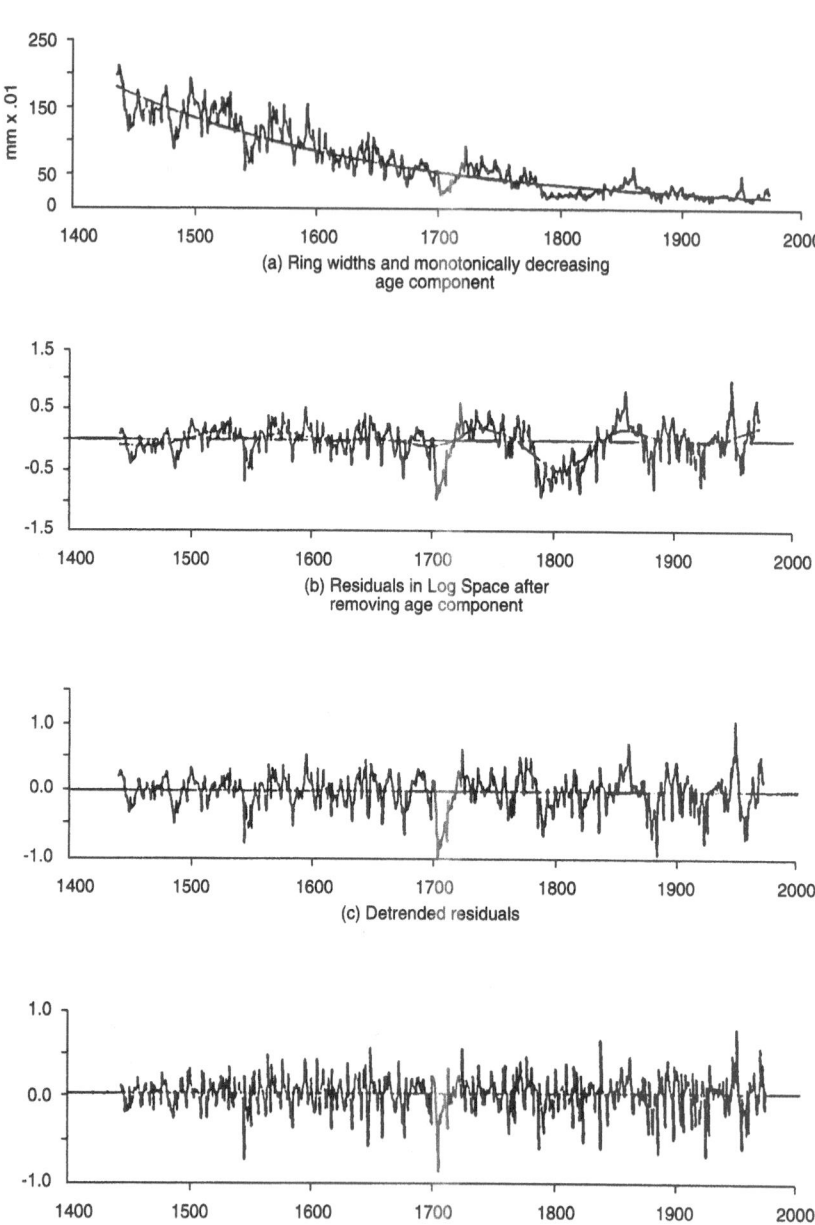

(a) Ring widths and monotonically decreasing
age component

(b) Residuals in Log Space after
removing age component

(c) Detrended residuals

(d) Residuals after fitting and removing
an AR(3) process

Figure 8.3 Examples of steps in the ring-width detrending procedure. Horizontal axis shows calendar years.

ORT example tree-specific components

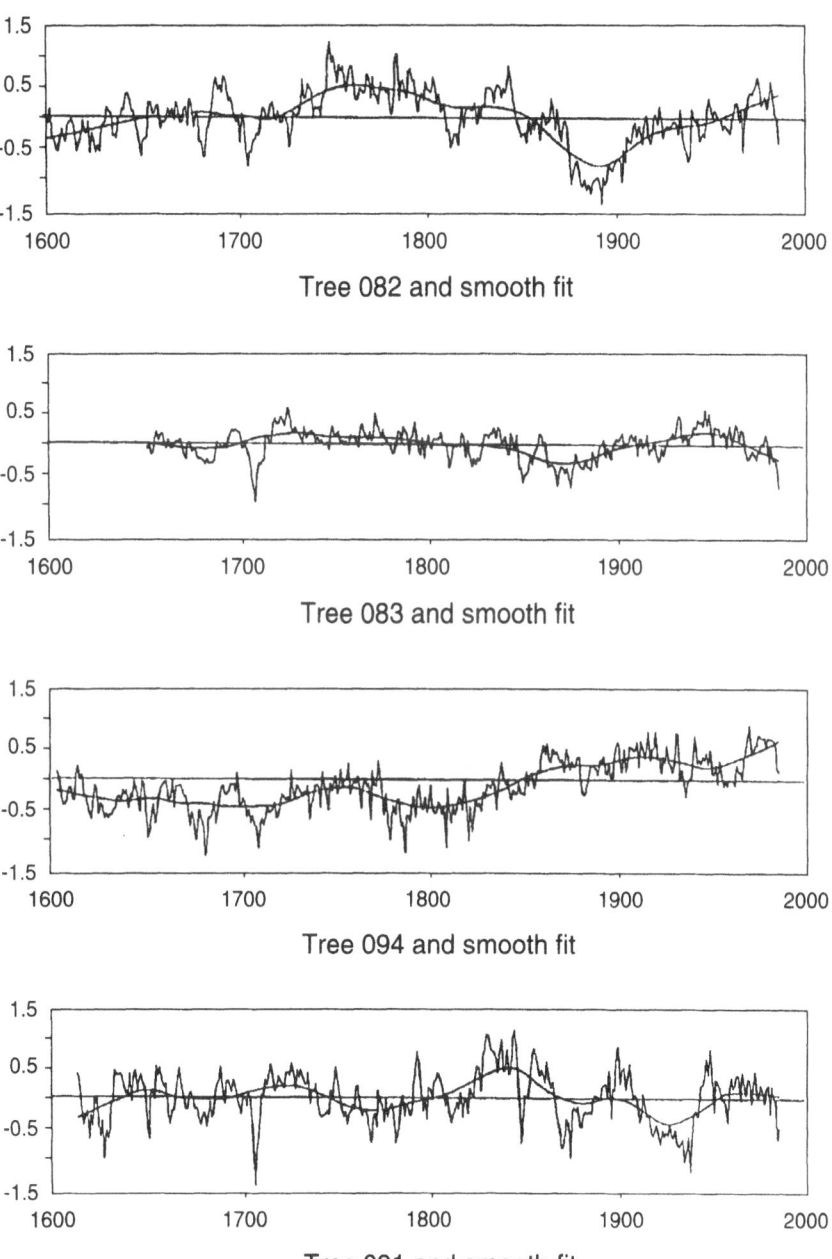

Tree 082 and smooth fit

Tree 083 and smooth fit

Tree 094 and smooth fit

Tree 091 and smooth fit

Figure 8.4 Examples of individual tree sequences after removal of age-related trends. The individualistic nature of the intermediate-term trends (smooth lines) suggest competitive effects.

are exceptional over the period of analysis, about 300 years. As an example, the mean growth index for the 1750–1984 stand average is 0.0107 (standard error (se) = 0.02) for ALT and -0.027 (se = 0.02) for SIL, but the mean growth indexes for the recent years 1931–1984 were 0.428 (se = 0.1) and 0.308 (se = 0.1) for these stands, respectively. All the trees sampled at ALT and SIL showed a growth increase, but at other sites the growth of some trees decreased even though the stand average increased.

Spatial patterns of variation

The values of the average cross-correlation coefficient (ρ) among the individual, prewhitened series for a given stand are high (Table 8.2), indicating that strong spatial homogeneity exists within stands across time. This consistency is potentially induced by stand-level factors, such as climate, that primarily affect the high-frequency component of ring-width variation.

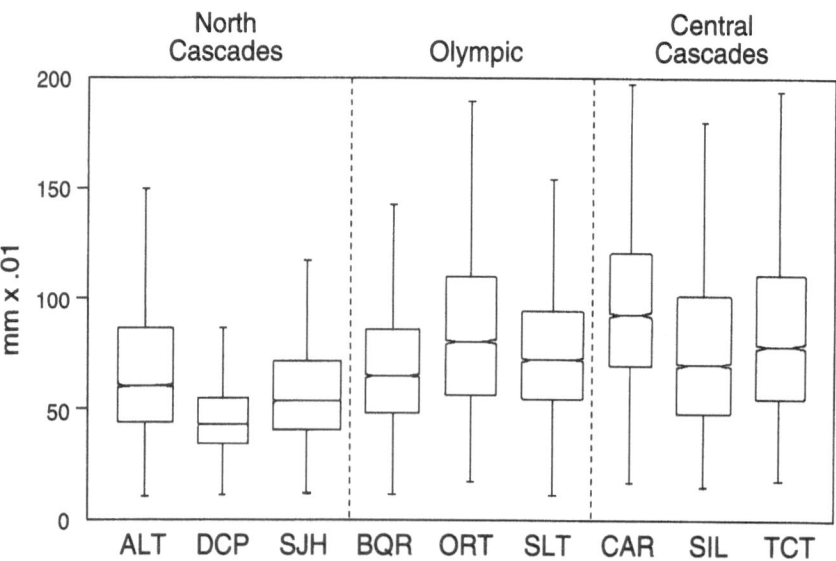

Figure 8.5 Boxplots comparing the total variation of ring widths among sites. The variation in ring width is represented by a box. The middle of the box represents the median of the sequence. The size of the box is the interquartile range with the upper and lower extremes corresponding to the upper and lower quartiles, respectively. The vertical lines show the full range of the data.

Table 8.2 Summary of site statistics and autoregressive fits.

Site	N	Mean length	Median width	Standard deviation	AR order (mode)	Mean			ρ
						ϕ_1	ϕ_2	ϕ_3	
ALT	21	408	54	40	2	0.46	0.07		0.40
DCP	19	387	37	13	3	0.33	0.14	0.13	0.46
SJH	26	521	45	24	2	0.29	0.15		0.35
BQR	20	456	63	24	2	0.42	0.11		0.41
ORT	21	372	76	36	1	0.60			0.46
SLT	31	353	67	32	1	0.57			0.46
TCT	21	345	72	34	1	0.54			0.44
SIL	27	370	57	39	1	0.47			0.43
CAR	20	270	91	34	1	0.60			0.49

N is the number of trees analyzed for the site.
Mean length is the average number of years per sequence.
Median width is in hundredths of a millimeter.
The *AR order* is the mode for all trees in the site.
ϕ_i represents the average value for the i-th AR coefficient.
ρ is the mean cross-correlation for the site, related to sensitivity.

Statistical associations among stands are shown by correlations among stand averages. Correlations are generally high between nearby stands (r = +0.7–0.8) and decay with distance (r = +0.3–0.4) (Figure 8.7). This indicates that tree growth data are spatially consistent at regional as well as at within-stand scales.

Pre-1960 vs post-1960 trend comparisons

Five sites showed an increase and four sites showed a decrease in growth after 1960 (Table 8.3). There are no patterns in these results that could be related to geographic features or pollution source areas. However, growth decreases since 1960 tend to be stronger than growth increases, as judged by both the magnitude of change and the number of trees changing at a given site.

These comparisons address only changes in magnitude and thus cannot be used to assess whether growth has recently become unusual with respect to climate variations. A change in the climate-growth relationship would suggest pollution effects, even if absolute growth rates were not unusual compared to long-term values. The investigation of this possibility is described in the following section.

Climate-Growth Relationships

We examined relationships between climatic data and both short-term, individual prewhitened residuals and intermediate-term, stand-average growth variations. Regression was used to establish climate-growth relationships for the prewhitened residuals, and simple correlation analysis was used to explore associations between local climate and stand averages. Ordinary regression models could not be fitted to the stand averages because the strong autocorrelation in these series violates the assumptions of ordinary least-squares fitting and makes inference from regression models invalid.

Methods

Prewhitened residuals

A regression model defined a priori and based on current understanding of climatic controls over growth of Douglas-fir in the Puget Sound Region (Salo 1974, Lassoie et al. 1985) was fitted to the individual prewhitened residual sequences for each site. Our objective was to

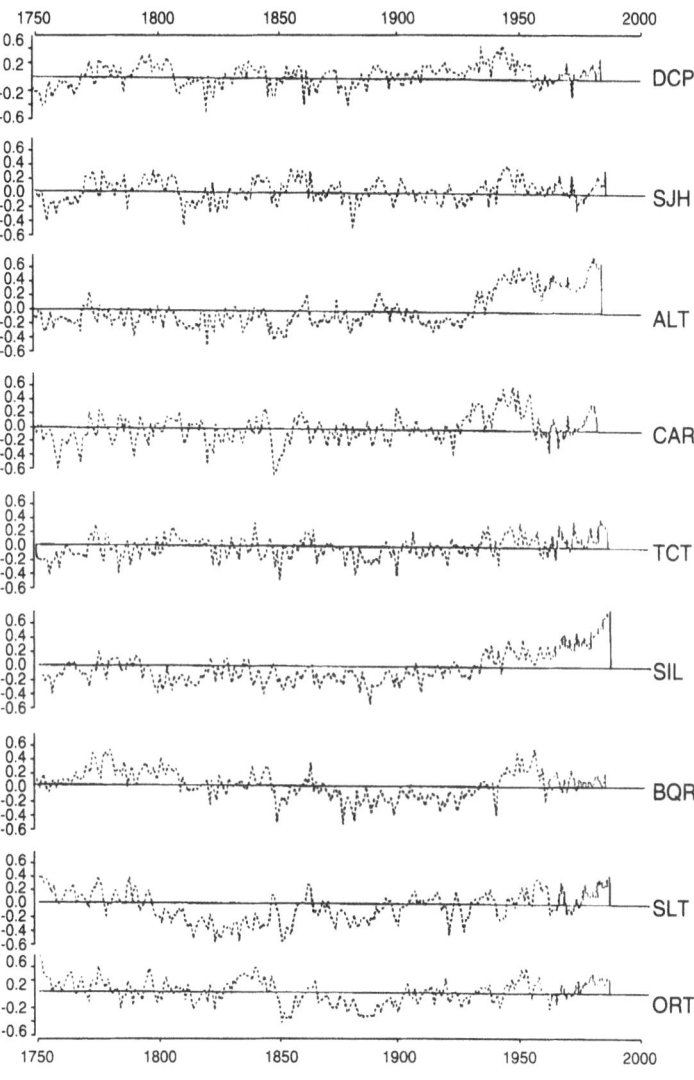

Figure 8.6 Stand averages of age-detrended tree-ring sequences for all sites (1750–1986).

1=ALT 2=BQR 3=CAR 4=DCP 5=ORT   6=SIL 7=SJH 8=SLT 9=TCT

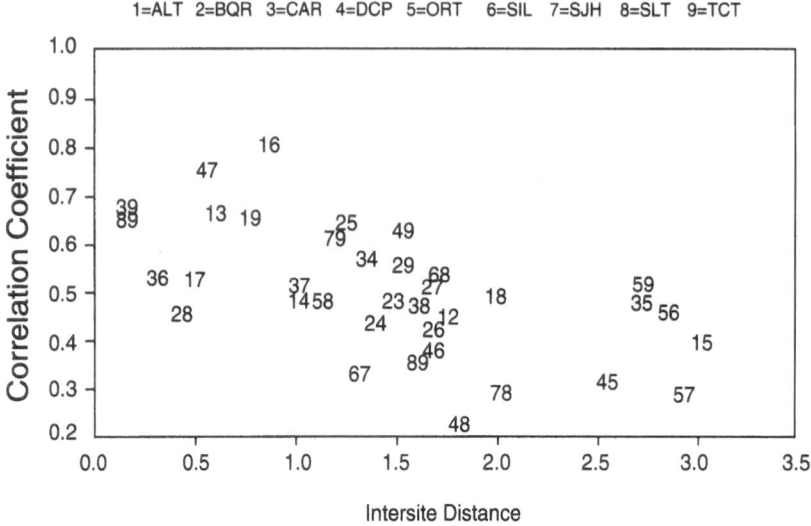

Figure 8.7 Correlations between stand averages plotted against an index of intersite distance based on latitudes and longitudes of site locations.

Table 8.3 Comparison of pre-1960 and post-1960 mean annual growth for each site. Means are standard units (growth adjusted for age effects). Also shown are the number of trees sampled per site (n) and the number of trees per site whose post-1960 mean growth differed from pre-1960 mean growth by more than .05 standard units; inc = increase, dec = decrease.

Site	pre 1960	post 1960	diff	SE (diff)	n	>0.050	<-0.050	main trend
ALT	0.442	0.460	0.018	0.051	21	9	7	inc
DCP	0.266	0.089	-0.177	0.055	18	4	14	dec
SJH	0.145	0.079	-0.065	0.035	21	5	11	dec
ORT	0.082	0.106	0.024	0.045	18	9	7	inc
BQR	0.165	0.045	-0.119	0.063	18	3	11	dec
SLT	0.082	0.140	0.058	0.067	28	14	11	inc
TCT	0.052	0.064	0.012	0.046	21	9	8	inc
SIL	0.187	0.463	0.276	0.057	25	19	3	inc
CAR	0.237	0.038	-0.198	0.046	17	2	13	dec

Pre-1960 corresponds to the period 1934–1960; post-1960 is 1961–1987

describe the relationship of tree growth and climate prior to 1960 and then use these models on the post-1960 data to see if they adequately represent post-1960 growth.

The basic methodology is a modification of the approach used by Cook et al. (1987), in which a pre-intervention series is analyzed, applied to a post-intervention series, and the adequacy of the model assessed. Our modifications are:

1. Application to individual tree series rather than to a stand average. This allowed us to examine whether trees differed substantially in their responses to short-term climatic variations.

2. Use of a priori defined models to describe relationships between tree growth and climate. This approach restricts the variables entering the analysis to biologically relevant climate variables expected to affect growth. All the a priori variables were simulta-neously entered into the regression equation, thus avoiding the limitations of step-wise procedures like those used by Cook et al. (1987) when variables are correlated (Draper and Smith 1981), as is the case with monthly climatic variables.

3. Use of all pre-1960 data to develop the model. We used residual analysis to identify lack-of-fit problems (Cook and Weisberg 1982) rather than leaving out a portion ("verification period") of the pre-1960 data. Our approach allows the use of all available climatic data, an important consideration in the Puget Sound Region where continuous climatic records are relatively short, starting after 1930 near most site locations.

4. Use of product-moment correlation between the actual and pre-dicted data (r), BIAS (mean of predicted values—mean of actual data), and mean absolute error (Makridakis et al. 1981) to assess forecasting success.

A priori models

Sites were classified as mid-elevation (600–1200 m) or low elevation (0–400 m) based on the presence of continuous, deep winter snowpack, and different a priori models were postulated for each group. Deep snow cover was considered an indication that trees were likely to be physi-ologically inactive. Most sites (seven) were from mid-elevations, where snow often persists from October–November to April–June. At such sites, spring and summer conditions were assumed to have greatest influence on growth. The two low elevation sites, not reported here, were fitted with fairly complex models that included all seasons.

The growth model for mid-elevation sites consisted of three variables:

1. Early spring temperature (DD1): Number of degree days above the mean maximum in March and April. Photosynthetic rates increase in response to rising air and soil temperatures in the spring. Early spring temperature should incorporate the direct effects of air temperature on photosynthesis as well as the indirect effects of snow melt on the physiological activity of trees.

2. Late spring and summer temperature (DD2): Number of degree days above the mean maximum for May through August. Warm temperatures accelerate cambial cell division and cell expansion is sensitive to warmer-than-average temperatures during the period of cambial activity (approximately May–August). Warm temperatures during this period should enhance the radial increment by increasing rates of cambial activity, cell expansion and photosynthesis.

3. Late summer precipitation: Total precipitation in July and August. The summer-dry climate of the Pacific Northwest results in low soil moisture in late summer. Rainfall during this period should improve tree water status and enhance growth via cambial effects (e.g., cell enlargement) or increased photosynthesis.

Stand averages

Product-moment correlation coefficients were estimated between the stand averages and monthly climatic variables. Stand averages were used because visual inspection of the individual tree data indicated some intermediate-term growth variations common to all trees at a site and thus potentially related to climatic variations. Individual tree series could not be used to analyze this level of dependence on climate, however, because intermediate-term variations in individual series also contain prominent individualistic trends related to changing competitive interactions (Figure 8.4).

As mentioned above, the strong temporal dependencies in the stand averages make regression analysis inappropriate for investigating climate-growth relationships of these series. Instead, we used correlation analysis and monthly climatic variables. This approach allowed us to explore associations between growth and a large number of climatic variables without losing degrees of freedom. In particular, we decided to explore the possible influence of climate in the prior winter and fall, which was not part of the a priori model. We used the entire length of climatic data, rather than dividing the analysis into pre- and post-pollution periods, because comparisons between these periods had not shown consistent differences (see pre-1960 versus post-1960 comparisons). Using the entire record increased the potential for identifying statistically significant associations.

Results

Prewhitened residuals

Pre-1960 Data. The significance of the model fit to the pre-1960 data was generally low in terms of explained variance (R^2)(Table 8.4). However, the relative importance and sign of different regression variables were remarkably consistent for individual trees within sites. In addition, climate-growth regression relationships were generally similar across sites (Table 8.4). The consistent sign of regression coefficients among trees and sites indicates that, despite low R^2 values, the regression results were not spurious.

Degree days in March-April were associated most strongly with short-term variations in tree growth. This variable showed a negative relationship with growth for more than 80% of the trees at all sites (Table 8.4). July-August precipitation was positively related to individual tree growth (> 80% of trees) at all but one site. Results for summer temperature were least consistent within and among sites: trees at half the Cascades sites showed a consistent positive relationship with summer temperature, trees at the remaining Cascades sites were inconsistent, and trees at the Olympic site showed a negative association between growth and summer temperature.

Pre-1960 versus Post-1960 Comparisons. The a priori model fit the pre-1960 data better than the post-1960 data (Table 8.4) at five sites. No sites showed a better fit for post-1960 data and two sites showed no change. In general, trees with worse fits for post-1960 data exhibited a positive bias and negative correlations for comparisons between predicted and actual data. That is, tree growth in the post-1960 period was less than predicted from the climate model developed for the pre-1960 period.

Stand averages

Variations in intermediate-term growth at the northernmost Cascades sites (DCP and SJH) showed relatively strong positive associations with temperature for most months (Figure 8.8), particularly in the prior fall and winter. ALT and the Central Cascades sites generally showed much weaker correlations for temperature, and relationships for late-spring and summer were often negative. Variations in intermediate-term growth are not strongly related to monthly precipitation at any site.

Climate-growth relationships: comparisons between temporal scales

Climate-growth relationships differed somewhat between the analyses of intermediate-term and short-term growth variations. For example, the strongest climatic relationship for prewhitened residuals, both within and across sites, was a negative association between growth and March-April degree days (Table 8.4), yet stand averages showed inconsistent relationships with March and April temperature (Figure 8.8). Similarly, short-term growth variations showed more consistent relationships across sites with precipitation than did intermediate-term trends.

Differences between the prewhitened residuals and the stand averages are expected on a purely statistical basis because trends (i.e., intermediate-term variations) dominate the stand averages (Table 8.2). The models are different as well. In the high-frequency models, the effects of the a priori defined variables are considered simultaneously, but for the intermediate-term variations, the climatic variables are examined separately. The high-frequency analysis did not consider the effects of prior fall and winter temperature, which dominate the intermediate-term component.

Table 8.4 Signs of regression coefficients for the seven mid-elevation sites. DD = number of degree days above the mean maximum. nc=not consistent in sign— where about half the trees had one sign, the others had the opposite. Relative fit is of a priori model to post-1960 data compared to pre-1960 data.

Site	Coefficients			Pre-1960		Relative
	Rain	DD1	DD2	max	fit	fit
	July-Aug	Mar-Apr	May-Aug	R^2		
Cascades						
DCP	nc	–	nc	0.27	poor	worse
SJH	+	–	+	0.33	poor	worse
ALT	+	–	+	0.31	poor	worse
TCT	+	–	nc	0.25	poor	no change
SIL	+	–	+	0.30	poor	worse
CAR	+	–	nc	0.47	good	worse
Olympic Peninsula						
BQR	+	–	–	0.63	good	no change

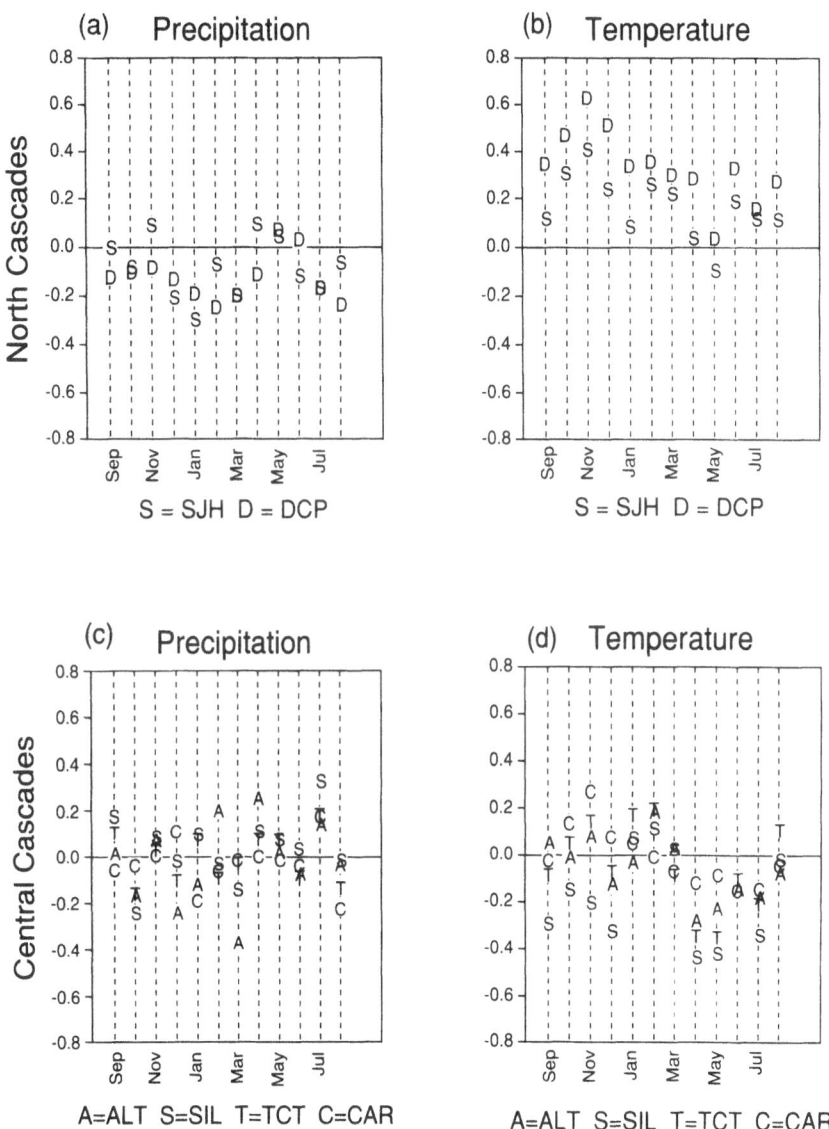

Figure 8.8 Correlations of Cascades stand averages (1930–1984) with average monthly temperature and total monthly precipitation, starting with the September preceding the growth year. ALT, a North Cascades site, is grouped in this figure with the Central Cascades sites. See text for details.

Discussion

Temporal Patterns of Ring-width Variability

Strong temporal dependencies are present within the growth series at all sites. AR(1) models fitted to the Olympic Peninsula and Central Cascades sites (Table 8.2) suggest that ecological processes driving radial growth have strong, but relatively short-term effects. The AR(2) and AR(3) models fitted to the North Cascades sites suggest greater lags in the processes controlling growth in this region. These geographic patterns are consistent with trends from other Western conifer forest regions that show the longest statistical persistence in tree-ring data from cold sites at upper treeline and high latitudes (Fritts 1976, Hughes et al. 1982). In cold environments, annual needle production is typically low, but individual needles are retained for several years. Under such conditions, needles may contribute substantially to the carbon economy of the tree long after they are produced (Teskey et al. 1984). Thus radial growth of trees may be influenced by processes (needle production) occurring several years prior to the growth year. Although there are no studies on the variation in needle longevity and physiological activity for Douglas-fir in western Washington, it is possible that the severe winters in the North Cascades cause longer needle retention and longer lag effects on growth than in the more southern and coastal sites.

Our data are the first evidence that suggests the growth of Douglas-fir has increased during the 20th-century at low to middle-elevations in western Washington (Brubaker et al. 1990). Prominent increases in the radial growth of trees have previously been found in a variety of high-elevation sites in the Cascade Mountains. For example, Graumlich and Brubaker (1986) found that the growth of mountain hemlock (*Tsuga mertensiana* [Raf.] Sarg.) and subalpine larch (*Larix lyallii* Donn) at upper treeline sites in the western and eastern Cascades, respectively, increased during the 20th-Century in response to increases in summer temperature. Similarly, increases in net productivity of mountain hemlock, Pacific silver fir and western hemlock are related to 20th-century climatic variations (Graumlich et al. 1989). Although the magnitude of 20th-century growth increases in Douglas-fir are not as great as for high elevation species, we suspect that climate caused the radial growth increases observed in our data.

Spatial Patterns of Ring-Width Variability

The prewhitened ring-width sequences are correlated among trees within sites. In terms of the conceptual model, these series are free of the effects of all factors influencing tree growth except for year-to-year variations in climate and random factors. Thus, the similarity of growth among trees suggests the importance of climatic controls over short-term tree-growth variations. Stand averages show important consistencies at larger spatial and longer temporal scales. The decreasing correlations between stand averages with distance (Figure 8.7) are consistent with the idea that nearby sites are influenced by similar weather conditions. Thus, analyses of spatial consistency in tree-growth data within and between stands support the hypothesis that climate is an important determinant of tree growth in the Puget Sound region.

Climate-Growth Relationships

Climate-growth relationships for the prewhitened residuals were consistent within sites and more than 80% of the trees at most sites showed the same sign for a given regression variable (Table 8.4). This finding and the strong cross correlations of prewhitened residuals within sites indicate the importance of climatic controls over short-term growth variations. Although the significance of the climate-growth model was never high for individual trees, the observed relationships are real. The generally poor fit for the pre-1960 data may be due to 1) limitations of using climate measurements taken at sites away from the tree sites, 2) limitations of the a priori climate-growth model, and 3) unexplained variation in individual tree growth records.

The interpretation of climatic controls over tree growth depends to some extent on the temporal scale of the analysis. For example, precipitation is more important in the analysis of short-term than intermediate-term growth variations. Graumlich et al. (1989) observed that precipitation was most strongly associated with net primary production of Cascades forests at short temporal scales and proposed that the physiological processes linking tree growth and precipitation (e.g., the control of stomatal conductance over photosynthesis, turgor controls over cambial cell expansion) are expressed primarily over short temporal frequencies.

The inconsistent results among trees and among sites for summer temperature in the a priori model suggests that growing season temperatures exert less control over growth than other variables included in the model. Correlations of intermediate-term growth variations also suggest that growing season temperatures are less important than temperatures during non-growth seasons. In fact, correlations for intermediate-term growth variations suggest that fall and winter temperature exert the most important influence over tree growth.

The strong negative association between March–April temperature and short-term growth variations was unexpected. We had anticipated a positive relationship, because we thought that warm temperatures should accelerate snowmelt and hasten the onset of physiological activity. March–April temperature is, as expected, positively related to intermediate-term growth variations at far northern Cascades sites (SJH and DCP), but results for ALT and the Central Cascades sites are inconsistent. We do not understand the reasons for these differences in the results for short- and intermediate-term variations.

More work is clearly needed to understand and model relationships between tree growth and climate in western Washington. The a priori model we explored represented a hypothesis to explain factors controlling tree growth at our sites. Although the various levels of spatial analyses demonstrated the importance of climatic controls over growth, our effort to model climate-growth relationships has not adequately characterized these relationships. Based on correlations observed between stand averages and monthly climatic data, winter and fall conditions should be incorporated into future models. The construction of future a priori models will also benefit from a better understanding of climate-growth relationships based on field measurements of physiological responses to environmental conditions.

Pre-1960 versus Post-1960 Comparisons

Growth losses since 1960 are more prominent than growth increases in terms of the average magnitude of growth change and the consistency of trends across trees within a site (Table 8.3). However, the geographic location of sites showing increased or decreased growth does not relate to pollution sources. Growth changes at sites downwind of Seattle and Tacoma (CAR, TCT, SIL), for example, are not consistently different from changes at sites in the North Cascades and Olympic Peninsula. These results suggest that pollution does not cause large-scale growth anomalies in the Puget Sound region.

In sum, there is little evidence that air pollution is causing significant regional changes in tree growth in the Douglas-fir forests of western Washington. The only evidence we have to suggest a pollution effect is the change in climate-growth relationships between the periods centering on 1960. Five of the six Cascades sites showed a poorer fit of the regression model for post-1960 data than for pre-1960 data (Table 8.4). Only one of the three Olympic Peninsula sites showed a poorer fit for the post-1960 data. However, sites with worse fits to post-1960 data do not show consistent relationships for other measures that might suggest a pollution effect. For example, sites with a poorer fit to the post-1960 data did not consistently show decreases in intermediate-term growth trends after 1960, as might be expected if pollution was causing changes in growth. Lastly, the interpretation of changes in climate-growth relationships is difficult because (i) the number of sites from each region is small, making inferences difficult; (ii) the length of the climatic record is short for both the pre- and post-1960 periods, further limiting inferences that can be made; and (iii) the fit of the regression model to the Cascades sites was relatively poor prior to 1960, so the meaning of a worse fit is not clear.

Conclusions

The most consistent temporal pattern in Douglas-fir ring-width data from the Cascade and Olympic Mountain Ranges is an increase in radial growth during the 20th-century. Although growth decreased at several sites since 1960 and climate-growth relationships have changed since 1960, there is no spatial pattern in these results that would indicate large-scale pollution effects in the Puget Sound region. On the other hand, our results do not exclude the possibility that pollution has affected tree growth at some individual sites. An important problem faced in our analysis was the lack of sufficient auxiliary data to assess causes of ring-width variations, particularly data on climatic variables and pollution exposure.

References

Agee JK (1981) Fire effects on Pacific Northwest forests: Flora, fuels and fauna. *Northwest Fire Council 1981 Conference Proceedings*, pp 54–66

Agee JK (1991) Fire history of Douglas-fir forests of the Pacific Northwest. In: Ruggiero LS, Aubry KB, Carey AB, Huff MH (eds) *Wildlife and Vegetation of Unmanaged Douglas-fir Forests*, USDA Forest Service General Technical Report, PNW-GTR-285, pp 23–34

Agee JK, Flewelling R (1983) A fire cycle model based on climate of the Olympic Mountains, Washington. In: *Proceedings: Seventh Conference on Fire and Forest Meteorology*. American Meteorological Society, Boston, pp 32–37

Basabe FA, Edmonds RL, Chang WL, Larson TV (1989) Fog and cloud water chemistry in western Washington. In: Olson RK, Lefohn AS (eds) (1989) *Effects of Air Pollution on Western Forests.* Transactions Series, No. 16, Air and Waste Management Association, Pittsburgh, PA, pp 33–49

Böhm M (1989) A regional characterization of air quality and deposition in the coniferous forests of the western United States. In: Olson RK, Lefohn AS (eds) (1989) *Effects of Air Pollution on Western Forests.* Transactions Series, No. 16, Air and Waste Management Association, Pittsburgh, PA, pp 221–224

Brubaker LB (1980) Spatial patterns of tree-growth anomalies in Pacific Northwestern forests. *Ecology* 61:798–807

Brubaker LB, Ford ED, Earle CJ, Vega-Gonzalez S, Ribic CA (1989) Growth variations in old-growth Douglas-fir forests of the Puget Sound area. Final Report EPA Project No. CR-814271-01-0, USEPA, Environmental Research Laboratory, Corvallis, OR

Brubaker LB, Graumlich LJ, Vega-Gonzalez S (1990) Long environmental records derived from tree-ring sequences in Washington and northern Oregon. *Proceedings of OCEANS 89 Symposium*, Seattle WA (in press)

Cook ER, Johnson AH, Blasing TJ (1987) Forest decline: Modeling the effect of climate in tree rings. *Tree Physiology* 3:27–40

Cook RD, Weisberg S (1982) *Residuals and Influence in Regression.* Chapman Hall, London

Draper NR, Smith H (1981) *Applied Regression Analysis.* John Wiley and Sons, New York

Edmonds RL, Basabe FA (1989) Ozone concentrations above a Douglas-fir forest canopy in western Washington, U.S.A. *Atmospheric Environment* 23:625–629

Franklin JF (1988) Pacific Northwest Forests. In: Barbour MG, Billings WD (eds) *North American Terrestrial Vegetation.* Cambridge University Press, New York, pp 103–130

Franklin JF, Dyrness DT (1973) *Natural Vegetation of Oregon and Washington.* General Technical Report PNW-8. US Department of Agriculture, Forest Service. Pacific Northwest Forest and Range Experiment Station, Portland, OR, 417p

Franklin JF, Hemstrom MA (1981) Aspects of succession in the coniferous forests of the Pacific Northwest. In: West D, Botkin D, Shuggart H (eds) *Succession: Concepts and Applications*. Springer-Verlag, New York, pp 212–229

DeBell DS, Franklin JF (1986) Old-growth Douglas-fir and western hemlock: A 36-year record of growth and mortality. *Western Journal of Applied Forestry* 2(4):111–114

Fritts HC (1976) *Tree Rings and Climate*. Academic Press, London

Graumlich LJ, Brubaker LB (1986) Reconstruction of annual temperature (1590–1979) for Longmire, Washington, derived from tree rings. *Quaternary Research* 25:223–234

Graumlich LJ (1987) Precipitation variation in the Pacific Northwest (1675–1975) as reconstructed from tree rings. *Annals of the Association of American Geographers* 77:19–29

Graumlich LJ, Brubaker LB, Grier CC (1989) Long-term trends in forest net primary productivity: Cascade Mountains, Washington. *Ecology* 70:405–410

Hemstrom MA, Franklin JF (1982) Fire and other disturbances of the forests in Mount Rainier National Park. *Quaternary Research* 18:32–51

Huff MH (1984) *Post-fire Succession in the Olympic Mountains, Washington: Forest Vegetation, Fuels, and Avifauna*. Doctoral Dissertation, University of Washington, Seattle, 223p

Hughes MK, Kelly PM, Pilcher JR, LaMarche VC Jr. (1982) *Climate from Tree Rings*. Cambridge University Press, Cambridge

Kulp JL (1986) Projections for the year 2020 in the Douglas-fir Region. In: Oliver CD, Hanley DP, Johnson JA (eds) *Douglas-fir: Stand Management for the Future*. University of Washington, Seattle, pp 3–7

Lassoie JP, Hinckley TM, Grier CC (1985) Coniferous forests of the Pacific Northwest. In: Chabot BF, Mooney HA (eds) *Physiological Ecology of North American Plant Communities*. Chapman and Hall, New York, pp 127–161

Makridakis SA, Anderson A, Carbone R, Fildes R, Hilbon M, Lewandowski R, Newton J, Parzen E, Winkler R (1981) The accuracy of extrapolation (time series) methods: Results of a forecasting competition. *Journal of Forecasting* 1:111–153

Meko DM (1981) *Application of Box-Jenkins Methods of Time Series Analysis to the Reconstruction of Drought from Tree-rings*. PhD thesis, University of Arizona, Tucson

Monserud RA (1986) Time series analyses of tree-ring chronologies. *Forest Science* 32:349–372

Oliver CD (1981) Forest development in North America following major disturbances. *Forest Ecology and Management* 3:153–168

Salo DJ (1974) *Factors Affecting Photosynthesis in Douglas-fir*. PhD thesis, University of Washington, Seattle

Spies TA, Franklin JF (1989) Gap characteristics and vegetation response in coniferous forests of the Pacific Northwest. *Ecology* 70:543–545

Spies TA, Franklin JF (1991) Composition, function and structure of old-growth Douglas-fir forests. In: Ruggiero LS, Aubry KB, Carey AB, Huff MN (eds) *Wildlife and Vegetation of Unmanaged Douglas-fir Forests*. USDA Forest Service General Technical Report, PNW-GTR-285, pp 71–89

Teskey RO, Hinckley TM, Grier CC (1984) Temperature induced change in the water relations of *Abies amabilis* (Dougl.) Forbes. *Plant Physiology* 74:77–80

Ulrich AH (1984) *US Timber Production, Trade, Consumption, and Price Statistics 1950–83*. United States Department of Agriculture Forest Service, Miscellaneous Publication No. 1442, 83 pp

Waring RH, Franklin JF (1979) Evergreen coniferous forests in the Pacific Northwest. *Science* 204:1380–1386

9

Coniferous Forests of the
Colorado Front Range

Part A: Mixed Species in Unmanaged
Old-Growth Stands
D. A. Graybill

Part B: Ponderosa Pine Second-Growth Stands
D. L. Peterson and M. J. Arbaugh

Forests along the Front Range of Colorado are exposed to elevated concentrations of ozone and other pollutants (see Chapter 3) due to emissions from the urbanized corridor stretching from Colorado Springs to Denver to Ft. Collins, and from localized mining activity and power plants. Although surveys for foliar ozone injury in the Front Range (James and Staley 1980) and Rocky Mountain National Park (Stolte, unpublished data from 1987) found no visible damage, the possibility of growth reductions was not explored.

In this chapter, we evaluate patterns of tree growth in old, unmanaged stands of several conifer species (Part A) and in younger, second-growth stands of ponderosa pine (Part B). Both approaches to investigating growth characteristics of conifers in the Front Range concentrated on widespread geographical and elevational coverage. Our objectives were to (1) determine if patterns in forest growth in the Front Range were other than could be attributed to typical trends and levels of variation, and (2) examine spatial patterns in forest condition in relation to expected patterns of pollutant exposure.

Forests of the Front Range

About one-third of Colorado is forested, with over 9 million hectares of forests (Miller and Choate 1964; Chapter 1, Table 1.1). Coniferous species occur at elevations ranging from about 1525 m to 3450 m, leading to one description of Colorado: "The second mile up is forested" (Deen 1945). Forests occur on soils derived from the gneiss, granites and schists that form this sector of the Rocky Mountains (see Chapter 4). The composition of forests depends primarily on elevation, topography, and disturbance history. The primary climatic controls on tree-growth here are typical of much of the West: dry soils in summer at low elevations, and short growing seasons at high elevations.

Rising from the plains, ponderosa pine (*Pinus ponderosa* var. *scopulorum*) forests dominate the lower foothills (beginning at about 2000 m), and comprise about 15% of Colorado's forests (USDA Forest Service 1981a). Ponderosa pine is one of the most widely distributed species in the western United States (Harlow and Harrar 1969), and is a dominant of many habitat types of the Front Range (Hess and Alexander 1986, Alexander 1987). The Rocky Mountain variety of ponderosa pine differs from that found in areas of California (*P. ponderosa* var. *ponderosa*) where pollutant injuries have been documented (see Chapters 11 and 12).

Ponderosa pine stands are found on north-facing slopes at lower elevations, and south-facing slopes at higher elevations. Many of the ponderosa pine forests east of the Continental Divide were clearcut in the mid-1800s to produce timbers for mines and ties for railroads. Between 1870 and 1960, ponderosa pine logs constituted about 40% of all timber harvested in Colorado (Miller and Choate 1964). In the middle of the ponderosa pine zone, stands tend to be dominated solely by ponderosa pine. Age class distributions tend to be bimodal, with an older cohort in the overstory, and a younger cohort beneath. Periodic surface fires (with return intervals of one to several decades) historically gave ponderosa pine forests an open, savannah-like appearance in some areas. In addition to occasional crown fires, the major natural disturbance in ponderosa pine forests is outbreaks of mountain pine beetles (*Dendroctonus ponderosae* Hopkins; Stevens et al. 1975). In the droughts of the 1920s and 1930s, beetle outbreaks killed 60 to 90% of the ponderosa pine over extensive areas of Colorado (Furniss and Carolin 1977). Dwarf mistletoe (*Arceuthobium vaginatum*) infections are widespread, causing reductions in tree growth (Hepting 1971, Hawksworth and Shaw 1984), and frequently tree death after several decades of infection.

In the upper elevations (about 2300 m) of ponderosa pine forests, mixtures with Douglas-fir (*Pseudotsuga menziesii*) are common. At higher elevations (up to about 2600 m), Douglas-fir forests contain large components of other conifers, particularly lodgepole pine (*Pinus contorta*). Douglas-fir forest types comprise about 10% of Colorado forests. Major disturbances include fire, and outbreaks of defoliating western spruce budworm (*Choristoneura occidentalis* Freeman) and Douglas-fir bark beetles (*Dendroctonus pseudotsugae* Hopkins). Budworm outbreaks in the late 1970s and early 1980s killed Douglas-fir trees throughout much of the Front Range, essentially removing it from many forests.

Lodgepole pine forests dominate landscapes between about 2600 and 3000 m, commonly occurring in even-aged, one-species stands. These forests account for about 15% of the forested area of Colorado, and supply about one-third of the current timber harvests (USDA Forest Service 1981a, Miller and Choate 1964). Many lodgepole pine trees retain cones attached to branches for more than a decade; following fire, these serotinous cones release enough viable seeds for abundant seedling production. In fact, many lodgepole pine forests are severely over-stocked, with thousands of small trees per hectare even in stands older than 50 years. At its upper elevational limits, lodgepole pine may comprise a substantial component of spruce-fir forests within the first century or two after major disturbances. In addition to wildfires, major disturbances include lodgepole pine beetle (*Dendroctonus murrayanae* Hopkins) outbreaks (Furniss and Carolin 1977) that routinely kill a large portion of trees in stands older than 60 years, and widespread infesta-tions of dwarf mistletoe (*Arceuthobium americanum*) that reduce growth and may cause mortality after several decades of infection (Hepting 1971).

Forests of aspen (*Populus tremuloides*) occur across the same range of sites as lodgepole pine, tending toward moister microsites within the range. Aspen forests are commonly successional to conifer forests (lodgepole pine or mixed conifer), but may be climax in some areas. About 20% of Colorado's forests are dominated by aspen, and aspen forests are espe-cially valued for aesthetics, wildlife habitat, and water production. The regeneration of decadent aspen stands is a major concern in Colorado, and high populations of elk (which eat young aspen suckers as well as bark of older aspen) are a major hindrance.

Mixed forests of Engelmann spruce (*Picea englemannii*) and subalpine fir (*Abies lasiocarpa*) extend from about 2800 m to 3400 m, often mixed with Douglas-fir, lodgepole pine, and limber pine (*Pinus flexilis*). Spruce-fir forests comprise about one-third of Colorado's forests, and Engelmann spruce is the most important commercial species in the state. Typical patterns of stand development after major disturbances (such as wild-fire) include thinning and transition stages (Chapter 6) where only subalpine fir is capable of regenerating in the dense shade of the over-story, and perhaps an old-growth stage where gap formation allows sufficient light for regeneration of spruce (Aplet et al. 1989). Aside from fire, the only major disturbance in this forest type is the spruce beetle (*Dendroctonus rufipennis* Kirby). Major outbreaks of spruce beetles commonly kill almost all large spruce trees across thousands of hectares. A large outbreak in the 1940s and 1950s killed most of the Engelmann spruce trees (about 11 million m^3) on over 100,000 hectares in the Colorado Rockies (Miller and Choate 1964, Furniss and Carolin 1977).

The Front Range is located largely within the Arapaho, Roosevelt and Pike National Forests (containing about 600,000 ha of land). About 20% of these forests have been set aside as wilderness areas, with over 300,000 visitor-days of use each year (USDA Forest Service 1981b). Recreational uses also include fishing (330,000 visitor-days), hunting (135,000 visitor-days), and other activities (more than 5 million visitor-days in total). Livestock grazing occurs on about 80,000 ha of mountain land within the forests, with about 20,000 animal-unit-months of utilization (one animal-unit-month is the amount of forage required by one cow and calf for one month). About 40,000 m^3 of timber are harvested annually, mostly through clearcut harvesting. The National Forests yield about 2.4×10^{10} m^3 (1.9 million acre feet) of water for irrigation.

Pollutants of the Front Range

Air pollution patterns in the Front Range are described in Chapter 3. In general, ozone concentrations in the Front Range are moderately high. Mean ozone concentration at Rocky Mountain National Park during the growing season (May to October) was 46 ppbv with the 90th percentile of 60 ppbv. Ozone levels at Manitou Experimental Forest were similar to those at Rocky Mountain National Park, with maximum values ranging between 40 and 80 ppbv (Zeller and McKinney 1989). Maximum ozone levels measured at Niwot Ridge (3050 m elevation) during 1981 (80 ppbv) were not uncommon (Fehsenfeld et al. 1983). The highest ozone concentrations usually occurred at valley and foothill locations adjacent to the central part of the Front Range. Mountain locations generally had lower concentrations, but were comparable to the northernmost and southernmost valley locations. There were only seven instances when ozone exceeded the national ambient air quality standard of 120 ppbv at a station in the monitoring network. Measurements of 100 ppbv or greater were made at all stations except Manitou and Colorado Springs.

Some information is available on patterns of bulk deposition of sulfate and nitrate across the Front Range; Figures 9.1a and 9.1b show the spatial trends for May 1982–May 1983 across 42 stations in Colorado (Lewis et al. 1984). Summer winds from the southeast (Figure 9.1c) are probably

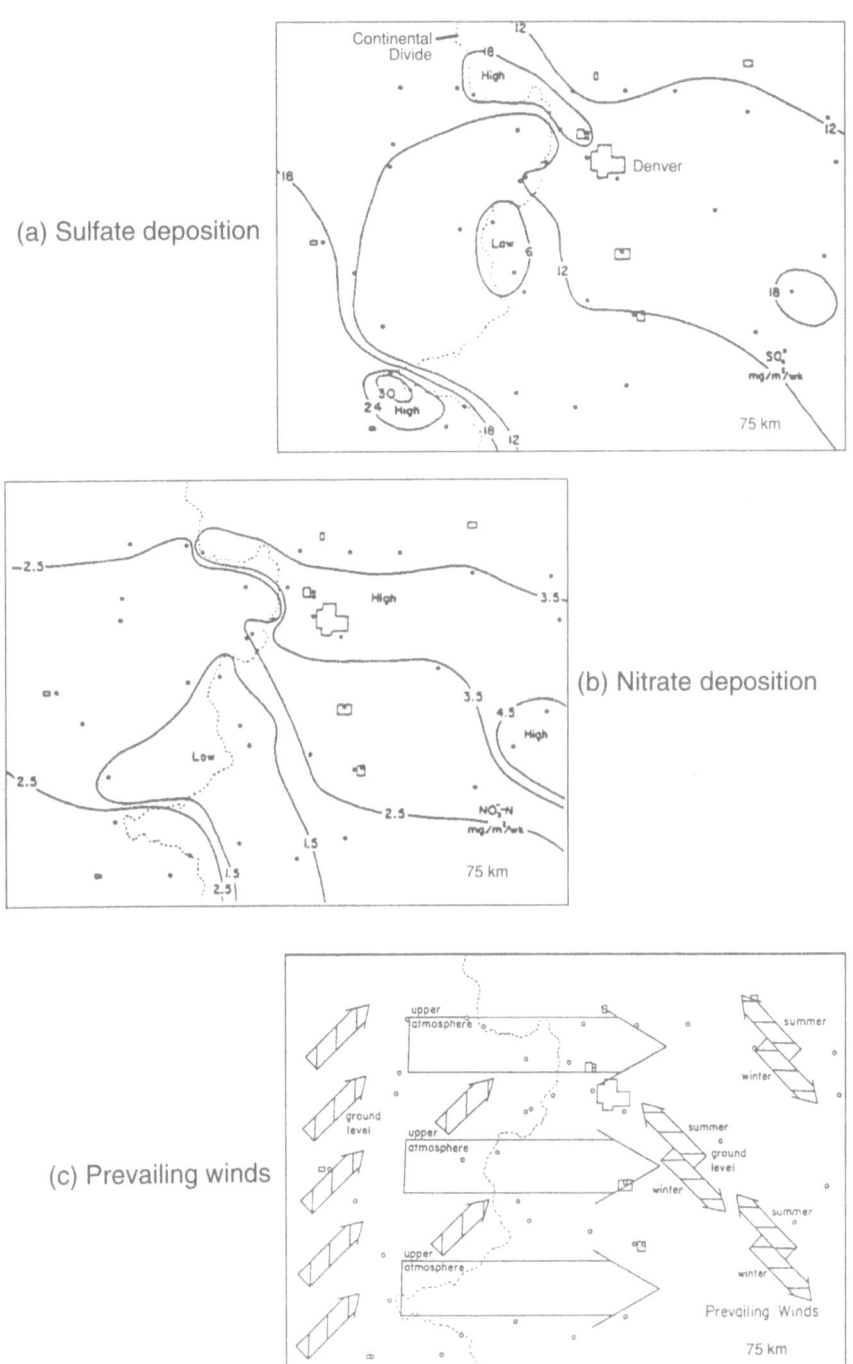

(a) Sulfate deposition

(b) Nitrate deposition

(c) Prevailing winds

Figure 9.1 Sulfate and nitrate deposition and prevailing winds in Colorado (from Lewis et al. 1984).

responsible for movement of urban and industrial pollutants from the greater Denver region to the higher elevations of the east facing sectors of the Front Range.

Inputs of pollutants such as lead to remote high altitude lakes in Rocky Mountain National Park began in the 1800s (Baron 1983). At that time it was probably a result of mining activity while more recently it could be a product of gasoline combustion. Amounts of other atmospheric pollutants such as PCB's and the PAH flouranthane have also been increasing in these lakes (Heit et al. 1984). In general, there appears to be a strong relationship between total acid deposition, the amount of wet precipitation, and elevation, resulting in greatest deposition to the higher forested and alpine areas (Lewis et al. 1984; Chapter 3, this volume).

Chapter 9A
Mixed Species in Unmanaged Old-Growth Stands

The Regional Sampling Approach

The approach in this study involved wide-ranging coverage of old-growth stands in the Front Range, in an exploratory mode. Obtaining a broad initial overview of this nature seemed the most useful strategy in the absence of a priori knowledge that conifers in any single sector of the region were experiencing abnormal growth. This regional sampling included 20 sites located along potential pollution gradients (Figure 9.1, Chapter 3). On north-south axes, they paralleled the Continental Divide, covering the latitudinal range from Fort Collins to Colorado Springs. An east-west axis included stands over the majority of elevations where tree-growth might be affected by pollutants. Paired sites on both sides of the Continental Divide offered the possibility of differences in pollutant exposure between sites, with the western sites more protected from urban emissions. Three pairs of sites were sampled, on either side of the divide and at about the same elevation: Engelmann spruce at upper treeline—MIS,TIS; lodgepole pine at intermediate elevations—ONL,HIL; and ponderosa pine at lower elevations—MOP,DMP (see Table 9.1, Figure 9.2).

Only stands considered unlikely to be harvested were selected for sampling in order that they might remain available for long-term air pollution research. This was feasible in most situations because the lands were controlled by various federal agencies such as the National Park Service and USDA Forest Service.

Site and sample selection procedures were in most cases attempts to identify, control and maximize or minimize the various signals in tree-ring series described in Chapter 7 (Equation 7.2—The Linear Aggregate Model). Control of these signals necessarily begins with the field sample selection strategy, both with respect to stand and site characteristics as well as to the selection of trees within a stand. A useful summary of sampling issues in dendrochronology has been provided by LaMarche (1982). The following paragraphs summarize the strategies that were used in collection of tree-ring series for this project.

A major issue in site and sample selection was avoidance of stands subjected to human or natural disturbances (such as harvesting, burning or insect epidemics) that may have in turn affected growth patterns. Open canopied stands were sought in order to minimize competition effects on ring width growth patterns. Sample trees were all rated for level of mistletoe infestation (Hawksworth 1977), and no tree with a rating of greater than 3.0 (moderately infected) was sampled. Other criteria that were used in selecting particular trees included a minimum age of approximately 100 years, dominant to codominant in canopy position, and no evidence of unusual defects. Two increment cores were taken, parallel to the topographic contour, from the lower bole of each tree.

From 10–20 trees were sampled at each site when only one age class was of interest. In six stands, I included trees in the 100 year age class as a younger sample and others of greater age for an older sample, to test whether trees of these different age classes might respond differently to climate or to possible pollution stresses. Thus, a total of 26 data sets were generated from the 20 sites.

Additional observations included the visual examination for injury of needles from each of three branches from five trees per stand. No visible injury such as chlorotic mottle associated with ozone damage was observed. Needle retention was not measured because the trees were from a variety of different elevations, and differences in time of needle retention might simply result from elevational differences (Ewers and Schmid 1981, Weidman 1939).

Chronology Development

Tree-ring index chronology development followed common procedures (Stokes and Smiley 1968, Robinson and Evans 1980; see Chapter 7). In brief, samples were first crossdated and ring widths were measured to the nearest 0.01 mm. All series were standardized and most were fit with deterministic models such as a negative exponential or straight line. Standardization in this manner removes natural biological trend due to

Table 9.1 Tree ring collections from unmanaged old-growth stands.

Chronology[a] Code	Species[b]	Elevation (m)	No. Trees	Dated Range	Signal:Noise Ratio1850–1987
ALO	PP	2560	13	1680–1988	22.64
ALY	PP	2560	10	1829–1988	16.43
DMO	PP	2621	11	1622–1988	6.20
DMY	PP	2621	11	1825–1988	13.48
EDO	DF	1981	8	1540–1988	12.14
EDY	DF	1981	10	1847–1988	12.63
EDP	PP	1890	17	1677–1988	27.86
FRL	LP	3219	14	1309–1988	17:00
HIL	LDP	2829	16	1854–1988	16.59
HTF	DF	2097	11	1794–1988	14.56
HTP	PP	1707	11	1658–1988	21.63
JCO	PP	1966	12	1547–1988	15.09
JCY	PP	1966	8	1802–1988	17.41
KAF	DF	1829	13	1791–1988	30.92
KAP	PP	1829	10	1699–1988	20.33
LYP	PP	1798	11	1657–1988	18.15
MIO	ES	3414	12	1564–1988	8.45
MIY	ES	3414	9	1802–1988	21.29
MOP	PP	2621	12	1428–1988	6.09
NIL	LP	3170	18	1321–1988	7.27
ONL	LDP	2804	18	1762–1988	11.68
RCL	LP	3353	19	1061–1988	2.55
TCP	PP	1939	14	1597–1988	20.65
TIO	ES	3505	12	1502–1988	8.99
TIY	ES	3505	11	1729–1988	15.15
VBP	PP	1920	13	1681–1988	33.72

a. Six pairs of codes designate chronologies of older and younger trees from some of the stands in Figure 9.2. The third character in the chronology code is respectively 'O' or 'Y'. ALO and ALY are from stand ALP; DMO and DMY are from stand DMP; EDO and EDY are from stand EDP; JCO and JCY are from stand JCP; MIO and MIY are from stand MIS TIO and TIY are from stand TIS.

b. PP = Ponderosa pine, DF = Douglas-fir, ES = Engelmann spruce, LDP= Lodgepole pine, LP = Limber pine

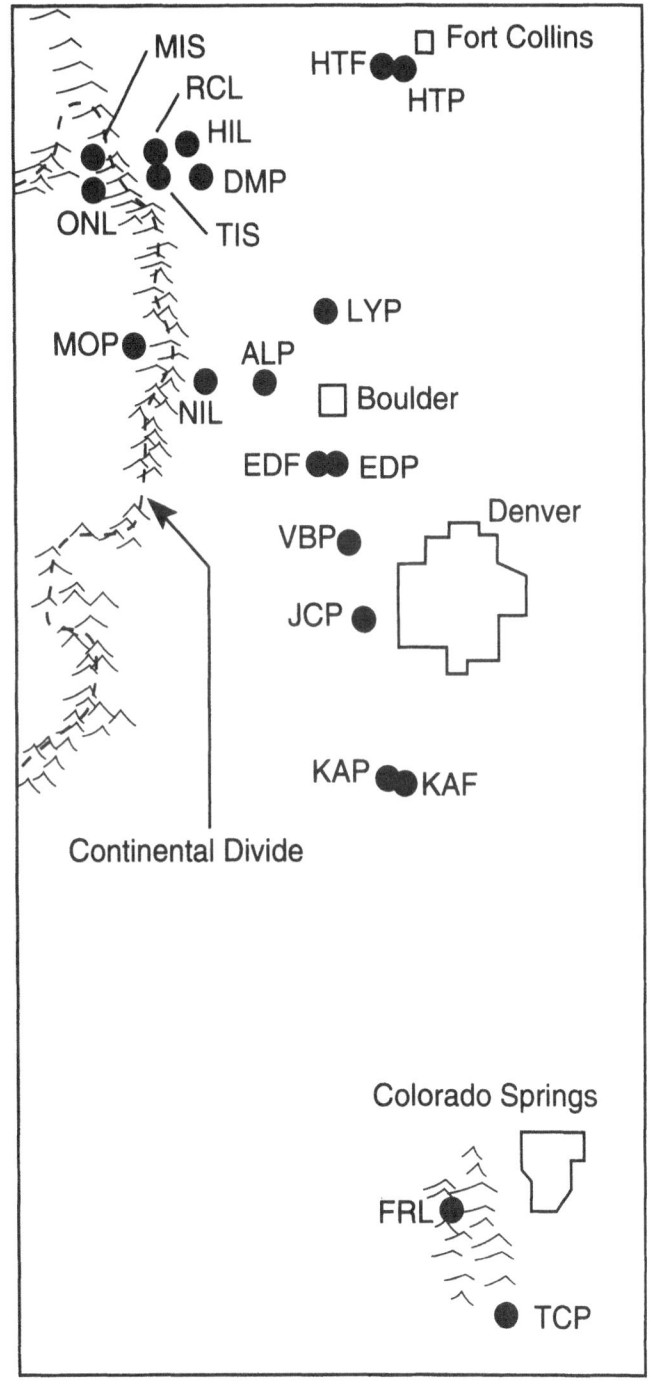

Figure 9.2   Location of unmanaged old-growth stands from which tree-ring chronologies were developed. See Table 9.1 for site characteristics.

aging, and preserves other trends that are of particular interest (Graybill 1979, 1982). Average index chronologies for the 26 data sets were then created.

Inspection of time series plots of the indices of individual tree averages and evaluation of analysis of variance (ANOVA) statistics (Fritts 1976) suggested that no multiple or highly divergent growth patterns occurred in any single data set. One consistent pattern was a tendency for signal-to-noise ratios (SNR) (Table 9.1) to be low for the wetter, high elevation stands (e.g. RCL) and for the more dense stands (e.g., ONL). In contrast, the SNR and common variance tended to be highest for stands on moisture-stressed sites at lower elevations (e.g., VBP). These differences in SNR and common variance are normally found along ecological and moisture gradients in the semi-arid, montane West (Fritts et al. 1965, LaMarche 1974, Graybill 1987). All of the index chronologies except one had significant autocorrelation, and this was removed by autoregressive moving average (ARMA) modelling and subsequent computation of residuals (Box and Jenkins 1976). The rationale for this is summarized in Chapter 10. The final series are referred to here as residual chronologies.

Climate-Tree Growth Relations

Relationships between climate and tree growth have been widely used to evaluate major changes in environmental conditions (Innes 1990). Pollution stress can lead to disruption of physiological activity through many potential routes (Kozlowski and Constantinidou 1986a,b), which would likely change the normal relationships between climate and growth. If such deviations were found in tree rings from recent decades when pollution levels have been highest, then further research on pollution type, level and concomitant physiological changes might be in order. Alternatively, other explanations can and should also be sought.

The first analytical procedure was to derive a model relationship between climate and tree growth for time periods thought to be relatively free from air pollution. Both 1950 and 1960 were selected as approximate turning points after which major increases in pollutant emissions occurred in Colorado. Simple correlations were computed between residual chronologies and climate data for the pre/post 1950 and pre/post 1960 time periods. For any one year of tree growth, these relations were reviewed over the preceding 16 months, or, May of the prior year through August of the year of growth. The climate variables included monthly and various seasonal and annual precipitation sums and temperature averages for the growing season months of May through August. The data sets were regional averages for Colorado climate divisions 1, 2 and 4 (see Chapter 2). Additionally, response function analysis (Fritts 1976) was used as an aid in determining the sensitivity of tree growth to temperature and precipitation.

Sixteen of the chronologies from elevations below 2625 m (MOP was the exception) had a relatively strong positive growth response to precipitation of the current spring and summer and the previous late summer and early fall. Tree growth responses to current spring and summer temperatures were generally negative but slight. These responses reflect the importance of soil moisture for growth. Relationships of tree growth to precipitation of the winter months were weak in most cases, probably because recharge of soil moisture with snowmelt is limited more by soil water holding capacity than by the quantity of snowmelt.

The remaining ten chronologies, all from elevations at 2600 m or greater, provided a diversity of growth responses to climate that were in some cases time dependent. Those chronologies were not used in the succeeding analysis but summaries of some of their important climate responses are provided below.

In the case of the 16 drought-sensitive series, I used simple linear regression (Draper and Smith 1981) to establish predictive relationships of tree-ring growth from climate. The independent variable in each analysis was an annualized precipitation sum, and the dependent variable was a residual tree-ring chronology. These analyses were accomplished over baseline periods from 1897 to 1950 and from 1897 to 1960. First, it was necessary to obtain some idea of the stability of the tree-growth climate relationship during these time periods. The issue was whether there were strongly or even qualitatively different relationships at different times. Successive portions of the data were used for calibration (model development) and verification (model testing). The intervals from 1897 to 1950 or 1960 were divided into thirds. Two-thirds of the data were used for calibration while the remaining third was used for verification. In another set of three trials, two thirds of the data were randomly selected for calibration while the remaining third was used for verification. The climate-tree growth relationships reviewed here were not completely stable through time because the period from about 1932–1942 was substantially warmer and drier than the preceding 25–30 years, as illustrated by Palmer Drought Severity Indexes (Palmer 1965) for June (Figure 9.3). This index of meteorologic drought integrates both temperature and precipitation, and low values indicate warm, dry conditions (see Chapter 2). Correlations of the tree growth series with precipitation and drought severity indexes were somewhat higher during times of drought than during wetter periods. This is a common phenomenon and is reasonably well understood from physiological perspectives (Fritts 1976).

The changes in the climate-tree growth relationships from period to period during the twentieth century were not however so dramatic that the relationships could be considered different classes or kinds of processes. Therefore, final calibrations were established with the full baseline period data sets. This was necessary and prudent because the climatic data sets available during the baseline periods covered only 55 to 65 years. All of the data were used to insure that the final calibration model was developed on the fullest possible range of data covariation.

Information under the Calibration heading in Table 9.2 describes the nature and strength of the relationships between the precipitation sums and tree growth in the baseline periods. The original correlations were all positive and significant ($\alpha = 0.05$), indicating strong relationships between tree growth and variation in precipitation. The equations derived from regression were then used to estimate tree growth from precipitation during recent decades, from both 1951 and 1961 to 1987. Major changes in the climate-tree growth relationship would indicate novel stresses on the trees. A lack of change would be indicated if the means of the predicted and actual growth were similar, and if the covariance between the predicted and actual growth were similar to that of the model calibration period, and if there were no anomalous trends in the actual growth which were not also present in the predicted growth.

Alternatively, there may be (1) a significant difference between the means of the actual and predicted tree growth series, (2) no significant covariance of the actual and predicted series, or (3) different trends in the actual and predicted series. Inability to reject any of these alternative hypotheses indicates that the actual and predicted growth empirically differ in some manner. The difference may result from anthropogenic impacts, or increases or decreases in naturally occurring stresses.

Differences of means and central tendency were tested with the t-test and the Wilcoxon matched-pairs signed-ranks test. Covariance was evaluated with Pearson's correlation coefficient and its square (the coefficient of determination), and a non-parametric sign test. Trends were sought in visual examination of time series plots and in evaluation of differences of actual and predicted growth values.

Results of these exercises (Table 9.2) indicate that the covariation (r^2) of actual and predicted growth in the post-1950 test periods was significant with one exception (HTF)(DF), although the sign test indicated poor climate prediction of the LYP(PP) growth relationship for both time periods and of that for JCY(PP) in the 1961–1987 time period. A review of our time series plots of these chronologies does not indicate substantial increases or decreases in tree growth, but simply a failure to closely track the climate data. A review of site notes does not suggest that anything

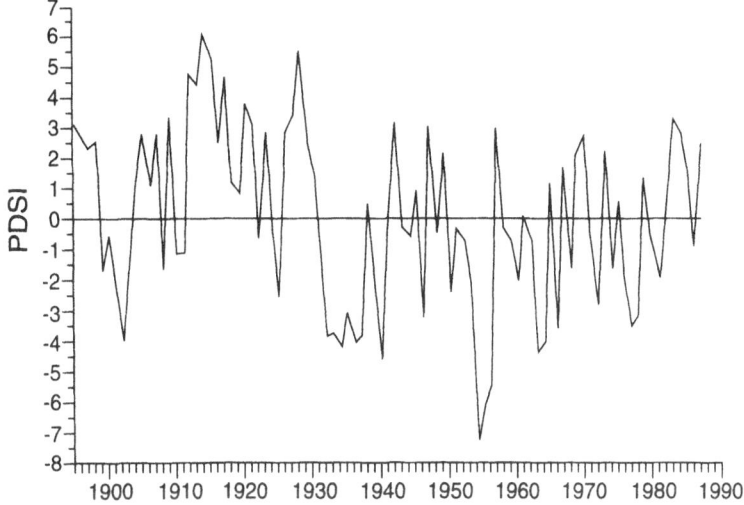

Figure 9.3 June Palmer Drought Severity Index (PDSI) time series for Colorado climatic division 4.

observed in the field would provide a plausible hypothesis for site or tree disturbance. It is however possible that Douglas-fir stands had been subject to spruce budworm infestation during the 1950s or 1960s.

Only a few major differences were found in the covariances, when comparing those from the baseline calibrations with those for the respective predictions, that are on the order of doubling or halving. One of those was the HTF stand noted above while others involved are ALO(PP), EDP(PP), and LYP(PP). I offer no explanation for these differences.

The t-test results suggest that the mean of each predicted series did not differ significantly from that of the actual growth in all cases, although the more stringent Wilcoxon test suggested that means differed in four cases. However, a review of time series plots of actual and expected values revealed no strong change in growth for these stands.

In brief summary, the results of these analyses suggest no major or widespread change in climate-tree growth relations in the decades following 1950 when various pollutants have reached historically high levels. However, the amounts of covariance between the actual growth

Table 9.2 Climate-tree ring model calibrations and predictions, old-growth unmanaged stand chronologies (see text and notes below).

CRN. CODE[a]		Season[b]	r^2	r^2	t-test	Wilcoxon test	sign test
		CALIBRATION		PREDICTION[c]			
ALO	1	P5	.19	.45	A	R	A
	2	P5	.26	.51	A	R	A
ALY	1	P4	.32	.40	A	A	A
	2	P4	.35	.37	A	R	A
DMO	1	P5	.27	.44	A	A	A
	2	P5	.32	.33	A	A	A
DMY	1	P5	.25	.39	A	A	A
	2	P5	.30	.32	A	A	A
EDO	1	P3	.26	.24	A	A	A
	2	P3	.27	.27	A	A	A
EDY	1	P3	.23	.30	A	A	A
	2	P3	.25	.38	A	R	A
EDP	1	P4	.12	.31	A	A	A
	2	P4	.17	.29	A	A	A
HTF	1	P2	.21	.04R	A	A	R
	2	P2	.17	.04R	A	A	R
HTP	1	P4	.30	.35	A	A	A
	2	P4	.30	.26	A	A	A
JCO	1	P5	.25	.43	A	A	A
	2	P5	.31	.35	A	A	A
JCY	1	P5	.25	.33	A	A	A
	2	P5	.30	.23	A	A	R
KAF	1	P4	.34	.48	A	A	A
	2	P4	.38	.43	A	A	A
KAP	1	P5	.19	.28	A	A	A
	2	P5	.24	.20	A	A	A
LYP	1	P4	.31	.33	A	A	R
	2	P4	.40	.20	A	A	R
TCP	1	P1	.48	.52	A	A	A
	2	P1	.46	.57	A	A	A
VBP	1	P4	.48	.52	A	A	A
	2	P4	.51	.46	A	A	A

a. '1' indicates calibration from 1897–1950, prediction from 1951–1986. '2' indicates calibration from 1897–1960, prediction from 1961–1986.

b. Annualized precipitation sums that were used in the analyses include the following:

Season	Months of precipitation
P1	Prior August through current July
P2	Prior September through current August
P3	Prior July through current June
P4	Prior September through November, current March through July
P5	Prior September, October, current March through July

c. 'R' indicates rejection of null hypothesis, 'A' indicates null hypothesis cannot be rejected. See text for specific hypotheses.

series and precipitation, and actual and predicted growth are generally less than 50%. The mean r^2 value for all pre-1950 and -1960 calibrations is 0.31 and the mean r^2 value for the post-1950 and -1960 predictions, excluding the HTF series, is 0.39. We therefore need to be cautious about generalizations.

Other Climate-Tree Growth Relationships

Given the initial difficulty of discerning any strong and consistent climate response in the ten high elevation chronologies, and the discovery of some time instability in the preliminary analyses of the low elevation chronologies, another approach was taken. The data sets were broken into three periods. This analysis used 1896–1931 for the first period (35 years) and divided the remaining 56 years equally into two 28 year segments, 1932–1959, 1960–1987. The earliest period is dominated by relatively wet, cool conditions, the second by warm, dry conditions, or drought, and the third by what might be termed recovery from drought (Figure 9.3). Correlation analyses indicated a varying time dependency in the response of several of these chronologies to some climate variables and that in several cases it is pronounced. One of the ten chronologies (RCL) showed no strong climate relationships in any period in these analyses. This record had no gross abnormalities such as strong upward or downward trend, or drastic discontinuities in growth level or variance.

Two major and different climate response patterns were commonly found among the remaining chronologies. One trend that occurred in all of the Engelmann spruce chronologies was a decrease in the correlation of growth with precipitation, moving from low to somewhat negative and significant in the case of MIO and MIY, and from moderate and significant to low and nonsignificant in the case of TIO and TIY (Table 9.3). This pattern also occurred in the ONL(LDP) and MOP(PP) series. A visual example of this is found in Figure 9.4, which shows the first and second principal component scores (PC1,2) (Jolliffe 1986) of the four residual spruce chronologies and the residual chronologies for ONL(LDP) and MOP(PP), plotted with an annualized precipitation series. The component scores summarize the major patterns of variability that are common to the individual chronologies. These stands are all at higher elevations, ranging from 2620 to 3505 m near upper treeline, and four of them are just at the western edge of the continental divide. One hypothesis for the trend toward a decreased correlation of growth with precipitation at these sites is that the trees have recently begun to respond in a negative fashion (or, a less positive fashion) to something that is associated with precipitation. This may include a variety of pollutants.

Table 9.3 Tree growth and climate correlations for selected time periods, unmanaged old-growth stand chronologies.

| Tree Ring Series | Climate Series[a] | Time Periods | | |
		1895–1931	1932–1959	1960–1987
PCA1[b] (Mix)	P1	.25 (.13)[c]	.18 (.35)	-.35 (.07)
PCA2 (Mix)	P1	.31 (.07)	.23 (.25)	.18 (.35)
MIO[d](ES)	P1	.27 (.10)	.02 (.92)	-.39 (.04)
MIY (ES)	P1	.09 (.59)	-.13 (.53)	-.48 (.01)
MOP (PP)	P1	.33 (.04)	.28 (.15)	.18 (.64)
ONL (LDP)	P1	.32 (.05)	.21 (.29)	-.04 (.85)
TIO (ES)	P1	.51 (.00)	.21 (.29)	.18 (.65)
TIY (ES)	P1	.35 (.04)	.13 (.53)	.03 (.89)
PCA1[e] (PP)	P5	.46 (.01)	.68 (.00)	.47 (.00)
PCA1[f] (DF)	P5	.39 (.02)	.63 (.00)	.61 (.00)
PCA1[g] (LP)	PD6	.16 (.35)	.54 (.00)	.38 (.04)

a. All climate codes but one are the same as those in Table 9.2. PD6 is the June Palmer Drought Severity Index.

b. The tree ring data used in the first two rows for PCA1 and PCA2 (principal components one and two) are the six chronologies just below them in the table, MIO through TIY. See comments under Figure 9.4 regarding this analysis.

c. Pearson correlation coefficients are followed by their two-tailed probability (alpha) in brackets.

d. The chronology identifiers such as MIO correspond to entries in Table 9.1.

e. This principal component is from an analysis that included Douglas-fir residual index chronologies EDO, EDY, KAF.

f. This principal component is from an analysis that included all residual ponderosa pine index chronologies except MOP.

g. This principal component is from an analysis that included the residual index chronologies FRL, HIL, NIL.

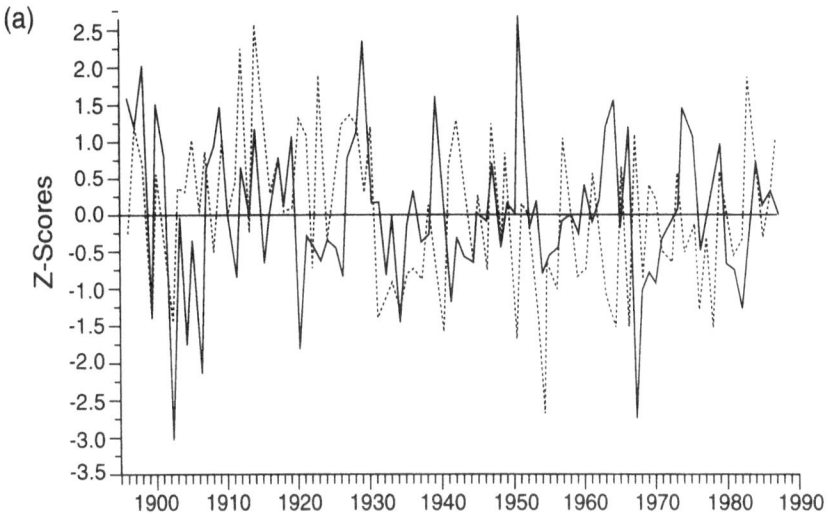

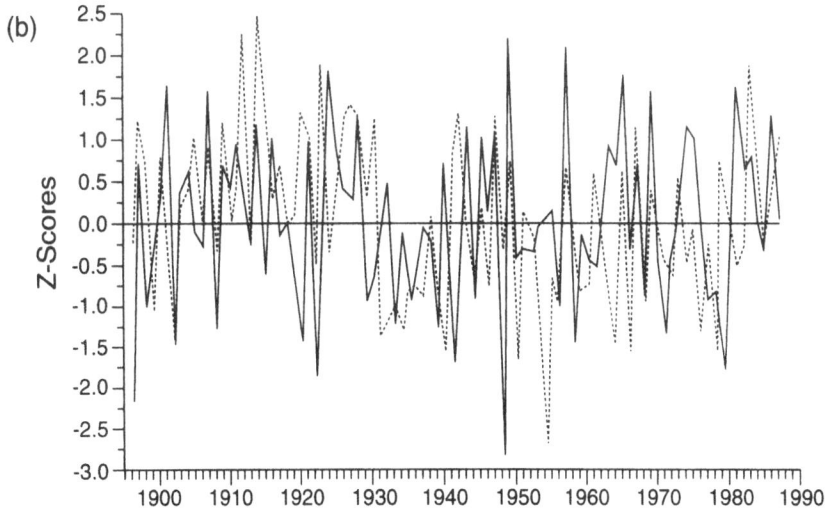

Figure 9.4 Principal component scores of chronologies (—) with time-dependent declines in response to annualized precipitation time series (····). The tree-ring chronologies in this analysis included MIO, MIY, MOP, ONL, TIO, TIY—see Table 9.2 and Figure 9.2. (a) The first principal component accounted for 53.4% of the variance while (b) the second accounted for 21.1%. The first component is weighted heavily on the spruce chronologies and the second by the limber pine and ponderosa pine chronologies. The precipitation values are the sum of monthly totals for the period from the August prior to growth through July of the year of growth.

The remaining chronologies share a single pattern whereby the correlations with moisture availability increase from the early cool-wet period to the middle, or driest period, and then decrease somewhat in the 1960–1987 period that was less stressful for growth (Table 9.3). Figures 9.5–9.7 illustrate this for PC1 of various groupings of the chronologies by species. This pattern is most common at the lower and drier elevations but is also seen in two of the limber pine series from high elevations (FRL, NIL). Limber pine, especially at these elevations, typically occurs on drier sites, and the lack of response to climate (low and nonsignificant r) in the early portion of this century may indicate that available moisture was not limiting to growth. The more recent positive response to moisture may then be understood in terms of the physiological issues noted above (Fritts 1976). This inference may also hold for a similar pattern of temporal correlations in one of the high elevation lodgepole pine chronologies (HIL). Records of field observations indicate that this is a relatively dry site for higher elevations. Conversely, our records indicate that the other lodgepole stand that had a decreasing positive response to precipitation (ONL) is in a relatively mesic setting.

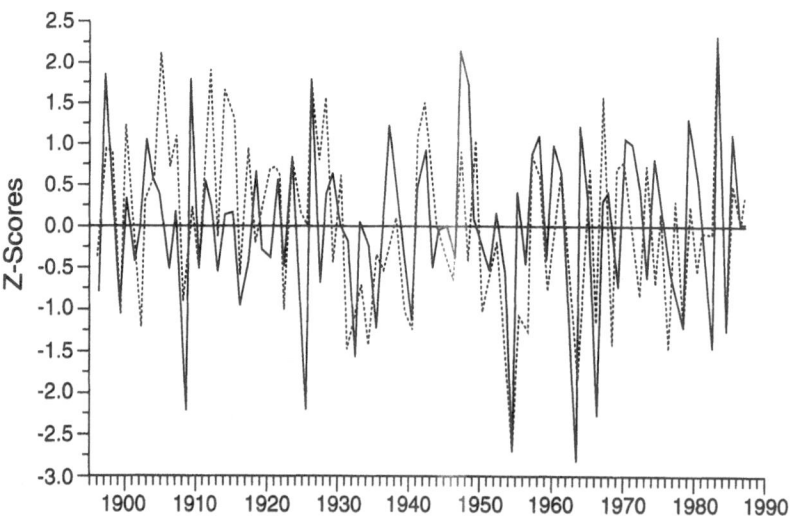

Figure 9.5 Principal component score one of ponderosa pine chronologies (—) and precipitation time series (····). The tree-ring chronologies in the analysis included all of the ponderosa pine except MOP—see Table 9.2 and Figure 9.2. This principal component accounted for 60.2% of the variance. The precipitation data is a sum of monthly totals for the September and October prior to growth and March through July of the year of growth.

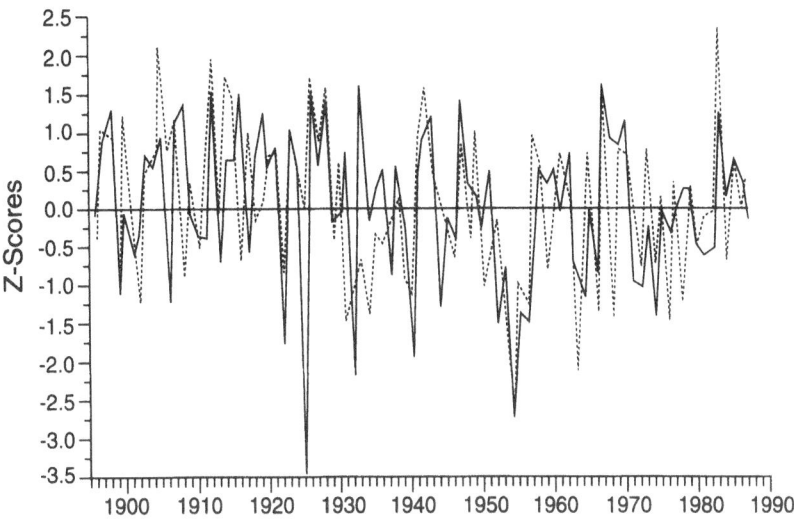

Figure 9.6 Principal component score one of Douglas-fir chronologies (—) and precipitation time series (⋯). The tree-ring chronologies in the analysis include EDO, EDY, KAF—see Table 9.2 and Figure 9.2. This principal component accounted for 75.6% of the variance. The precipitation data are the same as that for Figure 9.5.

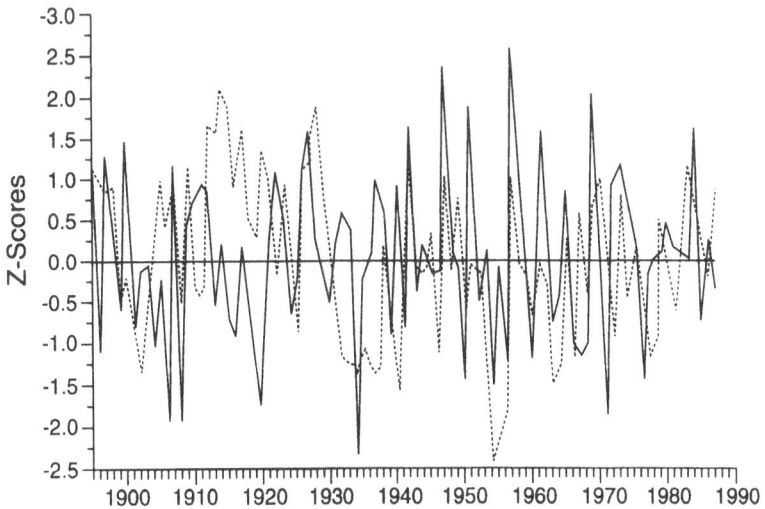

Figure 9.7 Principal component score one of limber pine chronologies (—) and June Palmer Drought Severity Index, Colorado Division four (⋯). The tree-ring chronologies in the analysis included FRL and NIL—see Table 9.2 and Figure 9.2. This principal component accounted for 65.8% of the variance.

Summary on Old-Growth Forests

Drought sensitive tree-ring chronologies from both youthful and older individuals at elevations below 2625 m show a reasonably consistent response to precipitation during the 20th century. This implies that pollution has not affected the relationship of tree growth to climate in these cases. Correlational analyses showed some time-dependent relationships between climate and tree growth. With the exception of the decreasing correlations with moisture availability at high elevations, none are considered to be so unusual that they require further investigation.

Geographic differences in tree growth response to climate are indicated by the fact that four of the six chronologies that showed decreasing response to precipitation over time are from trees on the west side of the Continental Divide, and most are at relatively high elevations. Reasons for this time-dependent growth response are not immediately obvious. The hypothesis was offered that this may reflect a negative response to some pollutant(s) that are associated with precipitation. No alternative hypotheses have been formulated.

Chapter 9B
Ponderosa Pine Second-Growth Stands

Research and Sampling Approach

Study sites for second-growth stands were initially selected with an on-the-ground survey of potential sites. Only areas dominated by second-growth ponderosa pine that showed minimal evidence of thinning or other disturbance were considered. Sampling was stratified in two groups: exposed and protected. Twenty exposed sites were located across the eastern edge of the Front Range facing the metropolitan areas stretching from Fort Collins to Colorado Springs (Figure 9.8, Table 9.4). We thought this area was exposed to relatively high concentrations of ozone. Ten protected sites were located to the west of Colorado Springs, across the Rampart Range, where we thought less exposure to ozone would occur due to topographic barriers and distance from pollution sources. Although this assumption was not supported by 1988 monitoring data from Manitou Experimental Forest (Zeller and McKinney 1989), we retain the exposed and protected terminology in the following discussion for convenience and consistency.

Trees were sampled in closed-canopy stands of ponderosa pine on slopes of 0–50%, between 2000 and 2700 m elevation. Sampling was conducted along 20 m wide transects, with 24 trees sampled per site. Trees were included in the sample if they had the following characteristics: (1) at least 50 years old at breast height (1.4 m above the ground), (2) no major crown or stem deformities or scars, (3) crown classification was dominant, codominant, or intermediate (sensu Spurr and Barnes 1980). Two cores were removed at breast height, the diameter (at 1.4 m) and height were measured, dominance was categorized based on crown classification, and needle retention was evaluated for five randomly selected trees per site, (three randomly selected branches per tree).

Chronology Development

Ring-width series were included in the database only if they could be confidently crossdated (Stokes and Smiley 1968, Fritts 1976, Swetnam et al. 1985). The ring widths were measured to the nearest 0.01 mm and were used to calculate basal area increments (BAI) over time. BAI time series for each tree were used in subsequent analyses.

Tree-growth trends were analyzed with Kalman filter procedures (Kalman 1960, Kalman and Bucy 1961). The Kalman filter is a recursive procedure using state-space formulation of a linear system that allows

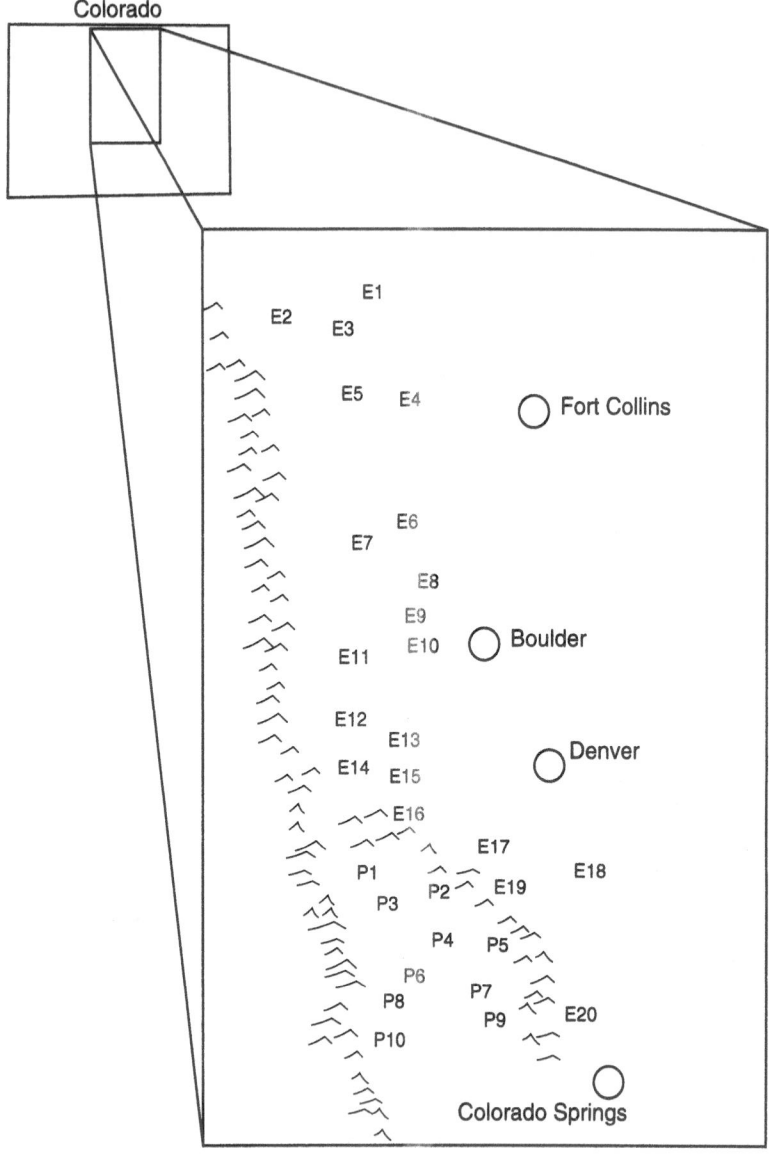

Figure 9.8 Location of second-growth ponderosa pine stands from which tree-ring chronologies were developed. See Table 9.4 for site characteristics.

Table 9.4 *Ponderosa pine tree-ring collections, second-growth stands.*

Site Code	Elevation (m)	Number of Samples	Beginning Year
E01	2400	23	1917
E02	2500	23	1916
E03	2500	23	1911
E04	2100	22	1936
E05	2200	24	1936
E06	2300	23	1920
E07	2600	23	1917
E08	2200	23	1930
E09	2500	23	1936
E10	2100	23	1930
E11	2700	23	1929
E12	2400	22	1932
E13	2500	23	1931
E14	2600	24	1934
E15	2500	24	1937
E16	2400	24	1908
E17	2300	23	1937
E18	2000	21	1920
E19	2400	24	1927
E20	2400	24	1923
P01	2400	24	1911
P02	2000	23	1923
P03	2100	24	1936
P04	2200	24	1923
P05	2600	24	1911
P06	2600	23	1921
P07	2500	24	1931
P08	2700	24	1905
P09	2500	24	1927
P10	2800	24	1913

parameters to be nonstationary with time. It has been used in other studies on averages of residual series (Visser and Molenaar 1986, Van Deusen 1987, see Chapter 7).

We used a version of the Kalman filter that would allow us to accurately detect long term changes in individual tree growth. The model formulation was:

$$Y_t = a_0(t) + a_1(t)Y_{t-1} + e_t$$

Where: Y_t = basal area increment at year t

$a_0(t)$, $a_1(t)$,... are parameters analogous to regression coefficients, also referred to as response functions

e_t = measurement error at time t

Parameter fluctuations through time were assumed to follow a random walk:

$$a_i(t) = a_i(t-1) + w_i(t) \qquad i = (0,1,2)$$

where a_i's are model parameters, and $w_i(t)$ is independent normally distributed random disturbance. Interpretation was based primarily on the parameter a_0, because it was the parameter most indicative of a trend component (see Chapter 7). A change in growth trend was defined to be a point at which $a_0(t)$ estimates changed such that a significant trend resulted. A trend was defined as a consistent direction of $a_0(t)$ increase or decrease of 10 years or more duration. Corresponding significant changes in $a_1(t)$ and $a_2(t)$ at the same time were used as supporting evidence. Series without significant departures from the stationarity hypothesis, or with growth patterns without clear trends were considered to have no change. Growth changes were recorded as an increase, decrease, or no change for each tree. They were summarized by site and expressed as frequencies per decade.

Growth trends were examined at a different level of resolution by dividing the basal area time series of each tree into three time segments: 1900–1929, 1930–1959, and 1960–1986. The proportional change between periods was calculated as:

DIFF1 = (BA2 - BA1) / BA1

where BA1 is total basal area for 1900–1929, and BA2 is total basal area for 1930–1959. DIFF2 was calculated as (BA3-BA1)/BA1, where BA3 is total basal area for 1960–1986. DIFF3 was calculated as (BA3-BA2)/BA2. Mean changes in DIFF1, DIFF2, and DIFF3 were calculated, and the hypothesis that the mean proportional changes were equal to 0 was tested with a two-sided student's t-test. Correlations of the mean differences with tree age were calculated.

The final part of the analysis examined trends and correlation of detrended basal area residuals with the nearest climatic station data. Monthly total precipitation and mean temperature for four weather stations (Waterdale, Boulder, Kassler, Cheesman) were used. These stations covered the length of the Front Range from north to south and were at lower elevation than most of the study sites. Variables used in the analysis were: annual precipitation (based on a precipitation year of October through September), spring temperature (April through July average) and summer temperature (June through August average). Basal area increment trends and correlations with climatic variables were summarized by site.

Site Characteristics

There were no major differences between exposed and protected sites with respect to elevation, slope, and aspect. Density varied greatly among sites, and ranged from 300 to 2820 stems/ha, while basal area varied from 11 to 35 m^2/ha. Ponderosa pine comprised the majority of stem density in most cases, and over 80% of stand basal area in all cases. Stands with high densities had large numbers of seedlings or suppressed individuals of ponderosa pine or quaking aspen in the understory. Mean diameter of sample trees ranged from 235 to 413 mm, and mean height ranged from 10 to 19 m; there was low variation in tree size at most sites. Mean earliest crossdated tree rings ranged from 1851 to 1921, with the oldest tree dated to 1765 and the youngest tree to 1937. Most trees at most sites had initial tree ring dates between 1880 and 1920. Most trees were relatively small, despite an average age of about 80 years, owing to the relatively short growing season in the Front Range. We measured needle retention to see if any spatial patterns were evident in relation to ozone exposure. The average age of the oldest class of needles retained on the trees ranged from 5.6 to 6.8 years, with lowest needle retention at exposed sites directly west of Denver. Highest needle retention was to the west of Fort Collins. Sites west of Denver were the only exposed sites to average less than 6 years needle retention, although some of the protected sites also had less than 6 years needle retention.

It is difficult to relate needle retention to ozone injury because ozone exposure is poorly characterized at mountain locations, and because there was no visible chlorotic injury. Morphological features such as needle retention may be related to genetic and environmental factors not measured in this study. The relatively low standard deviation measured for this variable suggests that needle retention was quite uniform within a site.

Lack of ozone-induced chlorotic injury corroborates the results of previous surveys of ponderosa pine in the Front Range (James and Staley 1980; K. Stolte, unpublished data from 1987). Pollution exposure has apparently not been high enough to produce observable injury symptoms even in a species as sensitive as ponderosa pine. This contrasts with other areas of the West where visible ozone injury is widespread (Pronos and Vogler 1981; Peterson et al. 1989; see Chapters 11, 12). The variety of ponderosa pine found in the Front Range (var. *scopulorum*) may not be as susceptible to ozone injury as varieties in other areas of the West (James and Staley 1980, Aitken et al. 1984).

Growth Trends in Second-Growth Stands

A wide variety of growth trends was evident within and among stands in the Front Range. Principal component analysis (PCA) was used to see if there was any homogeneity in growth trends. The variance accounted for by principal component axis 1 (PC1) ranged from 0.44 to 0.62, and the proportion of the variance accounted for by the first two axes ranged from 0.62 to 0.78. Additional axes accounted for only small portions of the variance. Furthermore, the growth trend of at least half the trees at each site (range of 10 to 19) correlated well with PC1 in all but two sites. This indicated that a large number of trees at any given site had similar growth trends.

A typical graphical result of PCA is displayed in Figure 9.9 to illustrate basal area growth trend of trees from a site in which most trees regenerated about 1900. There was no overall trend in PC1, but there was a prominent growth decrease starting in the second half of the 1940s. The graph in Figure 9.10 illustrates the growth trend of trees from a site in which many trees regenerated in the early to mid 1800s. The overall PC1 trend was decreasing with prominent periods of decrease during the 1920s and 1940s. The overall decrease in PC1 was typical of Front Range stands with older trees in which the period of time represented by PC1 included the normal reduction in basal area growth later in the tree's life.

Kalman filter results allowed us to examine individual growth trends in greater detail. These trends are displayed in Figures 9.11 and 9.12. One of the most striking trends was the large number of growth decreases in the 1940s. This decrease started in the latter half of the decade and continued into the 1950s. This decrease was found in at least some trees at each site and occurred in 83% of the trees in one site (Figure 9.11). This growth decrease in the 1940s was also observed in ponderosa pine in northern New Mexico (Swetnam 1987). A large number of growth decreases were also found in the 1920s, with up to 46% of the trees at a site having this trend. This trend was also found in ponderosa pine in northern New

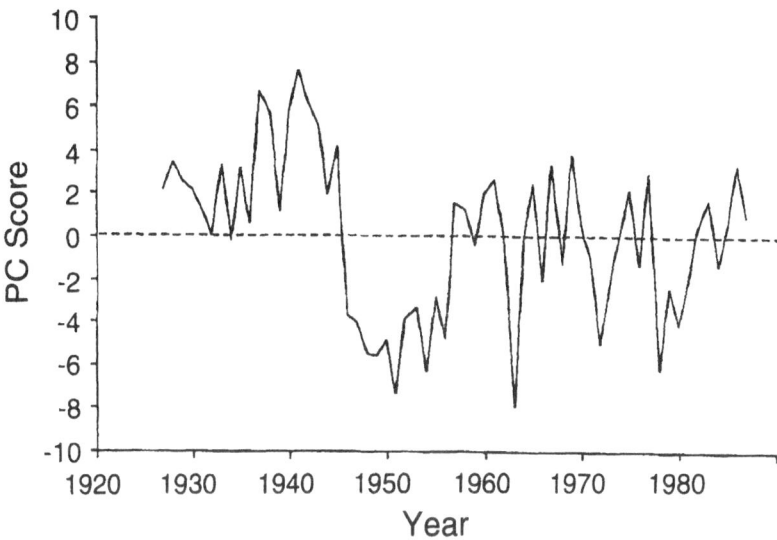

Figure 9.9 The first principal component for the time series of basal area for second-growth ponderosa pine from a site with average age of 80 years.

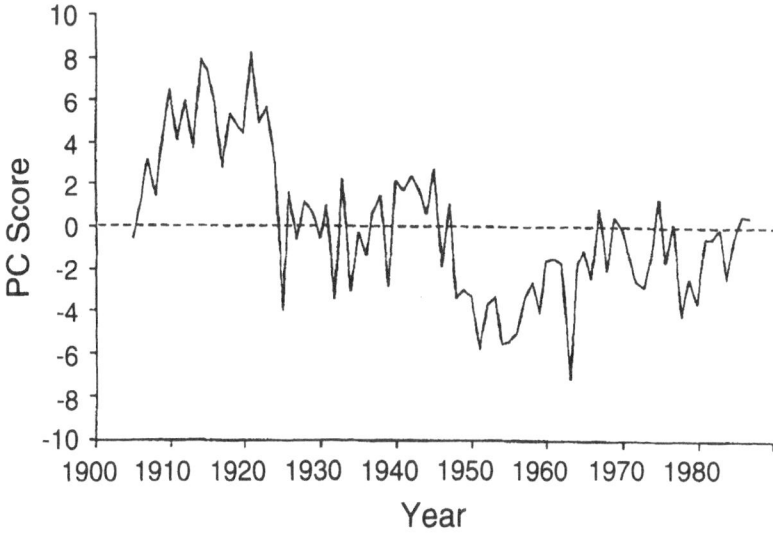

Figure 9.10 The first principal component for the time series of basal area for second-growth ponderosa pine from a site with average age of 100 years.

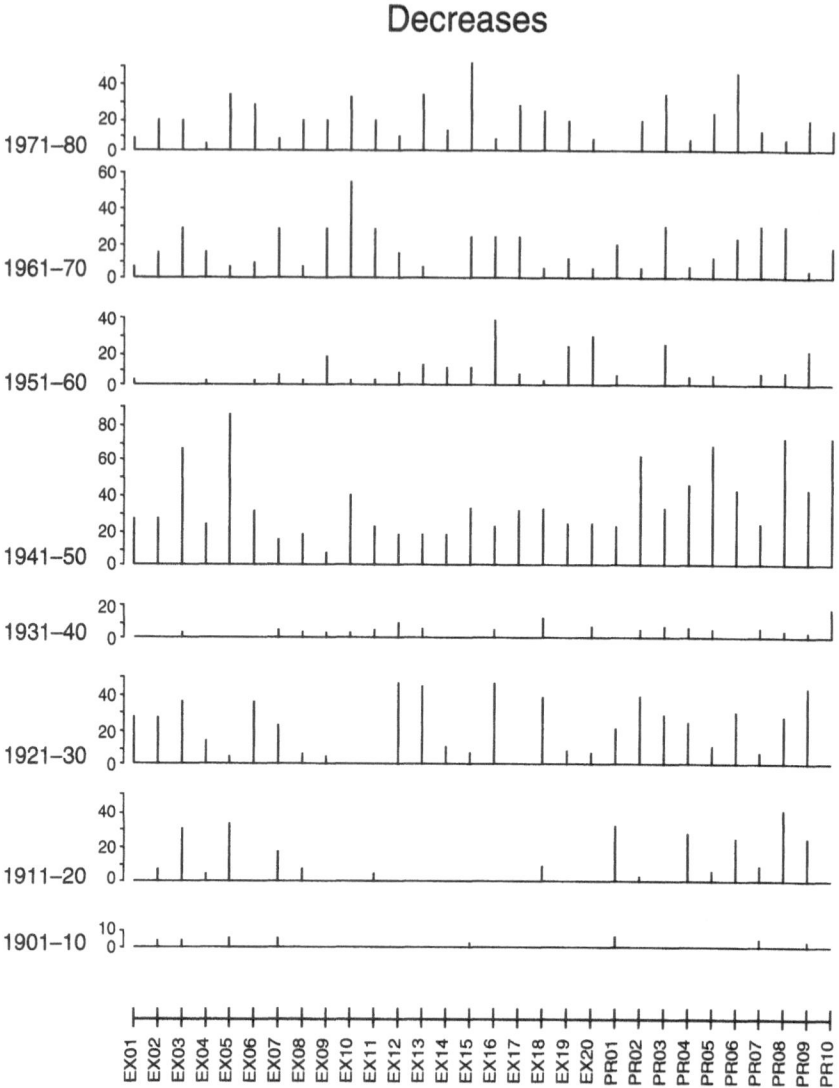

Figure 9.11 Percentage of second-growth ponderosa pine at each study site that had growth decreases is indicated by decade. The point at which the decrease started is indicated, rather than the duration of the decrease.

Increases

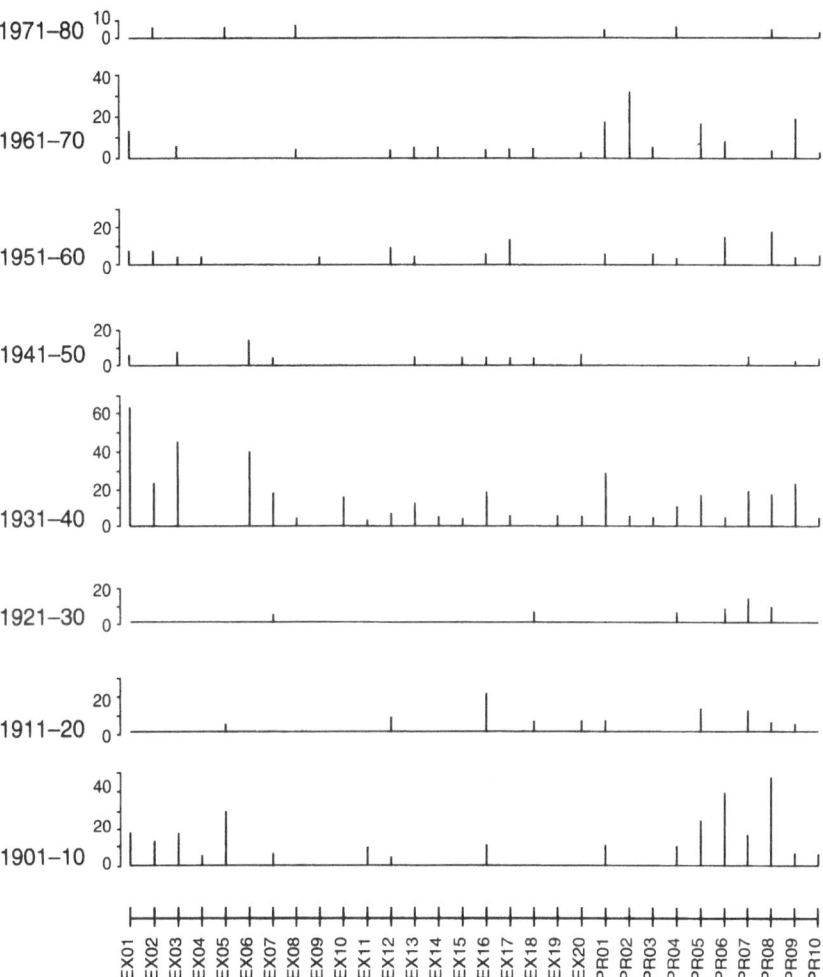

Figure 9.12 Percentage of second-growth ponderosa pine at each study site that had growth increases is indicated by decade. The point at which the increase started is indicated, rather than the duration of the increase.

Mexico (Swetnam 1987). There were very few trees with increases during this decade or the 1940s (Figure 9.12). Such decreases and increases in growth rates in the Front Range seem to have a periodicity of 7 to 24 years, with a mean period of about 14 years (Schweingruber et al. 1990).

The only prominent growth increase at any time in this century started in the late 1930s, despite relatively low precipitation during the early part of the decade (Figure 9.3). This period of increase was found in some trees at all but two sites, and occurred in 61% of trees at one of the exposed sites. In general, the Kalman filter analysis detected far more decreases than increases (Figures 9.11 and 9.12), which suggests that decreases tended to be of greater duration when they occurred. There did not appear to be any spatial patterns to decreases or increases.

Of greatest interest to this study was the possibility of any recent growth decreases that may be related to air pollution or other causes. There were a large number of sites that had growth decreases in the 1960s (maximum of 52% of trees at a site showed a decrease) and 1970s (maximum of 50%). If the number of decreases in the 1960s and 1970s are added, 12 to 82% of the trees at a given site had decreased growth during these decades. This may initially seem to be a significant result, but most trees had a sharp growth increase in the 1980s that compensated for lower growth in the previous two decades.

Basal area growth was also evaluated by comparing proportional growth changes (DIFF1, DIFF2, DIFF3) for the time segments 1900–29, 1930–59, and 1960–86. The mean values of DIFF1, DIFF2, and DIFF3 for all sites were 1.03, 0.84, and 2.73 respectively. All values of DIFF() were positive for all sites except two (for DIFF1 only), and nearly all values differed significantly from zero. This indicates that basal area increment generally increased through time, as one might expect from a second-growth stand of young-to-moderate age. DIFF1 and DIFF3 had significant negative correlations with tree age, but DIFF2 did not. This is probably the result of the steep basal area increment increase normally found in the years following germination compared to later growth.

Residuals of basal area growth correlated positively with annual precipitation and negatively with spring and summer temperatures (Table 9.5), similar to old-growth forests. Correlations were significant for nearly all sites and time periods for precipitation, and were significant for most sites for temperature with the exception of spring temperature in 1900–29 and 1960–86. Correlations were highest for all variables in the period 1930–1959. The value of the correlation coefficient increased between the first and second time periods at 20 sites for precipitation, 22 sites for spring temperature, and 20 sites for summer temperature. The correlation coefficient decreased between the second and third time periods at 24 sites for spring temperature.

Climatic analysis indicated a stronger relationship between growth and climate variables in 1930–59 than in the earlier or later time period. Growth correlated positively with precipitation and negatively with spring and summer temperature. The 1900–29 period was relatively cool and wet, while 1930–59 was relatively warm and dry. Tree growth was reduced somewhat during this latter period as evidenced by the large number of decreases in the 1940s (Figure 9.11). Furthermore, growth was affected to a greater degree by a combination of low precipitation and high summer temperatures at this time, which led to lower soil moisture availability during the growing season. Precipitation and temperatures were moderate in 1960–86, although relatively high precipitation in the 1980s may have influenced the high basal area growth in this decade.

Summary for Second-Growth Forests

The overall growth trend of ponderosa pine in the Front Range was typical of second-growth stand dynamics. Basal area increment increased rapidly during the early part of the life cycle and the rate of increase then slowed but still remained positive at most sites. Many older residual trees that were left from earlier timber harvests generally had decreasing basal area increment during much of this century. We found no evidence that ponderosa pine growth in these stands was affected by ozone exposure. Needle retention was slightly lower to the west of the Denver metropolitan area, but there was no evidence of chlorotic injury. Ozone levels in the Front Range are not much lower than in some parts of California where visible injury symptoms have been noted on ponderosa pine (Pronos and Vogler 1981, Peterson and Arbaugh 1988, see Chapters 11, 12). Possible explanations for the lack of symptoms in the Front Range could be differences in environmental stresses, in the diurnal and

Table 9.5. Mean correlation between basal area growth of second-growth ponderosa pine and climatic variables for three different time intervals. Correlation coefficients are mean values for all 30 study sites.

Time interval	Climate variable		
	Annual precip.	Spring temp.	Summer temp.
1900–1929	0.29	-0.12	-0.22
1930–1959	.44	-.30	-.36
1960–1986	.34	-.12	-.31

seasonal variations in ozone concentrations, and in the genetics of the different varieties of ponderosa pine. Physiological effects that are not visible could also be occurring in the Front Range. The large number of growth decreases in the 1960s and 1970s suggest a possible growth trend, but such decreases were common in earlier periods, and these recent decreases were largely negated by increases in the 1980s.

Conclusions from Old-growth and Second-growth Studies

Extensive geographical and elevational sampling of tree growth in the Colorado Front Range in 1988 resulted in dendrochronological samples from 50 coniferous forest stands. Twenty six tree-ring index chronologies were developed from unmanaged old-growth stands, and 30 basal area increment chronologies were developed from second-growth stands. Ponderosa pine is represented by 40 of these chronologies, with four each from Douglas-fir and Engelmann spruce, three from limber pine and two from lodgepole pine.

Visual examination of needle characteristics indicated no chlorotic mottling indicative of ozone damage. No visual symptoms of poor or declining growth were seen that were remarkable or suggestive of pollution damage.

Growth trends in second-growth stands indicated that most trees in any one stand had similar growth patterns over time, and the patterns of basal area growth were typical for stands of this age class and type. Several periods of growth increase and decrease were found but there did not appear to be any spatial pattern, and none appear related to pollution. Most trees in each of the old-growth stands also exhibited similar growth patterns, and no major differences were found in those patterns for index chronologies from younger individuals in the 100 year age class vs. older individuals.

Correlation analyses of all chronologies with climate time series discovered reasonably consistent and expected patterns in regional growth trends for most stands of ponderosa pine, Douglas-fir and limber pine, particularly those growing under conditions of moderate to strong moisture stress. There is some time dependency in these relationships that is caused by moisture variability. The period from about 1930 to 1959 was markedly drier than preceding or succeeding periods, and tree growth is more highly correlated with climate during that period. Most do however continue to respond in predictable or expected ways to climate in the recent decades of highest pollution, thereby suggesting that pollution has not affected the climate–tree growth relationship in most cases.

A second pattern of growth response to climate was discovered that was not understood. Six stands (four of Engelmann spruce, one of ponderosa pine and one of lodgepole pine) showed strong time-dependent changes in response to annual precipitation during the course of the 20th century. The pattern of change is from moderate and sometimes significant correlation with precipitation before 1932 to declining correlations that are sometimes negative and significant during the past three decades. Four of the six stands are near the western edge of the Continental Divide and all are above 2620 m in elevation. This kind of tree growth response could be due to the effects of precipitation borne pollutants.

References

Aitken WM, Jacobi WK, Staley JM (1984) Ozone effects on seedlings of Rocky Mountain ponderosa pine. *Plant Disease* 68:398–401

Alexander RR (1987) *Classification of the Forest Vegetation of Colorado by Habitat Type and Community Type.* USDA Forest Service Research Note RM-478

Aplet G, FW Smith, RD Laven (1989) Stemwood biomass and production during spruce-fir stand development. *Journal of Ecology* 77:70–77

Baron J (1983) Comparative water chemistry of four lakes in Rocky Mountain National Park. *Water Resources Bulletin* 19(6):897–902

Box GEP, Jenkins GM (1976) *Time Series Analysis: Forecasting and Control.* Holden-Day, San Francisco

Deen JL 1945. The second mile up is forested. *American Forests* 51:284–287, 308–310

Draper N, Smith H (1981) *Applied Regression Analysis.* Wiley, New York

Ewers FW, Schmid R (1981) Longevity of needle fascicles of *Pinus longaeva* (bristlecone pine) and other North American pines. *Oecologia* 51:107–115

Fehsenfeld FC, Bollinger MJ, Liu SC, Parrish DD, McFarland M, Trainer M, Kley D, Murphy PC, Albritton DL, Lenschow DH (1983) A study of ozone in the Colorado mountains. *Journal of Atmospheric Chemistry* 1:87–105

Furniss RL, Carolin VM (1977) *Western Forest Insects.* USDA Forest Service Miscellaneous Publication #1339, US Government Printing Office, Washington, DC

Fritts HC (1976) *Tree Rings and Climate.* Academic Press, New York

Fritts HC, Smith JW, Budelsky CA, Cardis JW (1965) Tree-ring characteristics along a vegetation gradient in northern Arizona. *Ecology* 46(4):393–401

Graybill DA (1979) Revised computer programs for tree-ring research. *Tree-Ring Bulletin* 39:77–82

Graybill DA (1982) Chronology development and analysis. In: Hughes MK, Kelly PM, Pilcher JR, LaMarche VC Jr (eds) *Climate from Tree-Rings.* Cambridge University Press, London, pp 21–28

Graybill DA (1987) A network of high-elevation conifers in the western United States for detection of tree-ring growth response to increasing carbon dioxide. In: Jacoby GC, Hornbeck JW (compilers) *Proceedings of the International Symposium on Ecological Aspects of Tree-Ring Analysis.* Department of Energy. National Technical Information Service Publication CONF-8608144, US Department of Commerce, Springfield, Virginia pp 463–474

Harlow WM, Harrar ES (1969) *Textbook of Dendrology.* McGraw-Hill, New York

Hawksworth FG (1977) *The Six-class Dwarf Mistletoe Rating System.* USDA Forest Service General Technical Report RM-48

Hawksworth FG, Shaw CG (1984) Damage and loss caused by dwarf mistletoe in coniferous forests of western North America. In: Wood RKS, Jellis GJ (eds) *Plant Diseases: Infection, Damage and Loss.* Blackwell Scientific Publications, Oxford, UK, pp 285–297

Heit M, Kluser C, Baron J (1984) Evidence of deposition of anthropogenic pollutants in remote Rocky Mountain lakes. *Water, Air, and Soil Pollution* 22:403–416

Hepting GH (1971) Diseases of forest and shade trees of the United States. *USDA Forest Service Handbook #386.* US Government Printing Office, Washington, DC

Hess K, Alexander RR (1986) *Forest Vegetation of the Arapaho and Roosevelt National Forests in Central Colorado: A Habitat Type Classification.* USDA Forest Service Research Paper RM-266

Innes J (1990) General aspects in the use of tree-rings for environmental impact studies. In: Cook ER, Kairiukstis LA (eds) *Methods of Dendrochronology: Applications in the Environmental Sciences.* Kluwer, Boston, pp 224–229

James RL, Staley JM (1980) *Photochemical Air Pollution Damage Survey of Ponderosa Pine Within and Adjacent to Denver, Colorado: A Preliminary Report.* USDA Forest Service Forest Insect and Disease Management Biological Evaluation R2-80-6

Jolliffe IT (1986) *Principal Components Analysis.* Springer-Verlag, New York

Kalman RE (1960) A new approach to linear filtering and prediction problems. *Transactions of the American Society of Mechanical Engineering Journal of Basic Engineering (Series D)* 82:35–45

Kalman RE, Bucy RS (1961) New results in linear filtering and prediction problems. *Transactions of the American Society of Mechanical Engineering Journal of Basic Engineering (Series D)* 83:95–108

Kozlowski TT, Constantinidou HA (1986a) Responses of woody plants to environmental pollution. *Forestry Abstracts* 47(1):5–51

Kozlowski TT, Constantinidou HA (1986b) Environmental pollution and tree growth. *Forestry Abstracts* 47(2):105–132

LaMarche VC Jr (1974) Frequency-dependent relationships between tree-ring series along an ecological gradient and some dendroclimatic implications. *Tree-Ring Bulletin* 34:1–20

LaMarche VC Jr (1982) Sampling strategies. In: Hughes MK, Kelly PM, Pilcher JR, LaMarche VC Jr (eds) *Climate from Tree-Rings.* Cambridge University Press, London, pp 2–6

Lewis WM, Grant MC, Saunders JF,III (1984) Chemical patterns of bulk atmospheric deposition in the state of Colorado. *Water Resources Research* 20(11):1691–1704

Miller RL, Choate GA (1964) *The Forest Resource of Colorado.* USDA Forest Service Resource Bulletin INT-3, Ogden, UT

Palmer WC (1965) *Meteorological Drought.* US Weather Bureau Research Paper 45, Washington, DC

Peterson DL, Arbaugh MJ (1988) Growth patterns of ozone-injured ponderosa pine (*Pinus ponderosa*) in the southern Sierra Nevada. *Journal of the Air Pollution Control Association* 38:921–927

Peterson DL, Arbaugh MJ, Robinson LJ (1989) Ozone injury and growth trends of ponderosa pine in the Sierra Nevada. In: Olson RK, Lefohn AS (eds) *Effects of Air Pollution on Western Forests.* Transactions Series, No. 16, Air and Waste Management Association, Pittsburgh, pp 293–307

Pronos J, Vogler DR (1981) *Assessment of Ozone Injury to Pines in the Southern Sierra Nevada, 1979/1980.* USDA Forest Service Pacific Southwest Forest Pest Management Report 81-20

Robinson WJ, Evans R (1980) A microcomputer-based tree-ring measuring system. *Tree-Ring Bulletin* 40:59–64

Schweingruber FH, Aellen-Rumo K, Weber U, Wehrli U (1990) Rhythmic growth fluctuations in forest trees of Central Europe and the Front Range in Colorado. *Trees: Structure and Function* 4:99–106

Spurr SH, Barnes BV (1980) *Forest Ecology.* John Wiley and Sons, New York

Stevens RE, Myers CA, McCambridge WF, Downing GL, Laut JG (1975) *Mountain Pine Beetle in Front Range Ponderosa Pine.* USDA Forest Service General Technical Report RM-7

Stokes MA, Smiley TL (1968) *An Introduction to Tree-Ring Dating.* University of Chicago Press, Chicago

Swetnam TW (1987) Western spruce budworm outbreaks in northern New Mexico: Tree-ring evidence of occurrence and radial growth impacts from 1700 to 1983. In: Jacoby GC, Hornbeck JW (compilers) *Proceedings of the International Symposium on Ecological Aspects of Tree-Ring Analysis.* Department of Energy. National Technical Information Service Publication CONF-8608144, Springfield, Virginia, pp 130–141

Swetnam TW, Thompson MA, Sutherland EK (1985) *Using Dendrochronology to Measure Radial Growth of Defoliated Trees.* USDA Forest Service Agricultural Handbook 639, Washington, DC

USDA Forest Service (1981a) *An Assessment of the Forest and Range Land Situation in the United States.* Forest Resource Report 22, Washington, DC

USDA Forest Service (1981b) *Arapaho and Roosevelt National Forests Land and Resource Management Plan.* USDA Forest Service, Rocky Mountain Region, Fort Collins, CO

Van Deusen PC (1987) Some applications of the Kalman filter to tree-ring analysis. In: Jacoby GC, Hornbeck JW (compilers) *Proceedings of the International Symposium on Ecological Aspects of Tree-Ring Analysis.* Department of Energy, National Technical Information Service Publication CONF-8608144, Springfield, Virginia, pp 566–578

Visser H, Molenaar J (1986) *Time Dependent Responses of Trees to Weather Variations: An Application of the Kalman Filter.* N. V. Tot Keuring Van Elektrotechnische Materialen, Research and Development Division, Arnhem

Weidman RH (1939) Evidence of racial influence in a 25-year test of ponderosa pine. *Journal of Agricultural Research* 59:855–887

Zeller KF, McKinney WR (1989) *USFS Manitou Experimental Forest 1988 Ambient Ozone and Meteorological Data Report.* Rocky Mountain Experiment Station Report, Fort Collins, CO

10

Coniferous Forests
of Arizona and New Mexico

D. A. Graybill and M. R. Rose

Forests in central and southern Arizona are exposed to relatively high concentrations of ozone and other pollutants (Chapter 3) that result from point sources (such as copper smelters) and urban areas (such as Phoenix and Tucson). To the north and east of the urban areas, air quality is better and forest exposure to pollutants is lower. We used a series of analytical strategies to evaluate the nature of variability in the annual growth increments of 41 tree-ring chronologies in forests from across central and southern Arizona and southern New Mexico.

Although there are many components of forest growth, radial growth is a single variable that is highly representative of forest growth in general (Avery and Burkhart 1983). Tree-ring series are especially useful because they provide the only long-term annual records of forest condition that begin before urban and industrial pollution of the 20th century and continue to the present. This permits the development of baseline characterizations of natural variability in growth during times prior to suspected pollution effects, and then a search for deviations of recent growth from those norms (Innes 1990).

The macroclimatic signal can be relatively strong in Southwestern trees, can usually be isolated, and relationships with tree growth can be interpreted. The interpretive framework is based on substantial physiological research and on models relating tree-growth to climate that have been developed in this region (Brown 1968, Budelsky 1969, Fritts 1976). Analytical techniques for working with those relationships are now well developed (Hughes et al. 1982, Cook and Kairiukstis 1990), and the normal relationships between climate and tree growth might change under conditions of pollution stress that lead to disruption of normal physiological activity (Kozlowski and Constantinidou 1986a,b).

Forests of Arizona and New Mexico

Extensive conifer forests arc across central Arizona into western New
Mexico, in and near the mountainous Mogollon Rim that separates the
upland Colorado Plateau from the dry Basin and Range Province (Figure
10.1). The two dominant tree species are ponderosa pine (*Pinus ponderosa*) and Douglas-fir (*Pseudotsuga menziesii*), with other species including southwestern white pine (*Pinus strobiformis*), Engelmann spruce
(*Picea engelmannii*), white fir (*Abies concolor*), blue spruce (*Picea pungens*),
limber pine (*Pinus flexilis*) and subalpine fir (*Abies lasiocarpa*). In the Basin
and Range Province of south central and southeastern Arizona, forests
occur on several isolated mountains surrounded by shrublands, grass-
lands, and deserts. Large sectors of the mountainous and upland regions
are also forested with varying proportions of pinyon pine (primarily
Pinus edulis), junipers (*Juniperus* spp.) and oaks (*Quercus* spp.); pinyon-
juniper woodlands are more extensive than all other forest types combined in both Arizona and New Mexico (USDA Forest Service 1981).

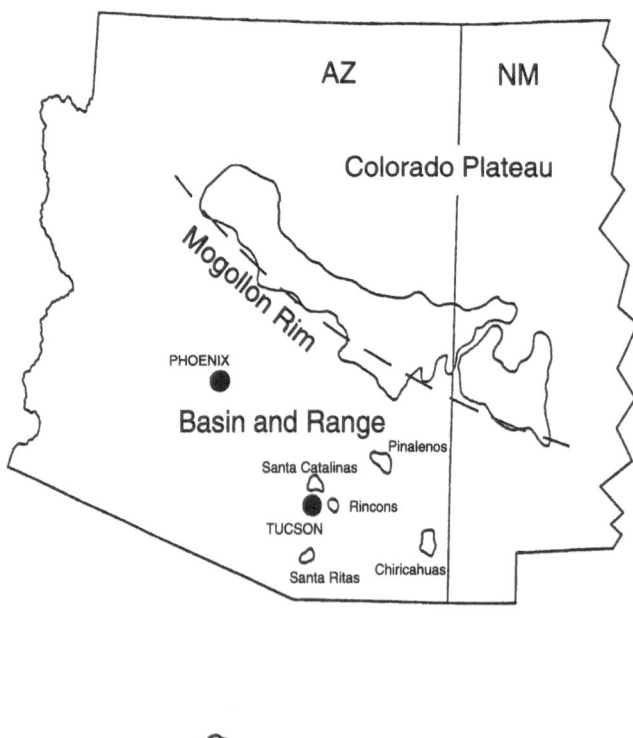

Figure 10.1 Physiographic features and coniferous forests of the research area.

Pinyon-Juniper Woodland

The pinyon-juniper woodlands occur at the lower elevational limits of forests in this region, typically occurring from 1370 to 2440 m. Estimates of the areal extent of these woodlands in Arizona and New Mexico range from about 9 million ha (USDA Forest Service 1981) to 12 million ha (Springfield 1976, Meeuwig and Basset 1983). This association is a source of fuel wood, fenceposts, Christmas trees, pinyon nuts, and important wildlife habitat. Its greatest economic importance may be as rangeland for cattle grazing. Lanner (1981) presents an intriguing introduction to the natural history and cultural use of pinyon pine, and a comprehensive summary of recent research and an extensive bibliography is available in Everett (1987).

Pinyon-juniper woodland exists in a variety of topographic and edaphic conditions. Low precipitation and high evaporative demand lead to moisture stress, modest leaf area indexes (1.0 to 3.6 m^2/m^2 on a projected basis; Schuler and Smith 1988) and low biomass productivity (maximum of about 1 m^3/ha annually; Buckman and Wolters 1987, Schuler and Smith 1988). Regeneration is intermittent and sparse (Samuels and Betancourt 1982, Ronco 1987). Within the woodland type, pinyon pine is the more common species at upper elevations and higher latitudes while juniper species are more common at the lower elevations and latitudes. The common juniper species include Utah juniper (*Juniperus osteosperma*), one-seed juniper (*Juniperus monosperma*), Rocky Mountain juniper (*Juniperus scopulorum*) and alligator juniper (*Juniperus deppeana*). Pinyon-juniper woodland is often bordered by grassland, oak woodland, or desert scrub at its lower limit, and by ponderosa pine or Gambel oak (*Quercus gambelii*) at upper limits.

Ponderosa Pine Forests

The ponderosa pine type forest is found throughout the region at elevations of 1525 to 2750 m, with its prime habitat between 2150 and 2450 m. It covers about 3.5 million ha in Arizona and New Mexico (Shupe 1965; USDA Forest Service 1981). There is an extensive literature regarding this major timber species in the Southwest (see also Chapter 9), such as the classic management monograph by Pearson (1950), and pioneering papers on ecology by Weaver (1951) and Cooper (1960, 1961). More recent silvicultural and fire ecology summaries are Schubert (1974), Ronco and Ready (1983), and Swetnam and Dieterich (1985).

At the lower elevations of its range, ponderosa pine is associated with juniper, pinyon, and Gambel oak. At upper limits, mixtures include Douglas-fir, aspen (*Populus tremuloides*) and various conifers, depending on latitude. Across most of its range, ponderosa pine occurs in extensive and pure stands, especially along the Mogollon Rim. In southeastern Arizona and southwestern New Mexico there is a ponderosa pine variety with five needled bundles (var. *arizonica* (Engelm.) Shaw) that may co-occur and interbreed with the three needled variety (*scopulorum* Engelm.).

Productivity of ponderosa pine forests is limited by low leaf area indexes (2.0 to 3.5 m^2/m^2 on a projected basis, Whittaker and Niering 1975, Peet 1988) that result from low supplies of water and nutrients. Rates of stem growth average about 2 to 4 m^3/ha in well-stocked stands (Ronco and Reddy 1983).

Cool and moist climatic conditions in 1919 allowed dense thickets to be established across much of the Southwest. Fire suppression and especially heavy grazing in preceding decades also helped insure survival of this large cohort of trees (White 1985).

Mixed Conifer Forests

Mixed conifer forests are complex assemblages of up to eight overstory conifers that cover over 500,000 ha of Arizona and New Mexico (Ronco et al. 1983) at elevations of 2450 to 3050 m. Mixed conifer forests occur as low as 1830 m in cool, moist canyon bottoms and on north-facing slopes (Pissot 1965, Jones 1974, Ronco et al. 1983). This forest type requires moister conditions than ponderosa pine, into which it grades at lower elevations. At upper limits, the mixed conifer type grades into the more cold-tolerant spruce/fir type. Douglas-fir is the dominant species across most of the mixed conifer type, mixed with ponderosa pine, white fir (*Abies concolor*), Engelmann spruce, blue spruce, limber pine, subalpine fir, southwestern white pine and corkbark fir (*Abies lasiocarpa* var. *arizonica* (Merriam) Lemm.).

Limitations on growth have not been experimentally determined in this forest type; leaf area indexes range from about 6 to 8 m^2/m^2 on a projected basis (Whittaker and Niering 1975), and stem growth of about 5 m^3/ha annually (Ronco et al. 1983).

Spruce-Fir Forests

Spruce-fir forests cover about 370,000 ha in Arizona and New Mexico (USDA Forest Service 1981), above the mixed conifer forests (about 2900 m) up to timberline at about 3350 m. Near timberline, subalpine and corkbark firs drop out of the stands, leaving pure stands of Engelmann spruce (Alexander and Engelby 1983). We did not sample this type because of its limited occurrence in the study region.

Pollutants of the Region

Parts of this region have recently experienced high levels of pollutant emissions, with spatial gradients in deposition amounts (Figure 10.2, USEPA 1978, Roth et al. 1985). Sulfur was primarily derived from nonferrous smelters in an area ranging from near the Arizona-Mexico border up to the base of the Mogollon Rim in central Arizona, and then eastward along the Mogollon Rim to the copper mining areas of eastern Arizona and southwestern New Mexico. While some of the mining activity has an extensive history that reaches back into the 19th century, the heaviest smelting activity occurred in the four decades following World War II (Dunning and Peplow 1959, Arizona Department of Economic Security 1983). Trijonis (1979) found a declining trend in visibility from the middle 1950s to the early 1970s, and hypothesized that this was most likely due to corresponding increases in sulfur dioxide, nitrogen oxides and hydrocarbon emissions. Nitrogen oxides and hydrocarbon emissions vary in source but changes in them can be inferred from increases in population and attendant increases in total vehicle and other sources of emissions since mid-century in Phoenix and Tucson. Because time series of ozone concentrations from those cities are less than two decades in length, and vary substantially in terms of monitoring frequency and number of stations, we did not use them in this project. Existing data on ozone patterns in this region are described in Chapter 3.

The Regional Sampling Approach

The forest types that we evaluated included pinyon-juniper woodland, ponderosa pine forests and mixed conifer types. These forests often form complex mosaics rather than extensive homogeneous communities, particularly in the southern mountain ranges. Major differences in moisture availability and evaporative demand occur on scales of hundreds of meters to a few kilometers, resulting from differences in elevation, aspect, and soils in this geologically complex region (see Chapters 1 and 4). Much of the annual precipitation (about half) occurs in winter, and water stored in the soil from winter drives tree growth early in the growing season. High temperatures and variable precipitation during the growing season are also important (Graybill 1989; Chapter 2).

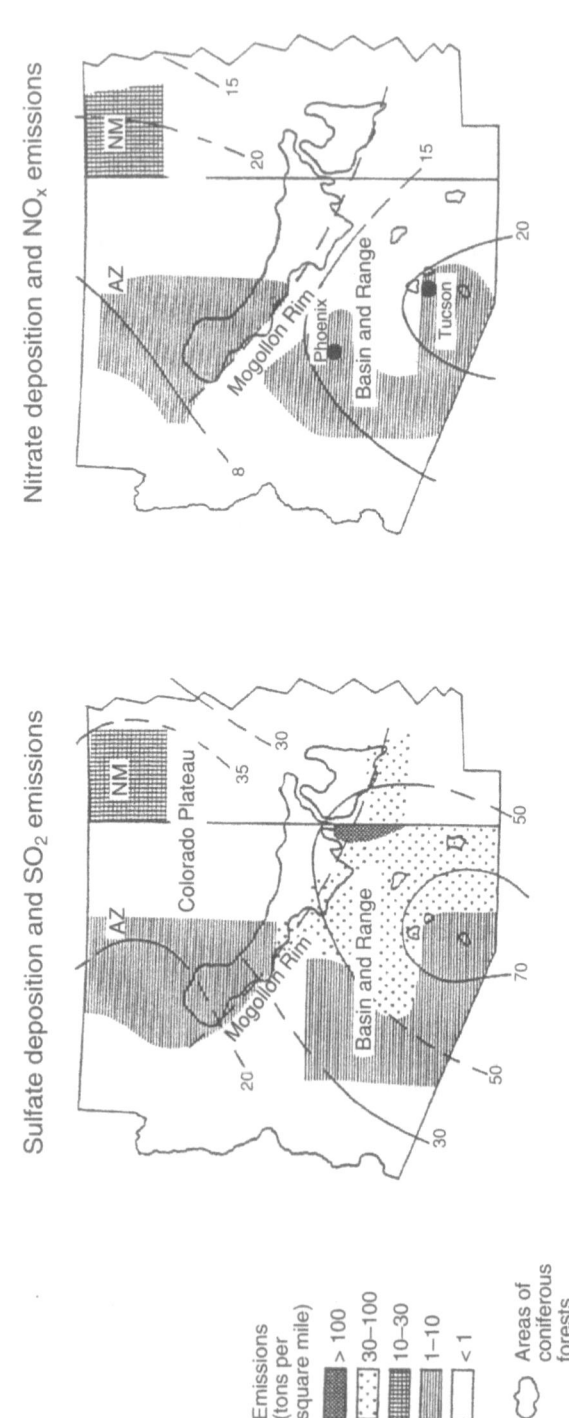

Figure 10.2 *Sulfur and nitrogen deposition and emissions in the southwest US, 1975 (after Roth et al. 1985, USEPA 1978). Contoured lines and associated numerical units are deposition as µeq/l.*

In designing our collection strategy we recognized the geographical differences in pollutant emissions and depositions that have a history of being greater in the southern than the northern area (Figure 10.2). These observations resulted in the stratification of stands into a group north of the Mogollon Rim, and a group to the south (Figure 10.3). Our coverage was wide ranging in the north, and focused primarily on ponderosa pine and Douglas-fir (Table 10.1). Coverage was also widespread in the south, with a more intensive sampling near the rapidly growing cities of Tucson and Phoenix (Figure 10.3). Although smaller than Phoenix in population (about 650,000 vs 1,000,000, Tucson and Phoenix Chamber of Commerce estimates, 1989), Tucson's pollution may potentially have the highest impact on forests at present or in the near future. The city has expanded to the base of two of the ranges (Santa Catalinas and Rincons), is not far from a third (Santa Ritas) and lies less than 1200–1500 m below and 16 km distant from ponderosa pine and mixed conifer forest types. Phoenix is farther from forested areas. We sampled the closest accessible ponderosa stand to Phoenix, about 80 km northeast of the city center.

Some stands in the south were selected because they were in the general vicinity (30–80km) of large copper smelters. We developed tree-ring chronologies from stands near Douglas, Morenci and Tucson, Arizona, and Hurley, New Mexico. The only area with smelter activity and coniferous trees within a few kilometers is near Miami and Globe, Arizona. Nash et al. (1975) have used dendrochronological procedures to examine the possibility of pollution effects on tree growth there.

We developed pinyon tree-ring chronologies from four different locations in the northern sector of the study region and from one in the Rincon Mountains. Junipers were not sampled because most cannot be readily crossdated. We sampled thirteen ponderosa stands in the north and eight in the south, representing a variety of stand conditions from relatively mesic and dense to relatively dry and open. We sampled three stands of southwestern white pine and one of white fir in the mixed conifer forests of southern Arizona. Douglas-fir was collected from four stands in the north and seven in the south.

Many of the sampling decisions at the stand and tree level were made to maximize a climatic signal and remove or control some of the other signals described in Chapter 7 (The Linear Aggregate Model; Equation 7.2). Open canopied stands on relatively well drained sites were most commonly sought to minimize the kinds of disturbance signals that can arise from competition and to maximize any drought signals. A few stands that exhibited some competition and were on relatively mesic sites were also sampled for comparative purposes.

Table 10.1 Tree-ring collections.

Site	Code	Species[1]	Elevation (m)	No. Trees	Dated Range	Signal:Noise Ratio 1850–1986
AGUA FRIA, NM	AFN	PN	2225	17	1361–1987	53.448
BEAVER CREEK, AZ	BCR	PP	2393	21	1559–1987	30.335
BLACK MTN, NM	BKF	DF	2697	39	1462–1987	34.499
BLACK MTN, NM	BKP	PP	2697	26	1555–1987	29.130
BLACK RIVER, AZ	BRF	DF	2438	24	1565–1987	25.682
DRY CREEK, AZ	DCP	PN	1378	23	1608–1987	40.070
EAGLE CREEK, AZ	EPN	PN	1695	20	1639–1987	28.501
G.C.DWELLINGS, NM	GCP	PP	1768	15	1528–1987	—[2]
GUS PEARSON, AZ	GPN	PP	2255	35	1571–1987	39.448
GIRL'S RANCH, AZ	GRP	PP	1969	23	1594–1987	46.315
GRASSHOPPER, AZ	GRS	PP	1798	32	1661–1986	46.951
MULETANK, AZ	MTP	PP	2309	40	1595–1987	6.169
MIMBRES JUNCT, NM	MJN	PN	1951	23	1604–1987	30.678
S.F. PEAKS, AZ	PDF	DF	2682	20	1762–1987	9.767
ROCKY GULCH, AZ	RGP	PP	1966	20	1659–1987	32.474
ROBINSON MTN, AZ	RMP	PP	2225	39	1620–1987	26.965
ROSE PEAK, AZ	RPP	PP	2316	16	1641–1987	52.667
SHOWLOW, AZ	SGP	PP	2073	22	1603–1987	60.870
SLATE MTN, AZ	SMP	PP	2194	19	1589–1987	61.684
WALNUT CYN, AZ	WCP	PP	2057	12	1413–1987	26.239
WALNUT CYN, AZ	WDF	DF	2027	20	1668–1987	20.530
BEAR WALLOW, AZ	BWF	DF	2484	22	1600–1987	11.282
DEVIL'S BATHTUB, AZ	DBN	PN	2286	19	1769–1987	7.968
GREEN MTN, AZ	GMF	DF	2194	25	1547–1986	17.358
GREEN MTN, AZ	GMT	PP	2194	43	1412–1986	19.681
HELEN'S DOME, AZ	HDE	PP	2536	14	1625–1987	18.564
HELEN'S DOME, AZ	HDW	WP	2536	13	1718–1987	16.861
MT. HOPKINS, AZ	MHP	PP	2133	12	1714–1987	17.650
NOON CREEK, AZ	NOO	PP	2347	20	1764–1987	15.456
NORTH SLOPE, AZ	NSA	WF	2438	10	1778–1987	18.390
NORTH SLOPE, AZ	NSE	PP	2438	10	1678–1987	9.456
NORTH SLOPE, AZ	NSF	DF	2438	17	1637–1987	19.067
NORTH SLOPE, AZ	NSW	WP	2438	12	1692–1987	13.853
ORD MTN, AZ	ORD	PP	2133	16	1560–1987	17.145
PINERY CYN, AZ	PCD	DF	2286	22	1641–1987	18.903
POST CREEK, AZ	PST	DF	2731	21	1516–1987	17.278
RHYOLITE CYN, AZ	RCF	DF	1829	27	1620–1987	27.190
RHYOLITE CYN, AZ	RPE	PP	1829	18	1629–1987	14.122
SANTA RITA, AZ	SRH	DF	2400	21	1486–1987	29.888
TUCSON SIDE, AZ	TSE	PP	2365	23	1654–1987	24.041
TUCSON SIDE, AZ	TSW	WP	2365	9	1663–1987	8.630

[1] PN = *Pinus edulis* (pinyon pine)
 PP = *Pinus ponderosa* (ponderosa pine)
 WP = *Pinus strobiformis* (southwestern white pine)
 DF = *Pseudotsuga menziesii* (Douglas-fir)
 WF = *Abies concolor* (white fir)
[2] Inadequate constant specimen depth during 1850–1986 period. Signal information not calculated.

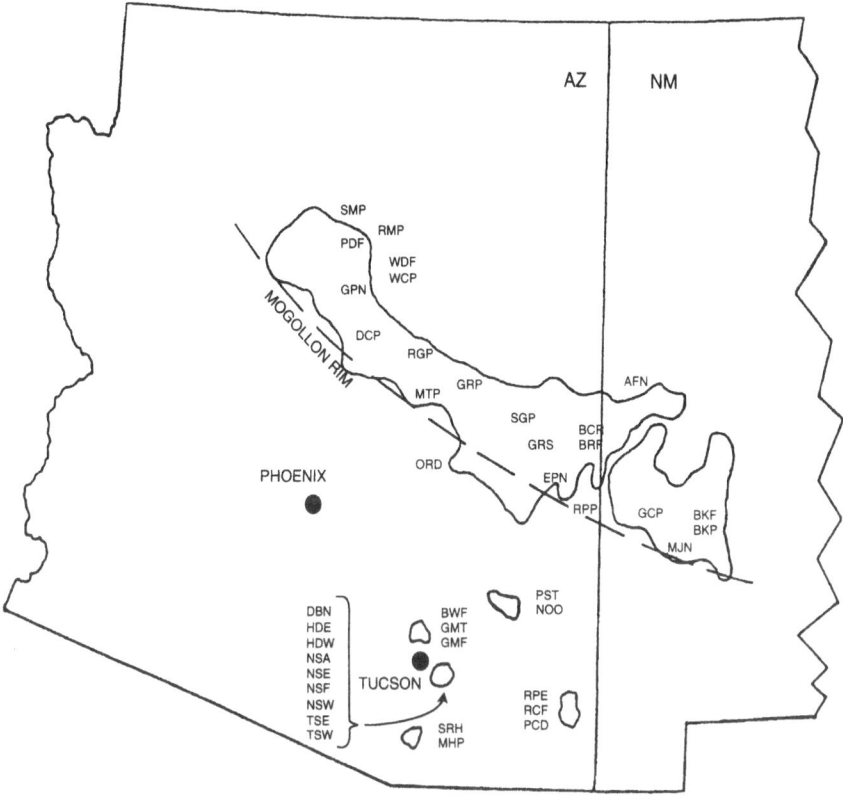

Figure 10.3 Location of stands selected for sampling. See Table 10.1 for site names and characteristics.

Stands with evidence of thinning or other anthropogenic disturbance such as road construction were avoided in order to limit disturbance signals unrelated to pollution. We also avoided trees with obvious defects such as heavy lightning damage, extreme lean, or dwarf mistletoe (*Arceuthobium* spp.) infection that was rated 3.0 (moderate infection) or greater with the Hawksworth (1977) system. We extracted two increment cores at breast height, transverse to the slope, from 15–25 dominant to codominant trees in each collection. Trees older than 150 years were sought. The final chronologies are from stands in a variety of environmental settings that range from the middle-elevation ranges of forests to lower elevation limits of forests.

Chronology Development

First, each ring width series was carefully crossdated and the widths were measured to the nearest 0.01 mm (Stokes and Smiley 1968, Robinson and Evans 1980). This was followed by personal examination as well as by computer aided procedures (Holmes 1983) to check both the dating and measurement accuracy. The final ring width series were then individually standardized and converted to tree-ring indices (Graybill 1979, 1982; Chapter 7). This procedure removed the biological growth trend and resulted in a new time series that was more homoscedastic than the ring widths. Importantly, standardization also allows direct comparison of the variation in growth patterns of all trees, regardless of age or rate of growth.

We primarily used simple deterministic curves such as negative exponentials or straight lines to model age-related growth trends. These curves minimized the chance of removing recent growth trends of interest that were not age related. Indices for individual ring width series were computed and averaged to form a mean index chronology for each collection. The averaging process has the effect of reducing the unique error associated with each component series and maximizes the common variance signal used in subsequent analyses.

The strength of the common signal in a tree-ring series can be estimated by the signal-to-noise ratio (Wigley et al. 1984). While not all of that signal is necessarily determined by climate, it is a useful descriptive measure for comparing the potential merits of different tree-ring chronologies for climate related research. In general, when that ratio drops below about 10:1, the common signal is minimal and the trees do not collectively provide strong estimates of factors that might control growth. The lowest values are normally found in settings where climate (or other factors) does not particularly limit growth and each tree has a relatively wide latitude for growth response. In this study the lowest values were found in relatively mesic stands that were not closed-canopied but where competition was more apparent than in others. The highest values were found in chronologies from trees on xeric sites with poor soils. These were open-canopied stands, presumably with limited or no competition among trees.

The persistence (autocorrelation) structure is one additional characteristic of tree-ring index chronologies that requires consideration. Temporal persistence in a tree-ring chronology is present when successive values in the series are not independent, that is, when each value can be partially predicted from past observations (Chatfield 1980). This component must be identified and treated appropriately before using the chronologies in regression-based analyses. If this is not done, variance estimates can become unreliable and uninterpretable (Wonnacott and Wonnacott 1981, Monserud 1986).

Persistence in a tree-ring series for any year can result from dependence on carbohydrates stored in prior years, from other physiological factors (Fritts 1976), or from persistence in a growth forcing climatic factor. We found that all but one of the tree-ring index chronologies in the current data set had significant autocorrelation. It was then necessary to determine whether this characteristic was primarily biological or climatic in origin so succeeding analyses might be conducted appropriately. We checked the five annual divisional precipitation sums and found them free of autocorrelation for their period of record, 1895–1986.

We next used standard Box and Jenkins (1976) Autoregressive Moving Average (ARMA) protocol to identify a low order parsimonious model that might be used in removing biological persistence from the tree-ring index chronologies. This first required definition of the appropriate time span for model evaluation. We chose the end date to use periods for model fitting that we thought preceded any period of probable abnormal growth. Both 1950 and 1960 were selected and used in all cases. They provide approximate turning points or brackets for post World War II industrial expansion and increased pollution. The choice of a beginning date was determined by the presence of an adequate number of sample index series in each final chronology. The number of component series normally decreases in the earliest years of a chronology to a point where variability increases due simply to sample size. This early portion of a chronology could therefore exert undue influence on the model fitting procedure and lead to unstable estimates of a best-fit model. We therefore determined the earliest year in which a subsample of series contained 90% of the common signal expressed in a chronology of all the trees (Wigley et al. 1984). Best-fit ARMA models were then developed using the length of series from that year to 1950 and again to 1960. We then removed autocorrelation from the entire length of each index chronology via appropriate transformations with those best-fit models established for the pre-1950 and pre-1960 time periods. These time series are referred to as residual chronologies.

Analytical Approaches

The first stage of analysis was model establishment. We performed a climate-tree-ring transfer function analysis to develop reliable predictive relationships of natural tree growth variation from climatic variation over the baseline periods of 1897 to 1950, and 1897 to 1960. Simple linear regression analyses were used and evaluated according to common criteria (Draper and Smith 1981). The dependent variable was always a residual tree-ring chronology while the independent variable was an annual precipitation sum of divisional data (NOAA 1988) for the period of August prior to growth through July of the year of growth. This annual sum was determined to be the strongest and most common predictor of annual growth after extensive screening of various monthly and seasonal values of temperature, precipitation and the Palmer Drought Severity Index (Palmer 1965).

We established statistical models in each baseline period over two subintervals for each of the 41 chronologies; the intervals from 1897 to 1950 and to 1960 were divided into halves. The first interval was initially used for calibration (model development) while the second interval was used for verification (model testing). Then the process was reversed and the second interval was employed in calibrating the model while the first segment was used for verification. This split period calibration/verification approach during the baseline period was adopted to test for time stability of the relationships between climate and tree growth. Only after the split period results were evaluated was a final calibration established with the full baseline period data set. This was important because the climatic data available during the baseline periods spanned only 55 to 65 years. All of the data were used to insure that the final calibration models were developed on the fullest possible range of data covariation.

The equations derived in the model development phase were used to estimate tree growth series in the 1951–1986 and 1961–1986 periods. The null hypothesis was that the estimated growth should be reasonably similar to actual growth in the recent intervals. Failure to reject the hypothesis would support the contention that there has been no change in the relationship between climate and tree growth. If so, the covariance between the predicted and actual growth should not be significantly different, and should be similar to the covariance between actual growth and annual precipitation in the baseline periods. In addition, the means of the predicted and actual growth after mid century should not be significantly different, and they should not exhibit divergent trends.

Alternatively, the covariance between the actual and predicted series may be poor or drop precipitously from that seen in the calibration period. In addition, significant differences may be seen between the means and variances of the actual and the predicted series or the actual and predicted series may have different trends. Inability to reject any of the alternative hypotheses implies that the actual and predicted growth differed in some manner. It does not necessarily mean that an anthropogenic impact on tree growth is the cause of the difference, merely that a difference exists and that some factor(s), including pollution, could be the causal agent(s). That pollution is actually the cause of any tree growth change would remain to be demonstrated.

Covariance was evaluated with Pearson's correlation coefficient and its square, the coefficient of determination. Differences of means and central tendency were tested with the t-test and the Wilcoxon matched-pairs signed-ranks test, while differences in variances were evaluated with an F ratio test. An alpha level of .05 was set for all tests and these were two-tailed. Descriptions of these statistical procedures are found in many basic texts (cf. Snedecor and Cochran 1979, Bradley 1968).

Results

Indications of Recent Growth Abnormalities

During the early stages of research we discovered a strong trend in recent growth that was not typical or natural, owing to divergence from the trend in annual precipitation. In the dating process we found that a moderate to substantial portion (up to about 50 percent) of the tree-ring series from ponderosa pine in seven stands could not be dated after times ranging from about 1920–1950 on to the final year of collection in 1986. In some cases more than 40 rings were absent from samples during the period of 1920–1986. The most extreme suppressions of growth were found in the post-1950 years. This was unexpected. Precipitation trends throughout the region show particularly high values near the beginning of the century with decline from the 1920s into drought conditions at mid-century (Chapter 2). This drought was followed by large increases in moisture in the 1970s and 1980s. Most tree-ring series we have dealt with in Arizona and New Mexico tracked this overall U-shaped pattern (Rose et al. 1981, Graybill 1989). However, the cores in the present study are the first series from the Basin and Range area, and six of the seven chronologies exhibiting this period of growth suppression are from that southern sector of Arizona. Discussion of possible reasons for these growth anomalies is presented below in the Summary and Discussion section. We next consider some of the quantitative aspects of climate and tree growth covariation.

Relationships Between Climate and Tree Growth

Predictive relationships between divisional climate data (August to July total precipitation) and each of the 41 residual tree-ring chronologies were established during the two baseline periods (Table 10.2). The final

Table 10.2 Model calibrations and predictions. See Table 10.1, Figure 10.3 for site information.

CODE	[1]Period	[2]Calibration r^2	[3]Prediction r^2	[4]t-test	[5]F-test
Northern group					
AFN	1	.42	.45	Rd	A
	2	.43	.44	Rd	A
BCR	1	.13	.41	A	Ri
	2	.18	.35	A	Ri
BKF	1	.43	.54	Rd	A
	2	.44	.60	Rd	A
BKP	1	.37	.58	A	Ri
	2	.40	.58	A	Ri
BRF	1	.43	.54	A	Ri
	2	.46	.51	A	Ri
DCP	1	.51	.50	A	Ri
	2	.47	.59	A	A
EPN	1	.37	.17	A	A
	2	.33	.20	A	A
GCP	1	.35	.42	Rd	Ri
	2	.40	.35	A	Ri
GPN	1	.31	.33	Rd	Ri
	2	.36	.34	Rd	A
GRP	1	.43	.19	A	A
	2	.47	.09R	A	A
GRS	1	.50	.30	A	Ri
	2	.47	.28	A	Ri
MTP	1	.16	.02R	A	Ri
	2	.18	.00R	A	Ri
MJN	1	.25	.64	A	Ri
	2	.33	.57	A	A
PDF	1	.35	.47	A	Ri
	2	.37	.41	A	Ri
RGP	1	.45	.34	Rd	A
	2	.46	.34	A	A
RMP	1	.40	.37	Ri	Ri
	2	.39	.31	A	Ri
RPP	1	.41	.33	Rd	A
	2	.40	.32	A	A
SGP	1	.48	.48	Rd	A
	2	.50	.46	Rd	A
SMP	1	.37	.50	Rd	A
	2	.43	.46	Rd	A
WCP	1	.55	.49	A	Ri
	2	.59	.42	A	Ri
WDF	1	.38	.52	A	Ri
	2	.39	.53	A	Ri

Southern Group

BWF	1	.23	.09R	A	Ri
	2	.16	.18	A	Ri
DBN	1	.09	.09R	A	Ri
	2	.11	.06R	A	Ri
GMF	1	.35	.22	Rd	A
	2	.31	.31	Rd	A
GMT	1	.27	.12	Rd	A
	2	.23	.16	Rd	A
HDE	1	.10	.03R	Rd	Ri
	2	.11	.01R	Rd	Ri
HDW	1	.24	.34	A	A
	2	.24	.30	A	A
MHP	1	.29	.30	A	Ri
	2	.29	.27	A	Ri
NOO	1	.33	.27	A	Rd
	2	.31	.22	Ri	Ri
NSA	1	.37	.48	A	A
	2	.36	.51	A	A
NSE	1	.17	.09R	Rd	Ri
	2	.16	.06R	Rd	Ri
NSF	1	.25	.24	Rd	A
	2	.24	.25	Rd	Rd
NSW	1	.16	.36	Rd	A
	2	.15	.45	Rd	A
ORD	1	.30	.20	Rd	Ri
	2	.28	.20	Rd	Ri
PCD	1	.37	.45	Rd	A
	2	.40	.43	Rd	A
PST	1	.29	.33	A	Ri
	2	.25	.43	A	Ri
RCF	1	.41	.50	Rd	A
	2	.43	.46	Rd	A
RPE	1	.23	.31	Rd	Ri
	2	.27	.27	Rd	A
SRH	1	.42	.41	Rd	Rd
	2	.40	.50	Rd	A
TSE	1	.25	.14	Rd	A
	2	.22	.09R	A	A
TSW	1	.16	.26	A	Ri
	2	.17	.19	A	Ri

Table 10.2 footnotes

[1] 1 indicates period of calibration is 1897 -1950, period of prediction is 1951 - 1986. 2 indicates period of calibration is 1897 - 1960 and period of prediction is 1961 - 1986.

[2] Covariation of prewhitened tree-ring indices and the sum of August through July precipitation for the period of 1897 - 1950 and 1897 - 1960. R indicates nonsignificant values (alpha = .05).

[3] Covariation of actual and predicted tree growth measured by prewhitened index chronologies. R indicates nonsignificant values (alpha = 0.05).

[4] T-test results of mean differences in actual and expected growth. Throughout the remainder of the table, A indicates no significant differences (alpha = .05) in test statistics, R indicates there are significant differences, d indicates actual values were less than predicted, i indicates actual values were greater than predicted.

adjusted r^2 values for the full calibrations across the baseline periods were significant in all cases, but two of the relationships in the south were weak and not clearly interpretable (DBN, HDE). Those covariances were however relatively constant per chronology for each of the two baseline periods. The range of r^2 values for the 20 chronologies in the south was slightly lower than for the 21 in the north (0.09–0.42 vs 0.13–0.55, 1896–1950 period) and the means were different (0.26 in the south, 0.38 in the north, t = 3.69, p<.001). The lower mean value in the south is primarily due to low calibration values for chronologies in the mountains near Tucson.

Results of the comparison of actual and estimated tree growth are summarized in Table 10.2. During the recent test periods, the covariation between actual and climatically predicted tree growth was not significant in 11 cases. Eight of the non-significant covariances were associated with chronologies in the southern area although some were not surprising because their climatic sensitivity in the calibration period was low. With the exception of the GRP chronology, all tests yielding non-significant covariances were associated with chronologies from dense stands on relatively mesic sites.

Normally, we would not expect to find major changes in the covariance of actual and predicted tree growth after mid-century, compared to the covariance of actual growth and climate before mid-century. Exceptions would probably be due to changes in the relationship between climate and tree growth, or to stand level disturbances that we did not detect in either the baseline or test periods. In most cases (28 chronologies) no major changes occurred in covariance when moving from the calibration to the recent test periods. Two kinds of exceptions were noteworthy, showing either a decrease or an increase in r^2 of 50% or more in either recent test period when that value was compared to the calibration r^2 of the respective baseline period. Eight chronologies showed a decrease, with three in the north (EPN, GRP, MTP) and five in the south (BWF, GMT, HDE, NSW, TSE). All of the southern chronologies were from the two mountain ranges nearest Tucson (Rincons and Santa Catalinas) and four were ponderosa pine sites with extreme growth suppression in many trees after 1950 (although those particular cores had too many missing rings to use in the chronologies). The site and stand characteristics of all but EPN and GRP tend toward mesic and dense rather than xeric and open.

Five chronologies showed substantially increased r^2 values in the recent periods; three in the north were from relatively dry and open settings (BCR, BKP, MJN), and two in the south were from relatively wet sites and dense stands (NSW, TSW).

The results of F-ratio tests showed that for about half of the series (22 of 41, 1951–1986; 19 of 41, 1961–1986) the variance of the expected series both in the north and south was significantly less than that of the actual growth series (Table 10.2, F-test, Ri). This usually indicates that not all of the variance in the actual series was estimated by the original calibration models, which is not uncommon. More surprisingly, three cases showed variances of the estimated series that exceeded the variances of the actual growth series (Table 10.2, F-test, Rd), indicating that the variance of the actual series decreased after mid-century. All instances were in the south (NOO, NSF and SRH, 1951–1986, NSF, 1961–1986). Figure 10.4 provides a good visual example of this kind of pattern that is also associated with a difference in trend and a difference of means in the actual and predicted growth for chronology SRH near Tucson.

Differences between the estimated and actual mean growth are of some interest. Both the t-test (Table 10.2) and the Wilcoxon test (results not presented) show many instances where the two means, or central tendencies, are significantly different, with the mean of the actual tree growth usually being less than that predicted from climate (denoted by a "d" in Table 10.2). For a few chronologies the means of actual growth were greater than predicted (denoted by an "i" in Table 10.2). The results of the t and Wilcoxon tests were similar, with just three of 82 possible disagreements; therefore only the t-tests are considered in further discussion.

The spatial distribution of chronologies with significant differences of actual and predicted mean growth in the 1951 to 1986 period are shown in Figure 10.5. Dark dots following each code indicate that actual growth is less than predicted from August to July divisional precipitation. Taking the data set as a unit, and generously permitting chance to operate 10% of the time, we might only expect that four of the chronologies per time period would exhibit significant differences in actual and expected mean growth. Instead, 20 of 41 series (51%) in the 1951–1986 period and 16 of 41 series (39%) in the 1961–1986 period show significantly less than expected growth.

Actual and predicted growth levels differed on a spatial basis, and to some degree by species. This involved sites in and near the Mogollon Rim (northern sector) vs those in the southern Basin and Range area, and regional differences in ponderosa pine and Douglas-fir growth predictions. These differences will next be considered from the standpoint of the entire data set and from one of a reduced data set. Some chronologies had low calibration r^2 values in one or both baseline periods, indicating a low climatic response. They were excluded from some comparisons when that value was less than 0.25 (or, the simple correlation was 0.5), thus creating the reduced data set.

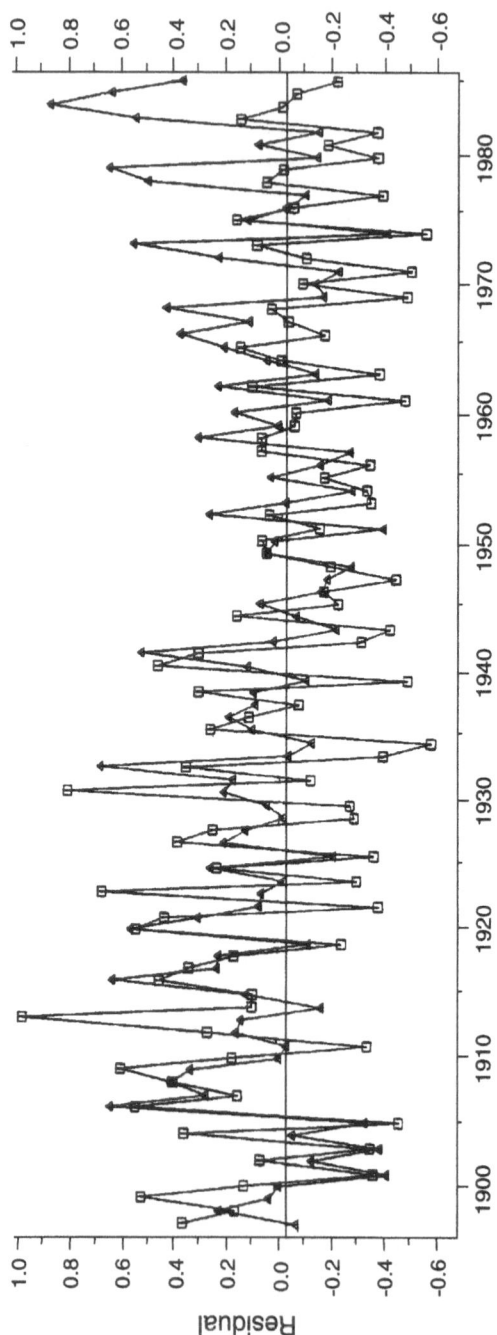

Figure 10.4 Time series plots of actual (squares) and predicted (triangles) tree growth for chronology SRH (ponderosa pine).

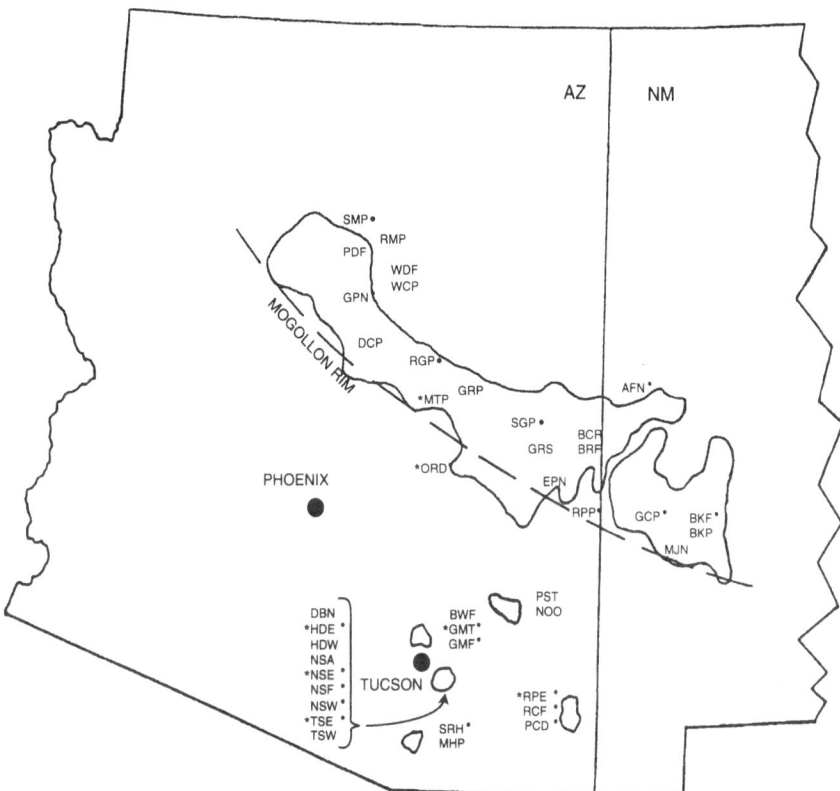

Figure 10.5 Geographic distribution of t-test results. A black dot following a chronology code indicates significantly lower mean growth in the actual series than in the predicted series for the 1951–1986 test period. An asterisk before a code indicates that approximately 50% of the trees sampled had extreme growth suppression after the mid-20th century.

Fisher exact tests and one-tailed probability levels for some of the most basic breakdowns of variation in mean growth response for both the full and reduced data set are provided in Table 10.3. The categories for each table are region (north, south) and evidence of significantly less than expected mean growth (fail) or of significantly greater than expected mean growth (pass). Cell values are simple counts of chronologies by category. However, these tests can only be considered gross descriptive statistics and cannot be used as estimates of population parameters for two reasons. First, the samples were not randomly selected. Second, not all chronologies are members of the same kind of population because the

Table 10.3 Fisher exact tests on mean growth variation.

	Full Data Set				Reduced Data Set		

Ponderosa pine, 1951–1986

A					**B**		
	Pass[1]	Fail				Pass	Fail
North	7	6		North		5	6
South	2	6		South		2	3
p = 0.200				p = 0.635			

Ponderosa pine, 1961–1986

C					**D**		
	Pass	Fail				Pass	Fail
North	10	3		North		8	3
South	3	5		South		2	2
p = 0.090				p = 0.682			

Douglas-fir, 1951–1986

E					**F**		
	Pass	Fail				Pass	Fail
North	3	1		North		3	1
South	2	5		South		1	5
p = 0.197				p = 0.119			

Douglas-fir, 1961–1986

G					**H**		
	Pass	Fail				Pass	Fail
North	3	1		North		3	1
South	2	5		South		1	4
p = 0.197				p = 0.167			

All species 1951–1986

I					**J**		
	Pass	Fail				Pass	Fail
North	13	8		North		11	8
South	8	12		South		4	6
p = 0.138				p = 0.168			

All species 1961–1986

K					**L**		
	Pass	Fail				Pass	Fail
North	16	5		North		14	5
South	9	11		South		4	6
p = 0.042				p = 0.085			

[1]Pass counts include chronologies with actual growth that is equal to or greater than expected. Fail counts include chronologies with growth that is significantly less than expected based on t-test results.

collections were made over gradients of growth conditions that range from open canopied and xeric to relatively dense and mesic for the region. In essence, each chronology represents a case study.

A review of the results for ponderosa pine (Table 10.3A–D) suggests that in the full data set there are some hints at north-south differences. This is not apparent in the reduced data set. However, the chronologies removed in this case were all from sites where many series could not be dated after mid-century. Four are from the mountains near Tucson (GMT, HDE, NSE, TSE), one is from the Chiricahuas (RPE) and only one is from the north (MTP—a dense forest interior site). Thus, removal of sites with relatively low calibration r^2 values, and for the most part, dramatically lower r^2 values between actual and estimated growth post-1950, may be obscuring the overall relationships best seen in the full data set.

The cell values for Douglas-fir (Table 10.3E–H) are low but there are limited indications of differences in mean expected growth by region. When all species are lumped (Table 10.3I–L) the strongest regional difference is in the 1961–1986 period in both the full and reduced data sets.

Reviewing the test results for other species, the only white fir chronology in the collection (NSA) showed no difference of means. It also has a large amount of explained variance during the verification period. Additionally, only one of the five pinyon chronologies (AFN) and only one of the southwestern white pine series (NSW) showed a difference between the predicted and actual mean growth.

Caution must be exercised in evaluating these empirical tests. Some tests, such as t-tests, are sensitive to the presence of one or a few large differences or outliers. A review of the time series plots of actual and predicted growth suggested that for three of the northern chronologies, rejection of the null hypothesis was due to this problem, and not to any aberrant level of growth (AFN, BKF, GCP). For example, Figure 10.6 illustrates the actual and predicted growth values for a Douglas-fir chronology from stand BKF in a remote area of southwestern New Mexico (see Figure 10.3 for site location). The covariance of actual and predicted growth was among the highest in the data set. However, the year 1973 was one of the wettest in the century and apparently the trees did not respond to that high value. This is not unexpected in the Southwest, especially when one or more of the preceding years had been relatively dry and stressful for growth, such as 1972. Removal of the 1973 values from the t-test computation dropped the significance of the t-value from .03 to .05 and removal of the 1984 values further dropped the significance to .09. In contrast, Figure 10.7 illustrates another chronology from the north (GPN-PP) where there was a striking change in both level

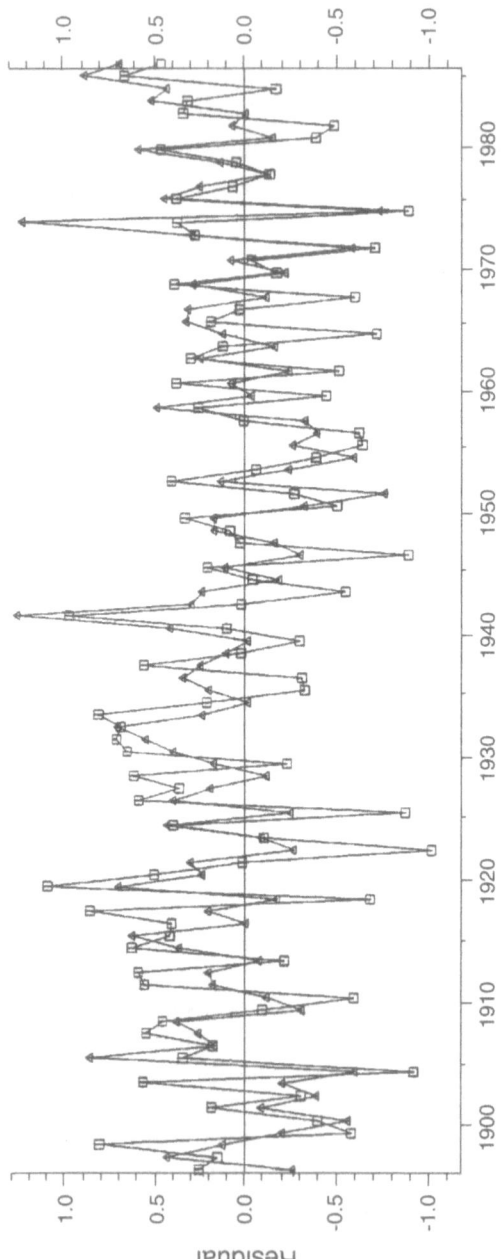

Figure 10.6 Time series plots of actual (squares) and predicted (triangles) tree growth for chronology BKF (Douglas-fir).

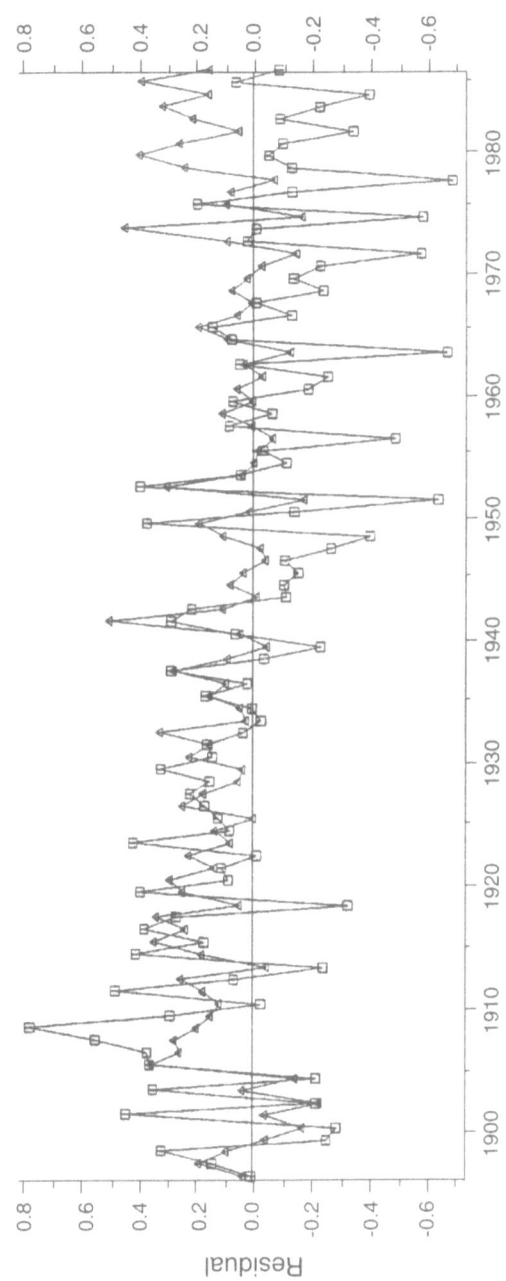

Figure 10.7 Time series plots of actual (squares) and predicted (triangles) tree growth chronology for GPN (ponderosa pine).

and trend of actual and predicted growth beginning in about 1965. In the southern area, the stands with significantly less than predicted growth in recent decades have time series plots that are similar to Figure 10.4. Significant t values in these cases appear to stem from a lack of growth response to increased moisture of the 1970s and early 1980s.

Summary And Discussion

Two types of anomalous tree growth have been discovered in the mid to late 20th century that are perhaps different in degree rather than kind. In the most extreme cases, some but not all trees in seven ponderosa pine stands exhibited growth patterns during recent decades that suggest mortality is imminent. Six of the seven stands are in the southern mountain ranges. In other less extreme cases there are statistically significant differences in actual and climatically predicted tree-ring growth in terms of covariance, trend, and particularly in mean value over the past few decades. These differences are most pronounced in southern Arizona and are most reliably documented throughout the region for ponderosa pine and Douglas-fir.

Another dendrochronologically based attempt to evaluate possible pollution effects on coniferous tree growth in central Arizona was reported by Nash et al. (1975). They focused on ponderosa pines within 7 km of one large copper smelter and 10 km of another. They first removed climatic variation from their tree-ring indices and then searched for non-climatic aberrations in growth. The results show some non-climatic decreases in tree growth from about 1908–1920, a period correspondent with early production at the two copper smelters. However, growth reductions were not apparent in succeeding years when the smelters remained in operation. The authors concluded that factors such as fire or competition might be responsible for some growth decreases and that air pollution could not be definitively identified as the cause. In contradistinction to some of the results of our study, they did not find aberrant growth from 1950–1966, the final years of their chronology. However, those years are the peak of 20th century drought and precede moisture increases that figure prominently in our analyses.

Although empirical tests such as those undertaken here may be statistically significant, they can only suggest that meaningful biological changes are occurring. These changes could result from a variety of processes:

- Lack of growth response to improved moisture conditions of the late 20th century relates to different intrinsic (genetic) abilities of individuals to rebound after the shock of drought stress they experienced in mid-century. It is possible, however, that younger trees than we sampled would have greater vigor and might better respond to recent moisture changes.

- Anthropogenic activity resulted in chemical deposition that affected relationships between climate and tree growth.

- Natural biological competition may be responsible for growth changes in some stands.

- Any combination of the above may be possible.

Investigation of two of those alternatives is now in progress. The cohorts of trees established in the cool, wet portion of the early 20th century may now be providing strong competition with older individuals for water, soil nutrients and light. A study has therefore been initiated to develop competition indices for trees of varying ages in many of the stands that we have evaluated throughout the region. This should allow us to determine if competition could be at least partially responsible for some of the anomalous growth patterns discerned here in the late 20th century. Younger trees in the 100 year age class are also being sampled in many of the same stands and their chronologies will be subjected to the same kinds of analyses used for the older trees. All of these lines of evidence will provide a more complete understanding of the issues addressed here but will probably also raise new questions.

References

Alexander RR, Engelby O (1983) Engelmann spruce-subalpine fir. In: Burns RM (ed) *Silviculture Systems for the Major Forest Types of the United States.* USDA Forest Service Agriculture Handbook 445:59–62

Arizona Department of Economic Security (1983) *Arizona Occupational Profile.* Arizona Department of Economic Security, Phoenix

Avery TE, Burkhart HE (1983) *Forest Measurements.* McGraw-Hill, New York

Box GEP, Jenkins GM (1976) *Time Series Analysis: Forecasting and Control.* Holden-Day, San Francisco

Bradley JV (1968) *Distribution-Free Statistical Tests.* Prentice Hall, Englewood Cliffs

Brown JM (1968) *The Photosynthetic Regime of Some Southern Arizona Ponderosa Pine.* PhD dissertation, University of Arizona, Tucson

Buckman R, Wolters GL (1987) Multi-resource management of pinyon-juniper woodlands. In: Everett RL (comp) *Proceedings Pinyon-Juniper Conference.* USDA Forest Service General Technical Report INT-215, pp 2–4

Budelsky CA (1969) *Variation in Transpiration and its Relationship with Growth for Pinus Ponderosa in Southern Arizona.* PhD dissertation, University of Arizona, Tucson

Chatfield C (1980) *The Analysis of Time Series.* Chapman and Hall, London

Cook ER, Kairiukstis LA (1990) *Methods of Dendrochronology: Applications in the Environmental Sciences.* Kluwer, Boston

Cooper CF (1960) Changes in vegetation, structure and growth of southwestern pine forests since white settlement. *Ecological Monographs* 30:129–164

Cooper CF (1961) Pattern in ponderosa pine forests. *Ecology* 42:493–499

Draper N, Smith H (1981) *Applied Regression Analysis.* Wiley, New York

Dunning CH, Peplow EH Jr (1959) *Rock to Riches.* Southwest Publishing, Phoenix

Everett RL (comp) (1987) *Proceedings Pinyon-Juniper Conference.* USDA Forest Service General Technical Report INT-215

Fritts HC (1976) *Tree Rings and Climate.* Academic Press, New York

Graybill DA (1979) Revised computer programs for tree-ring research. *Tree-Ring Bulletin* 39:77–82

Graybill DA (1982) Chronology development and analysis. In: Hughes MK, Kelly PM, Pilcher JR, LaMarche VC Jr (eds) *Climate from Tree-Rings.* Cambridge University Press, London, pp 21–28

Graybill DA (1989) The Reconstruction of Prehistoric Salt River Stream Flow. Chapter 3 in *The 1982–1984 Excavations at Las Colinas: Environment and Subsistence,* by Graybill DA, Gregory DA, Nials FL, Fish SK, Gasser RE, Miksicek CH, Szuter CR. Archaeological Series 162, Volume 5, Part 1, Cultural Resource Management Division, Arizona State Museum, University of Arizona, Tucson

Hawksworth FG (1977) *The Six-class Dwarf Mistletoe Rating System.* USDA Forest Service General Technical Report RM-48

Holmes RL (1983) Computer-assisted quality control in tree-ring dating and measurement. *Tree-Ring Bulletin* 43:69–78

Hughes MK, Kelly PM, Pilcher RJ, LaMarche VC Jr (1982) *Climate from Tree-Rings.* Cambridge University Press, Cambridge

Innes J (1990) General aspects in the use of tree-rings for environmental impact studies. In: Cook ER, Kairiukstis LA (eds) *Methods of Dendrochronology: Applications in the Environmental Sciences.* Kluwer, Boston, pp 224–229

Jones JR (1974) *Silviculture of Southwestern Mixed Conifers and Aspen: The Status of Our Knowledge.* USDA Forest Service Research Paper RM-122

Kozlowski TT, Constantinidou HA (1986a) Responses of woody plants to environmental pollution. *Forestry Abstracts* 47(1):5–51

Kozlowski TT, Constantinidou HA (1986b) Environmental pollution and tree growth. *Forestry Abstracts* 47(2):105- 132

Lanner RM (1981) *The Pinyon Pine: A Natural History.* University of Nevada Press, Reno

Meeuwig RO, Bassett RL (1983) Pinyon-juniper. In: Burns (ed) *Silviculture Systems for the Major Forest Types of the United States.* USDA Forest Service Handbook 445:84–86

Monserud RA (1986) Time series analyses of tree-ring chronologies. *Forest Science* 32(2):349–372

Nash TH, Fritts HC, Stokes MA (1975) A technique for examining non-climatic variation in widths of annual tree rings with special reference to air pollution. *Tree-Ring Bulletin* 35:15–24

NOAA (1988) United States divisional averages of temperature and precipitation, time corrected through September, 1987. Magnetic tape, National Climatic Data Center, Asheville, North Carolina.

Palmer WC (1965) *Meteorological Drought.* US Weather Bureau Research Paper 45, US Weather Bureau, Washington, DC

Pearson GA (1950) *Management of Ponderosa Pine in the Southwest.* USDA Forest Service Agricultural Monograph No. 6

Peet RK (1988) Forests of the Rocky Mountains. In: Barbour MG, Billings WD (eds) *North American Terrestrial Vegetation.* Cambridge University Press, Cambridge, pp 63–104

Pissot HJ (1965) *New Mexico's Forest Area and Timber Volume.* USDA Forest Service Research Note INT-32

Robinson WJ, Evans R (1980) A microcomputer-based tree-ring measuring system. *Tree-Ring Bulletin* 40:59–64

Ronco F (1987) Stand structure and function of pinyon-juniper woodlands. In: Everett RL (compiler) *Proceedings Pinyon-Juniper Conference.* USDA Forest Service General Technical Report INT-215, pp 12–22

Ronco F Jr, Ready KL (1983) Southwestern ponderosa pine. In: Burns RM (ed) *Silviculture Systems for the Major Forest Types of the United States.* USDA Forest Service Agriculture Handbook 445:70–72

Ronco F Jr, Gottfried R, Shaffer R (1983) Southwestern mixed conifers. In: Burns RM (ed) *Silviculture Systems for the Major Forest Types of the United States.* USDA Forest Service Agriculture Handbook 445:73–76

Rose MR, Dean JS, Robinson WJ (1981) *The Past Climate of Arroyo Hondo, New Mexico, Reconstructed from Tree Rings.* School of American Research Press, Arroyo Hondo Archaeological Series, Volume 4, Santa Fe, New Mexico.

Roth P, Blanchard C, Harte J, Michaels H, El Ashry MT (1985) *The American West's Acid Rain Test.* World Resources Institute and Energy and Resources Group, University of California, Berkeley

Samuels ML, Betancourt JL (1982) Modeling the long-term effects of fuelwood harvests on pinyon-juniper woodlands. *Environmental Management* 6:505–515

Schubert GH (1974) *Silviculture of the Southwestern Ponderosa Pine: The Status of Our Knowledge.* USDA Forest Service Research Note RM-123

Schuler TM, Smith FW (1988) Effect of species mix on size/density and leaf-area relationships in Southwest pinyon/juniper woodlands. *Forest Ecology and Managment* 25:211–220

Shupe DG (1965) *Arizona's Forest Area and Timber Volume.* USDA Forest Service Research Note INT-33

Snedecor GW, Cochran WG (1979) *Statistical Methods.* The Iowa State University Press, Ames

Springfield HW (1976) *Characteristics and Management of Southwestern Pinyon-Juniper Ranges: The Status of Our Knowledge.* USDA Forest Service Research Paper RM-160

Stokes MA, Smiley TL (1968) *An Introduction to Tree-Ring Dating.* University of Chicago Press, Chicago

Swetnam TW, Dieterich JH (1985) Fire history of ponderosa pine forests in the Gila Wilderness, New Mexico. In: Lotan JE, Kilgore BM, Fischer WC, Mutch RW (tech coords) *Proceedings: Symposium and Workshop on Wilderness Fire.* Nov 15–18, 1983 Missoula, Montana. USDA Forest Service General Technical Report INT-182

Trijonis J (1979) Visibility in the southwest—an exploration of the historical data base. *Atmospheric Environment* 13:833–843

USDA Forest Service (1981) *An Assessment of the Forest and Range Land Situation in the United States.* Forest Resource Report #22, Washington, DC

USEPA (1978) *National Air Quality, Monitoring, and Emissions Trends Report.* EPA450/2-78-052

Weaver H (1951) Fire as an ecological factor in the southwestern ponderosa pine forests. *Journal of Forestry* 49:93–98

White AS (1985) Presettlement regeneration patterns in a southwestern ponderosa pine stand. *Ecology* 66:589- 594

Whittaker RH, Niering WA (1975) Vegetation of the Santa Catalina Mountains, Arizona. V: Biomass, production, and diversity along the elevation gradient. *Ecology* 56:771–790.

Wigley TML, Briffa KR, Jones PD (1984) On the average value of correlated time series with applications in dendroclimatology and hydrometeorology. *Journal of Climate and Applied Meteorology* 23(2):201–213

Wonnacott TH, Wonnacott RJ (1981) *Regression: A Second Course in Statistics.* Wiley, New York

11

Mixed Conifer Forests of the Sierra Nevada

D. L. Peterson and M. J. Arbaugh

The mixed conifer forest of the Sierra Nevada in California is one of three areas in the western United States with visible symptoms of ozone injury to conifers (the other areas are the San Bernardino Mountains and other locations in southern California [see Chapter 12] and the Rincon Mountains of Arizona [Graybill and Rose 1989]). The Sierra Nevada contains the largest forest area in the world with documented damage from a nonpoint source pollutant, with an area approximately 500 km long having ozone exposure high enough to cause visible injury. Although ozone exposure is not as high as in the Los Angeles Basin (Chapters 3 and 12) or the pine-fir forests near Mexico City (Ciesla and Macias Samano 1987, Cibrian Tovar 1989), the mixed conifer forest of the Sierra Nevada has suffered air pollution stress since at least the early 1970s (Miller and Millecan 1971).

Ponderosa pine (*Pinus ponderosa* var. *ponderosa*) and Jeffrey pine (*Pinus jeffreyi*) are highly susceptible to ozone injury and associated stress under both field and experimental conditions (Miller et al. 1983). In this chapter, we focus on these species as sensitive indicators of stress associated with chronic ozone exposure in the Sierra Nevada. Forest condition was evaluated by quantifying spatial trends in ozone injury, and spatial and temporal patterns of tree growth.

Mixed Conifer Forest: Biogeographic Setting

Soils and Climate

Mixed conifer forest is found at elevations of 1000 to 2500 m throughout the Sierra Nevada and Transverse Ranges of California. It occurs mostly on soils derived from granitic parent material of the Sierra batholith. Soil development varies depending on parent material and geomorphology, although a large proportion of soils in the mixed conifer zone are inceptisols. Soils and vegetation in most areas have developed since the last glaciation. The climate is mediterranean, with warm, dry summers and cool, wet winters; it is not unusual to have little or no rainfall for several consecutive months in the summer (see Chapter 2). There is a strong rain shadow effect, with a gradient of increasing precipitation at higher elevations. Most streamflow occurs during the snowmelt runoff period of April-July (Stephenson 1988).

Species

The mixed conifer forest ecosystem is one of the most diverse forest types in the western United States. Dominant tree species include ponderosa pine, Jeffrey pine, white fir (*Abies concolor*), sugar pine (*Pinus lambertiana*), incense-cedar (*Libocedrus decurrens*), Douglas-fir (*Pseudotsuga menziesii*), and California black oak (*Quercus kelloggii*) (see Chapter 1). Giant sequoia (*Sequoiadendron giganteum*) is locally common at many locations in the southern and central Sierra Nevada. The distribution of these species varies greatly by aspect; ponderosa pine and Jeffrey pine are more common on south and west aspects, and white fir and incense-cedar are more common on north and east aspects. Douglas-fir is more common in the northern half of the Sierra Nevada where precipitation is higher. Jeffrey pine grows on shallower soils, steeper slopes, and at higher elevations than ponderosa pine (Griffin and Critchfield 1972), although these species hybridize in some locations and are often confused in the field. Ponderosa pine and Jeffrey pine are early successional species in the mixed conifer forest, and white fir and incense-cedar are later successional species. Ponderosa pine and Jeffrey pine are also considered more sensitive to ozone than any of the other species (Davis and Wilhour 1976).

Effects of Natural Factors and Management on Stand Structure

Both natural factors and management affect stand structure in mixed conifer forest. Fire is an important factor in the Sierra Nevada, with a higher incidence of fire at increasing elevations up to 2000 m (Parsons 1981). The early successional species ponderosa pine has thick bark, large buds, and open crown structure that confer a high degree of fire resistance (Brown and Davis 1973). It requires mineral soil exposure for seed germination, a condition normally found after fire. There is some speculation that fire suppression during this century has reduced the size of fires and therefore encouraged increasing dominance of shade tolerant species such as white fir in unmanaged stands (Parsons and DeBenedetti 1979).

There are numerous species of bark beetles associated with tree species of the mixed conifer forest (Furniss and Carolin 1977). Bark beetles are most common after fire or during periods of drought. There was extensive mortality in several tree species of the Sierra Nevada in 1989–91 as the result of a prolonged drought and subsequent bark beetle attack. There are several common fungal pathogens that damage mixed conifer forest species, the most prominent being annosus root rot (*Heterobasidion annosum*) (Parmeter et al. 1978, Scharpf 1978b). All Sierra Nevada tree species are subject to infection by various species of mistletoes and dwarf mistletoes (Scharpf 1978a).

Management practices are also a critical factor in determining stand structure. Many of the Sierra Nevada mixed conifer forests were harvested between 1880 and 1920, primarily with various types of selective cuts in which the best timber, predominantly pines, was removed. The degree to which stands have been managed varies considerably. Stands in National Parks are largely undisturbed, resulting in uneven-aged stands and large amounts of woody litter. The intensity of current management varies considerably in National Forests. Some managed stands are uneven-aged, resulting from selection harvests. The more intensively managed stands are normally even-aged, resulting from clearcutting blocks 15 ha or smaller. Management practices on private lands range from passive management (no cutting) to clearcutting.

Air Pollution

Ozone is a major air pollution threat in the Sierra Nevada, where exposure is higher than in any other forest region of the United States except the Los Angeles Basin (see Chapter 3). Nitrogen oxides and hydrocarbons from fossil fuel combustion in the Central Valley and San Francisco Bay regions are photochemically oxidized and transported eastward (Miller et al. 1972, Carroll and Baskett 1979, Ewell et al. 1989a) (Figure 11.1). There is a gradient of ozone exposure in the Central Valley and the Sierra Nevada, with highest concentrations in the south and lowest in the north (California Air Resources Board 1987, Ewell et al. 1989a). This gradient can be seen by examining data for exceedances of 100 ppbv from the California Air Resources Board database (Figure 11.2). The few ozone data available from mountain locations are correlated with data from the valley database (Pronos and Vogler 1981, California Air Resources Board 1987, Miller et al. 1988).

Figure 11.1 Photochemical haze (smog) moves into the mixed conifer forest zone of the southern Sierra Nevada on summer afternoons, as seen here near Board Camp Dome in Sequoia National Park. Ozone is a colorless component of this polluted air mass. (Photo by P. R. Miller)

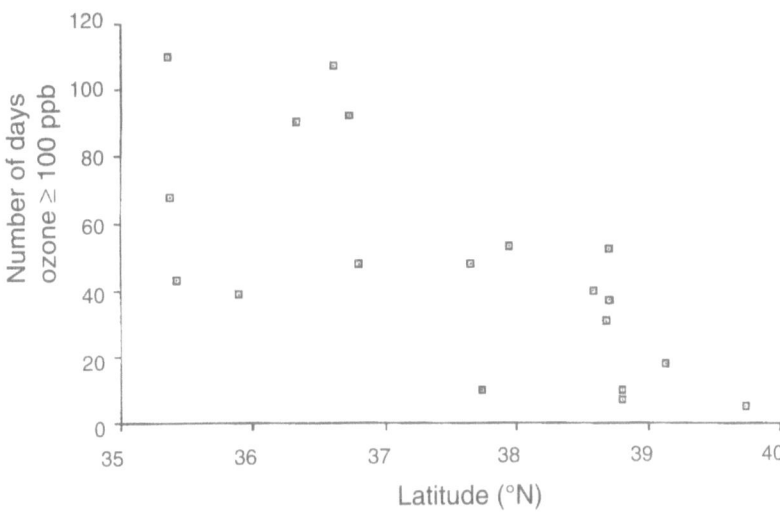

Figure 11.2 Number of days per year with ozone concentration ≥ 100 ppb at locations in and adjacent to the Sierra Nevada (California Air Resources Board 1987, Miller et al. 1988).

Highest ozone exposure occurs during the summer months and lowest exposure in winter, with ozone concentration peaks in the mid to late afternoon (see Chapter 3). Ozone concentrations are relatively high throughout the summer at some sites in the southern Sierra Nevada. Furthermore, minimum values are relatively high at mountain sites in the southern part of the range (see Chapter 3), and summer minimums are rarely less than 50 ppbv (Miller et al. 1988), because there is an insufficient concentration of nitrogen oxides to combine with and break down ozone at night. Plants are therefore exposed to moderate concentrations of ozone throughout the day, including the morning hours when gas exchange and photosynthesis are at their peak.

Regional Trends of Ozone Injury and Growth

Until recently most information on the effects of air pollution on mixed conifer forests of the Sierra Nevada has been limited to surveys of visible injury. Permanent plots designed to monitor tree condition with respect to ozone injury in the National Forests and National Parks of the Sierra Nevada are a valuable long term database (Pronos et al. 1978, Pronos and Vogler 1981, Warner et al. 1982, Allison 1982, 1984a,b, Duriscoe and Stolte 1989). Data from these plots can be used to document spatial and temporal trends of visible injury. It is unclear how ozone exposure and injury are related to growth and productivity of Western conifers, although there are some data available from controlled fumigation studies of seedlings (e.g., Grulke et al. 1989, Hogsett et al. 1989). There are relatively few studies on the effects of ozone on growth of mature trees in the mixed conifer forest of the Sierra Nevada (Peterson and Arbaugh 1988, Peterson et al. 1987a,b, Ewell et al. 1989b, Patterson and Rundel 1989).

Open Grown Jeffrey Pine and Ponderosa Pine in the Southern Sierra Nevada

Previous studies of air pollution effects on tree growth have focused on mixed conifer forest stands of the southern Sierra Nevada because ozone exposure and ozone injury are relatively high in this area (Pronos and Vogler 1981). We studied long term growth trends in both Jeffrey pine and ponderosa pine because these species are sensitive to elevated levels of ambient ozone (Davis and Wilhour 1976, Miller et al. 1983). We initially sampled only in open grown stands in Sequoia National Forest and Sequoia National Park, at sites where intertree competition was assumed to be minimal.

Trees were cored along the western edge of the Sierra Nevada, where there was visible ozone injury (symptomatic sites), and in the interior where injury was absent (asymptomatic sites). We evaluated long term growth trends by: (1) comparing the time series of radial growth between symptomatic and asymptomatic sites and (2) comparing growth trends before and after the inception of elevated ozone concentrations.

Mean annual radial increment (at dbh) of Jeffrey pine with symptoms of ozone injury was 11% less in the last 20 years of growth than that of trees at sites without ozone injury (Peterson et al. 1987a,b). Larger diameter trees (> 40 cm) and older trees (> 100 years) had greater decreases in growth than smaller and younger trees. Differences in radial growth patterns of injured and uninjured trees were prominent after 1965. Winter precipitation accounted for the largest portion of variance in growth of all trees (positive correlation), and summer temperature was negatively correlated with growth. Growth of ozone-injured trees was correlated more strongly with interannual variation in precipitation and temperature since 1965.

Although open grown ponderosa pine had the same level of visible injury as Jeffrey pine, annual radial increment showed no significant reduction for symptomatic ponderosa pine since 1965 compared to past growth or growth of asymptomatic trees (Peterson and Arbaugh 1988). First order autocorrelation and climatic variables accounted for a large proportion of the variance. Winter precipitation was again the most important climatic variable and was positively correlated with growth for all size and age classes. Summer temperature was also negatively correlated with growth in most cases. Both Jeffrey pine and ponderosa pine clearly respond strongly to soil moisture, as evidenced by the strong positive correlation with winter precipitation and strong negative correlation with summer temperature. The effects of ozone injury on growth may not have been as prominent in ponderosa pine because it grows on deeper soils with less potential for moisture stress.

Extensive Analysis of Ponderosa Pine in the Sierra Nevada

Research approach

Previous work on the effects of air pollution on tree growth in the mixed conifer forest of the southern Sierra Nevada was restricted to the southern portion of the range (Peterson et al. 1987a,b, Peterson and Arbaugh 1988), and evaluated only open grown trees in a relatively small geographic area. It was necessary to sample mixed conifer forest over a much larger area to characterize regional growth trends in denser, more typical stands. Ponderosa pine was the subject species because of its ozone sensitivity, wide distribution, ecological importance in mixed conifer forest, and economic importance.

Trees were sampled in seven federal administrative units in the Sierra Nevada: (from north to south) Tahoe National Forest, Eldorado National Forest, Stanislaus National Forest, Yosemite National Park, Sierra National Forest, Sequoia–Kings Canyon National Parks, and Sequoia National Forest (Figure 11.3). Ozone injury symptoms in ponderosa pine were previously documented in all these areas.

Four sites in areas with ozone injury symptoms (symptomatic) and four sites in areas without injury (asymptomatic) were identified in each National Forest and Park. Symptomatic sites were in the western portion of the Sierra Nevada with potential exposure to air pollutants from the Central Valley, and asymptomatic sites were in the interior of the range

protected from high air pollutant exposure (Figure 11.3). Not all trees
from symptomatic sites necessarily had visible injury symptoms at the
time of the study.

Sampling was conducted in mixed conifer forest stands, with only
ponderosa pines greater than 50 years old included in the sample. Trees
were sampled on south and west aspects, on slopes of 0 to 80%, and at
elevations of 900 to 2000 m. Twenty-four trees were sampled per site
during July and August of 1987. All symptomatic sites were sampled in
August to insure that crown ratings would be comparable.

Crown condition was recorded as an estimate of needle chlorosis and
number of years of needle retention, based on a sample of five randomly
selected branches from the lower crown of each tree. Chlorosis rating
was based on ozone injury only, using the system of Muir and

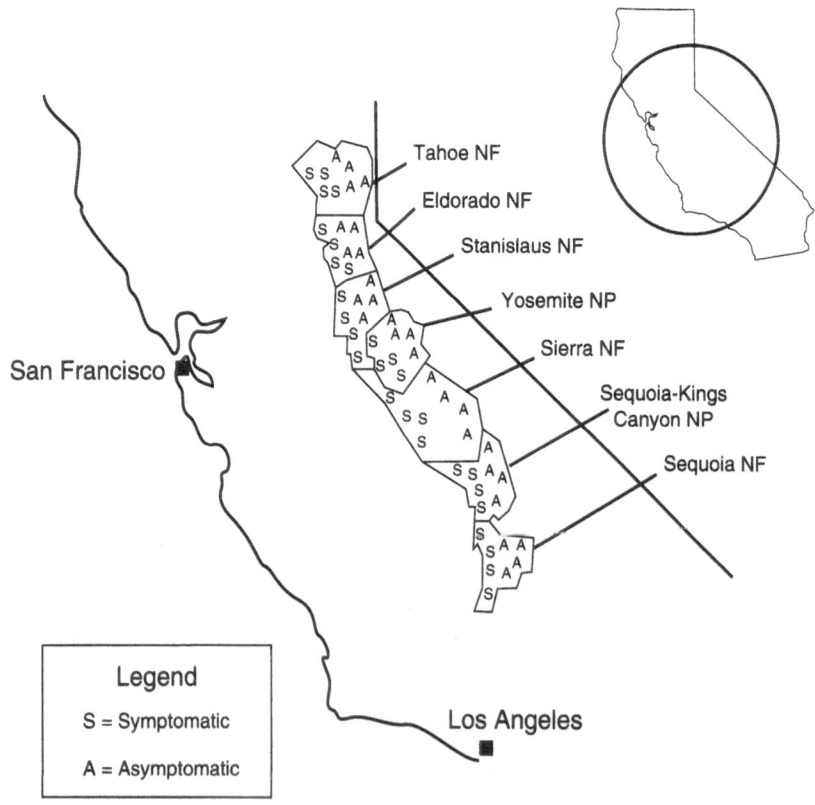

*Figure 11.3 Locations of study sites for the extensive study of ponderosa pine
growth in mixed conifer forest of the Sierra Nevada. The map is diagrammatic
and does not indicate exact locations. From Peterson et al. (1991).*

Armentano (1988), which assigns the following categories based on percent chlorosis: 0 = 0%, 1 = 1–5%, 2 = 6–25%, 3 = 26–50%, 4 = 51–75%, 5 = 76–95%, 6 = 96–99%, 7 = 100%. Chlorosis rating was done for current year, one-year old, and two-year old needles.

Two cores were removed from the cross-slope sides of each sample tree at breast height with an increment borer. Cores were stored in paper straws until processed. Cores were mounted in wood blocks and sanded until individual tracheids were visible with the microscope.

Ring widths were measured to the nearest 0.01 mm with standard dendrochronological methods (Stokes and Smiley 1968, Fritts 1976, Swetnam et al. 1985). Ring width measurements were used to calculate basal area increments (BAI) over time. BAI time series for each tree were used in subsequent analyses.

One of the fundamental requirements of our statistical analysis was that results could be interpreted on an individual tree basis. Results could then be compared to individual tree characteristics measured at the time of sampling. This approach prevents significant results from being obscured by averaging techniques used to aggregate data for many trees (Schweingruber 1986, Kienast et al. 1987, Peterson et al. 1990).

Analysis of tree growth trends was conducted with Kalman filter procedures, with a state-space formulation of a linear system that allows parameters to be nonstationary with time (Kalman 1960, Kalman and Bucy 1961, Visser and Molenaar 1986, Van Deusen 1987). It has been used previously in dendrochronology on averages of residual series (see Chapter 7). We used a version of the Kalman filter that would allow us to accurately detect long term changes in individual tree growth.

Interpretation was based primarily on the parameter a_0, because it was the parameter most indicative of a trend component (see Chapter 9b for Kalman filter formula.) A change in growth trend was defined to be a point at which $a_0(t)$ estimates changed such that a significant trend resulted. A trend was defined as a consistent direction of $a_0(t)$ increase or decrease of 20 years or more duration. Corresponding significant changes in $a_1(t)$ and $a_2(t)$ occurring at the same time were used as supporting evidence. Series without significant departures from the stationarity hypothesis, or that displayed growth patterns without clear trends were considered to have no change. One of the asymptomatic sites from Sequoia NF was eliminated from the analysis because its time series of ring widths was of insufficient length for the Kalman filter procedure.

The date and direction of growth changes were recorded, with changes recorded as an increase, no change, or a decrease in growth trend for each tree. Results were compiled by site and expressed as frequencies of increase and decrease per decade. Two equal time periods were selected, one prior to and the other during the time when ozone levels are assumed to have been elevated. These periods were 1921–50 and 1951–80 respectively. Frequencies of declines for the two periods, and between symptomatic and asymptomatic areas were summarized. Data were arranged in contingency table form, and differences in frequencies were analyzed with a chi-square procedure (Peterson et al. 1989).

In addition, we conducted an analysis to determine the relationship of growth with ozone injury. Growth ratio was calculated as the ratio of median basal area increment growth for 1982–1986 to median growth for 1951–1960. Growth ratio was then compared to needle retention, ozone chlorosis, and other variables using correlation analysis.

Stand and sample tree characteristics

Mixed conifer forest stands sampled in this study included a range of conditions in stem density, basal area, tree age, and management conditions. Stand density ranged from 175 to 2770 stems/ha, and basal area from 27 to 163 m^2/ha. Ponderosa pine comprised over 50% of basal area in all but seven stands and was never less than 35% of basal area, which indicates that it was a dominant species in all stands. Details of stand characteristics are found in Peterson et al. (1989). Mean diameter ranged from 32.3 to 104.0 cm for all sites, and mean age of tree cores ranged from 52 to 173 years. Several cores were crossdated in the 1600s, and the oldest ring measured was 1672. Trees from National Park sites tended to be older than trees from National Forests.

Ozone injury to needles

Ozone chlorosis and reduced needle retention in conifers can indicate stress because trees with fewer needles may have reduced photosynthetic production if chlorosis is high enough (Patterson and Rundel 1989) or enough biomass is removed (Helms 1970, Hom and Oechel 1983). There were large differences in needle retention between symptomatic and asymptomatic sites throughout the Sierra Nevada (Table 11.1). Needle retention was lower for southern Sierra sites, although it was relatively low for some northern sites such as Stanislaus NF and Eldorado NF as well. Needle retention at all symptomatic sites was lower than corresponding asymptomatic sites in the same National Forest or Park.

Ozone chlorosis in needles was also most common at the southernmost symptomatic sites, with most of the measurable injury at Sierra National Forest, Sequoia–Kings Canyon National Parks, and Sequoia National Forest (Peterson et al. 1989). Chlorotic injury in these forests was rare in

Table 11.1 *Mean number of years of needles retained (±1 SD) by ponderosa pines from symptomatic and asymptomatic sites. Data are intended for comparison with all other data, not just within a row (from Peterson et al. 1991).*

Forest	Site	Symptomatic	Asymptomatic
Tahoe NF	1	4.5 (0.6)	4.3 (0.5)
	2	4.4 (0.6)	4.8 (0.5)
	3	4.6 (0.5)	4.9 (0.6)
	4	4.3 (0.5)	4.8 (0.5)
Eldorado NF	1	3.0 (0.3)	4.5 (0.5)
	2	3.0 (0.2)	4.4 (0.6)
	3	3.7 (0.5)	4.0 (0.4)
	4	2.9 (0.4)	3.9 (1.2)
Stanislaus NF	1	2.6 (0.5)	4.8 (0.5)
	2	2.8 (0.4)	4.7 (0.5)
	3	3.6 (0.5)	4.3 (0.5)
	4	3.4 (0.5)	4.9 (0.3)
Yosemite NP	1	4.1 (0.2)	4.5 (0.2)
	2	3.9 (0.3)	4.8 (0.6)
	3	4.0 (0.2)	4.2 (0.4)
	4	4.0 (0.3)	4.3 (0.4)
Sierra NF	1	3.0 (0.2)	4.8 (0.6)
	2	3.4 (0.6)	4.6 (0.5)
	3	3.0 (0.2)	4.1 (0.3)
	4	3.0 (0.3)	4.4 (0.5)
Sequoia–Kings Canyon NP	1	3.2 (0.9)	4.0 (0.4)
	2	3.4 (0.5)	3.4 (0.6)
	3	3.4 (1.3)	3.2 (0.6)
	4	3.2 (0.7)	4.8 (0.4)
Sequoia NF	1	3.4 (0.5)	4.9 (0.2)
	2	3.6 (0.6)	5.0 (0.4)
	3	3.4 (0.6)	5.0 (0.5)
	4	3.4 (0.5)	—

current year needles, moderate in one-year-old needles (Figure 11.4), and common in two-year old needles (Figure 11.5). Correlations between needle retention and chlorosis were relatively weak but consistently negative, which indicates that trees with lower needle retention also had higher chlorosis. Neither needle retention nor chlorosis were correlated with diameter or age.

Injury was highest in Sequoia–Kings Canyon NP, where nearly all two-year old needles were damaged (Figure 11.5). Two-year-old needles from half of the trees were in injury class 2 (6–25% chlorosis), with many in higher injury classes as well. Over 50% of one-year-old needles had chlorotic injury. Injury of two-year-old needles from trees in Sequoia NF and Sierra NF was similar to Sequoia–Kings Canyon NP, but one-year-old needles were injured less severely. Forest sites north of Sierra NF had low injury with the exception of one site in Yosemite NP and two sites in Eldorado NF. There was almost no injury at Tahoe NF.

The spatial pattern of ozone chlorosis indicates that injury is clearly most severe in the southern Sierra Nevada and least severe in the north. This general trend corroborates previous observations of ozone injury at other locations in the Sierra Nevada (Pronos et al. 1978, Pronos and Vogler 1981, Warner et al. 1982, Allison 1982, 1984a,b, Duriscoe and Stolte 1989). Severity of ozone injury is also correlated with the ozone exposure gradient (Figure 11.2) previously discussed.

Growth trends in the 20th century

One of the difficulties in interpreting radial increment data is the high amount of variability normally present. For example, basal area incre-ment data for three trees from a symptomatic site in Sequoia–Kings Canyon National Parks are presented in Figure 11.6. One of the trees had a prominent recent growth decrease, one had a prominent increase, and one had no change. This example points out the advantage of assessing growth trends on an individual tree basis; if radial increment data for these trees had been averaged, the "average" graph would indicate no change, and potentially important trends for individual trees would not have been detected.

The Kalman filter procedure used in this study was helpful in detecting trends on an individual tree basis. For example, the basal area increment curve in Figure 11.7a indicates a trend of increasing growth at the end of the time series, although the start of this trend is difficult to identify. Analysis of the Kalman filter model parameters (Figure 11.7b) shows rather clearly that the change in growth trend occurred in the 1950s. Both basal area curves and time series of model parameters were used to estimate the occurrence of growth changes.

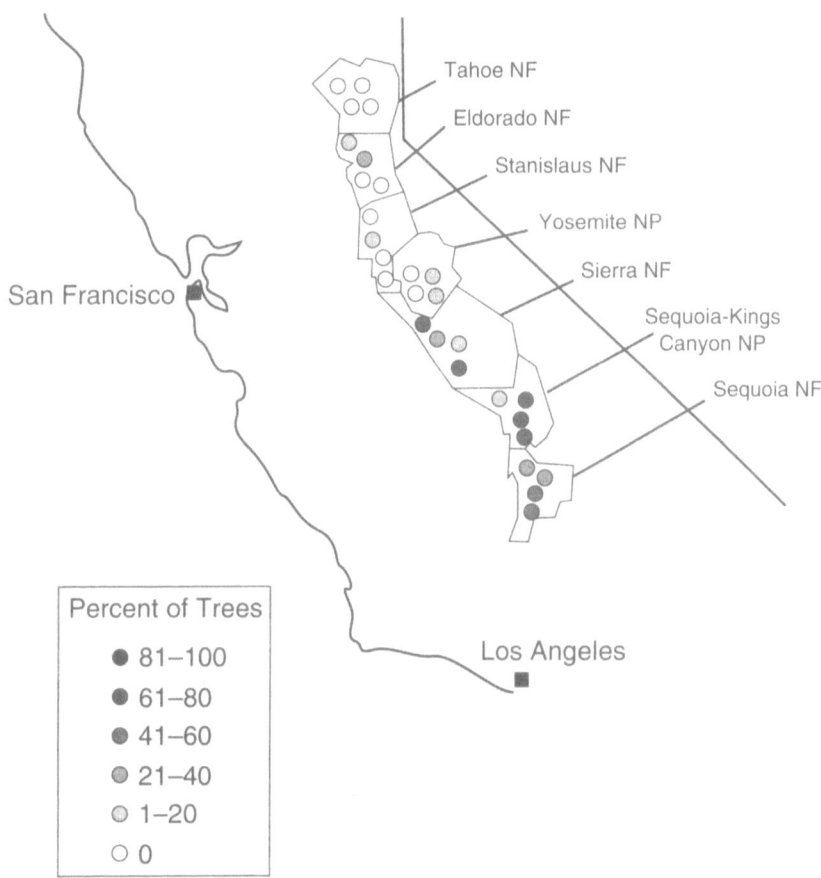

Figure 11.4 Percent of ponderosa pines with ozone injury in one-year-old needles at symptomatic sites.

There was wide variation in patterns of growth increases and decreases at both symptomatic and asymptomatic sites. One of the most prominent trends present in nearly all sites was a growth decrease beginning in the 1920s. We attribute this to low precipitation in the early 1920s, especially in 1924 which was associated with a very small ring in most tree cores. There were substantial growth increases starting in the late 1930s at many sites, however, which correspond with a period of relatively higher precipitation (Peterson et al. 1990). These two results indicate that regional growth trends were clearly related to climate.

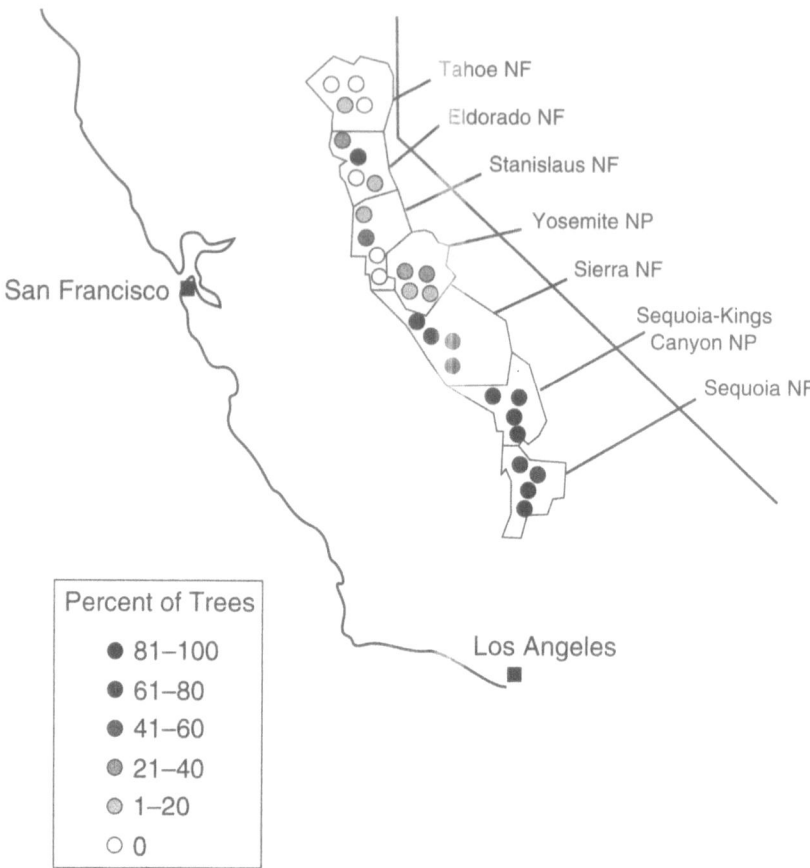

Figure 11.5 Percent of ponderosa pines with ozone injury in two-year-old needles at symptomatic sites.

Relationship of growth to air pollution

Growth since 1950 was of greatest interest to us, because of potential impacts from air pollution in recent years. Most symptomatic sites had trees with growth decreases in the 1950s or 1960s (Figure 11.8), although the number of decreases was never greater than 25% per site in either decade. The largest number of trees with decreased growth was in Sequoia–Kings Canyon NP. Most sites also had growth increases during the same time period. The number of trees with decreases and increases was similar for asymptomatic trees on a regional basis, with the largest number of recent growth decreases in the central Sierra (Figure 11.9).

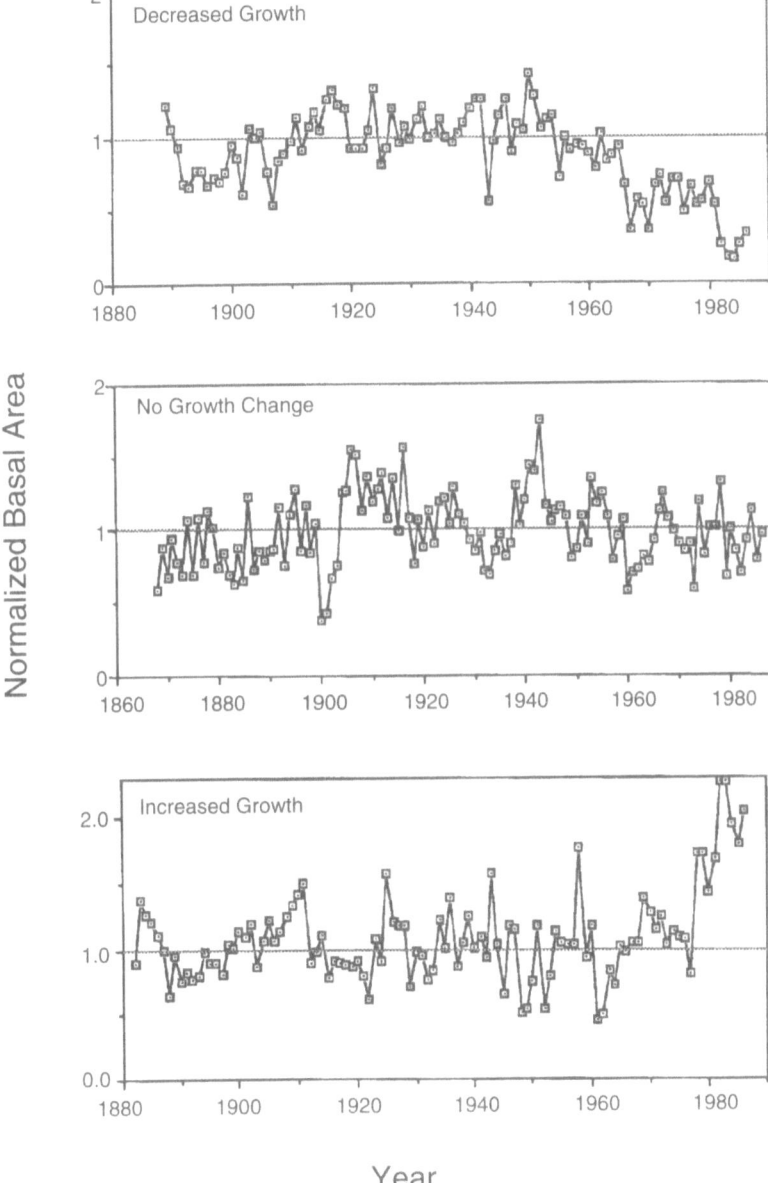

Figure 11.6 Time series of basal area increment growth for three ponderosa pines from a symptomatic site in Sequoia–Kings Canyon National Parks.

Core Number 1723

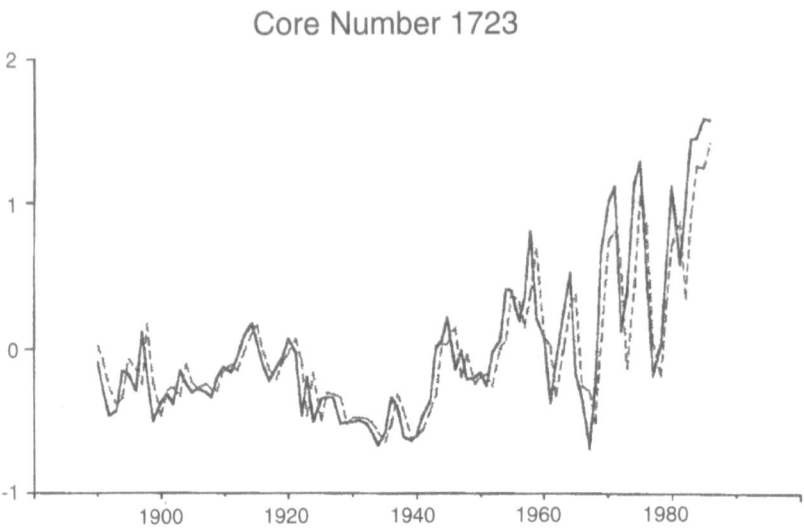

Residuals

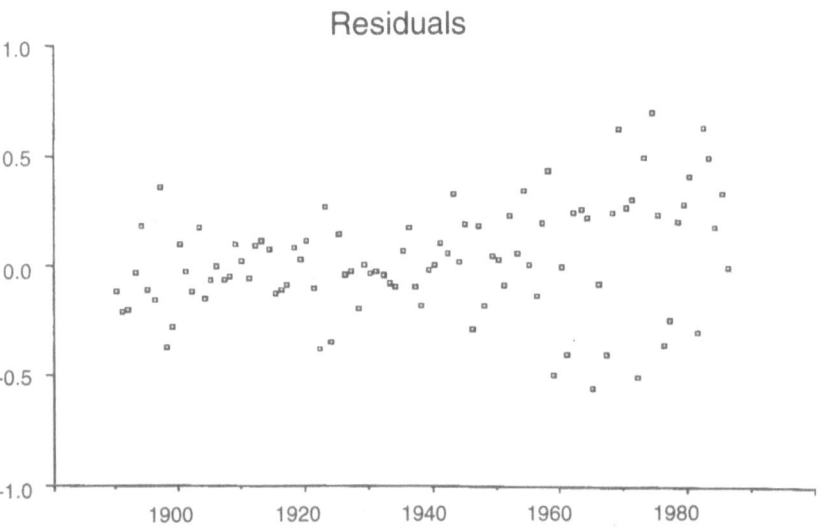

Figure 11.7a Time series of basal area increment growth and residuals for a ponderosa pine from an asymptomatic site in Sequoia National Forest. Values for the observed (solid line) and predicted (dashed line) basal area series are centered to 0.

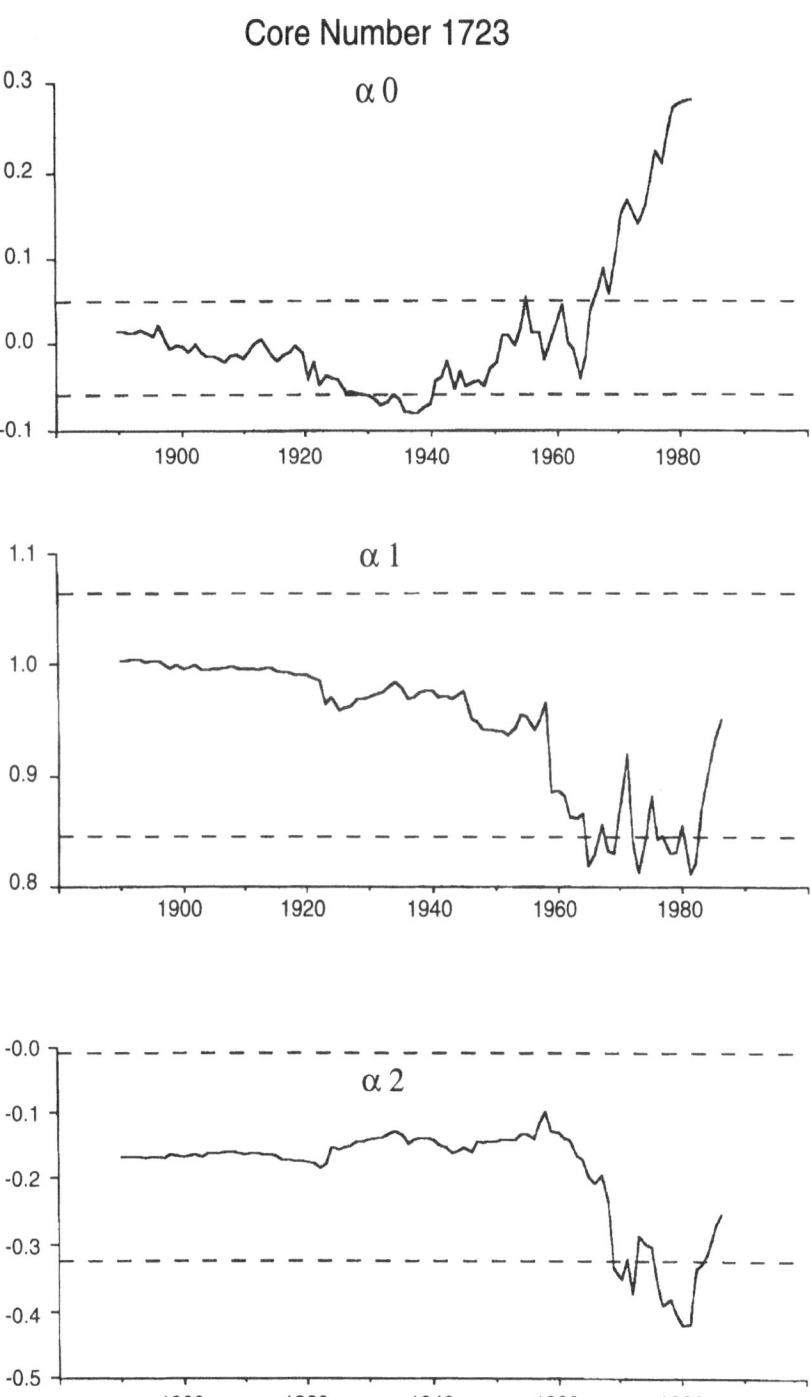

Figure 11.7b Kalman filter model parameters for the ponderosa pine in Figure 11.7a, indicating a change in growth trend in the 1950s. Dashed lines represent upper and lower 95% confidence intervals for stationary parameters.

No widespread pattern of either growth decreases or increases appeared during recent years in the Sierra Nevada. However, we analyzed some sites more closely in order to determine if growth decreases on a National Forest or Park level may have been caused by air pollution or other factors. It is important to identify such trends even if this response is not expressed over a large geographic area.

Recent growth changes (since 1950) at symptomatic sites were compared to decreases at asymptomatic sites and to past decreases at symptomatic sites. We found that symptomatic sites in Sequoia–Kings Canyon Na-

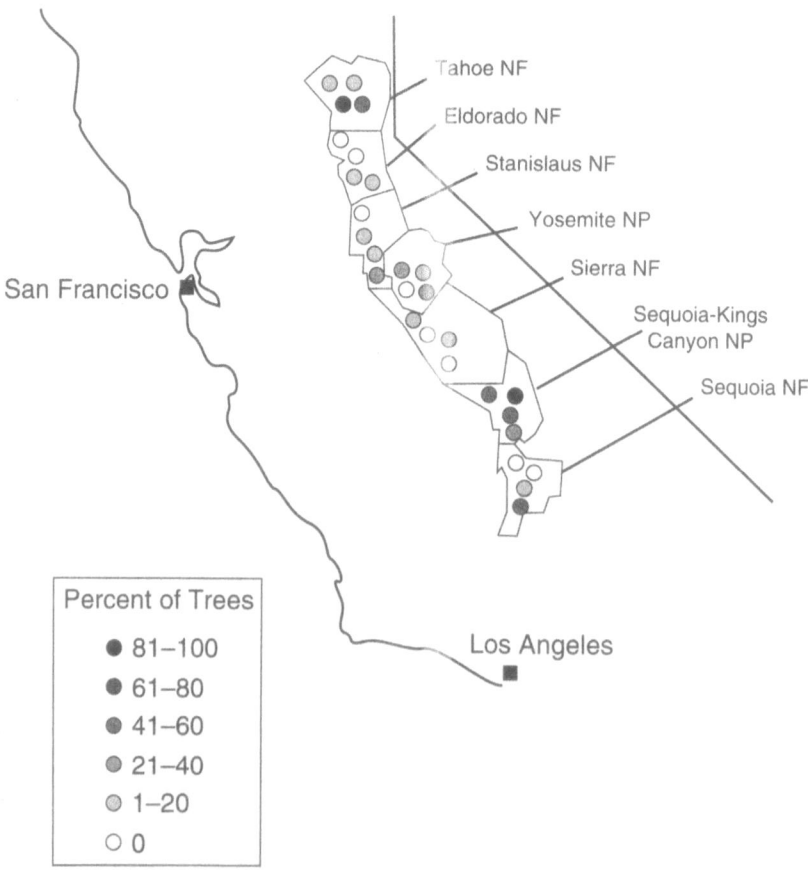

Figure 11.8 Decreases in basal area increment growth of ponderosa pine since 1950 at symptomatic sites.

tional Parks had a significantly greater number of both total growth decreases and net decreases (decreases minus increases) than most other forest locations. One site in Sequoia National Forest, one in Yosemite National Park, and two in Tahoe National Forest also had significant growth decreases.

We compared basal area growth with ozone injury in order to see if there was any quantitative relationship. Growth ratio (ratio of median basal area increment growth for 1982–1986 to median basal area growth for

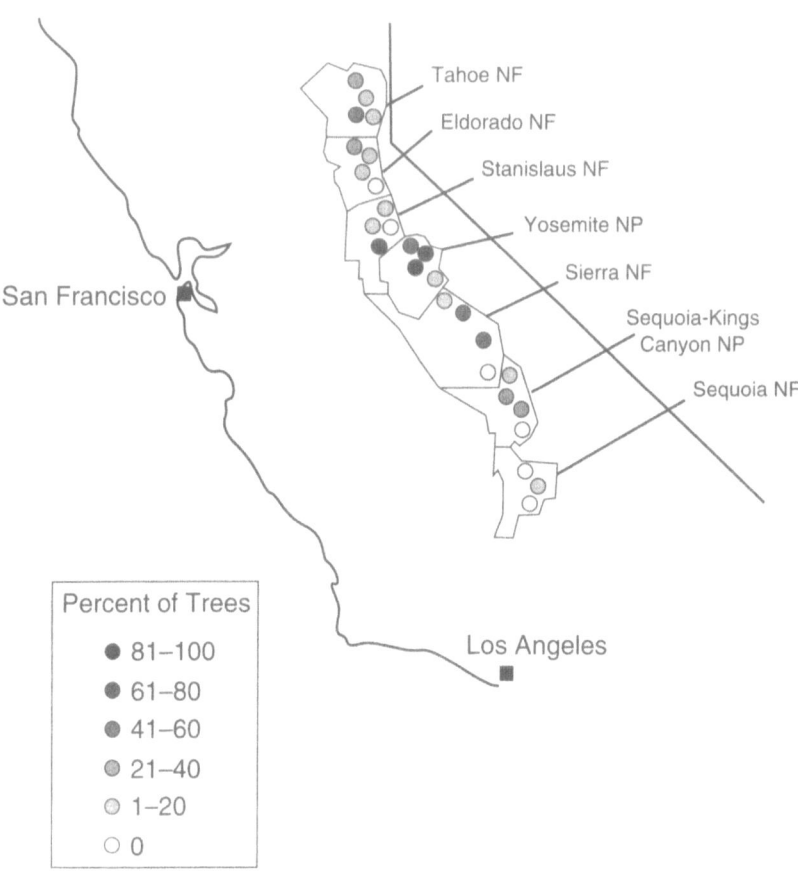

Figure 11.9 Decreases in basal area increment growth of ponderosa pine since 1950 at asymptomatic sites.

1951–1960) ranged from 0.89 to 3.16. Older and larger trees generally had lower growth ratio, but there were no apparent spatial patterns or major differences between symptomatic and asymptomatic sites.

Symptomatic sites from the three southernmost forests were used for the analysis because they were the only ones that had adequate levels of chlorosis (two-year old needles) to have any statistical meaning. Growth ratio was positively correlated with needle retention for all forests, although the correlation was relatively weak (Peterson et al. 1990); the correlation for Sequoia NF was considerably lower than for the other forests. This indicates that lower needle retention was correlated with lower growth. Correlations between growth ratio and ozone chlorosis were approximately zero in all cases. Growth ratio was correlated better with needle retention than with ozone chlorosis.

We investigated the relationship between injury and growth further by examining the correlations between growth ratio and needle retention for individual symptomatic and asymptomatic sites. We also calculated correlations for growth ratio with chlorosis in one-year old and two-year old needles of symptomatic trees. There was a weak positive correlation between growth ratio and needle retention for both symptomatic and asymptomatic sites (Peterson et al. 1990). Only a few of the positive correlations were highly significant, and none of the negative correlations were significant. Correlations between growth ratio and chlorosis varied between sites; there were no significant positive and negative correlations for both needle age classes, and no clear spatial pattern.

In summary, analysis on a stand basis corroborated the results of analysis on a forest basis. Growth ratio had a weak positive correlation with needle retention, but no clear relationship with ozone chlorosis. Needle retention was therefore a better indicator of growth in ozone-stressed trees, although the correlation is not strong enough to use for quantitative purposes. Tree diameter and age were not confounding factors because neither variable was correlated with needle retention.

Relationship of growth to other environmental factors

Previous studies in the Sierra Nevada have shown that interannual variation in winter precipitation accounts for the largest portion of variance in tree growth. Summers are warm and dry in the mediterranean climate of the Sierra, and soil moisture is dependent on winter snowfall and rainfall. Summer temperatures also affect the availability of soil moisture (Peterson et al. 1987a,b, Peterson and Arbaugh 1988).

As previously mentioned, many of the sites in this study had growth decreases in the 1920s, a period of low precipitation, and growth increases starting in the late 1930s, a period of higher precipitation. Growth reductions in the 1960s and 1970s may also be related to low precipitation. Tree rings were usually very small for the years 1961, 1976, and 1977, which were low precipitation years (see Chapter 2, Figure 2.5d). Only one or two years of high or low precipitation may have a long term impact on growth, because of a lag, or autoregressive effect.

The effects of stand dynamics in the Sierra Nevada are difficult to interpret because information on stand history is poor in most cases. We found that the time series of growth within stands in National Parks generally had greater variance than in National Forests. This pattern exists because undisturbed National Parks tend to have uneven-aged stands, and National Forests tend to have even-aged stands as the result of timber harvesting.

Thinning on National Forest lands is conducted to reduce competition and promote faster growth of the residual trees. This was almost certainly the cause of some of the growth increases observed since 1950 (few of these increases were observed in National Parks). These increases were generally synchronous and occurred in many trees per site.

Prominent growth reductions were found at two symptomatic sites in Tahoe National Forest. These growth reductions were probably due to high stem density (1350 stems/ha and 1680 stems/ha for symptomatic sites 2 and 3, respectively). These were extremely high densities compared to most other Sierra sites, and ozone exposure and injury were low in this area. We know of no other explanations for this recent change in growth.

The largest recent growth decreases in this study were found in asymptomatic sites in Yosemite NP; two of these sites had growth decreases throughout the 20th century. The probable cause of this pattern is annosus root rot, which is common throughout Yosemite Valley (Felix et al. 1973, Parmeter et al. 1978). This fungus causes wood decay, kills sapwood and cambium, and reduces uptake of water and nutrients (Scharpf 1978b). Growth trends at Yosemite NP indicate that pathogens have a large impact on ponderosa pine growth in some stands.

Conclusions

We found no evidence of large scale growth reduction in ponderosa pine throughout the Sierra Nevada. There were some regional growth trends, however, such as the period of growth decreases in the 1920s and the increases in the 1930s. Basal area increment growth tended to be similar within each site, and was often quite different from adjacent sites. There was higher variability among sites on National Forests than National Parks, which suggests that cutting and other management practices have affected growth trends. Within-site variability was higher in National Park sites.

Needle retention was lower at symptomatic sites than at asymptomatic sites throughout the Sierra. Growth was positively correlated with needle retention, although this correlation was too weak to develop a predictive relationship. Growth was not correlated with ozone chlorosis. Ozone stress may not have been severe enough or of long enough duration in the Sierra Nevada to produce quantifiable relationships between growth and ozone injury.

Most of our knowledge on the relationship between ozone dose and tree response is based on seedling studies (see Chapter 5), and little is known about dose-response for mature trees. Many growth decreases and increases occurred in both symptomatic and asymptomatic sites since 1950, with the largest number of decreases at symptomatic sites in the southern Sierra Nevada. These decreases were in the region of the Sierra with highest ozone exposure and highest ozone injury. It may never be possible to determine a cause-and-effect relationship, but we feel it is quite likely that ozone injury is part of a stress complex that has resulted in growth reductions at some sites in the southern Sierra Nevada.

We have characterized regional trends in ponderosa pine growth, but have no information on other tree species of the mixed conifer forest in the Sierra Nevada. All of the dominant trees in this forest type, such as Douglas-fir, white fir, incense-cedar, sugar pine, and California black oak, are generally considered to be more resistant to ozone injury than ponderosa pine (Davis and Wilhour 1976). We know little about stand dynamics in the mixed conifer forest, or about how a reduction in growth of one species might be compensated for by another. Ozone injury to ponderosa pine in mixed conifer forest adjacent to the Los Angeles Basin (Miller et al. 1989) has resulted in an accelerated rate of succession and dominance of white fir and incense-cedar (Miller 1983). Ozone injury is not as severe in the Sierra Nevada, but widespread injury and growth reductions at some sites indicate at least some loss of vigor at the present time.

We have previously documented a gradient of ozone exposure and visible injury in ponderosa pine in the Sierra Nevada (Peterson et al. 1989, 1991). We have also found that radial growth has declined in some stands with ozone injury (Peterson et al. 1987a,b, 1989, 1991). Ozone levels will probably remain the same or increase in the Sierra Nevada in the forseeable future, even if hydrocarbons and nitrogen oxides are reduced slightly (Innes 1981). The mixed conifer forest of the Sierra Nevada is one of the few forests in the world to exhibit widespread ozone injury and some growth reductions. Monitoring injury and growth in these forests to detect changes in their health and productivity should be a priority.

References

Allison JR (1982) *Evaluation of Ozone Injury on the Stanislaus National Forest.* USDA Forest Service, Pacific Southwest Region, Forest Pest Management Report 82-7

Allison JR (1984a) *An Evaluation of Ozone Injury to Pines on the Eldorado National Forest.* USDA Forest Service, Pacific Southwest Region, Forest Pest Management Report 84-16

Allison JR (1984b). *An Evaluation of Ozone Injury to Pines on the Tahoe National Forest.* USDA Forest Service, Pacific Southwest Region, Forest Pest Management Report 84-30

Brown AA, Davis KP (1973) *Forest Fire Control and Use.* McGraw-Hill, NY

California Air Resources Board (1987) *California Air Quality Data, Volume 18.* Aerometric Data Division, Sacramento, CA

Carroll JJ, Baskett RL (1979) Dependence of air quality in a remote location on local and mesoscale transports: A case study. *Journal of Applied Meteorology* 18:474–486

Cibrian Tovar D (1989) Air pollution and forest decline near Mexico City. *Environmental Monitoring and Assessment* 12:49–58

Ciesla WM, Macias Samano JE (1987) Desierto de los Leones: A forest in crisis. *American Forests* 93:29–31,72–74

Davis DD,Wilhour RG (1976) *Susceptibility of Woody Plants to Sulfur Dioxide and Photochemical Oxidants.* Report EPA-600/3-76-102, Environmental Research Laboratory, United States Environmental Protection Agency, Corvallis, OR

Duriscoe DM, Stolte KW (1989) Photochemical oxidant injury to ponderosa pine (*Pinus ponderosa* Laws.) and Jeffrey pine (*Pinus jeffreyi* Grev. and Balf.) in the National Parks of the Sierra Nevada of California. In: Olson RK, Lefohn AS (eds) *Effects of Air Pollution on Western Forests.* Transactions Series, No. 16, Air and Waste Management Association, Pittsburgh, pp 261–278

Ewell DM, Flocchini RG, Myrup LO (1989a) Aerosol transport in the southern Sierra Nevada. *Journal of Applied Meteorology* 28:112–125

Ewell DM, Mazzu LC, Duriscoe DM (1989b) Specific leaf weight and other characteristics of ponderosa pine as related to visible ozone injury. In: Olson RK, Lefohn AS (eds) *Effects of Air Pollution on Western Forests.* Transactions Series, No. 16, Air and Waste Management Association, Pittsburgh, pp 411–418

Felix LS, Parmeter JR, Uhrenholdt B (1973) *Fomes annosus* as a factor in the management of recreational forests. In: *Proceedings: 4th Conference on* Fomes annosus, University of Georgia, Athens, Georgia, pp 309–372

Fritts HC (1976) *Tree Rings and Climate.* Academic Press, New York

Furniss RL, Carolin VM (1977) *Western Forest Insects.* USDA Forest Service Miscellaneous Publication 1339, Washington, DC

Graybill DA, Rose MR (1989) Analysis of growth trends and variation in conifers from Arizona and New Mexico. In: Olson RK, Lefohn AS (eds) *Effects of Air Pollution on Western Forests.* Transactions Series, No. 16, Air and Waste Management Association, Pittsburgh, pp 395–407

Griffin JR, Critchfield WB (1972) *The Distribution of Forest Trees in California.* USDA Forest Service Research Paper PSW-82

Grulke NE, Miller PR, Wilborn RD, Hahn S (1989) Photosynthetic response of giant sequoia seedlings and rooted branchlets of mature foliage to ozone fumigation. In: Olson RK, Lefohn AS (eds) *Effects of Air Pollution on Western Forests.* Transactions Series, No. 16, Air and Waste Management Association, Pittsburgh, pp 429–441

Helms JA (1970) Summer net photosynthesis of ponderosa pine in its natural environment. *Photosynthetica* 4:243–253

Hogsett W, Tingey DT, Hendricks C, Rossi D (1989) Sensitivity of Western conifers to SO_2 and acid fog and ozone. In: Olson RK, Lefohn AS (eds) *Effects of Air Pollution on Western Forests.* Transactions Series, No. 16, Air and Waste Management Association, Pittsburgh, pp 469–491

Hom JL, Oechel WC (1983) The photosynthetic capacity, nutrient content, and nutrient use efficiency of different needle age-classes of black spruce found in interior Alaska. *Canadian Journal of Forest Research* 13:834–839

Innes WB (1981) Effect of nitrogen oxide emissions on ozone loads in metropolitan regions. *Environmental Science and Technology* 15:904–912

Kalman RE (1960) A new approach to linear filtering and prediction problems. *Transactions of the American Society of Mechanical Engineering Journal of Basic Engineering (Series D)* 82:35–45

Kalman RE, Bucy RS (1961) New results in linear filtering and prediction problems. *Transactions of the American Society of Mechanical Engineering Journal of Basic Engineering (Series D)* 83:95–108

Kienast F, Schweingruber FH, Bräker FH, Schar E (1987) Tree-ring studies on conifers along ecological gradients and the potential of single-year analyses. *Canadian Journal of Forest Research* 17:683–696

Miller PR (1983) Ozone effects in the San Bernardino National Forest. In: *Proceedings: Symposium on Air Pollution and the Productivity of the Forest.* Izaak Walton League of America, Arlington, Virginia, pp 161–197

Miller PR, Longbotham GJ, Longbotham CR (1983) Sensitivity of selected Western conifers to ozone. *Plant Disease* 67:1113–1115

Miller PR, McBride JR, Schilling SL, Gomez AP (1989) Trends of ozone damage to conifer forests in southern California. In: Olson RK, Lefohn AS (eds) *Effects of Air Pollution on Western Forests.* Transactions Series, No. 16, Air and Waste Management Association, Pittsburgh, pp 302–309

Miller PR, McCutchan MH, Milligan MP (1972) Oxidant air pollution in the Central Valley, Sierra Nevada foothills and Mineral King Valley of California. *Atmospheric Environment* 6:623–633

Miller PR, Millecan AA (1971) Extent of air pollution damage to some pines and other conifers in California. *Plant Disease Reporter* 55:555–559

Miller PR, Wilborn RD, Schilling SL, Gomez AP (1988) *Ozone Injury to Important Tree Species of Sequoia and Kings Canyon National Parks.* Final Report filed with the National Park Service Air Quality Division, Denver, CO

Muir PS, Armentano TV (1988) Evaluating oxidant injury to foliage of *Pinus ponderosa*: A comparison of methods. *Canadian Journal of Forest Research* 18:498–505

Parmeter JR, McGregor NJ, Smith RS (1978) *An Evaluation of* Fomes annosus *in Yosemite National Park*. USDA Forest Service, Pacific Southwest Region, Forest Pest Management Report 78-2

Parsons DJ (1981) The historical role of fire in the foothill communities of Sequoia National Park. *Madroño* 28:111–120

Parsons DJ, DeBenedetti SH (1979) Impact of fire suppression on a mixed-conifer forest. *Forest Ecology and Management* 2:21–33

Patterson MT, Rundel PW (1989) Ozone impacts on the photosynthetic capacity of Jeffrey pine in Sequoia National Park. In: Olson RK, Lefohn AS (eds) *Effects of Air Pollution on Western Forests*. Transactions Series, No. 16, Air and Waste Management Association, Pittsburgh, pp 419–427

Peterson DL, Arbaugh MJ (1988) Growth patterns of ozone-injured ponderosa pine (*Pinus ponderosa*) in the southern Sierra Nevada. *Journal of the Air Pollution Control Association* 38:921–927

Peterson DL, Wakefield VA, Arbaugh MJ (1987a) Detecting the effects of ozone pollution on growth of Jeffrey pine in the Sierra Nevada, California. In: Jacoby GC, Hornbeck JW (compilers) *Proceedings of the International Symposium on Ecological Aspects of Tree-Ring Analysis*. Department of Energy, National Technical Information Service Publication CONF-8608144, US Department of Commerce, Springfield, VA, pp 402–409

Peterson DL, Arbaugh MJ, Wakefield VA, Miller PR (1987b) Evidence of growth reduction in ozone-stressed Jeffrey pine (*Pinus jeffreyi* Grev. and Balf.) in Sequoia and Kings Canyon National Parks. *Journal of the Air Pollution Control Association* 37:906–912

Peterson DL, Arbaugh MJ, Robinson LJ (1989) Ozone injury and growth trends of ponderosa pine in the Sierra Nevada. In: Olson RK, Lefohn AS (eds) *Effects of Air Pollution on Western Forests*. Transactions Series, No. 16, Air and Waste Management Association, Pittsburgh, pp 293–307

Peterson DL, Arbaugh MJ, Robinson LJ (1991) Regional growth trends of ozone-injured ponderosa pine (*Pinus ponderosa*) in the Sierra Nevada, California, USA. *The Holocene* 1:50–61

Pronos J, Vogler DR (1981) *Assessment of Ozone Injury to Pines in the Southern Sierra Nevada, 1979/1980*. USDA Forest Service, Pacific Southwest Region, Forest Pest Management Report 81-20

Pronos J, Vogler DR, Smith RS (1978) *An Evaluation of Ozone Injury to Pines in the Southern Sierra Nevada*. USDA Forest Service, Pacific Southwest Region, Forest Pest Management Report 78-1

Scharpf RF (1978a) Mistletoes. In: Bega RV (ed) *Diseases of Pacific Coast Conifers*. USDA Forest Service Agricultural Handbook 521, Washington, DC, pp 121–141

Scharpf RF (1978b) Root diseases. In: Bega RV (ed) *Diseases of Pacific Coast Conifers*. USDA Forest Service Agricultural Handbook 521, Washington, DC, pp 142–156

Schweingruber FH (1986) Abrupt growth changes in conifers. *International Association of Wood Anatomists Bulletin* 7:277–284

Stephenson N (1988) *Climatic Control of Vegetation Distribution: The Role of Water Balance with Examples from North America and Sequoia National Park, California*. PhD dissertation, Cornell University, Ithaca, NY

Stokes MA, Smiley TL (1968) *An Introduction to Tree-ring Dating*. University of Chicago Press, Chicago

Swetnam TW, Thompson MA, Sutherland EK (1985) *Using Dendrochronology to Measure Radial Growth of Defoliated Trees*. USDA Forest Service Agricultural Handbook 639, Washington, DC

Van Deusen PC (1987) Some applications of the Kalman filter to tree-ring analysis. In: Jacoby GC, Hornbeck JW (compilers) *Proceedings of the International Symposium on Ecological Aspects of Tree-Ring Analysis*. Department of Energy Technical Information Service Publication CONF-8608144, US Department of Commerce, Springfield, VA, pp 566–578

Visser H, Molenaar J (1986) *Time Dependent Responses of Trees to Weather Variations: An Application of the Kalman Filter*. N. V. Tot Keuring Van Elektrotechnische Materialen, Research and Development Division, Arnhem, The Netherlands

Warner TE, Wallner DW, Vogler DR (1982) Ozone injury to ponderosa and Jeffrey pines in Sequoia–Kings Canyon National Parks. In: van Riper C, Whittig LD, Murphy ML (eds) *Proceedings: Conference on Research in California's National Parks*. University of California, Davis, CA, pp 1–7

12

Mixed Conifer Forests of the
San Bernardino Mountains, California

P. R. Miller

The San Bernardino Mountains are part of the Transverse Range Province that extends from west to east across parts of Santa Barbara, Ventura, Los Angeles, San Bernardino, and Riverside counties, California (Baily and Jahns 1954). The northern section of the San Bernardino National Forest, which is the subject of this chapter, is confined entirely to San Bernardino County. It is bounded on the north by the Mojave Desert, on the east by the Little San Bernardino Mountains, on the south by the upper Santa Ana Valley and Yucaipa-Beaumont Plains, and on the west by the San Gabriel Mountains. This chapter describes present forest conditions in the San Bernardino Mountains in relation to geology, soils, climate, biotic pests and diseases, fire, recreational pressure and resource management, and chronic exposure to photochemical oxidant air pollutants.

Biogeographic Setting

Geology

The San Bernardino Mountains were formed by an uplift along the San Andreas fault zone to the south and along several steeply dipping reverse faults to the north. Major topographic features include the prominent east-west ridge, with 13 peaks higher than 3000 m, that separates the upper Santa Ana River drainage basin from the San Gorgonio pass area to the south, and the subdued upland plain in the Crestline-Lake Arrowhead-Big Bear Lake regions (Figure 12.1). More than half of the mountain area is characterized by slopes greater than 50%. The major geologic materials include gneisses, schists, plutonic

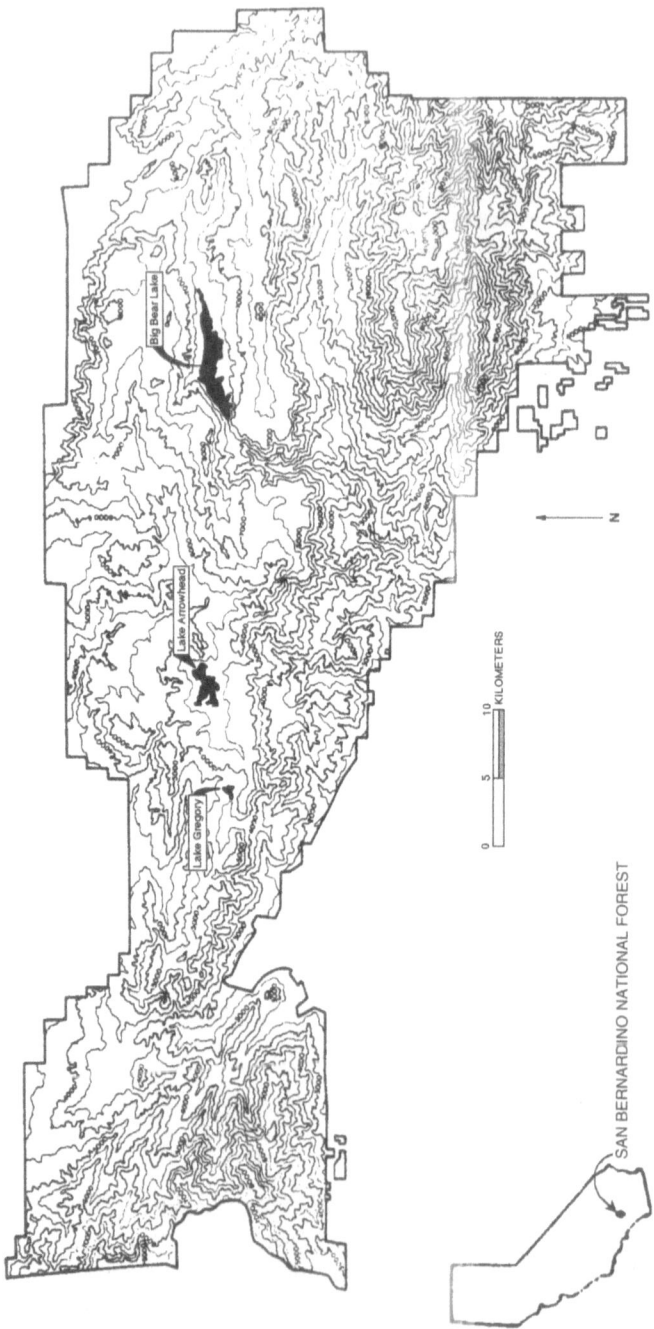

Figure 12.1 Topographic map of the San Bernardino Mountains (Taylor 1973). Contours shown in feet.

rocks, sediments and recent alluvium. The most prominent material exposed is a light colored quartz monzonite of the Mesozoic age called Cactus granite (Bailey and Jahns 1954).

Soils

Soil depths, profiles, and available water content were described at a number of forest vegetation plots established throughout the San Bernardino Mountains (Miller and Elderman 1977). Arkley (1981) described soils near Lake Arrowhead as coarse, loamy, mixed, mesic, Ultic Haploxerolls. To the east of Lake Arrowhead at elevations > 1900 m, soils were typically coarse-loamy mixed, frigid Xerumbrepts and Xerochrepts. Two locations were on fine-loamy Argixerolls. Strongly decomposed granite was generally encountered at depths of 1.3 to 2.0 m and probably extends downward to 5 m or more.

During 1974 through 1978, available moisture content of soils at the experimental plots ranged from 1.28 to 4.96 cm in the top 274 cm (the deepest sensor) of soil (Arkley 1981). Winter precipitation was almost always enough to recharge the top 274 cm, but available water was virtually gone from this layer by the end of September. Predawn xylem water potentials remained moderate even in September (Miller and Elderman 1977), demonstrating that trees were able to access water from sources in fractured rock deeper than 274 cm (Arkley 1981). Excessive drought conditions could develop after several consecutive years of diminished rainfall insufficient to recharge fractured areas in the underlying rock.

Climate

The San Bernardino Mountains experience a summer-dry mediterranean climate typical of southern California. More than 90% of precipitation occurs as rain or snow between October and April during large storms that originate in the Gulf of Alaska (see Chapter 2). The small amount of summer precipitation results from thundershower activity in August. Thunderstorms are mainly concentrated in the eastern third of the mountains. During spring and early summer the western half of the San Bernardino Mountains intercepts some moisture from the deep coastal fog that shrouds slopes and ridgetops below about 1800 m elevation.

A complete precipitation record from 1884 (Figure 12.2) exists for Big Bear Dam at 2077 m elevation. One way of assessing this record, in the context of understanding tree growth trends, is to determine the number of consecutive years with precipitation amounts above the long-term average (92.95 cm, 01 October to 30 September), and consecutive years

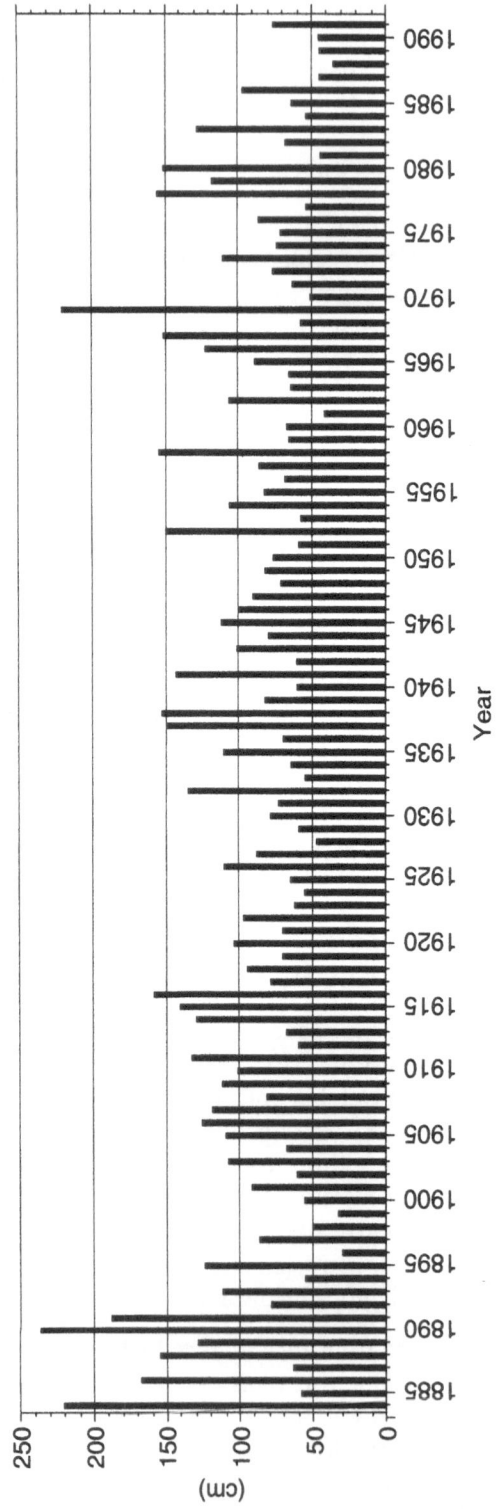

Figure 12.2 Precipitation record from 1884 to 1991 at Big Bear Dam. Totals shown are for water years 1 October – 30 September. For example, 1970 represents precipitation for October 1969 – September 1970.

below average. Another important variable is the timing of small amounts of moisture arriving during the dry season. The longest dry periods included 1896–1902, 1947–1951, and 1981–1991 (Table 12.1). Precipitation during June and July was almost always zero. During the 24 wet or dry intervals in the Big Bear Dam record, 30 years had total precipitation amounts for May, August and September that were higher than the long-term average. For example, during the years 1896 to 1902 there were seven consecutive years with below-average precipitation, of which three had higher than average summer (May, August, September) precipitation. Above average precipitation during May, August and September may partly compensate for lower-than-average winter precipitation (Arkley 1981). The significance of these trends in relation to stem growth is discussed below.

*Table 12.1 At Big Bear Dam, periods of two or more years, since 1884, during which every year had total precipitation greater than the long-term average for the site (92.95cm, 01 October to 30 September; water year date matches the September date) or every year had total precipitation less than the long-term average. All years in the periods marked in **bold** had greater than average precipitation. All years in the periods written in regular typeface had below average precipitation. The numbers in the single-digit columns to the right of each period are the number of years in the interval when the sum of May, August and September precipitation was greater than the long-term average of 2.75cm.*

Period	Wetter Than Average Summers	Period	Wetter Than Average Summers
1888–1891	4	**1945–1946**	0
1896–1902	3	1947–1951	2
1905–1907	2	1955–1957	2
1909–1911	2	1959–1961	0
1912–1913	1	1963–1965	2
1914–1916	2	**1966–1967**	1
1917–1919	0	1970–1972	1
1923–1925	0	1974–1977	2
1927–1931	2	**1978–1980**	2
1933–1934	0	1981–1982	0
1937–1938	0	1984–1985	0
1939–1940	1	1987–1991	1

A well known feature of summer climate in southern California is the formation of persistent, low-level temperature inversions that occur in conjunction with onshore flow and high pressure systems aloft. Air pollutants injected into the marine layer are inhibited from vertical diffusion by the warm air layer above (see Chapter 3). Photochemical reactions together with a lack of pollutant dispersion causes ozone to accumulate beneath the inversion as the entire marine air mass moves inland. Surface heating weakens the inversion as the day progresses and the polluted air begins to mix vertically. The sea breeze and afternoon upslope flow advect pollutants to forests along the western slopes of the mountains.

At least eight dominant air flow patterns are possible during March through November (Hayes et al. 1984; Figure 12.3). The percentage of time that individual patterns prevail at 0400, 1000, 1600, and 2200 hrs during the spring, summer and fall days is indicated in Table 12.2.

Table 12.2 South Coast Air Basin surface air flow types, including diurnal percentage of occurrence for spring, summer and fall months (Hayes et al. 1984).

Types	I	II	III	IV	V	VIa	VIb	VII
Time PST	On-shore	Sea Breeze	Off-shore	South-erly	Downslope/ Traditional	Weak Santa Ana	Full Santa Ana	Calm
Spring								
4:00 a	10	8	19	6	· 26	4	3	24
10:00 a	43	29	3	12	5	2	1	2
4:00 p	31	61	2	4	1	1	1	*
10:00 p	23	26	9	4	23	3	1	10
All	27	31	8	6	14	3	2	9
Summer								
4:00 a	10	5	4	4	34	1	1	37
10:00 a	51	41	1	6	1	*	*	0
4:00 p	26	73	0	1	0	0	0	0
10:00 p	34	39	2	2	18	1	*	5
All	30	40	2	3	13	1	*	11
Fall								
4:00 a	7	10	16	2	26	7	4	25
10:00 a	33	29	5	6	10	6	4	7
4:00 p	20	67	4	2	2	1	1	4
10:00 p	16	19	13	2	27	5	3	15
All	19	31	10	3	16	5	3	13

* < 0.5% of the time

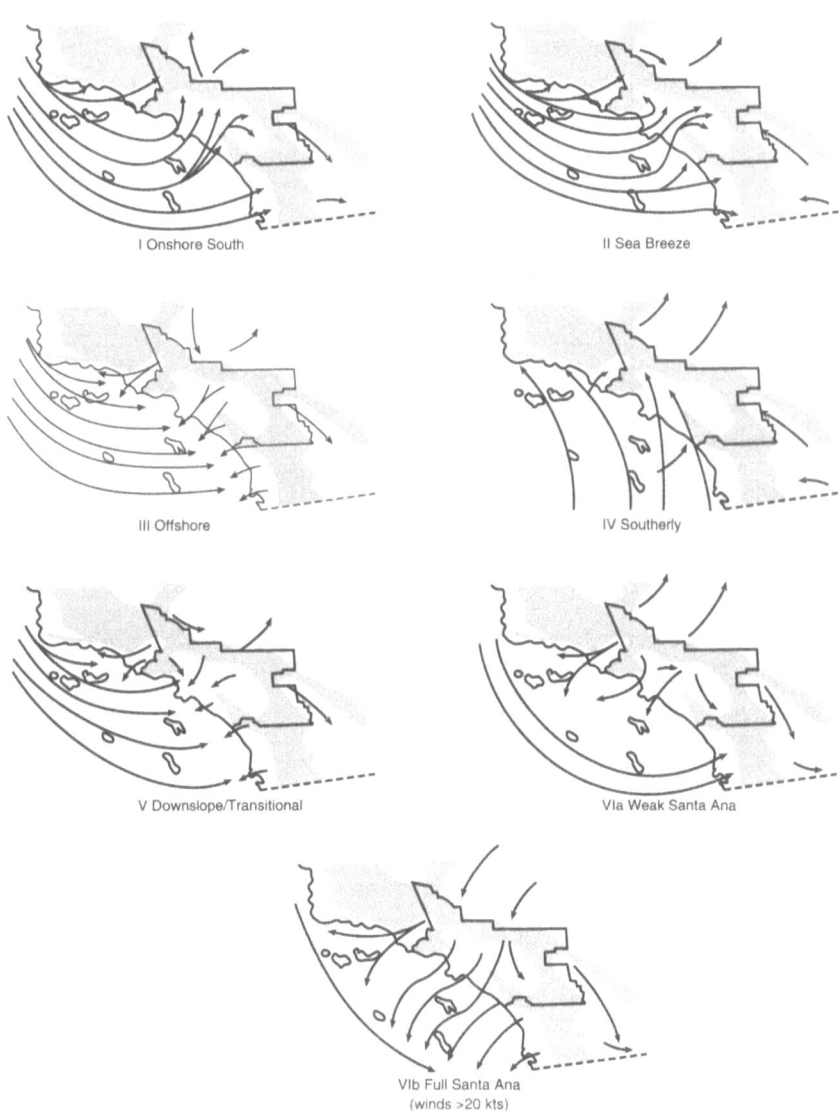

Figure 12.3 South Coast surface air flow types (Hayes et al. 1984), Class VII not shown (calm).

Vegetation

Vegetation in the San Bernardino Mountains is a complex combination of closed forest, woodland, and shrub formations. The general distribution of different vegetation types is shown in Figure 12.4 (Miller and Elderman 1977).

The ponderosa pine (*Pinus ponderosa*) and Jeffrey pine (*Pinus jeffreyi*) series has an overstory composed of either species (Paysen et al.1980). Sometimes both species mix along contact zones. The understory of ponderosa and Jeffrey pine forests is herbaceous or shrubby. The mixed conifer series includes sugar pine (*P. lambertiana*), white fir (*Abies concolor*), ponderosa pine or Jeffrey pine, incense-cedar (*Calocedrus decurrens*) and California black oak (*Quercus kelloggii*). The overstory is multistoried in dense stands and the understory growth is generally limited by excessive litter accumulation.

Elevation is the most obvious variable associated with distributions of the closed forest, woodland and shrub formations. Wright (1968) esti-mated elevational ranges for conifer species in the San Bernardino Mountains:

Pseudotsuga macrocarpa,	big-cone Douglas-fir	600–2150 m
Pinus attenuata,	knobcone pine	600–1375 m
Pinus coulteri,	Coulter pine	600–2150 m
Calocedrus decurrens,	incense-cedar	1225–2450 m
Pinus ponderosa,	ponderosa pine	1225–2300 m
Pinus jeffreyi,	Jeffrey pine	1375–2750 m
Pinus lambertiana,	sugar pine	1375–2750 m
Abies concolor,	white fir	1375–3050 m
Pinus contorta,	lodgepole pine	2450–3400 m
Pinus flexilis,	limber pine	2600–3500 m

Wright (1968) also examined the presence of knobcone pine, Coulter pine, and sugar pine along a single elevational transect in the San Bernardino mountains. Lower elevational limits were 860 m for knobcone pine, 1200 m for Coulter pine, and 1600 m for sugar pine. For established trees there was no evidence of growth decrease at the lower elevational limits of these species. This suggested that the lower elevational limit of each species was influenced by conditions required for seedling establishment. During the first two years after establish-ment, roots of seedlings still in the upper 40 cm of soil tended to dry out before the beginning of fall precipitation. Soils dry out faster at lower elevations. Furthermore, the rooting habit of established Coulter and knobcone pines was predominantly vertical while sugar pine roots were more shallow and horizontal. Further generalization from these results is limited by the variable, destructive effects of frequent fires that spread uphill from the chaparral shrub zone.

The species composition descriptions of mixed forests in southern California have shown slight differences over time, however, these differences may be partly attributable to survey methods. Parish (1917) reported 299 species of trees, shrubs, herbs and grasses in the forests dominated by either ponderosa or Jeffrey pine. McBride et al. (1975) collected 232 of these taxa, 67 fewer than Parish (1917); however, 83 species were collected that Parish did not report. Twenty exotics were found by McBride et al. (1975), while Parish reported only five. Twenty-six taxa are considered endemic to the San Bernardino Mountains; fifteen are considered rare or endangered. As to distribution in families, there were 34 genera of the Asteraceae and 19 genera of the Poaceae. For trees, the genus with the most species was *Pinus* with six, for shrubs it was *Arctostaphylos* with four species, and for herbaceous plants it was *Carex* with eight species (McBride et al. 1975).

Haase (1913) reported 16 species of foliose and fruticose lichens in the San Bernardino Mountains; these species were found mainly on the bark of conifers and oaks. Sigal and Nash (1980) found only eight of these species and four of them were present in very small quantities.

Fire History

One of the first accurate descriptions of unharvested forest vegetation in the San Bernardino Mountains emphasized that the forest was comprised mainly of large diameter yellow pines in very open stands with little or no understory growth (Leiberg 1900). This implies that the forests were largely in the natural fire climax phase. Repeated understory fires cleared out younger trees and other competing vegetation and fuel buildup between fires was not sufficient to cause much damage to the lower boles of most trees. Today this state of equilibrium has been replaced over large areas by a disclimax phase, i.e., a replacement or modification of the true climax often brought about by human activities (Clements 1936).

Man's hand in the disruption of the true climax began with extensive logging operations which started during the 1850s in the western section of the San Bernardino Mountains and extended eastward to the Running Springs-Green Valley area by the late 1890s (La Fuze 1971). In 1900, forest regrowth throughout the area was characterized as "dense young thickets 3–5 m high coming up after lumbering of 20 to 30 years ago"

Figure 12.4 *Vegetation type map of the San Bernardino Mountains (Miller and Elderman 1977).*

Ponderosa, Jeffrey, Coulter
and/or sugar pine, white fir,
incense-cedar, and/or big-cone Douglas-fir

Limber and lodgepole pine

Pinyon and/or juniper

Hardwoods (black oak)

Unstocked or non-commercial sites, barren and
waste areas and/or chapparal

Private land within National Forest boundary and/or lakes

Heart Bar State Park

San Gorgonio Wilderness Area

(Craig 1904). Some of the lands clear-cut by the Brookings Lumber and Box Company in the Running Springs area had burned repeatedly and the vegetation cover is characterized to this day as timberland chaparral (Horton 1960).

Development of second- and third-growth pine forest stands has been accompanied by an unprecedented buildup of dead fuel (Wilson and Dell 1971, Dodge 1972), and a change to dense understory growth, primarily due to fire suppression policies. Increased mortality of pon- derosa pine following ozone exposure and subsequent bark beetle attack has been a contributing factor to the increase in the understory of young trees of ozone-tolerant species such as incense-cedar (McBride et al. 1985).

McBride and Laven (1976) have determined, by means of fire scars and tree ring dating, that since 1905 (when fire suppression policy started) the interval between fires has more than doubled. For example, the fire interval for ponderosa pine stands was 10 years before 1860, 14 years between 1860 and 1904, and 32 years from 1904 to 1977. Management of this problem on Federal lands is included as a specific element in the San Bernardino National Forest Plan (USDA 1986), but there is no coordi- nated effort to manage fuel loading on extensive private lands contained within National Forest boundaries.

Major Causes of Tree Mortality

Periods of prolonged drought, e.g., 1896–1902, 1947–1951 (Table 12.1) or lower than average precipitation, e.g., from 1984 to 1988 (Figure 12.5), lead to extensive killing of stressed conifers by bark beetles. Pole-size to mature trees on entire hillslopes may be killed in a 1–2 year period (USDA 1989). Clear-cutting of forest stands in specific areas in the San Bernardino Mountains achieved a similar result (La Fuze 1971). Ozone- injured trees are also more vulnerable to immediate mortality from insect or disease complexes. Bark beetle attack is almost always the primary cause of mortality (Stark et al. 1968). Annual counts of trees killed by bark beetles in the Lake Arrowhead area for the period 1921 through 1951 (Taylor 1973) show maximum mortality during dry years and less mortality during moist years. Counts of killed trees remained high every year from 1951 until the end of the record in 1971. The sustained high level of mortality is coincident with the beginning and continuation of tree injury due to ozone.

Tree diseases acting alone do not usually cause extensive mortality in a short time span such as one or two years. Tree morbidity and mortality is spread out over a longer period of time, but the aggregated damage may ultimately reach levels comparable to beetle kills in the case of root

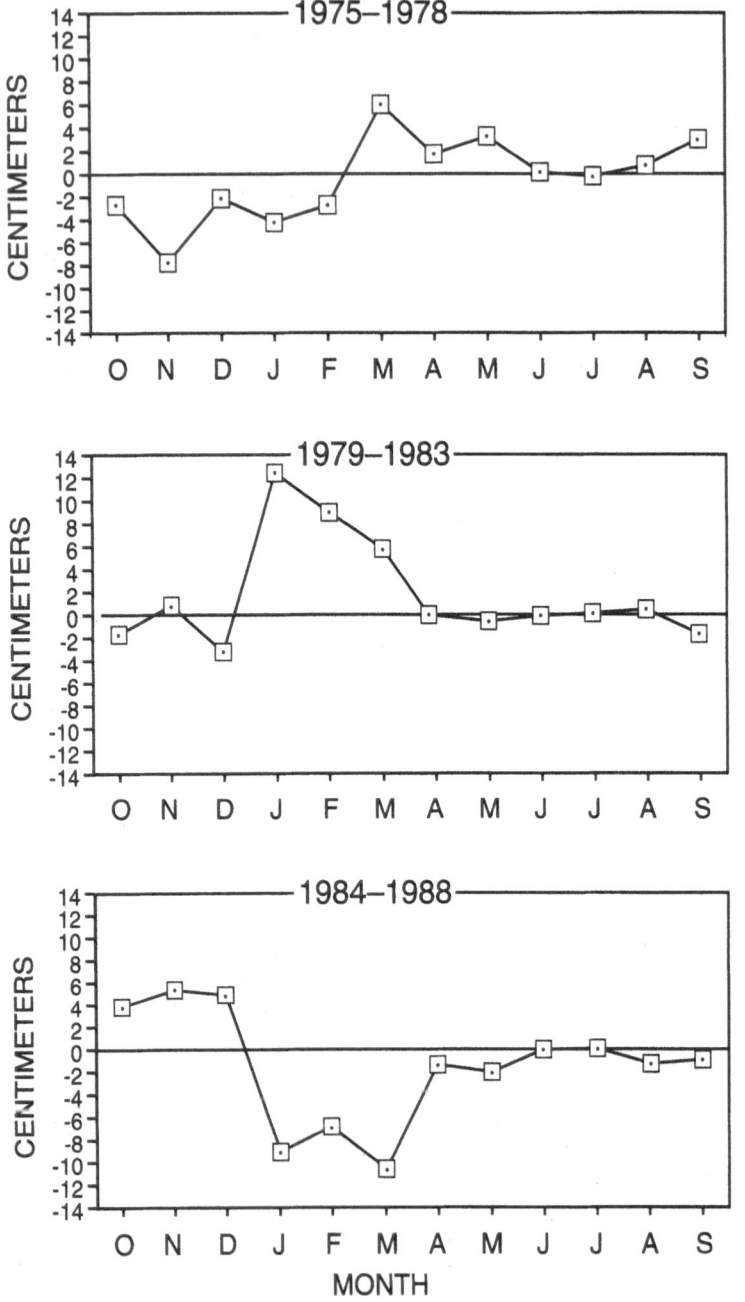

Figure 12.5 Departures of monthly precipitation amounts at Lake Arrowhead from the 1973–1987 monthly averages during three periods: 1975–1978, 1979–1983, 1984–1988. These periods correspond with the crown evaluation and stem growth measurement intervals followed by Miller et al. (1989). Water years are defined as October of previous year through September.

disease caused by *Heterobasidion annosum*. Dwarf mistletoe (*Arceuthobium campylopodum* var. *campylopodum*) is frequently present on ponderosa and Jeffrey pines where it causes stunting of small trees and stem deformations of larger trees.

History of Resource Use

Descriptions of forest distribution in the San Bernardino Mountains are available as early as 1769–1776 from diaries of explorers during the Mission period. All indications are that the extent of the forest corresponded to present day boundries (La Fuze 1971). In 1853–54, after California became a state, the Pacific Railroad Survey described the types of timber available. The first accurate maps and descriptions of forest types were reported by the U. S. Geological Survey to the Secretary of Agriculture (Leiberg 1899, 1900). New settlers from the nearby San Bernardino Valley built a crude road in 1852 to gain access to the abundant supply of timber in areas occupied today by the communities of Lake Gregory and Lake Arrowhead (Figure 12.1).

Logging operations increased as the timber harvest proceeded eastward into virgin timber. Beginning in 1891 some mills were reestablished in the western areas formerly cutover (La Fuz 1971). Timber harvesting peaked during the period from 1898 to 1912 with the establishment of a sawmill and narrow gauge railroad near the present location of Running Springs. Timber in drier, more sparse stands further to the east was in demand for use by local gold mining operations since the 1850s. During the early 1900s there were five sawmills in the Big Bear Lake area which provided enough lumber for local construction. Sawmills operated in the Mill Creek drainage between 1850 and 1870.

Timber on government land became less available, particularly after repeated fires and other abuses caused by logging operations in the Running Springs area. During the period extending from the 1870s to 1890s extensive sheep grazing occurred in the eastern half of the mountain area. Severe overgrazing resulted and sheepherders regularly set fires in an effort to improve forage. The mountains became a Forest Reserve in 1892 and were designated as the San Bernardino National Forest in 1925.

The annual cut of timber between 1948 and 1984 in Los Angeles and San Bernardino Counties (Hiserote et al. 1986) provides information on recent activities in the San Bernardino Mountains (Figure 12.6). The San Gabriel Mountains are located immediately to the west of the San Bernardinos; however, the forested area of the San Gabriel Mountains is much less than that of the San Bernardino Mountains. Between 1954 and 1984 the annual timber harvest averaged 11,162 thousand board feet with a high of 27,489 thousand board feet in 1963. The guidelines used for selecting trees to cut was a four category risk rating system based on crown condition. The classification aided a field decision as to whether a tree would succumb to bark beetle attack during the next decade. Category I and II trees were the healthiest and were not cut. Category III and IV trees were the least healthy and were marked for removal. In some heavily used recreation areas some category III trees were not removed. In 1953–54 a total of 5500 acres was harvested in the Barton Flats area (Hall 1958).

The impetus for this management action was the persistent loss of trees to bark beetles, which became more agressive when trees were weakened by other environmental stresses. Below-average precipitation from 1947–1951 and again from 1955–1957 (Table 12.1) was a major factor contributing to the decline of tree vigor. In 1956 the first report was filed by the Forest Service, and a description was published by Parmeter et al. (1962), of what later was proven to be extensive chronic ozone damage (Miller et al 1963, Richards et al. 1968). Thus, both drought and ozone injury were involved in predisposing ponderosa and Jeffrey pines to bark beetle attack.

Tourism gradually increased in popularity and permanent year-around communities began to appear about 1925. Today, privately owned lands, which were formerly timber or mining claims, are surrounded by larger

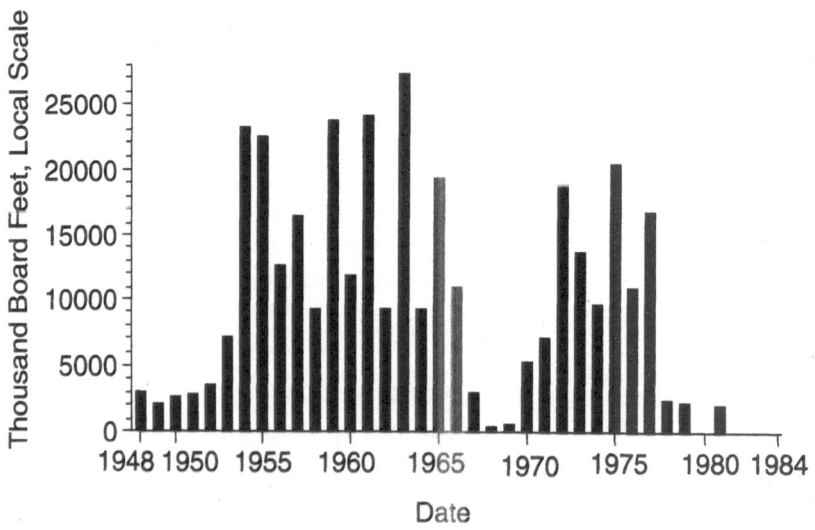

Figure 12.6 Timber harvest in Los Angeles and San Bernardino Counties from 1948 to 1984 (Hiserote et al. 1986).

areas of National Forest land in the Lake Gregory-Lake Arrowhead and Big Bear Lake areas. Larger unbroken blocks of National Forest land are found in the Santa Ana River drainage. Urban development on private lands in 1980 had already exceeded predictions for the year 2000. In addition, the Forest lies within two hours' driving time of the 14 million residents of the greater Los Angeles area. The size of the full-time population within its boundries, the amount of dense urban development contiguous to its boundries, and the sheer numbers of visitors and recreationists create tremendous pressures on the forest (USDA 1986).

Nature and Extent of Air Pollution

The accumulation of pollutants in the atmosphere of the Los Angeles Basin has been well documented and is of international fame (see Chapter 3). Pollutants such as ozone, nitrogen dioxide, peroxyacetyl nitrate, ethylene, sulfur dioxide, nitric acid, and fine particle matter, notably ammonium nitrate, are common (see Chapter 3, Bytnerowicz et al. 1987). For the most part, ozone concentrations are of greatest concern to forests surrounding the Los Angeles Basin as levels regularly exceed thresholds known to cause visible injury to vegetation. However, the concentrations of other chemical species appear to be too low to cause specific symptoms of leaf injury and no evidence is available to connect injury in the field with individual or combined actions of the other pollutants. Consequently, pollutant monitoring in the San Bernardino Mountains has focused principally on ozone.

The possible effects of acidic wet deposition have not been monitored in the San Bernardino Mountains. The pH of winter precipitation at 33 California sites between July 1984 and June 1986 was between 5.10 and 5.60 for 27 sites, and between 4.71 to 5.02 for the six remaining sites which were all in urban locations in southern California (California Air Resources Board 1986). The rainfall pH in the San Bernardino Mountains is probably in the 5.10 to 5.60 range, particularly since ammonia from dairy farms located between the urban source and the San Bernardino Mountains tends to neutralize acids.

The situation is different for fogs that typically occur in late spring and occasionally in the fall. Fog events in Riverside recorded pH values between 2.33 and 5.68 (Munger et al. 1990) and Waldman et al. (1985) found that the acidity of fogs elsewhere in the South Coast Air Basin dropped to as low as pH 1.69. No specific injury symptoms on forest vegetation have been observed that could be related to acid fog exposure.

Spatial gradient of ozone concentrations

Ozone concentrations on the valley floor peak around midday in July and August; at Sky Forest, a site in northwestern San Bernardino National Forest at 1709 m, peak ozone concentrations occur around 1700 hours; while at Barton Flats (1891 m), approximately 28 km to the east and downwind of Sky Forest, peak ozone concentrations occur closer to 1800 hours (Miller et al. 1986). Minimum ozone concentrations were low on the valley floor, often below detection limit of the instrument, whereas minimum concentrations are elevated to around 30 or 40 ppb at mountain sites. This spatial pattern is typical of ozone transported from a source area.

A spatial gradient in ozone exposure has been measured in the San Bernardino National Forest (Miller et al. 1986). Higher ozone concentrations were found in western portions of the forest than to the east (Figure 12.7). The 24-hour averages in the northwestern portion of the San Bernardino National Forest ranged between 90 and 140 ppb with average maximum values of between 200 and 240 ppb.

Temporal trends of ozone concentrations

A record of June through September ozone concentrations in the San Bernardino Mountains dates back to 1968 at Rim Forest, a site 2 km south of Lake Arrowhead (Miller and Elderman 1977). However, the reliability of the potassium iodide calibration procedure used for Mast total oxidant meters has been questioned and therefore this early record is not comparable at known levels of confidence with the data obtained after 1974 using ultraviolet ozone photometers.

During 1974 to 1978 ozone photometers were located at Camp Paivika and Sky Forest, and in 1978 the South Coast Air Quality Management District began to operate an ozone monitoring station at Lake Gregory. This station has continued operation up to the present. It bears the distinction of having the highest ozone concentrations of any station in California. During the single season (1978) that Camp Paivika and Lake Gregory (located about 5 km apart) instruments operated simultaneously, hourly ozone concentrations tracked very closely. The combined record for 1974 through 1987 shows a gradual decline in monthly average maximums. Hourly averages in ozone declined slightly during 1974 and 1980, but have remained relatively constant during the 1980s (Figure 12.8). This trend is supported by similar trends at most stations in the South Coast Air Basin (Davidson et al. 1985).

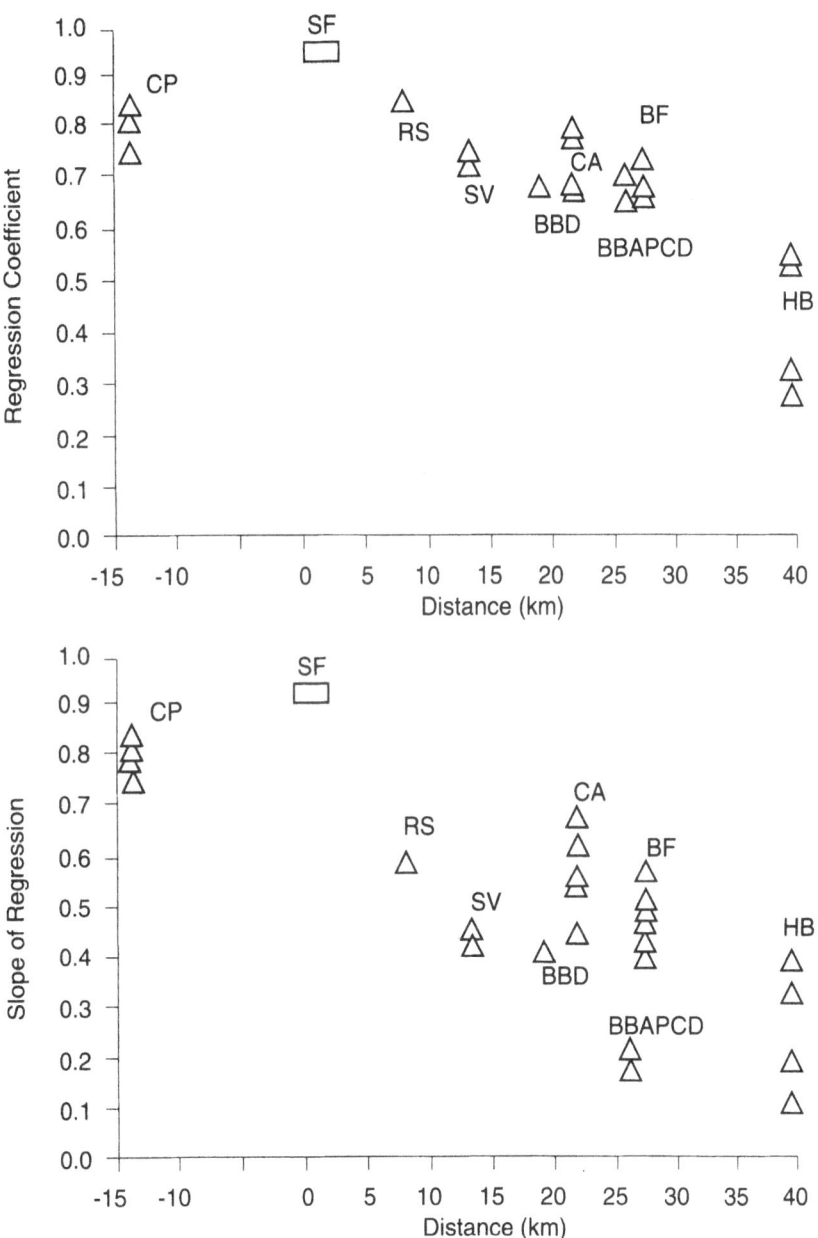

Figure 12.7 Regression coefficients (top) and slopes (bottom) from matching hourly average ozone concentrations at Sky Forest (SF) with similar data from 8 other stations in the San Bernardino Mountains indicate the west to east gradient of decreasing ozone concentrations (Miller et al. 1986).

Weather and ozone concentrations

Synoptic weather is a major factor determining transport of pollutants to inland mountains (see Chapter 3). Summer weather in southern California was categorized into five classes by McCutchan and Schroeder (1973) (Table 12.3). These classes range from the hot dry Santa Ana wind conditions, equivalent to patterns VIa and VIb in Figure 12.3, to cool moist conditions all day, equivalent to pattern I in Figure 12.3. The influence of weather on ozone concentrations in the San Bernardino Mountains was determined using the five weather classes and ozone data from May through October of 1974 and 1975 (Figure 12.9). Weather associated with moist, modified marine air in the morning and hot in the afternoon, or class 3, and moist modified marine air in the morning and warm in the afternoon, or class 4, resulted in sustained high levels of ozone when these types of days occurred consecutively (Miller and McCutchan 1977). During May through October in 1974 and 1975, the day-to-day combinations that occured most frequently were class 4 followed by class 4 (23% of the time), class 3 followed by class 3 (19%), and class 5 followed by class 5 (12% of the time).

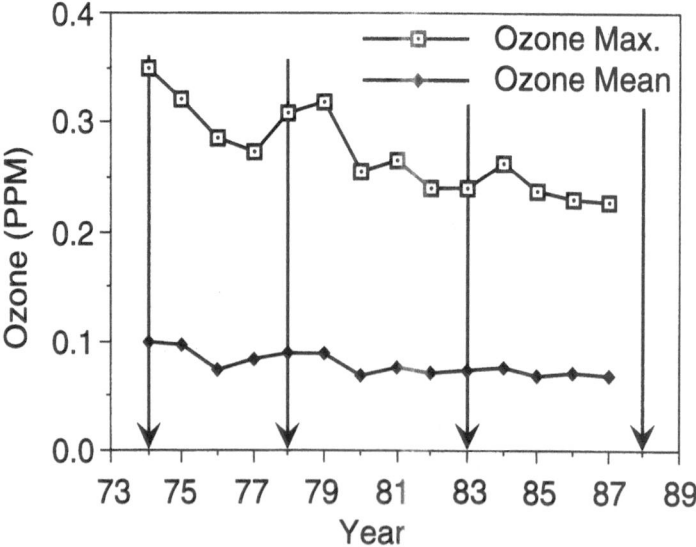

Figure 12.8 May through October 24-hour average ozone concentrations and average of monthly maximum ozone concentrations from 1974 to 1987 at Lake Gregory, a site representative of the western section of the San Bernardino Mountains (Miller and McBride 1989).

Table 12.3 Descriptions of meteorologic patterns for five classes of spring and summer days in southern California (McCutchan and Schroeder 1973).

Class	General Weather	Associated Synoptic Pattern at Surface	at 500mb
1	Hot, dry continental air throughout the day (Santa Ana)	Large high pressure over Great Basin	Strong northerly winds over the area
2	Relatively dry forenoon; modified marine air in afternoon, very hot	High pressure over Great Basin and thermal trough over desert	Subtropical high over area
3	Moist, modified marine air, hot in afternoon	Thermal trough over desert	High pressure over area
4	Moist, modified marine air, warm in afternoon	Thermal trough over desert	Trough over area
5	Cool, moist, deep marine air throughout the day	Synoptic-type low over desert	Deep trough or closed low over area

Pollutants and Forest Condition

Observations of unexplained foliar symptoms on ponderosa pine were described by Asher (1956) who named the condition "x-disease." The chlorotic mottling of needles resembled symptoms caused by black pine leaf scale (*Nuculaspis californica* Coleman), however, a survey by Stevens and Hall (1956) showed that trees with severe chlorotic mottle had little evidence of scale infestation and that there was poor geographic correspondence between scale infestations and the chlorotic mottle condition. The most severely affected stands were located along the ridgecrest overlooking the San Bernardino Valley. The nearest major source of "smokestack" pollutants was the Kaiser Steel plant at Fontana. George F. Edmunds, Jr. and B. L. Richards were employed as consultants by the

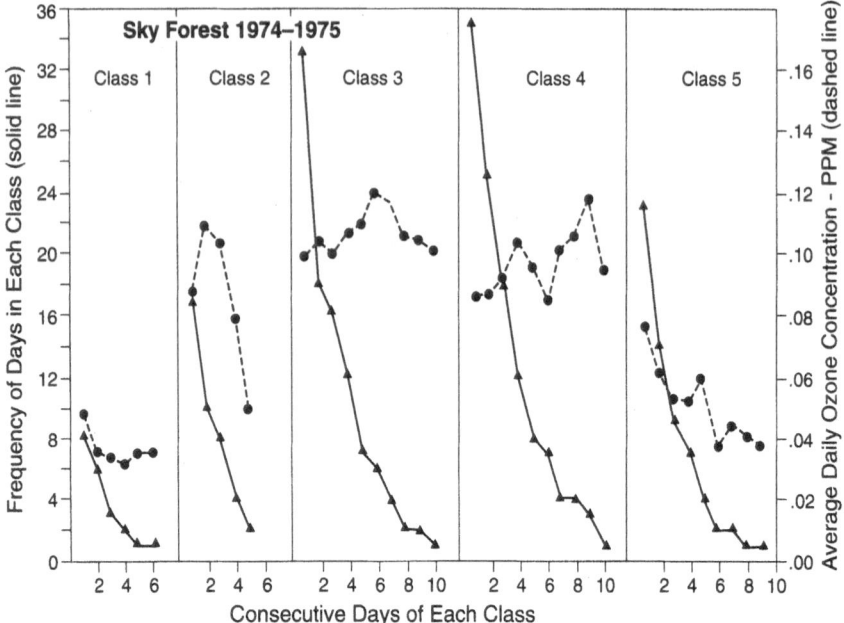

Figure 12.9 The frequencies with which each of five meteorological patterns (McCutchan and Schroeder 1973; Table 12.3) occurred on consecutive days (closed triangles), and associated daily average ozone concentrations (closed circles), Sky Forest, 1974–75.

Kaiser Steel Corporation to investigate the possibility that known pollutants, principally fluorides, might be responsible for the observed foliar injury. Field observations, summarized in a report to Kaiser Steel in 1959, did not implicate fluoride. Since chlorotic mottle symptoms were observed on pines at other locations in the Los Angeles Basin and the San Gabriel Mountains it was concluded that an area source such as "smog" might be the principal cause.

Fumigation of container-grown ponderosa pine seedlings with ozone (Richards et al. 1968) and treatments in branch enclosure chambers including carbon filtered air, ambient air, or filtered air plus ozone implicated ozone as the cause of the chlorotic mottle symptom (Miller et al. 1963). From 1968 through 1973, naturally regenerated ponderosa pine saplings were maintained from May through September in greenhouse enclosures. Some saplings were exposed to carbon-filtered air, the remaining to ambient air. A similar group of seedlings were exposed to ambient air under field conditions south of Lake Arrowhead where foliar injury was severe (Miller and Elderman 1977). During this period the weight of needles per internode (Figure 12.10) and both radial and height growth (Figure 12.11) increased in the filtered air treatment compared to ambient air treatments. During the last year of exposure, decreases in radial and height growth occurred. Increased competition by the seedlings for light and moisture is the most probable explanation for these decreases. This experiment provided additional evidence for the causal role of ozone. Exposure to other pollutants, both singly and in combination, did not cause the kind of chlorotic mottle observed in the field, whereas exposure of ponderosa pine seedlings to ozone in the laboratory yielded similar foliar injury to that experienced in forests surrounding the Los Angeles Basin.

Field observations of other conifer species native to the San Bernardino Mountains and exposure of container grown seedlings of other species to ozone indicate that the order of decreasing ozone susceptibility is ponderosa pine, Jeffrey pine, white fir, Coulter pine, incense-cedar, big-

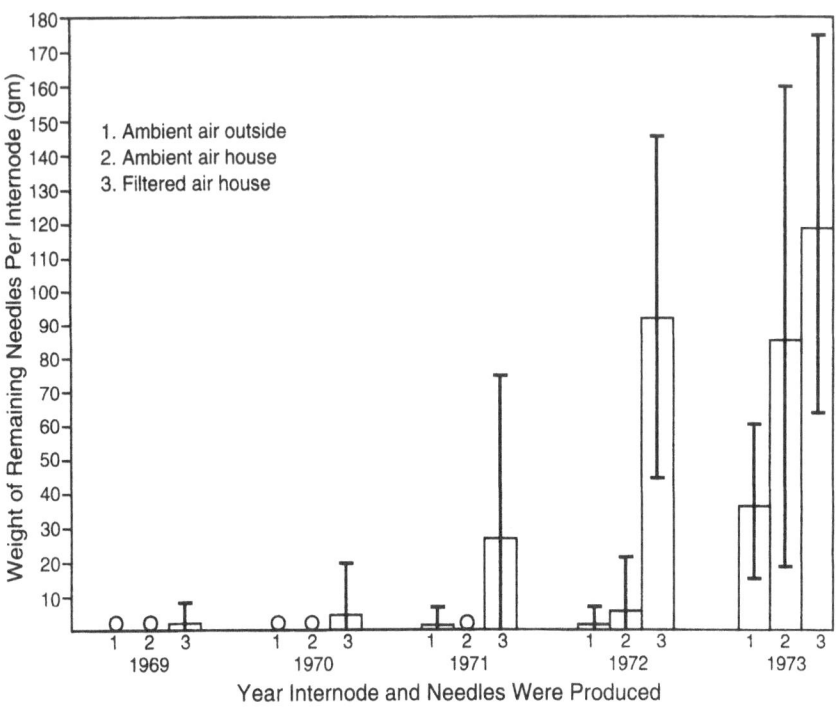

Figure 12.10 Mass of needles retained per internode by ponderosa pine saplings in 1973 after 5 years of exposure in a carbon-filtered air greenhouse, and to ambient ozone polluted air both inside and outside greenhouse enclosures (Miller and Elderman 1977).

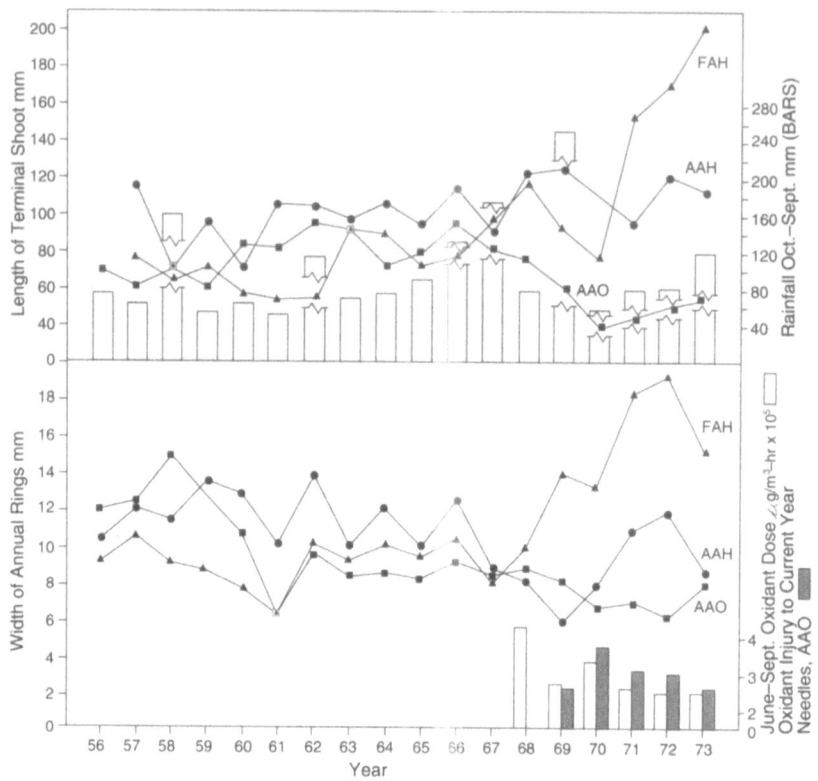

Figure 12.11 Height and radial growth of ponderosa pine saplings before and during exposure to carbon-filtered air in a greenhouse or to ambient, ozone-polluted air both inside and outside greenhouse enclosures (Miller and Elderman 1977). AAO = ambient air outside, AAH = ambient air house, FAH = filtered air house.

cone Douglas-fir, and sugar pine (Miller et al. 1983). The sensitivity of deciduous California black oak to ozone cannot be compared in the same terms, e.g. leaf retention, with that for evergreen conifers; however, California black oak develops leaf injury symptoms when present in the same stands as moderately to severely injured ponderosa pines (Miller et al. 1980).

Trends in Crown Condition, Stem Growth and Regeneration

The principal source of data for evaluating changes in crown condition and stem growth is a group of 18 plots in the San Bernardino National Forest that were established during 1973 (Miller and Elderman 1977,

Taylor 1980). The plots are laid out on parallel west to east transects (Figure 12.12) corresponding with gradients of decreasing ozone exposure (Figure 12.7) and characterize different series of the conifer forest formation (Figure 12.4). Observations and measurements were made annually between 1973 and 1978 (Taylor 1980) and at four to five year intervals since 1978. Ozone injury to pines was barely detectable where the 24 hour averages for May through September ranged between 50 and 60 ppb, whereas the highest level of injury was associated with averages of 100 to 120 ppb. Thus, most ozone injury was observed in the first 30 km of the 50 km long west-east transects (Miller et al. 1982).

Changes in ponderosa and Jeffrey pine crown condition, 1974–1988
Crown condition for individual trees was estimated from the number of annual needle whorls retained, the level of ozone injury on needles of each whorl, relative needle length, and relative amount of branch mortality in the lower crown. These elements were combined to form an ozone injury index (Miller et al. 1989). Ponderosa pines with crown injury comparable to different categories of the ozone injury index are shown in Figure 12.13. Ozone injury indices were compared for 1974, 1978, and 1988. Trees were divided into three classes based on stem

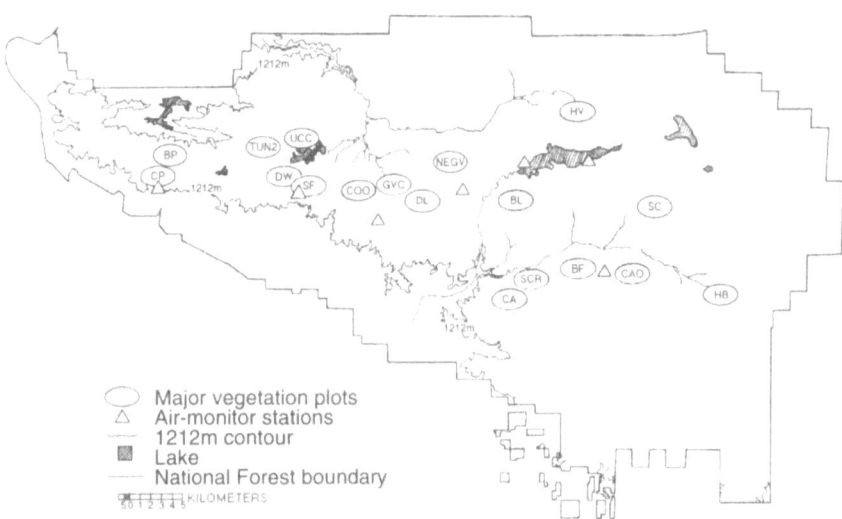

Figure 12.12 Locations of San Bernardino National Forest vegetation plots for observing long-term effects of ozone.

diameter at 1.3 m above the ground. The three diameter classes, 10 to 29.9 cm, 30 to 59.9 cm and > 60 cm, were chosen to investigate possible relationships between tree size and changes in crown condition.

Injury indices of trees in all size classes did not improve during the period 1974 through 1978 with the exception of Jeffrey pines with stem diameters greater than 60 cm. These Jeffrey pines were generally located in areas of lowest ozone exposure. Crown injury improved significantly between 1978 and 1988. The improved crown condition in recent years appears to be related to a general decline in both average and peak ozone concentrations at Lake Gregory (Miller et al. 1989, Figure 12.8).

Changes in crown condition at specific plots (Figure 12.14) were investigated using ozone injury indices for 1974, 1978, and 1988 at the plots located on the ozone gradient (Figure 12.12). In general, crown condition, expressed as ozone injury indices, improved with distance from the northwestern boundary of the forest; exceptions were Jeffrey pines at Barton Flats (BF) and Bluff Lake (BL). The two plots located nearest the northwestern boundary, Camp Paivika (CP), and Breezy Point (BP), were exposed to the highest levels of ozone and showed no improvement in crown condition, i.e., a decline in the ozone injury index was noted. Co-located ponderosa and Jeffrey pines at most sites exhibited similar injury indices as well as similar changes in ozone injury indices over time (Miller et al. 1989).

Tree diameter class did not appear to have any definite relationship to crown injury. There is some evidence to suggest that crown condition for trees with trunks greater than 60 cm in diameter at 1.3 m above the ground may have improved at a greater rate than did smaller trees (Miller et al. 1989).

Changes in ponderosa and Jeffrey pine stem diameter, 1974–1988

Tree growth is controlled by numerous internal and external variables. In particular, inter-annual variability in precipitation amounts can be detected in tree-ring records. Precipitation at Lake Arrowhead Fire Station was near average for the 1975 through 1978 study period, higher than average during 1979–1983, and below average from 1984–1988 (Figure 12.5).

Investigation of changes in stem diameter during the period 1974 through 1988 indicate two main patterns of growth response. First, steady improvements in growth were detected at most plots, specifically Camp O'Ongo (COO), Sky Forest (SF), Dogwood(DW), Green Valley Creek (GVC), Barton Flats, Camp Osceola (CAO), Deerlick (DL), Bluff Lake, and Holcomb Valley (HV). Growth at these sites improved during all three moisture periods, including periods of very low precipitation.

Figure 12.13 Crowns of four ponderosa pines approximating four ranges of chronic ozone injury include very severe to severe (upper left), severe to moderate (upper right), moderate to slight (lower left, center), and very slight to no visible injury (lower right). The precise index ranges for all categories are very severe (0–8), severe (9–14), moderate (15–21), slight (22- 28), very slight (29–35), no visible symptoms (36 and higher).

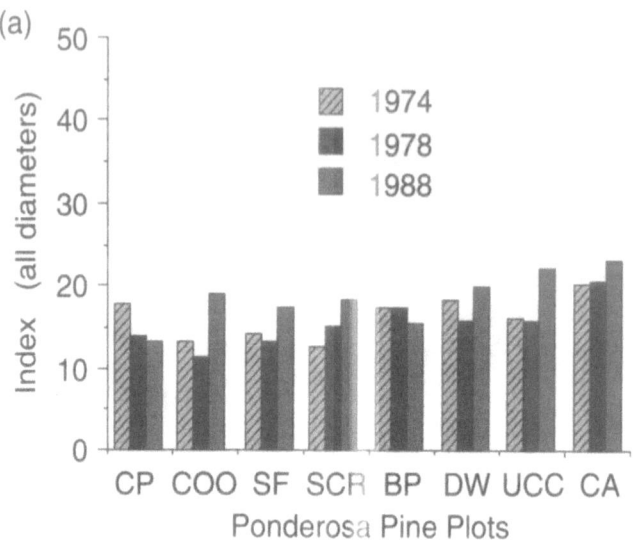

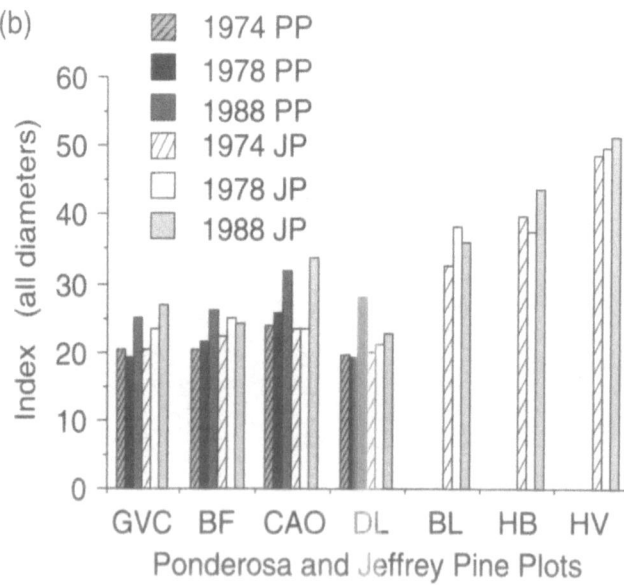

Figure 12.14 *Average ozone injury indexes in 1974, 1978, and 1988 at each of 15 forest plots, including (a) ponderosa pine plots, and (b) plots with both ponderosa and Jeffrey pine (Miller et al. 1989). Lower injury index values equal greater injury.*

Ozone exposure at these plots is moderate to high, but has decreased over the study period (Figure 12.8). Continued decrease in ozone concentrations may be sufficient to permit continuing improvement in growth similar to recovery of saplings maintained in carbon filtered air during exposure studies (Figure 12.11). Growth at Camp Paivika, Schneider Creek (SCR), Breezy Point, University Conference Center (UCC), and Heart Bar (HB) increased during the above-average precipitation period of 1979–1983 and decreased during below-average precipitation years of 1984–1988. The decreasing growth of surviving trees at Camp Paivika and Breezy Point is consistent with the ozone exposure gradient and a lack of improvement in crown condition. No growth response was detected at Camp Angeles (CA) during moist years and declines in growth occurred during dry years (Miller et al. 1989).

Ozone injury indices increased (crown condition improved) towards the east and lower ozone exposure. Growth on the other hand varied independently of ozone injury index when compared between plots. This implies that crown condition, as expressed by injury indices, was not the main factor controlling stem growth of ponderosa and Jeffrey pine. For example, trees with the best crown condition at Holcomb Valley had lower absolute amounts of stem growth than those at Camp Paivika and Breezy Point which had the worst crown condition. Available soil water content in the top 90 cm varied from 8 to 21% between plots. Regressions of growth and percent available water content yielded $r = +0.63$ for the above-average precipitation period 1979–1983 and $r = +0.67$ for the below-average precipitation period 1984–1988. Soil water content provides the best available explanation for differences in stem growth among plots (Figure 12.15). The relationship of stem growth with ozone injury index was tested at Barton Flats, where available soil water content (21%) was expected to be uniform throughout the plot. The regression yielded $r = +0.66$ for both precipitation periods (Figure 12.16).

In summary, the ozone injury index is a faithful indicator of crown health (number of annual whorls retained and condition of remaining needles) between plots receiving various levels of ozone exposure over several observation periods. The ozone injury index is a moderately good predictor of stem growth at a single plot. However, it fails to predict growth in comparisons between plots because different available soil water contents and perhaps other variables not yet recognized exercise a more profound control on stem growth. A system of forest plots for the purpose of monitoring changes in crown condition and stem growth over time should have comparable available soil water contents at plots situated on a gradient of ozone exposure.

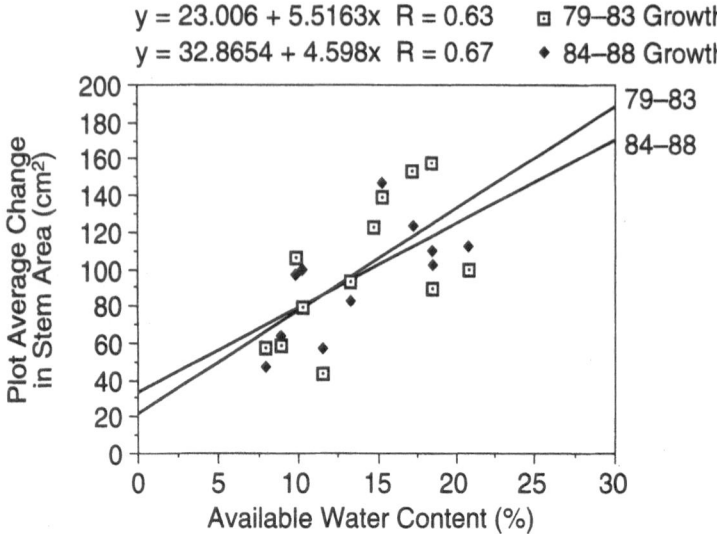

Figure 12.15 Relationship of average change in stem area during the "moist" 1979–1983 period, and "dry" 1984–1988 period to the percent available soil moisture in the 0–90 cm zone in soils at 13 plots (Miller et al. 1989).

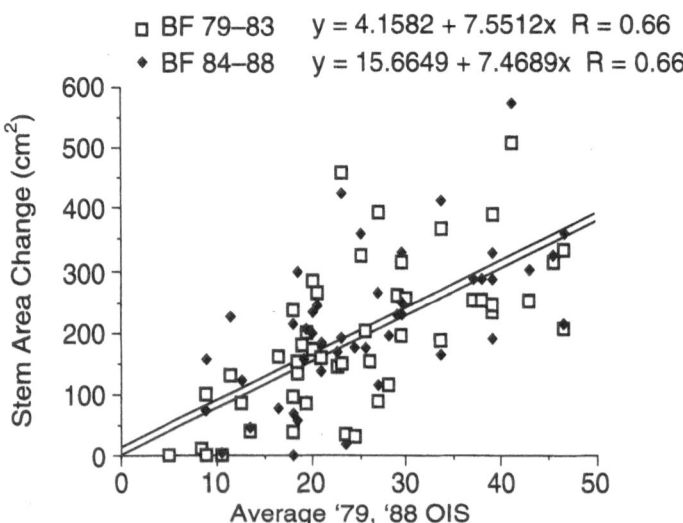

Figure 12.16 Relationship of ozone injury indexes in an individual plot (BF) to stem area changes during the 1979–1983 and 1984–1988 periods. OIS=Oxidant Injury Score; lower scores equal greater injury (Miller et al. 1989).

Factors Affecting Regeneration of Conifer Species

Reforestation by planting bare-root nursery stock is practiced at some intensively managed sites in the San Bernardino Mountains. However, natural regeneration is very important for the majority of the coniferous forests under control of federal land managers. The factors affecting cone production, seedling establishment, and seedling survival are discussed below.

Cone production by ponderosa and Jeffrey pines

The most important variable associated with cone production by ponderosa and Jeffrey pines in the experimental plots (Figure 12.12) was crown class, or position of a tree's crown relative to that of its neighbors (Luck 1980). Combined dominant and co-dominant crown classes of ponderosa pines represented 58% of all trees but produced 96% of the cones. The same was true for Jeffrey pines. During six years of observation, dominant ponderosa and Jeffrey pines 130 years and older with severe ozone injury to their crowns produced significantly fewer cones than uninjured trees. If co-dominant and dominant ponderosa and Jeffrey pines were combined, the severely injured trees also produced fewer cones than uninjured trees.

Trees with cones are actively sought by tree squirrels who feed on cones while they are still green (Miller and Elderman 1977). Squirrels tend to consume almost all of the cones on a single tree before moving on. Thus, if fewer cones are available on large, severely injured trees, squirrels will have a smaller number of trees to utilize. It would be undesirable, from a genetic standpoint, if seeds of uninjured trees were consumed, since a portion of the progeny from these individuals may possess genetic traits making them less sensitive to ozone injury.

Natural regeneration success of ponderosa pine

The average recruitment and average mortality of ponderosa pines of height about 1 m or less was determined over the following periods; 1975–1977 characterized by highest ozone and below-average precipitation; 1978–1983 with lower ozone than earlier but above-average precipitation; and 1984–1988 characterized with lowest ozone and below-average precipitation (Miller and McBride 1989). Recruitment was higher than mortality only during 1978–1983, and lowest recruitment occurred during 1984–1988 (Figure 12.17). Mortality was highest during 1975–1977 and lowest during the moist 1978–1983 period. Cone crop frequency is not known for the entire study period so it is impossible to evaluate the relative contributions of seed abundance and failure of seeds to germinate and successfully establish seedlings.

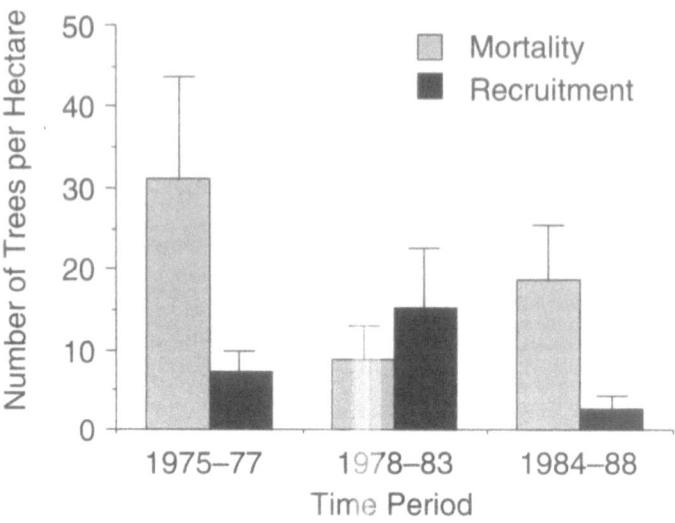

Figure 12.17 Averages of recruitment and mortality of ponderosa pine seedlings at 11 plots in the San Bernardino National Forest from 1975 through 1988 (Miller and McBride 1989).

The ideal seedbed for ponderosa pine is disturbed mineral soil resulting from fire or mechanical disturbance. Competing species such as white fir, incense-cedar, sugar pine and black oak establish more successfully by penetrating a litter layer. Establishment and mortality data for seedlings of all species is currently under investigation. Some preliminary estimates suggest that, during recent years, incense-cedar and sugar pine, all less sensitive to ozone than ponderosa pine, may be more successful than ponderosa pine in seedling establishment at plots with high ozone exposure (McBride et al. 1985).

Effects of Combined Stresses on Succession and Related Processes

Preceding sections evaluated the influences of ozone exposure and moisture availability on tree growth and seedling regeneration. Insect and disease complexes leading to morbidity and mortality are also an important factor to consider. A important goal is to understand how these variables fit into a hierarchical structure that influences community succession in the mixed conifer and pine forests of the San Bernardino Mountains.

Nutrient cycling is one of the most important processes to consider. Fire plays a major role in nutrient cycling. Following fire, nutrients are rapidly released from the litter, although some may be lost immediately to the atmosphere and to runoff. Nutrient availability is a "boom and bust" situation controlled by fire frequency. In the absence of fire, litter decomposition rates determine the supply of available nutrients. Chronic exposure to photochemical oxidant air pollution has important direct effects through atmospheric deposition and indirect effects through regulating rates of litter accumulation and litter decomposition.

There is good circumstantial evidence that bulk deposition of nitrogen compounds decreases with distance eastward as is the case with ozone. This assertion is supported by monitoring both air pollution (Bytnerowicz et al. 1987), and stream runoff after initial autumn rains (Riggan et al. 1985) in the San Gabriel Mountains, west of the San Bernardino Mountains. Studies of litter decomposition rates at plots located along the pollution gradient show more rapid decomposition rates at the western plots even when litter bags of Jeffrey pine needles from the eastern plots were exposed in the western plots (Bruhn 1980). Further, the nitrogen content of ponderosa pine litter was positively correlated with litter decomposition rates (Fenn and Dunn 1989). Thus, it appears that atmospheric deposition contributes more nitrogen to forests along the western end of the San Bernardino Mountains than along eastern portions.

Nitrogen availability in the litter layer is indirectly affected by rates of ozone-induced litter fall, as well as by the nitrogen content of needle tissue. The amount of litter fall was greatest from trees with moderate amounts of crown injury (Arkley and Glauser 1980), compared with severely injured and uninjured trees. Litter from ozone-injured trees has a higher proportion of needles from younger whorls that are inherently high in nitrogen content and low in calcium content. This undoubtedly contributes to the increased nitrogen content of litter at plots in the western portions of the San Bernardino Mountains (Fenn and Dunn 1989). The greater abundance of nitrogen in needle tissue of ponderosa pines at plots with severe crown injury may partly explain, along with the higher soil moisture holding capacity, the ability of trees with fewer annual whorls, as expressed by lower ozone injury indices, to exhibit improved stem growth.

Hypotheses proposed to explain long-term effects of ozone stress on ponderosa and Jeffrey pine forests must take into account the importance of largely unpredictable catastrophic events as these have the most dramatic long-term influence on the forest landscape (see Chapter 6). Fires that drastically modify or completely destroy existing forest cover have been a dominant force in the San Bernardino Mountains (Minnich

1988). The clock of forest succession can be set back to zero for thousands of acres in a matter of hours, and recovery may or may not lead to a similar forest cover over spans of 100 to 200 years. It is on the canvas of major fire effects that influences of other stress agents are painted as specific features or details. One such detail is the more rapid establishment of ozone-tolerant species (incense-cedar and white fir) on sites formerly occupied by ponderosa pine.

Summary

The accumulated effects of chronic ozone injury, and other man-caused disturbances, have driven extensive areas of conifer forest stands in the San Bernardino Mountains away from the fire climax phase, which existed for centuries before the arrival of settlers in the 1850s. Changes in stand composition set in motion by ozone exposure since the early 1950s are clearly evident, although the chronic effects of ozone injury appear to be decreasing at some forest plots. This trend is attributed to control strategies resulting in a general improvement of air quality in southern California. However, as population continues to increase in the South Coast Air Basin it remains to be seen if control measures will be able to maintain or effect additional improvements in air quality.

The inherent instability of the current disclimax phase of forest vegetation present over most of the ponderosa pine range can be improved by management strategies that simulate stand composition of the former fire climax. In the absence of stand improvement, including fuel reduction, there is a high probability that ignition of fires during dry, windy weather would have severe consequences for ponderosa and Jeffrey pine forests. Living foliage of ozone tolerant understory trees such as incense-cedar can be easily ignited by ground fires, and the accumulation of both needle and woody litter may result in crown fires that would destroy most remaining overstory pine trees. Such destruction may lead to a dominance of self-perpetuating, fire adapted, ozone tolerant shrub and oak species mixtures in place of the former pine and mixed conifer forests. This vegetation would resemble the timberland chaparral already present at some sites. Such vegetation would burn repeatedly and provide fewer commodity and amenity values than do the pine or mixed conifer forests of the San Bernardino Mountains.

References

Arkley RJ (1981) Soil moisture use by mixed conifer forest in a summer-dry climate. *Soil Science Society of America Journal* 45:423–427

Arkley RJ, Glauser R (1980) Effects of oxidant air pollutants on pine litter-fall and the forest floor. In: Miller PR (ed) *Proceedings of the Symposium on the Effects of Air Pollutants on Mediterranean and Temperate Forest Ecosystems.* USDA General Technical Report PSW-43

Asher JE (1956) *Observation and Theory on X-disease or Needle Dieback.* USDA Forest Service unpublished report, Arrowhead Ranger District, San Bernardino National Forest

Baily TL, Jahns RH (1954) Geology of the Transverse Range province, southern California. In: Jahns RH (ed) *Geology of Southern California, Bulletin 170.* California Division of Mines, Sacramento, pp 83–106

Bruhn JN (1980) *Effects of Oxidant Air Pollution on Ponderosa and Jeffrey Pine Needle Litter Decomposition.* PhD dissertation, University of California, Berkeley (Dissertation Abstract 82-00036)

Bytnerowicz A, Miller PR, Olszyk DM, Dawson PJ, Fox CA (1987) Gaseous and particulate air pollution in the San Gabriel Mountains of southern California. *Atmospheric Environment* 21:1805–1814

California Air Resources Board (1986) *Fourth Annual Report on Acid Deposition Research and Monitoring.* State of California, Sacramento, CA

Clements FF (1936) Nature and structure of climax. *Journal of Ecology* 24:252–284

Craig RD (1904) *Lumbering in the San Bernardino Mountains.* San Bernardino National Forest, Supervisors Office, Typescript

Davidson A, Hoggan M, Wong P (1985) *Air Quality Trends in the South Coast Air Basin, 1975–1984.* South Coast Air Quality Management District, El Monte, CA

Dodge M (1972) Forest fuel accumulation—A growing problem. *Science* 177:139–142

Fenn ME, Dunn PH (1989) Litter decomposition across an air-pollution gradient in the San Bernardino Mountains. *Soil Science Society of America Journal* 53:1560–1567

Haase HE (1913) The lichen flora of southern California. *Contribution U. S. National Herbarium* 17:1–132

Hayes TP, Kinney JJR, Wheeler NJM (1984) *California Surface Wind Climatology.* Aerometric Data Division, California Air Resources Board, Sacramento, 73p

Hiserote BA, Moen J, Bolsinger CL (1986) *Timber Resource Statistics for the San Joaquin and Southern California Areas.* USDA Forest Service Resource Bulletin PNW-132, 35p

Horton JS (1960) *Vegetation Types of the San Bernardino Mountains.* USDA Forest Service Technical Paper No 44, Pacific Southwest Forest and Range Experiment Station, Berkeley, CA

La Fuze PB (1971) *Saga of the San Bernardino Mountains.* San Bernardino County Museum Association, Redlands, CA

Leiberg JB (1899) San Gabriel, San Bernardino, and San Jacinto Forest reserves. In: Gannett H (ed) *19th Annual Report of the U.S. Geological Survey to the Secretary of Agriculture, Part 5, Forest Reserves.* GPO, Washington, DC, pp 359–370

Leiberg JB (1900) San Gabriel, San Bernardino, and San Jacinto Forest Reserves. In: Gannett H (ed) *20th Annual Report of the U.S. Geological Survey to the Secretary of Agriculture, Part 5, Forest Reserves.* GPO, Washington, DC, pp 411–479

Luck RF (1980) Impact of air pollution on ponderosa and Jeffrey pine cone production. In: Miller PR (ed) *Proceedings of the Symposium on the Effects of Air Pollutants on Mediterranean and Temperate Forest Ecosystems.* USDA General Technical Report PSW-43

McBride JR, Hill T, Milliken R, Laven N (1975) *A Checklist of the Vascular Plants of the Montane Coniferous Forest in the San Bernardino Mountains, California.* Unpublished typescript, Department of Forestry and Conservation, University of California, Berkeley, 45p

McBride JR, Laven RD (1976) Fire scars as an indicator of fire frequency in the San Bernardino Mountains, California. *Journal of Forestry* 74:439–442

McBride JR, Laven RD, Miller PR (1985) Effects of oxidant air pollutants on forest succession in the mixed conifer type of southern California. In: *Proceedings: Air Pollution Effects on Forest Ecosystems.* Acid Rain Foundation, St Paul, MN, pp 157–167

McCutchan MH, Schroeder MA (1973) Classification of meteorological patterns in southern California by descriminate analysis. *Journal of Applied Meteorology* 12:571–577

Miller PR, Parmeter JR, Taylor OC, Cardiff EA (1963) Ozone injury to the foliage of *Pinus ponderosa*. *Phytopathology* 53:1072–1076

Miller PR, McCutchan MH (1977) *Implications of Weather Patterns in the Regulation of Oxidant Air Pollutant Stress on Forest Vegetation.* Paper 77-15.3, Annual Meeting of the Air Pollution Control Association, Toronto, 16p

Miller PR, Elderman MH (eds) (1977) *Photochemical Oxidant Air Pollution Effects on a Mixed Conifer Forest Ecosystem.* US EPA Report No. EPA 600/3-77-104, 339p

Miller PR, Longbotham GJ, Van Doren RE, Thomas MA (1980) Effect of chronic oxidant air pollution exposure on California black oak in the San Bernardino Mountains. In: Plumb T (ed) *Proceedings of the Symposium on the Ecology, Management and Utilization of California Oaks.*USDA Forest Service General Technical Report PSW 44, pp 220–229

Miller PR, Taylor OC, Wilhour RG (1982) *Oxidant Air Pollution Effects on a Western Coniferous Forest Ecosystem.* US EPA, Research Brief, EPA 600/D-82- 276, Corvallis, OR, 10p

Miller PR, Longbotham GJ, Longbotham CR (1983) Sensitivity of selected western conifers to ozone. *Plant Disease* 67:1113–1115

Miller PR, Taylor OC, Poe MP (1986) *Spatial Variation of Summer Ozone Concentrations in the San Bernardino Mountains.* Paper 86-39.2, Annual Meeting of the Air Pollution Control Association, 14p

Miller PR, McBride JR (1989) Trends of ozone damage to conifer forests in the western United States, particularly southern California. In: Bucher JB, Bucher I (eds) *Air Pollution and Forest Decline: Proceedings of the 14th International Meeting for Specialists in Air Pollution Effects on Forest Ecosystems.* IUFRO P 2.05, Birmensdorf, Switzerland

Miller PR, McBride JR, Schilling SL, Gomez AP (1989) Trend of ozone damage to conifer forests between 1974 and 1988 in the San Bernardino Mountains of southern California. In: Olson RK, Lefohn AS (eds) *Effects of Air Pollution on Western Forests.* Transactions Series, No. 16, Air & Waste Management Association, Pittsburgh, pp 309–323

Minnich RA (1988) The Biogeography of fire in the San Bernardino Mountains of California—a historical study. *Geography 28,* University of California Press, Berkeley, 122p

Munger JW, Collett J, Daube B, Hoffmann MR (1990) Fogwater chemistry at Riverside, California. *Atmospheric Environment* 24B(2):185–205

Paysen TE, Derby JA, Black H Jr, Bleich BC, Mincks JW (1980) *A Vegetation Classification System Applied to Southern California.* USDA Forest Service General Technical Report PSW-45, Pacific Southwest Forest Range Experiment Station, Berkeley, CA, 33p

Parish SB (1917) An enumeration of the pteridophytes and spermatophytes of the San Bernardino Mountains, California. *Plant World* 20:208–223

Parmeter JR Jr, Bega RV, Neff T (1962) Chlorotic decline of ponderosa pine in southern California. *Plant Disease Reporter* 46:269–273

Richards BL Sr, Taylor OC, Edmunds GF Jr (1968) Ozone needle mottle of pine in southern California. *Journal of the Air Pollution Control Association* 18:73–77

Riggan PJ, Lockwood RN, Lopez EN (1985) Deposition and processing of airborne nitrogen pollutants in mediterranean-type ecosystems of southern California. *Environmental Science and Technology* 19:781–789

Sigal LL, Nash TH III (1980) Lichens as ecological indicators of photochemical oxidant air pollution. In: Miller PR (ed) *Proceedings of the Symposium on the Effects of Air Pollutants on Mediterranean and Temperate Forest Ecosystems.* USDA General Technical Report PSW-43, p 249

Stark RW, Miller PR, Cobb FW Jr, Wood DL, Parmeter JR Jr (1968) Photochemical oxidant injury and bark beetle (Coleoptera:Scolytidae) infestation of ponderosa pine. I: Incidence of beetle infestation in injured trees. *Hilgardia* 39:121–126

Stevens RE, Hall RC (1956) *Black Pine-leaf Scale and Needle Dieback, Arrowhead-Crestline area, San Bernardino National Forest, California: Appraisal Survey.* USDA Forest Service File Report, Pacific Southwest Forest Range Experiment Station, Berkeley, CA, 6p

Taylor OC (1973) Oxidant Air Pollutant Effects on a Western Coniferous Forest Ecosystem. Task B Report. Historical background and proposed systems study of the San Bernardino Mountains area. Statewide Air Pollution Research Center, University of California, Riverside, CA

Taylor OC (1980) *Photochemical Oxidant Air Pollution Effects on a Mixed Conifer Forest Ecosystem.* Final Report, Ecological Research Series, EPA 600/3-80-002, US EPA, Corvallis, OR, 196pp

USDA Forest Service (1986) *Draft Forest Plan.* USDA Forest Service, San Bernardino National Forest Supervisors Office, San Bernardino, CA

USDA Forest Service (1989) *California Forest Pest Conditions—1989.* Region 5 Forest Pest Management, San Francisco, CA

Waldman JM, Munger JW, Jacob DJ, Hoffman MR (1985) Chemical characterization of stratus cloudwater and its role as a vector for pollutant deposition in a Los Angeles pine forest. *Tellus* 37 (b):91–108

Wilson CC, Dell JD (1971) The fuels buildup in American forests. *Journal of Forestry* 69:471–475

Wright RD (1968) Lower elevational limits of montane trees. II. Environment-keyed responses of three conifer species. *Botanical Gazette* 129:219–226

Section III

Summary and Projections

13

Summary, Projections, and Recommendations

R. K. Olson, D. L. Peterson, and M. Böhm

Introduction

The first seven chapters of this book deal with the entire West. Chapters 8–12 then focus on forest condition in particular regions within the West. The goal of this chapter is to return to a West-wide perspective and in so doing (1) summarize the current condition of Western forests as affected by air pollution, (2) discuss future trends in forest condition, and (3) recommend research priorities for advancing our understanding of air pollution effects on Western forests.

Current Condition of Western Forests

Regional Studies Summary

Forest condition in five regions of the West was evaluated in Chapters 8–12. Each of these regions has elevated levels of air pollution, and ozone is the primary pollutant of concern. Forest condition was evaluated based on foliar symptoms of air pollution damage, and on trends in annual radial growth increment for one or more tree species in each region. Based on these indicators, large differences in air pollution effects were apparent between regions:

- No foliar damage attributable to air pollution was observed in western Washington (Chapter 8; Barnard et al. 1990) or in the Front Range of Colorado (Chapter 9). Both studies also found no regional abnormalities in growth trends that might be caused by air pollution.

- The Arizona study (Chapter 10) presents an intermediate case. Small amounts of ozone damage to ponderosa pine (*Pinus ponderosa*) foliage have been observed in stands near Tucson (Graybill and Rose 1989). Growth reductions or anomalies (including absence of annual growth rings in some years) were found in a number of stands, primarily in the southern portion of the study region where current and historic air pollution levels are highest. A number of factors may be responsible for these growth reductions; air pollution may or may not be a contributing factor.

- The two California studies—Sierra Nevada (Chapter 11) and San Bernardino Mountains (Chapter 12)—present clearer evidence of air pollution effects. Ozone chlorosis and needle loss on ponderosa pine was found throughout a 500 km length of the Sierra Nevada. Damage was most severe at the southern end of these mountains, and decreased northward as ozone levels declined. However, no corresponding pattern of regional growth decline was found.

There is also a gradient of foliar damage and ozone concentrations in the San Bernardino Mountains. Along this gradient, increased pollutant exposure was associated with reductions in radial growth and cone production by sensitive species, shifts in community composition, and alteration of nutrient cycles. These impacts have resulted from the interactions of air pollution, fire, insects, pathogens, and other stresses.

Across the West

Developing general statements about the response of Western forests to air pollution is hampered by a lack of information. There is little evidence of deleterious effects in most parts of the West, but many areas have not been studied. No systematic and comprehensive survey of the West for air quality or forest responses to air pollution has been conducted.

The regional case studies described in this book represent the broadest survey to date of the condition of Western forests with respect to air pollution, and encompass the regions of the West with the highest levels of air pollution. However, the case studies leave many gaps in coverage both within and between regions. Within regions, each study examined at most only a few of the major tree species, and measured only two main indicators of forest condition. Some regions with elevated pollution levels (e.g., Salt Lake City area, Chapter 3) were not surveyed.

Where available, data from other studies corroborate the findings of the regional case studies. Earlier surveys of the Colorado Front Range also failed to find any visible foliar damage from air pollution (James and Staley 1980). Patterns of foliar injury in the Sierra Nevada as described in Chapter 11 match the results of other surveys (e.g., Pronos and Vogler 1981, Duriscoe and Stolte 1989). A study of Jeffrey pine in the Sierra Nevada showed that trees with foliar damage from ozone had reduced radial growth compared to asymptomatic trees (Peterson et al. 1987). Surveys in the San Bernardino Mountains and other areas surrounding the Los Angeles basin have shown ozone damage to black oak (*Quercus kelloggii*) and other non-coniferous species (Miller et al. 1979, Westman 1985, Westman and Price 1988).

Some areas that lack information on forest condition do have information on air quality (Chapter 3). Although this data set has some major limitations, it does provide an additional overlay for evaluating the potential effects of air pollution on Western forests.

Bormann (1985) proposed a six part classification of the response of forests to air pollution (Table 13.1). His scheme describes sequential stages of increasing forest degradation, and serves as a useful framework for assessing the overall state of Western forests. Within the uncertainty stemming from dataset limitations, Western forests can be categorized as follows:

Stage 0 Very few Western forests qualify as Stage 0 because there are essentially no atmospherically pristine forests in the West. Long-range transport of sulfur (Cahill et al. 1981), ozone (Legge and Krupa 1989), and other pollutants reaches even remote forests. Coastal forests may receive polluted air from inland sources under certain meteorological conditions (Vong 1989). All forests are also exposed to elevated concentrations of carbon dioxide (Watson et al. 1990).

Stage I Most Western forests fall into this category. Air pollution is low and any effects are not detectable.

Stage IIA Much of the mixed conifer forest of the west slope of the Sierra Nevada is Stage IIA as evidenced by foliar damage on sensitive individuals and species. Hardwood forests on the west slope of the Sierra Nevada and forests around the southern and western edge of the lower Central Valley of California might be classified as IIA based on pollutant levels. Some forests in the Transverse and Peninsular Ranges of southern California may be at this stage as may certain forests in the mountains of southern Arizona.

Stage IIB Some forests in the most polluted portions of the San Bernardino Mountains have reached this stage. The most sensitive species, ponderosa pine, is declining in some areas but is unlikely to be lost from

Table 13.1 Summary of forest ecosystem response to increasingly severe air pollution stress. Adapted from Bormann (1985).

Stage 0	Pollution levels are insignificant. Ecosystems unaffected.
Stage I	Pollution levels are low. Ecosystems serve as a sink for some pollutants, but species and ecosystem functions are relatively unaffected.
Stage IIA	Pollutants affect some aspect of the life cycle of sensitive species or individuals, which are subtly and adversely affected. Sensitive plants may have reduced photosynthesis, a change in reproductive capacity, or a change in resistance to insects and pathogens.
Stage IIB	Populations of sensitive species decline with increased pollution stress as does their functional importance in the ecosystem.
Stage IIIA	High concentrations or long exposures of pollutants cause basic changes in the structure and productivity of the ecosystem. Tolerant species replace intolerant ones, and shrubby species replace trees. Regulation of energy flow and biogeochemical cycles decreases.
Stage IIIB	Ecosystem collapse. Severe losses of species, ecosystem structure, and nutrients. Damage is so severe that full recovery may never occur even if the pollution is eliminated.

the landscape. Air pollution data suggest that other forests in the Los Angeles airshed could be at this stage.

Stage IIIA–B There are no significant areas of Western forests at this stage of decline due to air pollution. Historically, there are examples of major localized forest damage around large point sources of sulfur or fluoride emissions (Lynch 1951, Shaw et al. 1951, Scheffer and Hedgecock 1955, Treshow et al. 1967, Miller and McBride 1975, Smith 1990), but these emissions have been reduced or eliminated and forest recovery is underway.

The conifer forests surrounding Mexico City are similar to many Western coniferous forests, and these forests are at Stage IIIA (Ciesla and Macias Samano 1987, Cibrian-Tovar 1989). Could these forests represent the future condition of some forests of the western United States? Or more generally, what are the likely trends in the condition of Western forests?

Future Condition of Western Forests

Confident predictions of the future condition of Western forests require two types of knowledge: (1) accurate predictions of changes in air quality and other environmental stresses, and (2) a solid understanding of the relationship between stresses and forest condition including the integrated effect of multiple stresses. Our understanding of both of these topics is limited. Predictions of future emissions are very uncertain (Streets 1991), and translating future emissions scenarios into deposition scenarios for Western forests adds an additional layer of potential error (Venkatram 1991). Our understanding of the effects of air pollution on forests still has many gaps at both the tree (Chapter 5) and stand or ecosystem (Chapter 6) level.

While air pollution is the primary focus of this book, other stresses must be considered because of possible interactions or synergistic effects. For example, the changes in community composition of some heavily polluted ponderosa pine forests in the San Bernardino Mountains (Chapter 12) include the effects of multiple factors such as air pollution, fire, insects, and competition between tree species. Air pollution alone would not have had the same overall effect.

Despite the uncertainty, decisions on forest management and regulation of pollutants will continue to be made (Peterson et al. 1992a,b). Acknowledging and understanding the uncertainties is an important component of the decision-making process. A brief review of current knowledge of future trends in air pollution and other key stresses of Western forests may shed some light on the complexity and difficulty of predicting the future condition of Western forests.

Future Changes in Forest Stresses

Air pollution

Air quality and emissions data from the 1980s are summarized for the West in Chapter 3. Levels of air pollution in the 1990s and beyond will depend in part on trends in emissions. Changes in emissions will be determined primarily by population growth, per capita energy consumption, the particular mix of industries and energy sources that evolves, and the rate of implementation of cleaner technologies. Energy prices and government policies on energy and pollution will have a major impact on these variables.

None of the variables that influence emissions trends in the West can be predicted with high certainty. Multiple variables and complex interactions make predictions of emissions and air quality very difficult, but some trends are clear. The population of the West is growing rapidly (Table 13.2). Western population more than doubled in the four decades following 1950, and will likely double again before 2050 as the result of high levels of immigration and rising fertility rates (Ahlburg and Vaupel 1990).

One effect of this population increase will be to increase the number and extent of large urban sources of emissions. As urban sprawl expands, pollution sources will become more regional. For example, the Tacoma–Seattle–Vancouver corridor is urbanizing, a process that will increase the area of Pacific Northwest forests directly downwind from urban areas.

Transportation is the West's largest source of volatile organic compounds and oxides of nitrogen; both are precursors for ozone formation (Chapter 3). Numbers of automobiles are increasing at a higher rate than population as is the total number of kilometers driven (Hagerman 1990). If continued, these trends will exacerbate the effect of population growth on emissions.

Total point source emissions of sulfur and nitrogen dioxide in the West are expected to increase gradually during the 1990s (Young et al. 1988). Increased industrialization and demand for electricity in response to population growth will be offset to some degree by improvements in combustion technology and the retirement of less efficient power plants (South 1991).

Carbon dioxide is not usually classified as an air pollutant, but anthropogenic emissions represent a major alteration of the atmosphere with potentially important consequences for Western forests. From a pre-industrial concentration of 280 ppmv, atmospheric carbon dioxide has increased to 353 ppmv in 1990 (Watson et al. 1990). If emissions remained constant at 1990 levels, atmospheric concentrations would reach about 450 ppmv in 2050 (Watson et al. 1990). However, the demands of a rapidly growing world population for energy and agricultural land could increase emissions; an increase of 2% annually would lead to an atmospheric carbon dioxide concentration of 575 ppmv in 2050 (Watson et al. 1990).

Table 13.2 Resident population by state (1000s) for 1950 and 1989 (US Bureau of the Census 1989, 1990) and prediction of total population of the West in 2050.

State	1950	1989	2050
Arizona	750	3556	—
California	10586	29063	—
Colorado	1325	3317	—
Idaho	589	1014	—
Montana	591	806	—
Nevada	160	1111	—
New Mexico	681	1528	—
Oregon	1521	2820	—
Utah	689	1707	—
Washington	2379	4761	—
Wyoming	291	475	—
West	19562	50158	122000[a]

[a]Prediction based on projected total US population of 430 million (Ahlburg and Vaupel 1990—medium projection) and a Western growth rate twice the national average. This corresponds to the relative growth rates for the period 1970–1987 (US Bureau of the Census 1989).

While the greatest effect of this increase on forests may be through changes in Western climate (see following section), elevated levels of carbon dioxide may have direct effects. Responses of C3 plants to increased carbon dioxide may include enhanced growth, increased rates of photosynthesis, reduced rates of photorespiration, increased water use efficiency, and reduced stomatal conductance (Melillo et al. 1990). Other potential effects on woody plants include accelerated needle abscission and chlorosis (Surano et al. 1986, Houpis et al. 1988), increased nodulation in nitrogen-fixing woody plants (Norby et al. 1987), and increased mycorrhizal colonization (O'Neill et al. 1987). At the ecosystem level, differential responses of species to increased carbon dioxide could alter competitive interactions, and changes in foliar carbohydrate levels could increase rates of herbivory and decrease rates of decomposition (Melillo et al. 1990).

Increased carbon dioxide could have the positive effect of reducing or partially offsetting the injurious effects of gaseous pollutants (Allen 1990). Reduced stomatal conductance would decrease rates of pollutant uptake by foliage, and increased rates of photosynthesis could increase the availability of photosynthate for repair of pollutant damage.

Climate change

Increased atmospheric concentrations of carbon dioxide and other greenhouse gases such as methane and nitrous oxide may cause an enhanced greenhouse effect and an increase in mean global air temperature (Houghton et al. 1990). Although predictions of global temperature changes have large uncertainties, estimates of future warming are within the range of 0.2°C to 0.5°C per decade (Houghton et al. 1990). This rate of increase would result by the year 2050 in a global mean air temperature 1.2°C to 3°C higher than in 1990.

Differences in climatic changes between and within regions are likely to be large, and predictions at regional scales are more uncertain than at the global scale (Mitchell et al. 1990). A region as large and complex as the West will have a large amount of variability in the magnitude and types of changes that may occur. In addition to temperature, amounts and patterns of precipitation may change. Western forests respond strongly to both variables, particularly as they interact to influence moisture availability (see Chapter 2).

An evaluation by Franklin et al. (1991) of the potential effects of climate change on forests of the Pacific Northwest serves as an example of the types of effects that global warming could have on Western forests. The study utilized a climate scenario representing potential changes under a doubled carbon dioxide environment: increases in mean temperature for the region ranging from 2°C to 5°C, and little change in precipitation, either in amount or seasonality. However, increased temperatures result in substantial decreases in available moisture and a longer summer dry period. Based on temperature alone, a climatic change of this magnitude would translate into a shift of climate zones upward in elevation 500–1000 m or northward 200–500 km (Franklin et al. 1991).

Any corresponding shifts in the spatial distribution of tree species will lag considerably behind changes in climate (Brubaker 1986). In the near-term, the effects of climatic warming in the Northwest will be felt on forests in their present locations; effects could be positive or negative depending on current site conditions relative to the ecophysiology of the species occupying the site. For example, moist sites where growth is limited by cold temperatures could benefit from warmer conditions.

Alternatively, water-limited sites could experience increased stress. Stressed trees are more susceptible to outbreaks of pests and pathogens (Stoszek 1988). A warmer climate may also increase the number of generations of a pest in a year and expand its overwintering range (Melillo et al. 1990). Franklin et al. (1991) suggest that a 2.5°C increase in mean temperature in the Northwest could allow the balsam woolly aphid (*Adelges piceae*) to expand upward into the subalpine zone where mature subalpine fir would be particularly vulnerable.

Eventually, shifts in competitive advantages among species may alter community composition in a particular area (Leverenz and Lev 1987, Melillo et al. 1990). This process could be accelerated if climatic warming were accompanied by an increased frequency of disruptive events such as fire and windstorms (Graham et al. 1990, Overpeck et al. 1990). In the Pacific Northwest, altered disturbance patterns including an increased magnitude and frequency of fires, storms, and pest/pathogen outbreaks could cause much more rapid changes in forest condition than the direct effects of increased temperature and moisture stress (Franklin et al. 1991).

Air pollution and climate change may interact in several ways. Increased air temperatures can cause the production of higher concentrations of ozone (Whitten and Gery 1986). Air pollution can increase the susceptibility of trees to pests and pathogens (see Chapter 12), and could accelerate the decline of individuals under climatic stress. Alternatively, drought may act to reduce the effects of gaseous pollutants on trees by reducing stomatal conductance and pollutant uptake.

Meteorological conditions conducive to the accumulation of pollutants may occur more frequently as climate warms (Schneider 1989). For example, stronger and more frequent valley inversions could lead to higher ozone exposure to forests near urban areas located in basins. Increased thermal convection could increase the intensity and duration of forest exposure to pollutants (see Chapter 3).

Ultraviolet-B radiation

Chlorofluorocarbons (CFCs) are components of air pollution that may indirectly affect Western forests. CFCs are greenhouse gases and contribute to global warming (Watson et al. 1990); they also catalyze the destruction of stratospheric ozone, one effect of which is to increase the flux of ultraviolet-B radiation (UV-B; 280–320 nm) to the earth's surface (Titus and Seidel 1986).

A 1% decrease in stratospheric ozone could result in about a 2% increase in the flux of UV-B (as weighted for biological effectiveness) through the stratosphere (Caldwell et al. 1989, Blumthaler and Ambach 1990). Loss of stratospheric ozone over the United States from 1978–1991 is estimated at 4–5% on average with about a 3% loss during the more critical (for plant response) summer season (Appenzeller 1991). Future rates of change will depend in part on rates of emissions of CFCs, which in turn depend on development of replacement chemicals and the effectiveness of the Montreal Protocol, an international CFC control agreement (Stetson 1990).

However, groundlevel UV-B flux may not increase proportionately to stratospheric ozone depletion due to interception of UV-B by tropospheric ozone and aerosols (Frederick and Lubin 1988, Scotto et al. 1988). Increased UV-B can accelerate rates of tropospheric ozone production (Whitten and Gery 1986, Liu and Trainer 1988), exacerbating one environmental problem while mitigating another.

The potential effects of increased UV-B on plants are initiated at the level of molecular photochemistry; for example, disruption of photosynthetic mechanisms (Caldwell et al. 1989). Impacts at this level may translate into effects at successively higher biological levels of organization such as reductions in photosynthesis, altered patterns of carbon allocation, and altered production of secondary chemicals; reduced growth and altered morphology (Kossuth and Biggs 1981, Sullivan and Teramura 1988); and changes in competitive relationships between species (Gold and Caldwell 1983).

However, UV-B response of plants is complicated by shielding, pigments, and photorepair mechanisms as well as interactions with other environmental stresses (Urbach 1989). Most research to date has not been with tree species (Tevini and Teramura 1989), and results from the limited studies of tree seedlings have been inconclusive (e.g., Kossuth and Biggs 1981, Sullivan and Teramura 1988). No research has been conducted on the response of whole ecosystems to enhanced UV-B. In general, the potential for deleterious effects of UV-B on Western forests is not clear.

Disturbances

Natural (e.g., insect and pathogen outbreaks, fire, windthrow, landslide) and human-caused (e.g., conversion to nonforest uses, harvest, fragmentation) disturbances to Western forests are discussed in Chapter 1. Over the next 50 years, most Western forests will be affected by at least one of these processes. Changes in the rate of any of these processes can have important ramifications for forest condition in the West.

Only 12% of Western forest land is legally reserved from commercial uses such as grazing, mining, and development as well as timber harvesting (Waddell et al. 1989), and recreational uses affect most reserved lands. Approximately 50% of Western forests are currently or potentially managed for timber production (Waddell et al. 1989). Annual softwood timber harvests (by volume) in the West are projected to be about 8% higher in 2030 than in 1986 (USDA 1990), although debates over issues such as the set-aside of additional forest land for the protection of endangered species make projections of future harvests very uncertain.

While cutting is clearly a major stress to forest ecosystems, management practices can have other implications for forest condition. Young managed forests may be more susceptible to wildfire than old-growth forests (Perry 1988), and monocultures may be more susceptible to pest outbreaks than more diverse natural forests (Schowalter 1988). Fragmented patches of forest may also be more susceptible to blowdown (Franklin et al. 1991). Alternatively, proper management of forests can increase stand vigor and resistance to stresses through selective thinning, supplemental fertilization, and other practices.

Continued population growth (Table 13.2) will drive continued conversion of forest lands to developments. Increased use of forests for recreation and other purposes also increases the chances of accidental fires. As discussed in the previous section on climate, global warming may cause an increase in severe windstorms and lightning strikes. Increased drought could also contribute to an increase in fire frequency (Clark 1988), and increased susceptibility of trees to insect outbreaks (Mattson and Haack 1987, Stoszek 1988).

While forest management practices can contribute to forest degradation, proper management is a key to mitigating the effects of global climate change and other stresses (Franklin et al. 1991). For example, managers can avoid planting species on sites that are marginal for them under current climate due to low water availability or high temperatures; regional warming would have the greatest effect on these sites. Pollutant tolerant genotypes can be planted in areas where air pollution is a current or potential problem. At a more complex level, ecosystem concepts can be used to develop forest management strategies that are more resistant to perturbations and more sustainable (Perry and Maghembe 1989).

Future Changes in Forest Condition

It is obvious that the future effects of air pollution on Western forests must be evaluated in the context of numerous other environmental, economic, and social variables. Interactions among stresses are important. The effects on a forest of one stress may be exacerbated or ameliorated by a second stress. Also, trends in one stress may influence trends in other stresses (Table 13.3).

Table 13.3 Effects of an increase in the magnitude of a stress on the magnitude of other stresses. (+, −): net effect is a slight increase or a slight decrease; (++, −−): net effect is a large increase or a large decrease; (0): net effect is minimal or unknown.

Increasing stress	Responding stress					
	AP	CW	UV-B	PG	ND	HD
Air pollution[1]		++	++	−	+	0
Climatic warming	+		0	−	++	0
UV-B	+	0		−	+	0
Population growth	++	++	+		+	++
Natural disturbance	+	+	0	0		0
Human-caused disturbance	+	+	0	0	+	

[1]Includes carbon dioxide and chlorofluorocarbon emissions.

It is also clear that forest condition is not a function of a single variable such as radial growth or chlorotic mottling of foliage. Biogeochemical cycles, species interactions, patterns of energy flow, and numerous other processes can be altered as part of a decline in forest condition. A complete prediction of the future condition of Western forests would have to consider all of these processes.

Predictive models

A fully integrated view of ecosystem impacts of pollution will probably require the use of computer simulation models that explicitly account for the main effects, the interactions among processes, and the feedback effects of both pollution and natural stresses. However, the availability of mechanistic models of air pollution effects on Western forests is limited, particularly models that can integrate effects at multiple levels of biological organization.

Kiester et al. (1990) reviewed existing models relevant to the National Acid Precipitation Assessment Program (NAPAP) assessment of the effects of acid rain and other pollutants on US forests. The twelve models included in the review address scales ranging from foliage to stands to watersheds. Only five of the models were classified by the reviewers as being potentially useful for predictions; of these five only one, the Plant-Growth-Stress Model (PGSM; Chen and Gomez 1990), addresses a Western species. PGSM is a single tree process/empirical model designed to simulate the effects of acid rain and ozone on ponderosa pine and other species. At the time of the NAPAP assessment, PGSM was described as not being sufficiently calibrated for use in projections (Kiester et al. 1990).

Process models of forest responses to stress continue to be improved (Dixon et al. 1990) as does the understanding of physiological processes necessary to support such models (see Chapter 5). The continuing development of models that simulate the impacts of pollution and other stresses at the level of leaves, branches and trees is likely to improve our overall understanding of these impacts. At this scale, validation of model projections may be possible. For models of whole forest ecosystems, poor opportunities for validation will probably limit the ability of models to make strong predictions of forest responses. While ecosystem-scale models can serve as formal statements of how forests work, and allow exploration of the possible effects of pollutants, their value is likely to be in increasing our understanding of system processes rather than in providing believable projections.

General assessments

Our current understanding of environmental trends and their effects on forest condition is not sufficient to support specific predictions about the future condition of Western forests. It is, however, possible to make some general statements about areas of forests at greatest risk from air pollution. These are the forests where increases in air pollution or continued high levels of air pollution will occur in combination with other stresses. In other words, areas with an overlap of:

- Air pollution: Forests in the airshed of a rapidly growing urban area are at greatest risk of increased or continued high exposure to air pollution. Examples include Puget Sound, Washington; Front Range, Colorado; Rincon Mountains, Arizona; Wasatch Mountains, Utah; and all California forests currently categorized as Stage II.

- Climate change: Species growing on sites toward the dry or hot end of their environmental tolerance will be the first to show stress from climatic warming. In the West, this will often be at the lower boundary of a species' elevational range. At a smaller scale, site factors such as soil water holding capacity will help to define areas of early stress from climatic warming.

- Disturbances: Human-caused disturbances, especially land conversion, will dominate at the urban edge. Natural disturbances such as insect outbreaks will be more evenly mixed with human effects such as logging at greater distances from the urban boundary. Obviously, displacing a forest with development can be classified as the most severe disturbance, but the relative severity of other disturbances is highly variable.

For example, ponderosa pine planted off-site at low-elevations of the southern Sierra Nevada in an area where road building has compacted the soil and disrupted drainage would be at risk of further decline in condition. An important point is that attention must be paid to small as well as large scales—the risk of pollution damage and general forest decline may vary within the boundaries of a larger region. It should also be kept in mind that air pollution is only one of several factors contributing to this risk.

Research Recommendations

Numerous research needs have been identified throughout earlier chapters. Many of these research questions are being addressed by ongoing research programs, and research priorities will change as projects are completed and new results (and questions) become available. There are, however, three research priorities likely to stand the test of time, and to maintain their fundamental importance to research and regulatory programs addressing air pollution and Western forests.

Long-Term Monitoring of Forest Condition and Stresses

Repeated measurements of forest condition and important stresses, taken as part of a spatial framework in a manner that can produce regional estimates of condition with known confidence, are essential to an understanding of the relationship between air pollution and forest health. Because of the high spatial and temporal variability of both stresses and forests, one-time surveys are not sufficient. Baseline and trend data are essential for determining the relationships between air pollution, other stresses, and the condition of forest ecosystems. Monitoring data are also critical for determining the effectiveness of regulatory or mitigation efforts to protect forest health (Silsbee and Peterson 1991).

An essential task in developing a monitoring system is the identification of indicators of forest condition—variables that can be measured and interpreted in terms of forest condition and the effect of particular stresses. The complexity of a forest ecosystem, the many points at which pollutants can impact the system, and the similarity in ecosystem response to different stresses make this a difficult task.

Efforts to develop a forest monitoring program are underway. The USDA Forest Service is designing a Forest Health Monitoring Program (Brooks et al. 1991) in collaboration with the US Environmental Protection Agency's Environmental Monitoring and Assessment Program (EMAP; Messer et al. 1991). These types of efforts should be continued.

Improved Predictive Capabilities

By the time a monitoring system detects a change in forest condition, it may be too late for effective mitigation, and it is certainly too late for prevention. Small changes in forest condition are not easy to detect relative to the normal variability of forests, and detectable changes may occur only after thresholds have been passed (Kiester et al. 1990). Trends in stresses such as climate change have large momentum, and are not quickly reversed.

Improved mechanistic models that predict both trends in stresses and the responses of forests need to be developed. A mechanistic basis for these models is necessary because they will have to predict beyond the bounds of existing data (Ford and Kiester 1990). Concurrent research on the physiological mechanisms of forest response to environmental stresses is essential. Modeling and experimental work go hand-in-hand; models are used to identify the most critical information gaps and research needs, while experimental results assist in validating and improving the models.

Policy/Science Links

Strong links need to be forged between science and policy, and between scientists and policy-makers. If government officials who set policies for controlling emissions and regulating forest management practices do not receive research results in a useful form, then research will contribute little to the condition of Western forests. Because government officials often respond to public demands, public education must also be a priority of scientists.

References

Ahlburg DA, Vaupel JW (1990) Alternative projections of US population. *Demography* 27(4):639–652

Allen LH Jr (1990) Plant responses to rising carbon dioxide and potential interactions with air pollutants. *Journal of Environmental Quality* 19:15–34

Appenzeller T (1991) Ozone loss hits us where we live. *Science* 254:645

Barnard JE, Lucier AA, Brooks RT, Johnson AH, Dunn PH, Karnosky DF, Brandt CJ, Richter DD (1990) Changes in forest health and productivity in the United States and Canada. NAPAP SOS/T Report 16, In: *Acidic Deposition: State of Science and Technology, Volume III*. National Acid Precipitation Assessment Program, 722 Jackson Place NW, Washington, DC 20503

Blumthaler M, Ambach W (1990) Indication of increasing solar ultraviolet-B radiation flux in alpine regions. *Science* 248:206–208

Bormann FH (1985) Air pollution and forests: An ecosystem perspective. *Bioscience* 35:434–441

Brooks RT, Miller-Weeks M, Burkman W (1991) *Forest Health Monitoring: New England, 1990 Summary Report*. USDA Forest Service NE-INF-94-91, Northeast Forest Experiment Station, Radnor, PA

Cahill TA, Kusko BH, Ashbaugh LL, Barone JB, Eldred RA, Walther FG (1981) Regional and local determinations of particulate matter and visibility in the southwestern United States during June and July, 1979. *Atmospheric Environment* 15:2011–2016

Caldwell MM, Teramura AH, Tevini M (1989) The changing solar ultraviolet climate and the ecological consequences for higher plants. *Tree* 4:363–367

Cibrian-Tovar D (1989) Air pollution and forest decline near Mexico City. *Environmental Monitoring and Assessment* 12:49–58

Ciesla WM, Macias Samano JE (1987) Desierto de los Leones: A forest in crisis. *American Forests* 93:29–31,72–74

Clark JS (1988) Effects of climate change on fire regimes in northwestern Minnesota. *Nature* 334:233–235

Duriscoe DM, Stolte KW (1989) Photochemical oxidant injury to ponderosa (*Pinus ponderosa* Laws.) and Jeffrey pine (*Pinus jeffreyi* Grev. and Balf.) in the national parks of the Sierra Nevada of California. In: Olson RK, Lefohn AS (eds) *Effects of Air Pollution on Western Forests.* Transactions Series, No. 16, Air & Waste Management Association, Pittsburgh, pp 261–278

Franklin JF, Swanson FJ, Harmon ME, Perry DA, Spies TA, Dale VH, McKee A, Ferrell WK, Means JE, Gregory SV, Lattin JD, Schowalter TD, Larsen D (1991) Effects of global climatic change on forests in northwestern North America. *Northwest Environmental Journal* 7(2):233–254

Frederick JE, Lubin D (1988) Possible long-term changes in biologically active ultraviolet radiation reaching the ground. *Photochemistry and Photobiology* 47:571–578

Gold WG, Caldwell MM (1983) The effects of ultraviolet-B radiation on plant competition in terrestrial ecosystems. *Physiologia Plantarum* 58:435–444

Graham RL, Turner MG, Dale VH (1990) How increasing CO_2 and climate change affect forests. *BioScience* 40:575–587

Graybill DA, Rose MR (1989) Analysis of growth trends and variation in conifers from Arizona and New Mexico. In: Olson RK, Lefohn AS (eds) *Effects of Air Pollution on Western Forests.* Transactions Series, No. 16, Air & Waste Management Association, Pittsburgh, pp 395–407

Hagerman E (1990) California's drive to mass transit. *World Watch* 3(5):7–8

Houghton JT, Jenkins GJ, Ephraums JJ (eds) (1990) *Climate change: The Intergovernmental Panel on Climate Change Scientific Assessment.* Cambridge University Press, Cambridge, UK

James RL, Staley JM (1980) *Photochemical Air Pollution Damage Survey of Ponderosa Pine Within and Adjacent to Denver, Colorado: A Preliminary Report.* USDA Forest Service Forest Insect and Disease Management Biological Evaluation R2-80–6

Kossuth SV, Biggs RH (1981) Ultraviolet-B radiation effects on early seedling growth of Pinaceae species. *Canadian Journal of Forest Research* 11:243–248

Legge AH, Krupa SV (1989) Air quality at a high elevation, remote site in western Canada. In: Olson RK, Lefohn AS (eds) *Effects of Air Pollution on Western Forests*. Transactions Series, No. 16, Air & Waste Management Association, Pittsburgh, pp 193–206

Leverenz JW, Lev DJ (1987) Effects of carbon dioxide-induced climate changes on the natural ranges of six major commercial tree species in the western United States. In: Shands WE, Hoffman JS (eds) *The Greenhouse Effect, Climate Change, and US Forests*. The Conservation Foundation, Washington, DC, pp 123–156

Liu SC, Trainer M (1988) Responses of the tropospheric ozone and odd hydrogen radicals to column ozone change. *Journal of Atmospheric Chemistry* 6:221–233

Lynch DW (1951) Diameter growth of ponderosa pine in relation to the Spokane pine-blight problem. *Northwest Science* 25:157–163

Mattson WJ, Haack RA (1987) The role of drought in outbreaks of plant-eating insects. *BioScience* 37:110–118

Messer JJ, Linthurst RA, Overton WS (1991) An EPA program for monitoring ecological status and trends. *Environmental Monitoring and Assessment* 17:67–78

Melillo JM, Callaghan TV, Woodward FI, Salati E, Sinha SK (1990) Effects on ecosystems. In: Houghton JT, Jenkins GJ, Ephraums JJ (eds) *Climate Change: The Intergovernmental Panel on Climate Change Scientific Assessment*, Cambridge University Press, Cambridge, UK, pp 283–310

Miller PR, Longbotham GJ, VanDoren RE, Thomas MA (1979) Effect of chronic oxidant air pollution exposure on California black oak in the San Bernardino Mountains. In: *Proceedings of the Symposium on the Ecology, Management and Utilization of California Oaks*. Pacific Southwest Forest and Range Experiment Station Research Paper

Mitchell JFB, Manabe S, Meleshko V, Tokioka T (1990) Equilibrium climate change – and its implications for the future. In: Houghton JT, Jenkins GJ, Ephraums JJ (eds) *Climate Change: The Intergovernmental Panel on Climate Change Scientific Assessment*, Cambridge University Press, Cambridge, UK, pp 131–172

Overpeck JT, Rind D, Goldberg R (1990) Climate-induced changes in forest disturbance and vegetation. *Nature* 343:51–53

Perry DA (1988) Landscape pattern and forest pests. *Northwest Environmental Journal* 4:213–228

Perry DA, Maghembe J (1989) Ecosystem concepts and current trends in forest management: Time for reappraisal. *Forest Ecology and Management* 26:123–140

Peterson DL, Arbaugh MJ, Wakefield VA, Miller PR (1987) Evidence of growth reduction in ozone-injured Jeffrey pine (*Pinus jeffreyi* Grev. and Balf.) in Sequoia and Kings Canyon National Parks. *Journal of the Air Pollution Control Association* 37:906–912

Peterson DL, Schmoldt DL, Eilers JM, Fisher RW, Doty RD (1992a) Guidelines for evaluating air pollution impacts on class I wilderness areas in California. USDA Forest Service General Technical Report PSW-000 (in press)

Peterson J, Schmoldt DL, Peterson DL, Eilers JM, Fisher RW, Bachman R (1992b) Guidelines for evaluating air pollution impacts on class I wilderness areas in the Pacific Northwest. USDA Forest Service General Technical Report PNW-000 (in press)

Scheffer TC, Hedgcock GC (1955) *Injury to Northwestern Trees by Sulfur Dioxide from Smelters.* USDA Forest Service Technical Bulletin No. 1117, 49p

Schowalter TD (1988) Forest pest management: A synopsis. *Northwest Environmental Journal* 4:313–318

Shaw CG, Fischer GW, Adams DF, Adams MF, Lynch DW (1951) Fluorine injury to ponderosa pine: A summary. *Northwest Science* 25:156

Shriner DS, Heck WW, McLaughlin SB, Johnson DW, Irving PM, Joslin JD, Peterson CW (1990) Response of vegetation to atmospheric deposition and air pollution. NAPAP SOS/T Report 18, In: *Acidic Deposition: State of Science and Technology, Volume III.* National Acid Precipitation Assessment Program, 722 Jackson Place NW, Washington, DC 20503

Silsbee DG, Peterson DL (1991) Designing and implementing comprehensive long-term inventory and monitoring programs for National Park System lands. Natural Resources Report NRR-91/04. National Park Service, Denver, CO

South DW (1991) Technologies and other measures for controlling emissions: Performance, costs, and applicability. In: Irving PM (ed) *Acidic Deposition: State of Science and Technology, Summary Report of the US National Acid Precipitation Assessment Program.* National Acid Precipitation Assessment Program, 722 Jackson Place NW, Washington, DC, pp 217–227

Stetson M (1990) Who'll pay to protect the ozone layer? *World Watch* 3(4):36–37

Stoszek KJ (1988) Forests under stress and insect outbreaks. *The Northwest Environmental Journal* 4:247–261

Streets DG (1991) Methods for modeling future emissions and control costs. In: Irving PM (ed) *Acidic Deposition: State of Science and Technology, Summary Report of the US National Acid Precipitation Assessment Program.* National Acid Precipitation Assessment Program, 722 Jackson Place NW, Washington, DC, pp 229–234

Sullivan JH, Teramura AH (1988) Effects of ultraviolet-B irradiation on seedling growth in the Pinaceae. *American Journal of Botany* 75:225–230

Tevini M, Teramura AH (1989) UV-B effects on terrestrial plants. *Photochemistry and Photobiology* 50:479–487

Titus JG, Seidel S (1986) Overview of the effects of changing the atmosphere. In: Titus JG (ed) *Effects of Changes in Stratospheric Ozone and Global Climate, Vol. I: Overview.* US Environmental Protection Agency, Washington, DC, pp 3–19

Urbach F (1989) The biological effects of increased ultraviolet radiation: An update. *Photochemistry and Photobiology* 50:439–441

US Department of Agriculture (1990) *An Analysis of the Timber Situation in the United States: 1989-2040.* General Technical Report RM-199, US Department of Agriculture, Forest Service, Rocky Mountain Forest and Range Experiment Station, Ft. Collins, CO

United States Bureau of the Census (1989) *Statistical Abstract of the United States: 1989 (109th edition).* Washington, DC

United States Bureau of the Census (1990) Current Population Reports, Series P-25, No. 1058, State Population and Household Estimates: July 1, 1989, US Government Printing Office, Washington, DC

Venkatram A (1991) Relationships between atmospheric emissions and deposition/air quality. In: Irving PM (ed) *Acidic Deposition: State of Science and Technology, Summary Report of the US National Acid Precipitation Assessment Program.* National Acid Precipitation Assessment Program, 722 Jackson Place NW, Washington, DC, pp 85–88

Vong RJ (1989) Background concentration of sulfate in precipitation along the west coast of North America. In: Olson RK, Lefohn AS (eds) *Effects of Air Pollution on Western Forests.* Transactions Series, No. 16, Air & Waste Management Association, Pittsburgh, pp 117–135

Watson RT, Rodhe H, Oeschger H, Siegenthaler U (1990) Greenhouse gases and aerosols. In: Houghton JT, Jenkins GJ, Ephraums JJ (eds) *Climate Change: The Intergovernmental Panel on Climate Change Scientific Assessment*, Cambridge University Press, Cambridge, UK, pp 1–40

Westman WE (1985) Air pollution injury to coastal sage scrub in the Santa Monica Mountains, southern California. *Water, Air, and Soil Pollution* 26:19–41

Westman WE, Price CV (1988) Detecting air pollution stress in southern California vegetation using Landsat Thematic Mapper band data. *Photogrammetric Engineering and Remote Sensing* 54:1305–1311

Whitten GZ, Gery MW (1986) Effects on urban smog resulting from changes in the stratospheric ozone layer and in global temperature. *EPA Workshop on Global Atmospheric Change and EPA Planning: Final Report*, EPA-600/9-86/016, US Environmental Protection Agency, Washington, DC

Young JR, Ellis EC, Hidy GM (1988) Deposition of air-borne acidifiers in the Western environment. *Journal of Environmental Quality* 17:1–26

Index

Scientific and common names of conifer and hardwood tree species discussed in book. Presence of species in physiographic regions is indicated; x = major occurrence, m = minor occurrence. Species ranges from Little (1971; see Chapter 1).

CONIFERS

	PNW	CA	NR	SR	MH
Abies amabilis (Pacific silver fir)	x	m			
Abies concolor (white fir)	x	x		x	x
Abies grandis (grand fir)	x	x	x		
Abies lasiocarpa (subalpine fir)	x		x	x	x
Abies magnifica (California red fir)	x	x			
Abies procera (noble fir)	x	x			
Calocedrus decurrens (incense-cedar)	x	x	m	m	
Chamaecyparis lawsoniana (Port Orford cedar)	x	x			
Chamaecyparis nootkatensis (Alaska yellow cedar)	x	x	m		
Cupressus arizonica (Arizona cypress)		m			x
Juniperus deppeana (alligator juniper)				m	x
Juniperus monosperma (one-seed juniper)				x	x
Juniperus occidentalis (western juniper)	m	x	x	m	
Juniperus osteosperma (Utah juniper)			m	x	x
Juniperus scopulorum (Rocky Mt. juniper)	m		x	x	x
Larix lyalli (subalpine larch)	x		x		
Larix occidentalis (western larch)	x		x		
Picea engelmannii (Englemann spruce)	x	m	x	x	m
Picea pungens (blue spruce)				x	x
Picea sitchensis (Sitka spruce)	x	x			
Pinus albicaulis (whitebark pine)	x	x	x	x	
Pinus aristata (bristlecone pine)				x	
Pinus attenuata (knobcone pine)	m	x			
Pinus balfouriana (foxtail pine)		m			
Pinus cembroides (Mexican pinyon)					x
Pinus contorta (lodgepole pine)	x	x	x	x	
Pinus coulteri (Coulter pine)		x			
Pinus edulis (pinyon pine)				x	x
Pinus engelmannii (Apache pine)					x
Pinus flexilis (limber pine)		x	x	x	
Pinus jeffreyi (Jeffrey pine)		x		x	
Pinus lambertiana (sugar pine)	x	x	m		
Pinus leiophylla (Chihuahua pine)				x	
Pinus longaeva (Great Basin bristlecone pine)				x	
Pinus monophylla (singleleaf pine)		x			
Pinus monticola (western white pine)	x	x	x		
Pinus ponderosa (ponderosa pine)	x	x	x	x	x
Pinus sabiniana (digger pine)		x			
Pinus strobiformis (southwestern white pine)				m	x

CONIFERS—*continued*

	PNW	CA	NR	SR	MH
Pseudotsuga macrocarpa (big-cone Douglas-fir)		x			
Pseudotsuga menziesii (Douglas-fir)	x	x	x	x	x
Sequoia sempervirens (redwood)		x			
Sequoiadendron giganteum (giant sequoia)		x			
Taxus brevifolia (Pacific yew)	x	x	x		
Thuja plicata (western redcedar)	x	x	x		
Tsuga heterophylla (western hemlock)	x	x	x		
Tsuga mertensiana (mountain hemlock)	x	x	x		

HARDWOODS

	PNW	CA	NR	SR	MH
Acer circinatum (vine maple)	x	x			
Acer macrophyllum (bigleaf maple)	x	x	m		
Alnus rubra (red alder)	x	x	x		
Arbutus menziesii (Pacific madrone)	x	x			
Castanopsis chrysophylla (golden chinquapin)	x	x	m		
Juglans californica (California walnut)		x			
Juglans major (Arizona walnut)					x
Lithocarpus densiflorus (tanoak)	m	x			
Platanus wrightii (Arizona sycamore)					x
Populus trichocarpa (black cottonwood)	x	x	x	x	
Populus tremuloides (quaking aspen)	x	x	x	x	x
Quercus agrifolia (coast live oak)		x			
Quercus arizonica (Arizona white oak)					x
Quercus chrysolepis (canyon live oak)	x	x		m	x
Quercus douglasii (blue oak)		x			
Quercus emoryi (Emory oak)					x
Quercus engelmannii (Engelmann oak)		x			
Quercus gambelii (Gambel oak)				x	x
Quercus garryana (Oregon white oak)	x	x	m		
Quercus hypoleucoides (silverleaf oak)					x
Quercus kelloggii (black oak)	m	x			
Quercus lobata (valley oak)		x			
Quercus oblongifolia (Mexican blue oak)					x
Quercus wislizenii (interior live oak)		x			
Umbellularia californica (California-laurel)	x	x			

Physiographic regions (see Figure 1.2):

PNW   *Northern Cascades, Central Cascades, Puget Trough, Olympic Mountains, Oregon Coast Ranges*

CA   *Southern Cascades, Sierra Nevada, California Trough, Transverse and Peninsular Ranges, California Coast Ranges, Klamath Mountains*

NR   *Northern Rockies, Columbia Plateaus*

SR   *Southern Rockies, Central Rockies, Wyoming Basin, Colorado Plateaus, Great Basin*

MH   *Mexican Highlands*